# SCALE DEPENDENCE AND SCALE INVARIANCE IN HYDROLOGY

Whether processes in the natural world are dependent or independent of the scale at which they operate is one of the major issues in hydrologic science. In this volume, leading hydrologists present their views on the role of scale effects in hydrologic phenomena occurring in a range of field settings, from the land surface to deep fractured rock.

Self-contained and thought-provoking chapters cover both theoretical and applied hydrology. They provide critical insights into important topics such as general circulation models, floods, river networks, vadose-zone processes, groundwater transport, and fluid flow through fractured media. This book is intended as an accessible introduction for graduate students and researchers to some of the most significant questions and challenges that will face hydrologic science in the twenty-first century.

# SCALE DEPENDENCE AND SCALE INVARIANCE IN HYDROLOGY

*Edited by*

GARRISON SPOSITO

*University of California at Berkeley*

CAMBRIDGE UNIVERSITY PRESS
Cambridge, New York, Melbourne, Madrid, Cape Town, Singapore, São Paulo, Delhi

Cambridge University Press
The Edinburgh Building, Cambridge CB2 8RU, UK

Published in the United States of America by Cambridge University Press, New York

www.cambridge.org
Information on this title: www.cambridge.org/9780521571258



First published 1998
This digitally printed version 2008

*A catalogue record for this publication is available from the British Library*

*Library of Congress Cataloguing in Publication data*

Scale dependence and scale invariance in hydrology / edited by
Garrison Sposito
p. cm.
Includes bibliographical references.
ISBN 0-521-57125-1 hardback
1. Hydrology – Statistical methods. 2. Scaling laws (Statistical physics)
I. Sposito, Garrison, 1939– .
GB656.2.S7S33
551.48′072 – dc21 97-33252
CIP

ISBN 978-0-521-57125-8 hardback
ISBN 978-0-521-08858-9 paperback

FOR JAMES J. MORGAN

*And sensibly, though so much*
*later, the wandering Bloom*
*replied, "Ireland," said Bloom*
*"Iwas born here. Ireland."*

# Contents

# Contributors

Roger Beckie
Department of Earth and Ocean Sciences
University of British Columbia

Alberto Bellin
Department of Civil and Environmental Engineering
University of Trento

R. Cuenca
Department of Bioresource Engineering
Oregon State University

William E. Dietrich
Department of Geology and Geophysics
University of California at Berkeley

Vittorio Di Federico
D.I.S.T.A.R.T.
University of Bologna

Vijay K. Gupta
Hydrologic Sciences Program and C.I.R.E.S.
University of Colorado

R. Haverkamp
L.T.H.E.
Université Joseph Fourier

Jan W. Hopmans
Department of Land, Air, and Water Resources
University of California at Davis

Paul A. Hsieh
Water Resources Division
United States Geological Survey

Vivek Kapoor
School of Civil and Environmental Engineering
Georgia Institute of Technology

Peter Kitanidis
Department of Civil Engineering
Stanford University

David R. Montgomery
Department of Geological Sciences
University of Washington

Shlomo P. Neuman
Department of Hydrology and Water Resources
University of Arizona

Donald R. Nielsen
Department of Land, Air, and Water Resources
University of California at Davis

J. -Y. Parlange
Department of Agricultural and Biological Engineering
Cornell University

Klaus Reichardt
Department of Physics
University of São Paulo

Andrea Rinaldo
Institute of Hydraulics
University of Padova

Ignacio Rodríguez-Iturbe
Department of Civil Engineering
Texas A&M University

P. J. Ross
Division of Soils
Commonwealth Scientific and Industrial Research Organization

Yoram Rubin
Department of Civil and Environmental Engineering
University of California at Berkeley

David Russo
Department of Soil Physics
The Volcani Center

Garrison Sposito
Department of Civil and Environmental Engineering
University of California at Berkeley

T. S. Steenhuis
Department of Agricultural and Biological Engineering
Cornell University

Edward C. Waymire
Departments of Mathematics and
Oregon State University

Eric F. Wood
Department of Civil Engineering and Operations Research
Princeton University

T. -C. J. Yeh
Department of Hydrology and Water Resources
University of Arizona

# Preface

*The reverse side also has a reverse side.*
—sign on Telegraph Avenue (6.9.96)

In his Josiah Willard Gibbs Lecture[1] some years ago, the gifted theoretical physicist Elliott Montroll recounted a statistical analysis of the prices of merchandise offered in the Sears annual catalog during the first 85 years of the twentieth century. That catalog, which Montroll termed "a magnificent database of Americana," was prepared carefully to feature items that reflected current public taste at prices that were appropriate for competitive merchandising of the time. The price of an item that was sold by Sears over many years would, of course, be expected to change in the catalog as the cost of living changed, or as technology related to the manufacture of the item improved. Some catalog items (e.g., the buggy whips sold in the 1910 catalog) would disappear altogether and be replaced as new technologies made them obsolete.

Despite the many vicissitudes of American life and of company operation, Montroll found that the frequency distribution of the prices in any one of the Sears catalogs published since 1900 closely approximated a lognormal distribution. More remarkably, the single-catalog standard deviation of the logarithm of price remained essentially constant over the 85-year period, although the single-catalog mean of the logarithm of price generally increased annually, reflecting an increasing cost of living. Montroll was quick to point out that this observed constancy of the single-catalog standard deviation of log(price) was to be expected if price changes were primarily the results of inflation, for then the price of an item in a given catalog would differ only by a scale factor from its price in a previous catalog, and that factor would always drop out when the single-catalog standard deviation of log(price) was calculated. Thus the constancy of the latter statistic signaled the fact that the distribution of prices in the Sears catalogs was scale-invariant. Evidently, year after year, catalogs were created with the items in them priced so as to maintain constant the relative number of items available within any selected range of prices, provided only that this range was scaled annually for inflation.

No reader of the Montroll lecture can be left untouched by the evident simplicity and beauty of his analysis. Whatever may be the complexity of American socioeconomic behavior, or the vagaries in Sears merchandising activities, the catalog pricing structure that emerged exhibited an inherent symmetry that typified it profoundly: that of *scale invariance*. The search for scale invariance is, of course, not limited to economics. It has had a long and successful history in the engineering disciplines,[2] nucleated by a celebrated theorem of the soil physicist Edgar Buckingham.[3] It is also a flourishing industry within the earth sciences, particularly the geophysical branches. Stefan Machlup, in an entertaining reminiscence,[4] has described a number of temporal geophysical phenomena, ranging from the stages of rivers to the magnitudes of earthquakes, for which the frequency distribution is scale-invariant, or approximately so. He infers that these natural phenomena must be complex enough to possess "a large ensemble of mechanisms with no prejudice about scale." An important corollary he adds is that present human efforts to control them likely will eliminate part of this ensemble, thereby producing frequency distributions that do exhibit intrinsic scales.

Scale invariance in the spatial domain is now almost universally associated with fractal concepts that emerged 30 years ago when Benoit Mandelbrot[5] reinterpreted the exponent in the power-law relationship between coastline length and measuring increment that was discovered by the eccentric dilettante Lewis Fry Richardson,[6] who himself had been motivated by the notion that the extent of conflict between countries was positively correlated with the degree of irregularity in their common borders. Fractal geometry has since become the signature approach to both spatial-scale invariance and temporal-scale invariance, as epitomized by self-similarity in the patterns of hydrologic and other geophysical processes.[7]

This volume is a compendium of essays contributed by engineers and scientists who have spent much of their recent professional lives thinking about and investigating issues of scale dependence or scale invariance in terrestrial hydrologic phenomena. More precisely, as occurs in the more encompassing sister discipline of landscape ecology,[8] the fundamental question they have been asking is whether "a given phenomenon appears or applies across a broad range of scales, or whether it is limited to a narrow range of scales." Their research has thus been motivated by a desire for understanding and application, rooted in a common quest for simplicity in scientific explanation, and informed by successful insights communicated through three decades of intense study by physical scientists concerned with scale effects. What they have to say herein is not necessarily shared belief among their colleagues in hydrology, nor is it necessarily comprehensive in the sense of a normal review article. What they do have to impart to the reader is a considered distillate of their own professional experience, gained by grappling both experimentally and conceptually with scale issues as they arise in continental hydrologic processes. Whether the reader ultimately will find the result to be an engaging book of ideas or simply a volume that would be more aptly entitled (in the current spate of Austenmania) *Pride and Prejudice*, only time can resolve.

I am most grateful to my colleagues Roger Beckie, Vivek Kapoor, Peter Kitanidis, Keith Loague, Shlomo Neuman, Donald Nielsen, Jean-Yves Parlange, Andrea Rinaldo, Ignacio Rodríguez-Iturbe, Yoram Rubin, David Russo, Chin-Fu Tsang, and Eric Wood for their dedicated service as referees for the chapters of this book while in draft form. I am also much indebted to Catherine Flack for suggesting that I edit such a book, to Holly Johnson of the Cambridge University Press for superb management of all matters relating to the production of this book, and to my wife, Mary, for her continual encouragement and support, at all scales.

Garrison Sposito
Berkeley, California

## References

[1] Montroll, E. W. 1987. On the dynamics and evolution of some sociotechnical systems. *Bull. Am. Math. Soc.* 16:1–46.

[2] Barenblatt, G. I. 1996. *Scaling, Self-Similarity, and Intermediate Asymptotics.* Cambridge University Press.

[3] Buckingham, E. 1915. The principle of similitude. *Nature* 96:396–7.

[4] Machlup, S. 1981. Earthquakes, thunderstorms, and other $1/f$ noises. In: *Sixth International Conference on Noise in Physical Systems*, ed. P. H. E. Meijer, R. D. Mountain, and R. J. Soulen, Jr., pp. 157–60. Special publication 614. Washington, DC: National Bureau of Standards.

[5] Mandelbrot, B. B. 1967. How long is the coast of Britain? Statistical self-similarity and fractional dimension. *Science* 155:636–8.

[6] Richardson, L. F. 1961. The problem of contiguity: an appendix of statistics of deadly quarrels. *General Systems Yearbook* 6:139–87.

[7] Fleischmann, M., Tildesley, D. J., and Ball, R. C. 1989. *Fractals in the Natural Sciences.* Princeton University Press.

[8] Pickett, S. T. A., and Cadenasso, M. L. 1995. Landscape ecology: spatial heterogeneity in ecological systems. *Science* 269:331–4.

*Gratitude is expressed to The Board of Trinity College Dublin for permission to reprint TCD MS 58 (the Book of Kells), fol 33r, on the cover of this book.*

# 1

# Scale Analyses for Land-Surface Hydrology

ERIC F. WOOD

## 1.1 Introduction

This chapter discusses some research problems associated with the scaling of land-surface hydrologic processes involved in the land-surface water-and-energy budget. As pointed out by Dubayah, Wood, and Lavallée (1996), the term "scaling" has come to have multiple definitions, depending not only on the general discipline (e.g., hydrology, meteorology, geography, physics) but also on the application within a discipline. Dubayah et al. (1996) refer to a process as exhibiting scaling "if no characteristic length scale exists; i.e. the statistical spatial properties of the field do not exhibit scale-dependent behavior." Blöschl and Sivapalan (1995) refer to scaling as the transfer of information between different spatial (or temporal) lengths. The "transfer of information" may consist in mathematical relationships, statistical relationships, or observations describing physical phenomena. This definition is consistent with that of Dubayah et al. (1996), albeit more qualitative. The term "upscaling" is often used by the remote-sensing community to describe going from observations on a small spatial (or temporal) scale to observations on larger scales, whereas "downscaling" consists in going from large-scale observations to smaller scales.

Interest in scaling as it relates to the land-surface water-and-energy budget arose from the more general problem of land-surface parameterization in climate models. In the 1970s, a series of Atmospheric General Circulation Model (AGCM) climate studies demonstrated the importance of land hydrology for the earth's climate: the sensitivity of albedo to climate (Charney et al., 1977), and the influence of soil-moisture anomalies (e.g., Walker and Rowntree, 1977). Those studies and related work resulted in the establishment of a Joint Scientific Committee Working Group on Land Surface Processes, with Peter S. Eagleson as chairman, under the World Meteorological Organization/International Council of Scientific Unions (WMO/ICSU) Joint Scientific Committee for the World Climate Research Programme (WCRP). That working group organized the Study Conference on Land Surface Processes in Atmospheric General Circulation Models in January 1981 help assess the state of the art in modeling land-surface hydrologic processes on the scale of AGCMs, to recommend research activities to improve knowledge in this area, and to suggest data requirements for initialization, validation, and parameter evaluation (Eagleson, 1982). The work and influence of that committee helped establish climatologic experimental programs like HAPEX (Hydrology Atmospheric Parameterization Experiment) and ISLSCP (International Satellite Land Surface Climatology Programme) to provide experimental data (however limited they may be) that have been used for more than a decade for scaling studies of land-surface processes. In the late 1980s the ICSU established the core project Biospheric Aspects of the Hydrological Cycle (BAHC). One research focus is spatial and temporal integration of biospheric–hydrospheric interactions. Even though such programs are under way, in general the land-surface research by the climate-modeling community has focused primarily on improved process models at small scales, using data collected through experiments like HAPEX-MOBILY and FIFE '87, with less emphasis on scaling point observations and the development of mathematical descriptions of land-surface processes at larger scales. Thus there remain opportunities for researchers to make significant contributions to the field.

My own interest in investigating land-surface hydrologic parameterizations and scaling arose through the influence of the seminal paper by Dooge (1982) that is part of the proceedings of the Study Conference on Land Surface Processes in Atmospheric General Circulation Models (Eagleson, 1982). This paper remains required reading for those who wish to understand the early foundational work in the scaling of land-surface hydrologic processes for AGCMs. Dooge clearly identifies the key role of soil moisture and its variability in the coupling of land-surface hydrology to atmospheric models and reviews the progress in estimating soil-moisture changes and evaporation, from field scales on the order of 10–100 hectares (ha) to catchment scales (100–1,000 km$^2$) to regional and AGCM grid scales (10,000–100,000 km$^2$). He comments that "in linking [soil moisture and evaporation] phenomena on that scale (10–100 ha.) to the usual scale (100–1000 sq. km.)... a number of approaches have been tried [but] this problem is largely an unsolved one" (Dooge, 1982). Though we have learned much over the past 15 years and certainly have gained an appreciation of the scaling problem, Dooge's comment is as valid today as in 1982. Over the past 15 years, considerable effort has been expended in trying to solve Dooge's problem. Much of that research has been reported in special conference and workshop proceedings that offer good overviews of current thinking and progress. Such publications include those by Gupta, Rodríguez-Iturbe, and Wood (1986), Bolle, Feddes, and Kalma (1993), Kalma and Sivapalan (1995), and Stewart et al. (1996).

The remainder of this chapter will be organized as follows: In the next section, the scaling problem as it relates to land-surface hydrologic processes will be more formally stated. Section 1.3 discusses the important role of variability in scaling. Section 1.4 presents selected research findings, first looking at simple analytical modeling results, followed by insights from analyses using complex land-surface models, and finally inferences derived from climatologic field experiments. Section 1.5 addresses the "approach of equivalent parameters," which is widely used by remote-sensing and micrometeorology scientists in scaling land-surface processes and fields. As described in Section 1.4, most of the work to date has been by empirical analysis. Sections 1.6 and 1.7 explore the potential of using statistical self-similarity theory for scaling land-surface processes. The final section provides a discussion of the progress to date and a look toward the future. Throughout these sections, an attempt has been made to indicate the areas of greatest potential for further research.

## 1.2 Statement of the Scaling Problem

The terrestrial water-and-energy-balance equations can be written as follows:

$$\left\langle \frac{\partial S}{\partial t} \right\rangle = \langle P \rangle - \langle E \rangle - \langle Q \rangle \tag{1.1}$$

and

$$\langle R_n \rangle = \langle \lambda E \rangle + \langle H \rangle + \langle G \rangle \tag{1.2}$$

where $S$ is the moisture in the soil column, $E$ is the evaporation from the land surface into the atmosphere, $P$ is the precipitation from the atmosphere to the land-surface, $Q$ is the net runoff from the control volume, $R_n$ is the net radiation at the land-surface, $\lambda$ is the latent heat of vaporization, $H$ is the sensible heat, and $G$ is the ground heat flux. The spatial average for the control volume is denoted by $\langle \cdot \rangle$. Equations (1.1) and (1.2) are valid over all scales, and only through the parameterization of individual terms does the water-and-energy balance become a "distributed" or "lumped" model.

By a "distributed" model I mean a model that accounts for spatial variability in inputs, processes, and parameters. This accounting can be either explicit – in which case the actual patterns of variability are represented, as, for example, in the European hydrologic system model SHE (Abbott, 1986a,b) and the three-dimensional finite-element catchment model of Paniconi and Wood (1993) – or statistical – in which case the patterns of variability are represented statistically [examples being models like TOPMODEL (Beven and Kirkby, 1979) and its variants (Moore, O'Laughlin, and Burch, 1984; Famiglietti and Wood, 1994; Liang et al., 1994), in which topography, soils, and vegetation play important roles in the distribution of water within a catchment].

By a "lumped" model I mean a model that represents the catchment (or grid) as being spatially homogeneous with regard to inputs and parameters. In engineering hydrology these models include the well-known unit hydrograph and its variants; in climate modeling these include the complex atmospheric–biospheric models such as the Biosphere Atmosphere Transfer Scheme (BATS) (Dickinson et al., 1986) and the simple-biosphere (SiB) model (Sellers et al., 1986).

Why do scaling problems exist? That is, why cannot data or model output be simply scaled up or down to meet the task at hand? Because the gridded output is inherently scale-limited by the grid spacing of the model or observation sensor, little information can be inferred below the smallest grid scale. As pointed out by Dubayah et al. (1996), this can hamper modeling over large areas, especially when the processes are spatially autocorrelated, because most hydrologic processes scale nonlinearly; that is, the moments (e.g., the mean and variance) obtained at one spatial scale may be significantly different from those obtained at a larger or smaller scale. Efforts to parameterize subgrid-scale variability are means for adjusting the statistical properties of fields to incorporate unmodeled fine-scale variability. In fact, very few models participating in the WCRP's Project of Intercomparison of Landsurface Parameters Scheme (PILPS) consider subgrid variability, either implicitly or explicitly. The model of Liang et al. (1994) considers variability in infiltration capacity (and soil moisture) through a distribution function, and the model of Koster and Suarez (1992) uses a mosaic approach for different vegetation types.

The terrestrial water balance, including infiltration, evaporation, and runoff, is known to be a highly nonlinear and spatially variable process, especially because of the key role that surface-soil moisture plays and its spatial and temporal variabilities due to variabilities in soil properties (e.g., Freeze, 1980). Thus the problem:

> What is the linkage between the parameterization of point-scale hydrologic processes and the parameterization of catchment-scale processes?

Study of this problem began in 1984. Figure 1.1, adapted from that early work, provides the framework. In general terms, the problem was stated as follows: Given a point representation of infiltration, drainage, and evapotranspiration, $g\{\theta(x), i(x,t)\}$, that is a function of spatially varying parameters $\theta(x)$ subjected to spatially and temporally varying inputs $i(x,t)$, which results in water-balance (and energy-balance) fluxes $o(x,t)$, we want to know how the integrated function of $g\{\theta(x), i(x,t)\}$, represented by $G(\Theta, I)$, changes as we consider the integrated function of $o(x,t)$ defined by

$$O = \iint_{t\in T, x\in A} o(x,t)\,dx\,dt \tag{1.3}$$

Here, $o(x,t)$ can be viewed as point fluxes, and $O$ as the hillslope, catchment, or AGCM grid scale, depending on the size of the domain $A$. The degree to which

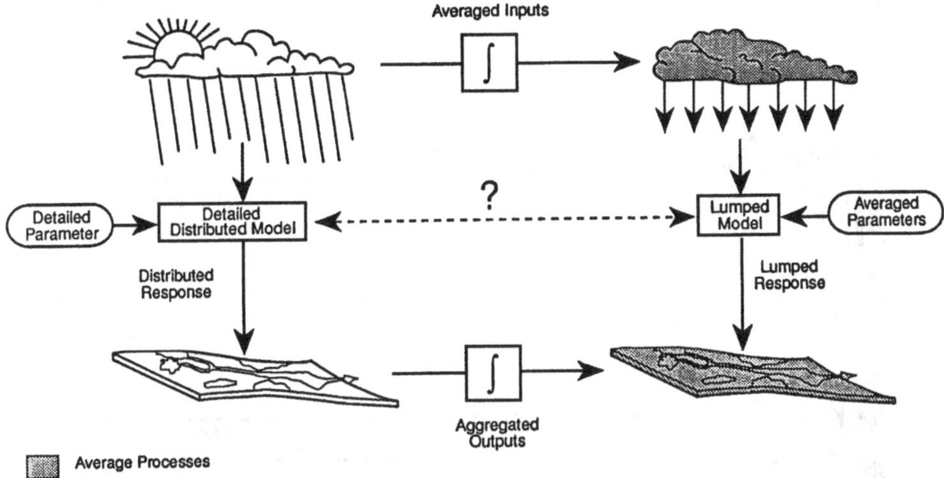


Figure 1.1. Schematic of aggregation and scaling in hydrologic modeling.

the aggregated outputs from distributed models will correspond to the output from lumped models will depend in part on the nonlinearity of the model process equations relative to the input fields, the spatial autocorrelation and scaling properties of those fields, and the amount of model spatial interaction.

The early research focused on the transition $g(\cdot) \to G(\cdot)$ under varying scales and stochastic spatial/temporal correlations. Perfect spatial correlation implies $g(\cdot) = G(\cdot)$. If $g(\cdot)$ is linear, then the stochastically varying variables can be represented by effective (or average) values $\langle O \rangle$ and $\langle I \rangle$, such that $g(\langle O \rangle, \langle I \rangle) = G(\langle O \rangle, \langle I \rangle)$. More discussion of effective parameters appears in Section 1.5. As nonlinearity in the point-scale processes increases, then averaging of the function $g(\cdot)$ occurs, and the functional form of $G(\cdot)$ will be significantly different from $g(\cdot)$.

## 1.3 Understanding the Role of Variability

Understanding the roles of spatial and temporal variabilities is central to understanding scaling. Variability in atmospheric forcings (rainfall and radiation), variability in land-surface characteristics (soil, vegetation, and topography), and variability in land-surface hydrologic processes all affect the transition $g(\cdot) \to G(\cdot)$. In an effort to understand the role of variability, Wood et al. (1988) carried out an empirical averaging experiment. Basically, their study consisted in averaging runoff over small subcatchments, aggregating the subcatchments into larger catchments, and repeating the averaging. By plotting the mean runoff against mean subcatchment area, they noted that the variance decreased until it was rather negligible at a catchment scale of about 1 $km^2$. That analysis has been repeated for the runoff ratio (Wood, 1995) and evaporation (Famiglietti and Wood, 1995) using data from Kings Creek, which was

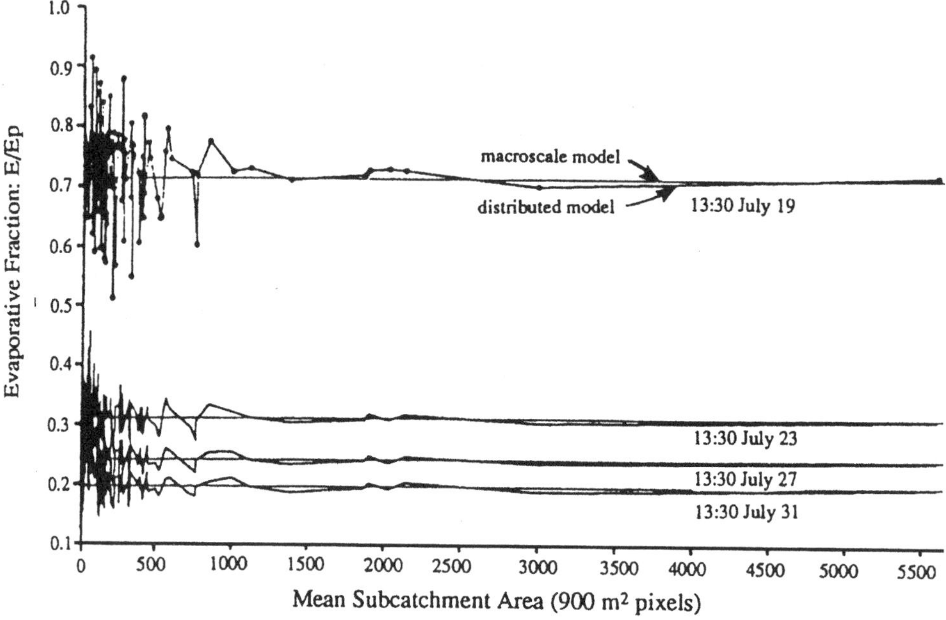


Figure 1.2. Comparison of inter-storm evapotranspiration values from a distributed hydrologic model and from a macroscale model for four times during the July 18–31, 1987, inter-storm period. The modeled catchment is the 11.7-km$^2$ Kings Creek catchment, which is part of the FIFE study area. The distributed model had a grid resolution of 30 m. The catchment was disaggregated into subcatchments, ranging from 66 to 5, depending on the scale of analysis. (From Wood, 1995, with permission.)

part of the FIFE '87 experiment. Figure 1.2 shows a typical result. The behavior of the catchment shows that at small scales there is extensive variability in both runoff and evaporation. This variability appears to be controlled by variabilities in soils and topography whose correlation length scales are on the order of $10^2$–$10^3$ m – typical hillslope scales. At an increased spatial scale, the increased sampling of hillslopes leads to a decrease in the difference between subcatchment responses. At some scale, the variance between hydrologic responses for catchments of the same scale should reach a minimum. Wood et al. (1988) suggest that this threshold scale reflects a representative elementary area (REA), which they propose as a fundamental building block for hydrologic modeling and scaling. As defined by Wood et al. (1988), "the REA is the critical scale at which implicit continuum assumptions can be used without explicit knowledge of the actual pattern of topographic, soil, [vegetation,] or rainfall fields. It is sufficient to represent these fields by their statistical characterization." As pointed out by Beven (1995), the REA concept does *not* denote the scale at which average or equivalent parameters can be used in the continuum (macroscale) descriptions of the fluxes, because at the REA scale the distribution of characteristics may still be important – it is only the pattern of those characteristics that is unimportant.

The REA concept has raised some controversy since it was proposed. It is worth noting that several other hydrologic runoff studies have supported REA dimensions on the order of 1 km$^2$. In addition, a semivariogram from NS001-derived evaporative fraction for the EFEDA/Barrax site has a correlation length on the order of 750 m, which is quite consistent with the REA of 1 km$^2$ (see Fig. 5 of Bastiaanssen et al., 1996). Raupach (1993) describes three levels of heterogeneity from the perspective of the convective boundary layer (CBL). The smallest is the microscale heterogeneity (1–5 km), at which the turbulence and mixed-layer properties in the CBL cannot adjust to variations in the surface conditions. This may very well correspond to a sub-REA scale where pattern is important. Raupach defines macroscale heterogeneity as existing when the CBL turbulence adjusts fully from one "patch" to another, and he suggests that a homogeneous representation of the patch, using equivalent parameters (Raupach and Finnigan, 1995), can be made. I contend that at this scale, the macroscale patch representation may have to include the statistical representation of subgrid variability, as suggested by the REA concept.

Shuttleworth (1988) suggests the existence of "organized" and "disorganized" variabilities in surface cover (cover types A and B, respectively), depending upon whether or not the variabilities in surface conditions are "reflected in, and responsible for, associated changes in the mesoscale meteorology." Organized variability (cover type B) provides an organized response to the CBL and will have correlation lengths greater than about 10 km. As one can see, there is a loose correspondence concerning spatial variability and the scaling of land-surface processes. Nonetheless, as discussed throughout this chapter, I believe that the handling of variability is a critical problem and that the progress to date has been rather limited.

## 1.4 Results from Empirical Model Studies

The use of simplified models that can be solved analytically, or more complex models that must be solved numerically, can provide insights into scaling when used within a sensitivity framework. This allows the effects of spatial variability to be explored in a "what if" framework. The general approach is to develop the results for the model in hand and then make the leap of faith that the results will hold for the real world. All too often authors have taken that leap with great confidence, without the necessary caveats. A critical evaluation of papers using that approach is mandatory. In particular, the following questions should be asked: Are the dynamics too simple? Have critical processes been left out? Are the conditions tested reflective of observed conditions? In many cases, models that are just too simplistic are used under the guise that they "catch the essential features." In other cases the modeler ignores some physical process. One example of this is the paper by Collins and Avissar (1994), in which they conclude that stomatal resistance is the most important factor whose variability must be accounted for in upscaling. Their model has only limited soil-moisture representation, and stomatal resistance accounts for any surface control on evapotranspiration. Readers may misinterpret the essential conclusion that spatial variability in the *soil–vegetation* control is the critical process in understanding the scaling of surface hydrology and that the stomatal resistance is the surrogate for this in the Collins and Avissar (1994) paper.

Have such analyses provided essential insights? Can this type of approach provide essential insights into the scaling of land-surface processes? It is an appealing approach, and my belief is that progress can be made through carefully designed sensitivity studies using simplified models. Unfortunately, too many researchers have carried out such studies because they are relatively easy, often with models that appear mathematically elegant. It is critical that the findings from these models be further verified through observations and/or analyses with more complex land-surface models that have been validated through observations.

### 1.4.1 Results Using Simplified Models

Let us review some results from sensitivity studies in which I have been involved. Although my students and I learned a great deal from these studies, there was the belief that fundamental improvements to our land-surface models were needed before the conclusions reached with the simplified, analytical models could be generalized. One problem that we analyzed (Sivapalan and Wood, 1986; Wood, Sivapalan, and Beven, 1986) was the effect of spatial heterogeneity in soils and rainfall, with and without topographic effects, on the infiltration response of a catchment. Quasi-analytical expressions were derived for the statistics of the ponding time and infiltration rate. Examples of the derived mean areal infiltration rates are shown in Figures 1.3 and 1.4 for the different cases. These results show a strong bias (high) in mean infiltration rates if only mean values of soil properties and rainfall intensities are used in conjunction with point-scale infiltration rates. The results also show that the critical variable in understanding the spatial response is the time to ponding, its space–time distribution, and the resulting proportion of the catchment that is saturated. The upward bias has the effect of overestimating soil moisture and underestimating runoff.

The spatial correlation of the infiltration rate is an important variable in understanding the scale effects of runoff production. The findings of Sivapalan and Wood (1986) and Wood et al. (1986) indicate that the inclusion of soil variability has the effect of reducing the correlation and scale of variability in the infiltration rate. Thus, the results suggest opposing roles for rainfall and soil variability, with the importance of each depending on the relative magnitude of its variance. Whereas these results were derived for idealized catchments, the findings can provide important insights for actual catchments. For example, the constant-rainfall/variable-soil case may relate to medium-size, geologically variable catchments subjected to large-scale cyclonic events of relatively low rainfall intensities, whereas the variable-rainfall/constant-soil case could be related to small homogeneous catchments subjected to convective rain events.

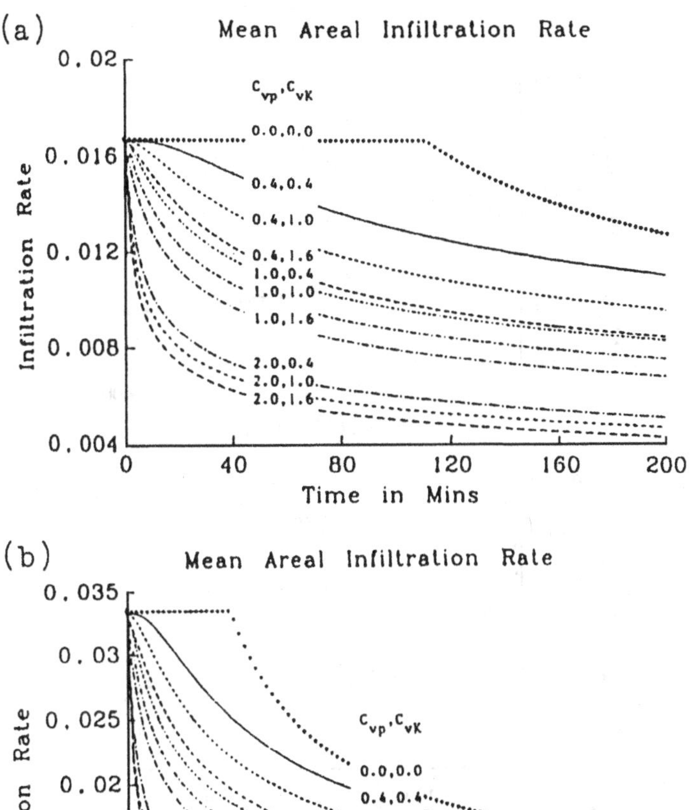


Figure 1.3. Mean areal infiltration rates due to spatially variable soils and rainfall. $C_{vp}$ and $C_{vK}$ refer to the coefficients of variation in precipitation and saturated-soil conductivity, respectively. The average saturated-soil conductivity was 0.008325 cm/min for both cases: (a) mean precipitation, 0.01665 cm/min; (b) mean precipitation, 0.0333 cm/min. (From Wood et al., 1986, with permission.)

This line of research was set aside (by myself) about 10 years ago to focus on improving the underlying process representation of the land surface, specifically the inclusion of vegetation and a surface energy balance, and to test these parameterizations against field data collected under FIFE and remote-sensing soil-moisture studies. As will be discussed further in a later section, Sellers, Heiser, and Hall (1992b), using analyses based on FIFE '87 data, suggested that land–atmosphere models are almost scale-invariant. Wood (1994) presented the counterargument, illustrating the mechanism for increased spatial variability in evapotranspiration under

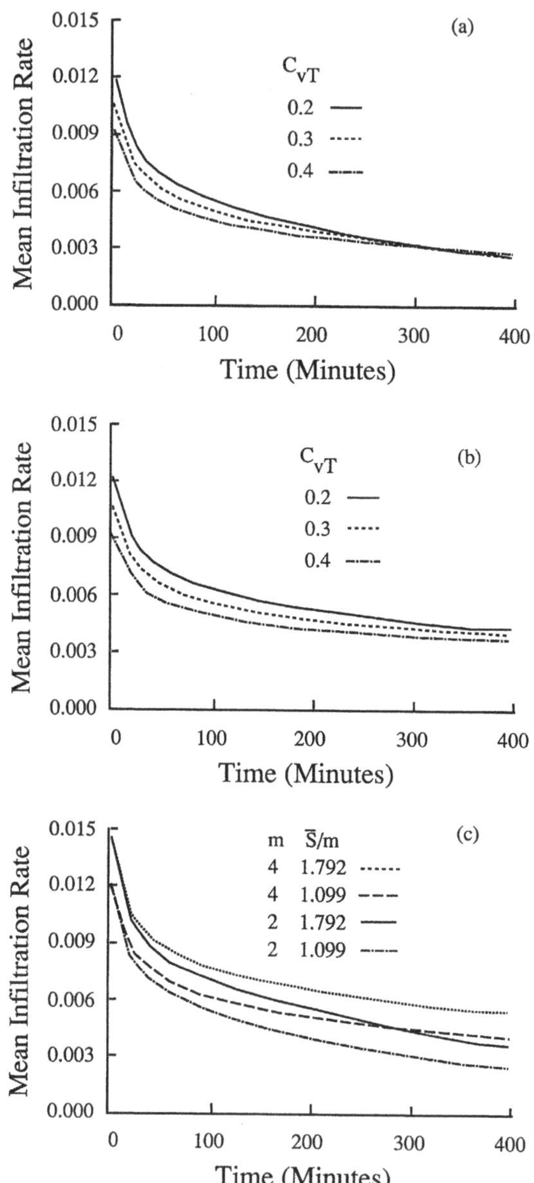


Figure 1.4. Mean areal infiltration rates due to spatially variable soils, rainfall, and topography. $C_{vp}$, $C_{vK}$, and $C_{vT}$ refer to the coefficients of variation in precipitation, saturated-soil conductivity, and topographic index, respectively. The average precipitation was 0.01665 cm/min, and average saturated-soil conductivity 0.008325 cm/min, for all cases: (a) $\bar{S}/m = 1.0986$, $m = 2$; (b) $\bar{S}/m = 1.0986$, $m = 4$; (c) $C_{vT} = 0.5$. The variable $\bar{S}$ refers to the catchment-mean soil-moisture deficit, and $m$ is a recession parameter within TOPMODEL (Beven and Kirkby, 1979). (From Wood et al., 1986, with permission.)

conditions of variable soil moisture and/or high atmospheric evaporative demand. More recently, an analytical model for the evaporative fraction (actual evapotranspiration/potential) has been developed (Wood, 1997) in order to gain insight into the effect of spatially aggregating soil moisture. The effect of aggregation is captured in an "evaporation scaling ratio," which is the ratio of the average evaporative fraction considering soil-moisture variability to the average evaporative fraction using aggregated soil moisture. The sensitivity of the evaporation scaling ratio depends on three parameters:

1. the ratio of the equilibrium soil moisture to the mean soil moisture, where the equilibrium soil moisture is defined as the soil-moisture level at which the atmospheric evaporative potential and the ability of the soil–vegetation system to satisfy this potential are in equilibrium (i.e., a land–atmosphere equilibrium with regard to soil moisture, which will vary temporally),
2. the soil–vegetation sensitivity factor to soil-moisture stress, which varies for different soils and vegetation types, and
3. the variability of soil moisture within a region, as measured by its coefficient of variation.

A series of sensitivity experiments was conducted by varying the soil–vegetation sensitivity factor and the soil-moisture variability. Figure 1.5 shows an example result. The analysis suggests that the practice of aggregating soil-moisture levels results in evapotranspiration calculations that overestimate actual evaporation during periods of low atmospheric demand (early morning, late afternoon, winter periods, etc.) and underestimate evapotranspiration during periods of high demand (midday summer periods). The conclusion is that the advisability of using aggregated soil-moisture levels (or "effective" parameters) in lumped macroscale models depends on the variability of soil moisture and the sensitivity of the soil–vegetation system to low moisture levels. This work, along with that of Collins and Avissar (1994) and others, emphasizes the need to better understand how vegetation reacts to environmental stress, such that the evapotranspiration rates are below the potential – an unresolved problem in the parameterization of land-surface schemes.

### *1.4.2 Results Using Complex Simulation Models*

These findings need to be verified through more detailed modeling over a wide range of conditions and (hopefully) through some observations. It is suggested that land-surface climatology experiments and soil-moisture experiments could provide the necessary input data for such studies. Recently there have been several studies looking at the roles of soil and rainfall variabilities within the context of understanding

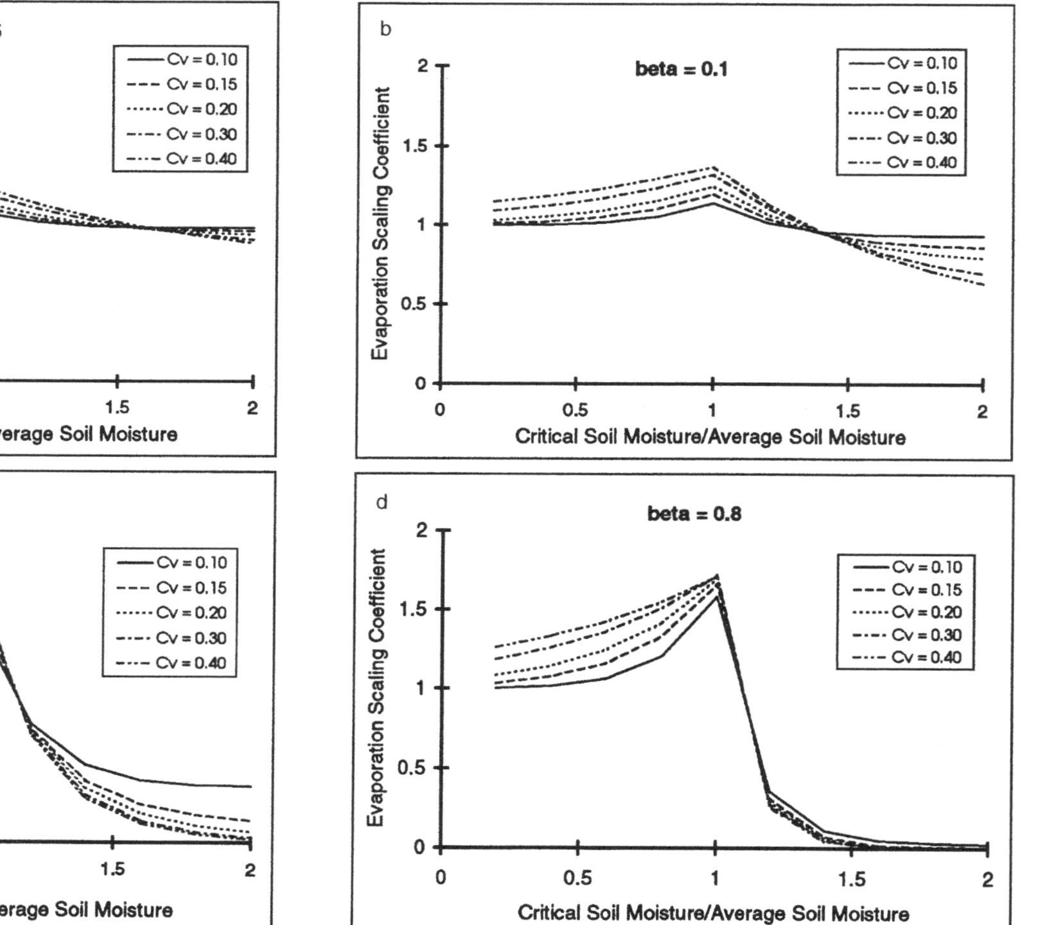


Figure 1.5. Evaporation scaling coefficient versus the ratio of critical soil moisture to areally averaged soil moisture for a soil-moisture coefficient of variation $C_v$ ranging from 0.10 to 0.40: (a) $\beta = 0.05$; (b) $\beta = 0.10$; (c) $\beta = 0.40$; (d) $\beta = 0.80$. The variable $\beta$ refers to the sensitivity of the soil–vegetation system to soil moisture, with higher values being more sensitive; the "critical" soil moisture refers to the soil-moisture level where the soil–vegetation system decreases evapotranspiration relative to the potential level. (From Wood, 1997, with permission.)

the importance of parameterizing the subgrid variabilities of soil, vegetation, and precipitation in AGCMs.

I shall now comment on certain aspects of three sensitivity modeling studies (Wood, Lettenmaier, and Arola, 1994; Liang, Lettenmaier, and Wood, 1996; Bonan, 1996). All of those studies supported the general insights gained from the simpler analyses, but the complexity of the land-surface model (or AGCM) severely limited the simulation analyses that could be carried out – both in domain size and in length of simulation. In the study by Wood et al. (1994), the land-surface hydrology model of Wigmosta, Vail, and Lettenmaier (1994) was used to simulate the water-and-energy balance over the 525-$km^2$ Little Washita, OK, catchment from September through December 1993 at a grid resolution of 120 m. Hourly precipitation data were from the then recently installed WSR 88D Doppler rain radar, with a product pixel resolution of $4 \times 4$ km. That study tested the sensitivity of catchment-aggregated moisture and energy fluxes and state variables (such as surface temperature), as well as their spatial distributions, to three aggregation levels of precipitation products: 4 km, 20 km, and catchment average. The results were that the lowest resolution (catchment-average precipitation) gave the highest diurnal peaks in latent heat, and the highest resolution ($4 \times 4$ km) gave the lowest diurnal peaks.

These effects are attributable to differences in the generation of direct runoff and differences in vegetation-interception storage. In particular, uniform precipitation results in all of the catchment experiencing evaporation of water intercepted by the canopy, whereas with higher-resolution (more spatially variable) precipitation, some parts of the catchment may remain dry. The role of canopy interception is expected to be most important in the summer, when solar radiation is high and the spatial variability of precipitation is greatest, and in areas (e.g., forest) where canopy interception is greater. It has also been found that the highest-resolution precipitation results in locally higher soil-moisture contents, and hence locally greater extents of saturated areas and more runoff, than does lower-resolution precipitation.

A more comprehensive simulation study by Liang et al. (1996) was carried out to develop a parameterization for subgrid fractional rainfall appropriate to the variable-infiltration-capacity two-layer (VIC-2L) land-surface model. In that study, the (simulated) AGCM grid was divided into 2,500 pixels, each independently modeled, which served as "ground truth." As in the previous work, the assumption of uniform rainfall resulted in readings of higher evaporation, lower sensible heat flux, much smaller runoff, and higher soil moisture, as compared with the case of subgrid precipitation variability. For the forest-site simulation, with fractional rainfall coverage of 30% (of the grid area), the higher evaporation was 36% on an annual basis. For the grassland case, similar results were obtained.

Bonan (1996) evaluated a subgrid parameterization of infiltration and surface runoff in a land-surface model coupled to an AGCM. Using an average over a 5-year simulation, he found that consideration of subgrid precipitation variability resulted in model predictions of lower soil moisture and drier soils, less evaporation, and higher surface air temperatures. Bonan found that some of the predictions (using the subgrid parameterization) for large continental basins agreed better with observations (and some worse) than did the results from the uniform-precipitation AGCM simulations. He attributed that to the relative importance of geologic controls on the different runoff mechanisms in the different basins.

I would suggest that there are more complex interplays between soil and rainfall variabilities, between soil-moisture and evapotranspiration variabilities, and between "potential" evapotranspiration (where atmospheric conditions control the evaporative demand) and actual evapotranspiration (where the soil–vegetation system controls the evaporative supply) than are generally recognized. An understanding of these competing conditions, both for differing climatic regions and over time for specific locations, will be central to an understanding of scaling – especially the transformation $g(\cdot) \rightarrow G(\cdot)$ discussed earlier.

### *1.4.3 Results from Empirical Studies Based on Large-Scale Field Experiments*

In the planning of the HAPEX and ISLSCP climate experiments there was an assumption that the collected data would be sufficient to carry out the scaling analyses implied by Eagleson's WCRP/ICSU Joint Scientific Committee Working Group on Land Surface Processes (Eagleson, 1982). For example, in the First ISLSCP Field Experiment (FIFE) the first goal of the study was an upscale integration of models, which Sellers et al. (1992a) described as being to "test the soil-plant-atmosphere models developed for small-scale applications and to develop methods to apply them at the larger scales (kilometers) appropriate to atmospheric models and satellite remote sensing." In the ISLSCP/BOREAS research announcement (NASA, 1992), a main objective was to "develop and validate remote sensing algorithms to transfer our understanding of the above processes [exchanges of energy and water between the boreal forest and the atmosphere] from local scales to regional scales." In addition, there have been experiments under the HAPEX program in France (HAPEX/MOBILY) and western Africa (HAPEX/SAHEL) and the EFEDA experiment in Spain (Bolle, 1993). What have the experiments yielded?

Sellers et al. (1992b) carried out a modeling study with the FIFE '87 data to test the scale invariance of the surface energy balance. The study divided up the domain ($15 \times 15$ km) into 225 subunits, interpolated atmospheric forcings and soil-moisture contents from station data, used high-resolution Thermatic Mapper (TM) data for vegetation, and calculated, using the simple-biosphere (SiB) model (Sellers et al., 1986), the spatial distributions of latent heat, sensible heat fluxes, and canopy conductance. The averaged fields were then compared to the fluxes determined by SiB using domain-averaged atmospheric forcings and soil-moisture inputs. They carried out four one-day simulations, using "golden days" that tended to be clear, well-watered (well-behaved) days. The modeling results showed agreement between the distributed and aggregated simulations to within 3%. Similar results, for a one-day simulation using HAPEX-MOBILY data, have been reported by Noilhan (1994).

An important question arises as to the reason for the difference between these reported findings and the suggested results from hypothetical modeling studies. Let us suggest two reasons: (1) the focus on rain-free golden days, with clear skies and moist conditions, days that tend to feature more scale invariance than rainy days; (2) the simulation of single days and the use of interpolated soil moisture to initialize those runs. It is important to note that the findings of Liang et al. (1996), Bonan (1996), Wood et al. (1994), and similar studies all arose from the subgrid variability of precipitation and the important role that rainfall interception by vegetation and infiltration plays in modeled variability. The single-day studies of dry golden days, initialized with station-interpolated soil moisture, seem to offer only very weak findings regarding scale invariance (as a colleague noted, "It never rains on golden days").

Peters-Lidard, Zion, and Wood (1997) carried out continuous water-and-energy-balance computations for the FIFE '87 experimental period using a land-surface model whose spatial variability in moisture arose from the topographic redistribution suggested by Beven and Kirkby (1979). Peters-Lidard et al. (1997) found that a partial-rainfall-coverage factor was needed for the interception storage, or else the latent-heat fluxes would have been too large. For the infiltration computations in the model, uniform net rainfall was assumed. That resulted in higher FIFE-area average soil-moisture values (consistent with the hypothetical modeling studies) for a day or so, by which time the modeled soil-moisture content for the top 10 cm was consistent with measured values. Although the FIFE-area station-averaged water- and energy-flux data were well matched with the area average estimated by the model, the model underestimated the flux variability across the stations.

The ability of a model to match the observed variability across an experimental domain is critical if model scaling is to be carried out. Using data from the ISLSCP/BOREAS experiment, Cooper et al. (in press) modeled the fine-scale surface fluxes over a 75,000-km$^2$ region within the large-scale BOREAS study area under cloud-free conditions. The input data included surface meteorological observations, soil and vegetation parameters, and satellite-retrieved surface-radiation fluxes, and the biosphere–atmosphere flux-exchange model has been described by Smith et al. (1993). A one-day (golden day) simulation of DOY 216 (August 4, 1995) was carried out to assess the errors that would arise in flux-aggregation schemes based on mean values in the forcing variables and input data (land-surface characteristics). In all, 14 model experiments were conducted to investigate 10 major aggregation options, which included using averaged forcings and land-surface characteristics and states. The most significant effects resulted from averaging soil properties and/or soil moisture. For the baseline run, the observed soil-moisture values obtained for all of the BOREAS tower flux sites were applied to grid points with similar vegetations and soil classes. They found the most significant sensitivity to be in the handling of spatial variability in soil properties and soil moisture. The range of variability, relative to the base case, went from an increase in latent heat of over 70% (constant soil moisture, meteorological inputs, and land characteristics) to a decrease in latent heat to 33% of the base case (constant soil moisture, but variable meteorological inputs and land characteristics). It appears that there were inconsistencies between the soil-moisture characteristics and land-surface characteristics that resulted in overestimation or underestimation of the latent heat fluxes. It is important to note that those findings underestimated the aggregation problems, because they were for only a one-day simulation.

For long-term simulations, input variability, especially for precipitation, must be maintained, or else the derived soil-moisture variability will be too low, which may bias the latent and sensible heat fluxes, as noted by Cooper et al. (in press). Figure 1.6 shows soil-moisture simulations for the May–October 1987 FIFE period carried out by Peters-Lidard et al. (1997). The macroscale land-surface model used a soil topographic index to represent and control soil-moisture variability in the root/transmission zone. Because of the lack of rainfall variability, the modeled

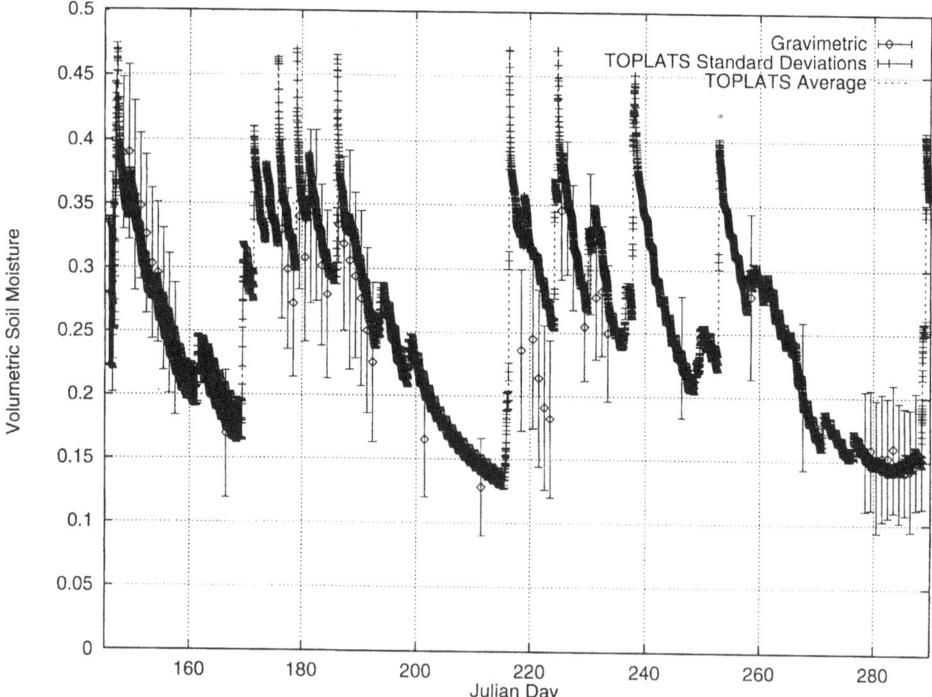


Figure 1.6. Time series of model-derived average soil-moisture values for the FIFE '87 study compared with observed gravimetric soil-moisture values from 37 sites.

surface-soil moisture underestimated the observed soil-moisture variability – to a great extent because the model underrepresented the variability in surface-soil characterization.

It appears to me that much of the soil-physics research into soil variability has not been effectively transferred (or understood) by those studying land-surface coupling. Work reported by Nielsen, Bigger, and Erh (1973), Beven (1983), and Schmugge and Jackson (1996) has provided a basis for appreciating the variability that is observed in the field and, subsequently, the spatial controls that can be exerted by the soil–vegetation system on evapotranspiration. In general, climate experiments like FIFE and BOREAS have failed to provide in their experimental designs the collection of spatial data necessary to develop and test scaling theories. One wonders if that was because of a lack of understanding, by the experiment organizers, of these land-surface controls. Part of that problem was addressed in the BOREAS experiment. Fortunately, the National Science Foundation (NSF), through its Earth Sciences Hydrology Program, and Environment Canada provided extra money for a rain radar for the BOREAS Southern Study Area (SSA). It is expected that the resulting data set, in conjunction with satellite-derived solar-radiation data, will provide the necessary distributed forcing data to explore, develop, and test scaling theories. Hydrologists working in the BOREAS study have argued, without much success, for improved soil data; the work of Cooper et al. (in press) suggests that this will be important for upscaling the tower observations for the BOREAS region.

## 1.5 Flux Matching and Equivalent Parameters

A scaling approach widely used by micrometeorologists for modeling fluxes over heterogeneous landscapes is often referred to as "flux matching." Examples of this approach include the work of Lhomme (1992), Raupach (1993), and Lhomme, Chehbouni, and Monteny (1994), as well as many of the papers in the volume edited by Stewart et al. (1996) – the latter being workshop proceedings on land-surface hydrologic scaling through remote sensing. The problem can be posed as follows:

> Given a point-scale model $g\{\theta(x), i(x,t)\}$ (of surface fluxes, for example), where the parameters $\theta(x)$ and inputs $i(x,t)$ vary spatially, find the equivalent parameters $\Theta$ (and inputs $I$) such that the "lumped" flux will equal the aggregated flux, assuming that the point-scale model still holds. That is,

$$g\{\Theta, I\} = \int_A g\{\theta(x), i(x,t)\}\, dA(x) \tag{1.4}$$

Then the parameters $\Theta$ are referred to as "equivalent" (or sometimes "effective") parameters, and equation (1.4) is referred to as "flux matching." This procedure is often referred to as "upscaling" (Stewart et al., 1996). What are the essential issues? Assuming a homogeneous surface, the sensible ($H$) and latent ($\lambda E$) heat fluxes can be described by the following expressions (Brutsaert, 1982):

$$H = \rho C_p \frac{(T_s - T_a)}{r_a} \tag{1.5}$$

$$\lambda E = \frac{\rho C_p}{\gamma} \left[ \frac{e^*(T_s) - e_a}{r_a - r_s} \right] \tag{1.6}$$

where $\rho$ is air density, $C_p$ is specific heat at constant pressure, $\gamma$ is the psychrometric constant, $T_s$ is surface temperature, $T_a$ and $e_a$ are the temperature and vapor pressure of the air at a reference height $z$ above the ground, $r_a$ is the aerodynamic resistance, $r_s$ is the surface (soil or vegetation) resistance, and $e^*(T_s)$ is the saturated vapor pressure at the surface temperature. Many investigations have attempted to find "equivalent" resistances such that equation (1.6) aggregated over the area $A$ will match observations of latent heat at a point. That is, by matching the aggregated latent heat $\langle \lambda E \rangle = \int_A \lambda E \, dA$, they define an equivalent (or effective) resistance from (1.6) as being

$$\langle r_a + r_s \rangle^{-1} = \int_A (r_a + r_s)^{-1} \, dA \tag{1.7}$$

A potential problem with this approach is that the resistances are themselves parameterizations of the heterogeneous landscape. For example, the surface resistance is often represented as (Jarvis, 1976)

$$(r_s)^{-1} = g_c^* f(\delta e) f(T_w) f(\theta) \tag{1.8}$$

where $g_c^*$ is the conductance under unstressed surface conditions and $f(\cdot)$ are reduction factors for the conductance because of vapor-pressure deficit gradients $\delta e$, soil temperatures $T_w$, and soil moisture $\theta$. All of these functions are in themselves parameterized (e.g., Sellers et al., 1992b).

There is an implicit assumption in the use of effective parameters that the form of the relationships, that is, the point-scale homogeneous relationships, is scale-invariant; only the constants in them need to be varied to account for heterogeneity. The published research, to my knowledge, has tested this approach only with idealized, hypothetical modeling studies (e.g., Lhomme et al., 1994; Raupach and Finnigan, 1995) or by testing the effective parameters on isolated (golden-day) data sets of limited heterogeneity (Stewart et al., 1996). There have been several examples in which researchers have tried to upscale models through the use of remote sensing in an attempt to match the areal, remotely sensed fluxes with point-flux tower data.

Surely this is going in the wrong direction, and it reminds me of attempts in the groundwater-transport literature to find "equivalent" dispersion parameters.

Further testing of the concept of flux matching and the use of equivalent parameters is critical. These tests must be over large areas ($> 10^6$ km$^2$) that feature significant contrasts and they should last for at least a year. The challenge is in understanding how the equivalent parameters vary, spatially and temporally, as surface conditions change.

## 1.6 Similarity Theory and Scaling

The use of similarity theory is well developed in fields related to hydrology, such as hydrodynamics and turbulence, but only limited research has been done in hydrology, and virtually none in the land-surface modeling described in this chapter. In hydrology, the concepts of the unit hydrograph embed the concepts of similarity: Two catchments with the same unit hydrograph will produce the same discharge hydrograph for the same excess-rainfall input. Rodríguez-Iturbe and Valdes (1979) developed a parameterization for the unit hydrograph based on a macroscale characterization of river-channel networks. The runoff-production model of Beven and Kirkby (1979), TOPMODEL, embeds a similarity principle in that two locations within a catchment with the same soil-topographic index will respond similarly, and two catchments with the same distribution of soil-topographic indices will respond in similar manners.

These ideas have been incorporated to investigate the similarities in flood-frequency responses of catchments (Hebson and Wood, 1986; Sivapalan, Wood, and Beven, 1990) and runoff production in catchments (Sivapalan, Beven, and Wood, 1987). Sivapalan et al. (1987) defined a set of similarity parameters that incorporated rainstorm intensity and variability, soil characteristics and variability, and catchment-response variability. Larson et al. (1994) extended that approach and tested it on a number of small agricultural catchments in Western Australia. Robinson and Sivapalan (1995) used the results of Larson et al. (1994) to develop a spatially lumped model of runoff generation at the catchment scale. The studies by Larson et al. (1994) and Robinson and Sivapalan (1995) made important contributions to the scaling of catchment runoff.

The foregoing approach appears promising for extension to the surface energy balance. What are the challenges? In hydrology there is a history of water-balance models that implicitly incorporate spatial variability. This is missing in the surface-energy models. Thus, the approach to date has been the development of equivalent parameters for upscaled point models. Until we can develop surface-energy models that incorporate the heterogeneity in surface-energy partitioning, I think that limited progress will be made in applying similarity concepts.

## 1.7 Statistical Self-similarity: Its Potential for Scaling Land-Surface Fluxes

As we have seen, almost all the scaling results for land-surface water and energy fluxes are from empirical analyses. The potential exists for applying statistical self-similarity to the scaling of surface fluxes. Wood (1997) and Dubayah et al. (1996) have applied statistical self-similarity to the scaling of soil moisture, and the potential exists for its application to the fluxes of surface water and energy. Consider the scaling of the variable $O(A)$, where $A$ refers to a spatial-scale parameter and where the parametric dependence of $O$ on $A$ implies statistical homogeneity within a region. The process $O$ is defined as spatially scaling if the following holds:

$$E[O^q(A_i)] = \left(\frac{A_j}{A_i}\right)^{K(q)} E[O^q(A_j)] \tag{1.9}$$

where $q$ is the order moment, $K(q)$ is a set of scaling exponents associated with the moments, and the ratio $A_j/A_i$ is the scale parameter $\lambda$. If the spatial scale $A_j$ corresponds to the coarsest scale (coarsest grid spacing), then finer scales (finer grid spacings) corresponding to scale $A_i$ will have scale parameters going from $\lambda = 1$ (for the coarsest grid) to $\lambda = \frac{1}{2}, \frac{1}{4}, \frac{1}{8}$, and so forth, as the resolution is increased from $A_j$, $4A_j$, $16A_j$, and $64A_j$ values, and so on, down to the finest resolution. The process is said to be simple scaling if the scaling exponents are linearly related to the moment order $q$, that is, $K(q) = qC$; otherwise it is multiscaling.

As discussed by Dubayah et al. (1996), if the scaling exponents are known or can be determined at the scale of observation, that implies that the process can be modeled or observed at one scale and its statistical properties inferred at another scale. Dubayah et al. (1996) applied a scaling analysis to both remotely sensed soil moisture (airborne ESTAR L-band, passive microwave radiometer) and modeled soil moisture over the 525-km$^2$ Little Washita, OK, catchment during the period June 10–18, 1992. The catchment started out very wet, and there was a consistent decline in the near-surface soil moisture during the nine days. The surface-soil moisture was modeled using a semidistributed water-and-energy-balance model with a spatial modeling scale of 30 m.

For the model-derived soil moisture, the largest scale factor had an aggregation size of 128 × 128 pixels (3,840 m), with the smallest scale factor equal to $1/128$. Results for the land-surface-model-derived soil moisture are shown in Figures 1.7 and 1.8. Figure 1.7 shows log plots of moments $q$ versus the scale factor $\lambda$ for eight days. The modeled fields show reasonable log-log linearity over a wide range of scales. Figure 1.8 plots the scaling exponents $K(q)$ against the moments $q$ to look for signals of multiscaling. There are signals of multiscaling for four days, June 10–13, where the curves are nonlinear and concave. Multiscaling was also observed in the analysis of the ESTAR-derived remotely sensed data.

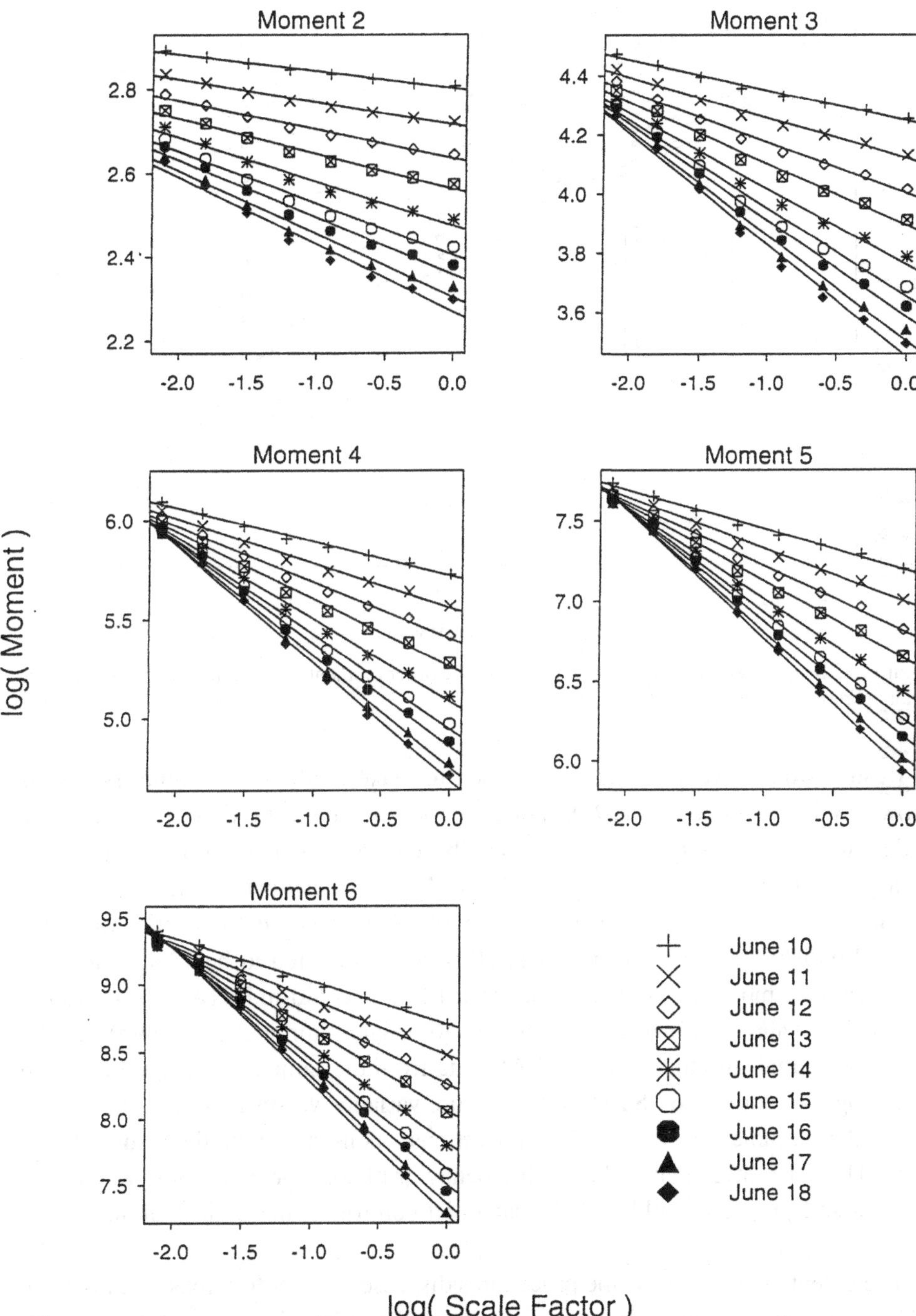


Figure 1.7. Log(moment) versus log(scale factor) for model-derived soil-moisture fields. The range of scale factors used aggegrations over areas from 30 m to 3,840 m. (From Dubayah et al., 1996, with permission.)

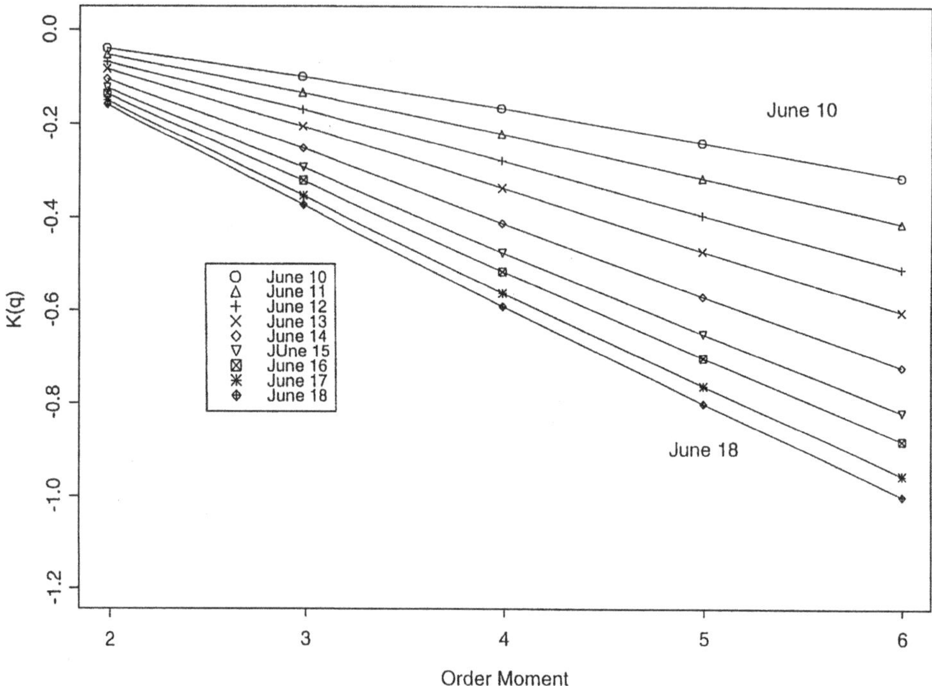


Figure 1.8. Scaling exponent $K(q)$ as a function of the order moment $q$ for the modeled soil-moisture fields. (From Dubayah et al., 1996, with permission.)

Results from these analyses can be used to make predictions at other scales, as shown by Dubayah et al. (1996). For example, they took the ESTAR data, which had a finest resolution of 200 m, and used those data to estimate soil-moisture variability at 30 m, the finest modeled resolution. Using the estimated scaling exponents for the second moment, $K(2)$, from the ESTAR analysis, the expected value for the soil-moisture standard deviation at 30 m was estimated to be 12.9%, as compared with the model-derived value of about 14%. The 200-m ESTAR soil-moisture standard deviation for that day was 6.6%. Rodríguez-Iturbe et al. (1995) investigated the spatial structure of the ESTAR data from the same experiment and found that a log-log plot of the ESTAR soil-moisture variance versus area was linear, with slopes that became steeper with drier conditions – consistent with the results shown here. They also investigated the scaling behavior of soil porosity to test whether or not that soil property could be a fundamental control on the scaling of soil moisture.

The potential exists to use the procedures discussed earlier for areas ranging from AGCM grids of $10^4$ km$^2$ to large catchments on the order of $10^6$ km$^2$. Limited studies across scales could be conducted to first determine scaling properties. Modeling (or remote-sensing observations) could then be conducted using a resolution on the order of 1 km$^2$, from which the scaling coefficients $K(q)$ could be estimated. Thus inferences about variability at finer grid spacings (subgrid variability) could be made.

In a best-case scenario, direct links might be developed between $K(q)$ and environmental conditions such as topography, vegetation, soils, and so forth (Dubayah et al., 1996). Such links will be critical to understanding the physical mechanisms by which surface water and energy fluxes are manifested across scales over the landscape.

## 1.8 Summary of Progress to Date

This chapter has attempted to review the progress in the scaling of land-surface hydrology, particularly the fluxes of water and energy. Recognition of the need for scaling dates back to the 1970s, when hydrologic processes were being incorporated into AGCMs. The desire to understand how land-surface hydrologic processes could be represented over grid sizes of $10^5$ km$^2$ motivated the climate-experiment programs like HAPEX, ISLSCP, and BAHC referred to earlier. Each of those programs has as a research focus the scaling of hydrologic processes. It is reasonable to summarize where progress has been made. My evaluation is as follows:

*Process Parameterization at Small Scales.* I think there has been significant progress because of the data collected during the climate experiments. Ten years ago, hydrologists did not think of coupled water-and-energy-balance models, let alone the coupling of the land surface to the CBL. The model experiments carried out under the PILPS program have enhanced the community's understanding of how different parameterizations behave and of the importance of land-surface controls on the surface water-and-energy balance.

*Scaling at the $10^3$-km$^2$ Scale.* Minor progress has been made. Recent climate experiments (e.g., FIFE, HAPEX/MOBILY, HAPEX/SAHEL) have collected data at this scale, but the designs of the experiments have not lent themselves to rigorous water-and-energy-flux scaling studies. The work that has been done to date has tended to feature empirical scaling studies on golden days. There is a need to develop improved theories and to make the observations to test them.

*Scaling of Hydrologic Remote Sensing.* There has been almost no progress. The potential for developing scaling theories for water and energy fluxes derived (or inferred) from remotely sensed data is significant. The reason for the lack of progress is the lack of data. It never fails to amaze me that given all of our satellites, we have produced so few usable data sets.

*Scaling at the Continental Scale ($\sim 10^6$ km$^2$).* Some progress has been made because of the Global Energy and Water Experiment (GEWEX) activities in the Mississippi River basin, the Mackenzie basin, and the Baltic Sea region (BALTEX). There have been demonstrations that land-surface models can describe water balances at continental scales over long periods (Abdulla et al., 1996; Pauwels, Wood, and Lettenmaier, 1996).

To further explore quantitative scaling of land-surface water and energy fluxes, I am convinced that we need to move up to large scales ($\sim 10^6$ km$^2$). To that end I am compiling a land-surface data base that will allow multiyear modeling of water and energy fluxes at a 1-km resolution over the 625,000-km$^2$ basins of the Red River and Arkansas River. Until data sets of this type are created, there can be no significant progress beyond the work reviewed in this chapter.

Is it possible to look down the path and see where developments may provide a way forward? Or at least potential areas for further research beyond the areas discussed earlier?

*The Limits of Reductionism.* Over the past 10 years I and my students and colleagues have incorporated additional processes into our water-and-energy-balance models. At some point this becomes a futile effort, and the added complexity does not add to the ability of models to represent land-surface exchanges at the scale of our measurements. This is especially true of measurements over heterogeneous areas. Recently, Keith Beven (University of Lancaster, U.K.) has clearly demonstrated the over-parameterization problem through his Generalized Likelihood Uncertainty Estimation (GLUE) analyses and "dotty" diagrams (Beven, 1995; Freer, Beven, and Ambroise, 1996; Franks et al., 1997). It is unclear at what scale the adding of complexity becomes futile, or at any particular scale which processes control land–atmosphere exchanges and how these controls may vary with land-surface conditions. I am fairly convinced that simpler (reduced-domain) models can perform as well as complex models, but how to represent scale-dependent process equations in the simpler models is still unresolved. It is my hope that large-scale detailed simulations will provide the necessary surrogate data sets to explore simpler models.

*Upscaling and Downscaling.* These terms have been widely and somewhat loosely used in the literature. To me, "upscaling" means relating measurements taken at small scales to estimates at large scales. Without being called by that name, this has been done for years with point data such as precipitation data – going from gauge data to catchment-area estimates. "Downscaling" is taking large-scale data (usually spatial) and determining the statistical characteristics at smaller scales. Sometimes "downscaling" is used to refer to the matching of large-scale data to point data. An unappreciated area is the interpretation and use of downscaled data, especially when point observations are available. This requires an understanding of the spatial statistical characteristics of the field in question.

The issues involved in upscaling and downscaling have been fairly well discussed in this chapter, and they require the development of techniques discussed in Section 1.7. Unfortunately, most analyses have been done rather empirically, especially with satellite data, where the focus has been on simply matching data between the satellite and point observations. I am convinced we can do better, but again consistent data sets are needed.

*Design of Large-Scale Experiments.* It is my belief that large-scale field experiments can be better designed through premeasurement hydrologic and ecosystem modeling to help design and guide the experiment. When the BOREAS experiment was in its early planning stages, Steve Running (University of Montana) suggested that preliminary ecosystem modeling be done, using the currently available data, to help design the experiment. Had that been done, the surprise of observing very low latent-heat fluxes might have been avoided, and the high Bowen ratios observed across the study area might have been accounted for in the experimental design. A fairly rough water balance should have indicated those low evapotranspiration rates and possibly could have led to different types of measurements that would have been of help in understanding the evapotranspiration controls by vegetation in that boreal environment.

The standard stratified-design approach (one set of measurements over each vegetation type, one over each soil type, one duplication, etc.) confuses the taking of measurements with the attempt to understand the underlying processes at the experimental-site scale. It seems to me that the important questions are related to the mean values for the flux and storage terms (water, energy, net primary production, etc.), their spatial and temporal variabilities, and the conditions that control both the mean and the variance. The question of how much variability is expected within a vegetation or soil class, versus across different classes, is central to the experimental design. I believe that modeling can certainly contribute in that area, and it will be interesting to determine the value of such "up-front" analyses.

During the past 15 years there has been a significant increase in the awareness that an understanding of the scaling properties of land-surface water and energy exchanges is an important hydrologic problem. Further understanding will be needed for improved climate- and weather-prediction models, for the design of satellite sensors for parameters such as soil moisture, and for the utilization of current satellite data (such as surface radiation from the Geostationary Operational Environmental Satellite, GOES) in models. The increased awareness of such needs has led to only marginal efforts to develop the data sets that are crucial for further progress. It behooves the hydrologic community to press forward toward such data sets so that progress on important scaling problems can continue.

## Acknowledgments

The knowledge I have gained and the findings I have presented in this chapter have been due to the wonderful students, postdoctorates, and colleagues who have interacted and contributed to this work and to the funding agencies whose support was instrumental to this research. In particular, I would like to recognize my former student Murugesu Sivapalan, who worked on the early scaling research and has continued to pursue extremely innovative scaling research at the University of Western Australia, my recently departed student Christa Peters-Lidard (Georgia Institute of Technology), and my research associate Mark Zion. Keith Beven (University of Lancaster, U.K.) and Dennis Lettenmaier (University of Washington) have been steadfast friends and colleagues from whose ideas I have benefited enormously through our discussions and joint work in hydrologic modeling, scaling, and field research. My early research in scaling has been supported through grants from the National Aeronautics and Space Administration (NASA) (NAG-5-491, NAG-1392) and the United States Geological Survey (14-08-0001-G1138); the scaling research with data from climate and remote-sensing field programs has been supported through NASA grants (NAG-5-899 and NAG-5-1628). This support is gratefully acknowledged and appreciated.

## References

Abbott, M. B., Bathurst, J. C., Cunge, J. C., O'Connell, P. E., and Rasmussen, J. 1986a. An introduction to the European hydrological system – systeme hydrologique European SHE. 1. History and philosophy of a physically-based distributed modeling system. *J. Hydrol.* 87:45–59.

Abbott, M. B., Bathurst, J. C., Cunge, J. C., O'Connell, P. E., and Rasmussen, J. 1986b. An introduction to the European hydrological system – systeme hydrologique European SHE. 2. Structure of a physically-based distributed modeling system. *J. Hydrol.* 87:61–77.

Abdulla, F. A., Lettenmaier, D. P., Wood, E. F., and Smith, J. A. 1996. Application of a macroscale hydrological model to estimate the water balance of the Arkansas–Red River basin. *J. Geophys. Res.* (*Atmos.*) 101:7449–61.

Bastiaanssen, W. G. M. 1995. Regionalization of surface flux densities and moisture indicators in composite terrain. Ph.D. thesis, Landouwuniversitait Wageningen, Report 109, DLO-Winand Staring Centre, Wageningen, The Netherlands.

Bastiaanssen, W. G. M., Pelgrum, H., Menenti, M., and Feddes, R. A. 1996. Estimation of surface resistance and Priestly-Taylor $\alpha$-parameter at different scales. In: *Scaling Up in Hydrology Using Remote Sensing*, ed. J. B. Stewart, E. T. Engman, R. A. Feddes, and Y. Kerr, pp. 93–112. Chichester: Wiley.

Beven, K. 1983. Surface water hydrology – runoff generation and basin structure. *Rev. Geophys. Space Phys.* 21:721–30.

Beven, K. 1995. Linking parameters across scales: subgrid parameterizations and scale dependent hydrological models. In: *Scale Issues in Hydrological Modeling*, ed. J. D. Kalma and M. Sivapalan, pp. 263–82. New York: Wiley.

Beven, K., and Kirkby, M. 1979. A physically based, variable contributing area model of basin hydrology. *Hydrol. Sci. Bull.* 24:43–69.

Blöschl, G., and Sivapalan, M. 1995. Scale issues in hydrological modeling: a review. In: *Scale Issues in Hydrological Modeling*, ed. J. D. Kalma and M. Sivapalan, pp. 9–48. New York: Wiley.

Bolle, H.-J. 1993. EFEDA: European field experiments in a desertification-threatened area. *Ann. Geophys.*11:173–89.

Bolle, H.-J., Feddes, R. A., and Kalma, J. D. (eds.). 1993. *Exchange Processes at the Land Surface for a Range of Space and Time Scales*. IAHS publication 212, Institute of Hydrology. Wallingford, Oxon, UK: IAHS Press.

Bonan, G. B. 1996. Sensitivity of a GCM simulation to subgrid infiltration and surface runoff. *Climate Dynamics* 12:279–85.

Brutsaert, W. 1982. *Evaporation into the Atmosphere*. Dordrecht: Kluwer.

Charney, J., Quirk, W., Chow, S., and Kornfield, J. 1977. A comparative study of the effects of albedo change on drought in semi-arid regions. *J. Atmos. Sci.* 34:1366–85.

Collins, D., and Avissar, R. 1994. An evaluation with the Fourier Amplitude Sensitivity Test (FAST) of which land-surface parameters are of greatest importance for atmospheric modeling. *J. Climate* 7:681–703.

Cooper, H. J., Smith, E. A., Gu, J., and Shewchuk, S. In press. Modeling the impact of averaging on aggregation of surface fluxes over BOREAS. *J. Geophys. Res. (Atmos.)*

Dickinson, R. E., Henderson-Sellers, A., Kennedy, P. J., and Wilson, M. F. 1986. Biosphere–atmosphere transfer scheme (BATS) for the NCAR community model. Technical note/TN-275+STR. Boulder, CO: NCAR.

Dooge, J. C. I. 1982. Parameterization of hydrological processes. In: *Land Surface Processes in Atmospheric General Circulation Models*, ed. P. S. Eagleson, pp. 243–88. Cambridge University Press.

Dubayah, R., Wood, E. F., and Lavallée, D. 1996. Multiscaling analysis in distributed modeling and remote sensing: an application using soil moisture. In: *Scale, Multiscaling, Remote Sensing and GIS*, ed. M. F. Goodchild and D. A. Quattrochi, pp. 93–111. Boca Raton: CRC Press.

Eagleson, P. S. (ed.). 1982. *Land Surface Processes in Atmospheric General Circulation Models*. Cambridge University Press.

Famiglietti, J. S., and Wood, E. F. 1994. Multiscale modeling of spatially variable water and energy balance processes. *Water Resour. Res.* 30:3061–78.

Famiglietti, J. S., and Wood, E. F. 1995. Effects of spatial variability and scale on areally averaged evapotranspiration. *Water Resour. Res.* 31:699–712.

Franks, S. W., Beven, K. J., Quinn, P. F., and Wright, I. R. 1997. On the sensitivity of soil-vegetation-atmosphere transfer (SVAT) schemes: equifinality and the problem of robust calibration. *Agric. Forest Meteorol.* 86:63–75.

Freer, J., Beven, K., and Ambroise, B. 1996. Bayesian estimation of uncertainty in runoff prediction and the value of data: an application of the GLUE approach. *Water Resour. Res.* 32:2161–73.

Freeze, R. A. 1980. A stochastic-conceptual analysis of the rainfall-runoff process on a hillslope. *Water Resour. Res.* 16:391–408.

Gupta, V. K., Rodríguez-Iturbe, I., and Wood, E. F. (eds.). 1986. *Scale Problems in Hydrology*. Dordrecht: Reidel.

Hebson, C., and Wood, E. F. 1986. A study of scale effects in flood frequency response. In: *Scale Problems in Hydrology*, ed. V. K. Gupta, I. Rodríguez-Iturbe, and E. F. Wood, pp. 133–58. Dordrecht: Reidel.

Jarvis, P. G. 1976. The interpretation of the variation in leaf water potential and stomatal conductances found in canopies in the field. *Phil. Trans. R. Soc. Lond. B.* 273:593–610.

Kalma, J. D., and Sivapalan, M. (eds.). 1995. *Scale Issues in Hydrological Modeling*. New York: Wiley.

Koster, R., and Suarez, M. 1992. A comparative analysis of two land-surface heterogeneity representations. *J. Climate* 5:1379–90.

Larson, J. E., Sivapalan, M., Coles, N. A., and Linnet, P. E. 1994. Heterogeneity and similarity of catchment responses in small agricultural catchments in Western Australia. *Water Resour. Res.* 30:1641–52.

Lhomme, J. P. 1992. Energy balance of heterogeneous terrain: averaging the controlling parameters. *Agric. Forest Meteorol.* 61:11–21.

Lhomme, J. P., Chehbouni, A., and Monteny, B. 1994. Effective parameters of surface energy balance in heterogeneous landscape. *Boundary-Layer Meteorol.* 71:279–309.

Liang, X., Lettenmaier, D. P., and Wood, E. F. 1996. A one-dimensional statistical-dynamic representation of subgrid spatial variability of precipitation in the two-layer VIC model. *J. Geophys. Res. (Atmos.)* 101:21403–22.

Liang, X., Lettenmaier, D. P., Wood, E. F., and Burges, S. 1994. A simple hydrologically based model of land surface water and energy fluxes for general circulation models. *J. Geophys. Res. (Atmos.)* 99:14415–28.

Moore, I. D., O'Laughlin, E. M., and Burch, G. J. 1984. A contour-based topographic model for hydrological and ecological applications. *Earth Surf. Proc. Landforms* 13:305–20.

NASA. 1992. Boreal ecosystem–atmosphere study (BOREAS). Research announcement NRA-92-0SSA-1, Office of Space Science and Application. Washington, DC: National Aeronautics and Space Administration.

Nielsen, D., Bigger, J. W., and Erh, K. T. 1973. Spatial variability of field measured soil-water properties. *Hilgardia* 42:215–60.

Nijssen, B., Lettenmaier, D. P., Wetzel, S. W., Liang, X., and Wood, E. F. 1997. Simulation of runoff from continental-scale river basin using a grid-based land surface scheme. *Water Resour. Res.* 33:711–24.

Noilhan, J. 1994. Defining area-average parameters in meteorological models for land surfaces with meso-scale heterogeneity. Presented at the ISLSCP-BACH Tucson Workshop on Aggregate Descriptions of Heterogeneous Land-Covers, Tucson, AZ, March 23–24.

Paniconi, C., and Wood, E. F. 1993. A detailed model for simulation of catchment scale subsurface hydrologic processes. *Water Resour. Res.* 29:1601–20.

Pauwels, V., Wood, E. F., and Lettenmaier, D. P. 1996. Large-scale hydrological modeling over high latitude basins. Presented at Second International Scientific Conference on the Global Energy and Water Cycle, Washington, D.C., June 17–21.

Peters-Lidard, C., Zion, M., and Wood, E. F. 1997. A soil-vegetation-atmosphere transfer scheme for modeling spatially variable water and energy balance processes. *J. Geophys. Res. (Atmos.)* 102:4303–24.

Raupach, M. R. 1993. The averaging of surface flux densities in heterogeneous landscapes. In: *Exchange Processes at the Land Surface for a Range of Space and Time Scales*, ed. H.-J. Bolle, R. A. Feddes, and J. D. Kalma, pp. 343–56. IAHS publication 212, Institute of Hydrology. Wallingford, Oxon, UK: IAHS Press.

Raupach, M. R., and Finnigan, J. J. 1995. Scale issues in boundary-layer meteorology: surface energy balances in heterogeneous terrain. In: *Scale Issues in Hydrological Modeling*, ed. J. D. Kalma and M. Sivapalan, pp. 345–68. New York: Wiley.

Robinson, J. S., and Sivapalan, M. 1995. Catchment-scale runoff generation model by aggregation and similarity analysis. In: *Scale Issues in Hydrological Modeling*, ed. J. D. Kalma and M. Sivapalan, pp. 311–30. New York: Wiley.

Rodríguez-Iturbe, I., and Valdes, J. 1979. The geomorphologic structure of hydrologic response. *Water Resour. Res.* 15:1409–20.

Rodríguez-Iturbe, I., Vogel, G. K., Rigon, R., Entekhabi, D., Castelli, F., and Rinaldo, A. 1995. On the spatial organization of soil moisture fields. *Geophys. Res. Lett.* 22:2757–60.

Schmugge, T. J., and Jackson, T. J. 1996. Soil moisture variability. In: *Scaling up in Hydrology Using Remote Sensing*, ed. J. B. Stewart, E. T. Engman, R. A. Feddes, and Y. Kerr, pp. 183–92. New York: Wiley.

Sellers, P. J., Hall, F. G., Asrar, G., Strebel, D. E., and Murphy, R. E. 1992a. An overview of the first international satellite land surface climatology project (ISLSCP) field experiment (FIFE). *J. Geophys. Res.* (*Atmos.*) 97:18345–71.

Sellers, P. J., Heiser, M. D., and Hall, F. G. 1992b. Relations between surface conductance and spectral vegetation indices at intermediate (100 $m^2$ to 15 $km^2$) length scales. *J. Geophys. Res.* (*Atmos.*) 97:19033–59.

Sellers, P. J., Mintz, Y., Sud, Y. C., and Dalcher, A. 1986. A simple biosphere model (SiB) for use within general circulation models. *J. Atmos. Sci.* 43:505–31.

Shuttleworth, W. J. 1988. Macrohydrology – the new challenge for process hydrology. *J. Hydrol.* 100:31–56.

Sivapalan, M., Beven, M., and Wood, E. F. 1987. On hydrological similarity: 2. A scaled model of storm runoff production. *Water Rosour. Res.* 23:2266–78.

Sivapalan, M., and Wood, E. F. 1986. Spatial heterogeneity and scale in the infiltration response of catchments. In: *Scale Problems in Hydrology*, ed. V. K. Gupta, I. Rodríguez-Iturbe, and E. F. Wood, pp. 81–106. Dordrecht: Reidel.

Sivapalan, M., Wood, E. F., and Beven, K. 1990. On hydrological similarity: 3. A dimensionless flood frequency model using a generalized geomorphologic unit hydrograph and partial area runoff generation. *Water Resour. Res.* 26:43–58.

Smith, E. A., Crosson, W. L., Cooper, H. J., and Wang, H.-Y. 1993. Estimation of surface heat and moisture fluxes over a prairie grassland. Part 3: Design of a hybrid physical/remote sensing biosphere model. *J. Geophys. Res.* (*Atmos.*) 98:4951–78.

Stewart, J. B., Engman, E. T., Feddes, R. A., and Kerr, Y. (eds.). 1996. *Scaling up in Hydrology Using Remote Sensing*. New York: Wiley.

Walker, J. M., and Rowntree, P. R. 1977. The effect of soil moisture on circulation and rainfall in a tropical model. *Q. J. Roy. Met. Soc.* 103:29–46.

Wigmosta, M. S., Vail, L. W., and Lettenmaier, D. P. 1994. A distributed hydrology-vegetation model for complex terrain. *Water Resour. Res.* 30:1665–79.

Wood, E. F. 1994. Scaling, soil moisture and evapotranspiration in runoff models. *Adv. Water Resour.* 17:25–34.

Wood, E. F. 1995. Scaling behaviour of hydrological fluxes and variables: empirical studies using a hydrological model and remote sensing data. In: *Scale Issues in Hydrological Modeling*, ed. J. D. Kalma and M. Sivapalan, pp. 89–104. New York: Wiley.

Wood, E. F. 1997. Effect of soil moisture aggregation on surface evaporative fluxes. *J. Hydrol.* 190:397–412.

Wood, E. F., Beven, K., Sivapalan, M., and Band, L. 1988. Effects of spatial variability and scale with implication to hydrology modeling. *J. Hydrol.* 102:29–47.

Wood, E. F., Lettenmaier, D. P., and Arola, A. 1994. Sensitivity of surface moisture and energy fluxes to the spatial distribution of precipitation. Presented at the GEWEX conference of the Royal Society, London, July 1994.

Wood, E. F., Sivapalan, M., and Beven, K. 1986. Scale effects in infiltration and runoff production. In: *Conjunctive Water Use* (proceedings of the Budapest symposium), pp. 375–87. IAHS publication. 156. Wallingford, Oxon, UK: IAHS Press.

# 2

# Hillslopes, Channels, and Landscape Scale

WILLIAM E. DIETRICH and DAVID R. MONTGOMERY

## 2.1 Are Landscapes Scale-invariant?

There is no doubt that channel networks display scale invariance in some respects, but this does not mean that landscapes lack distinctive scales, nor does it mean that the processes that evolve landscapes do not have scale dependence and do not impart distinctive scale-dependent morphology. Rodríguez-Iturbe and Rinaldo (1996) have summarized recent papers on channel networks and have argued for the importance of explaining the apparent scale invariance of landscape organization. Although the ubiquitous branching networks of valleys may indeed be scale-invariant or may possess attributes of multiscaling, knowledge of that cannot explain many of the fundamental issues involved in landscape form and evolution. In this chapter we focus on issues of scale and the importance of linking a process to an appropriate scale. We raise here more questions than answers, but in so doing perhaps make a case that as much as scale invariance is an appealing attribute of landscapes, understanding the controls on the actual scales of landscape features in both space and time can provide critical theoretical and practical insights into landscape processes and evolution.

Part of this discussion about scale in geomorphology is driven by the advent of computer-based analysis of digital topographic surfaces. In effect, our view of landscapes over the past 20 years has shifted from one of limited analysis of topographic contours (usually focusing on individual hillslope and river profiles or calculation of drainage density) to fully two-dimensional (or three-dimensional, depending on how one counts) grid-based investigations. Geomorphologists seem poised to realize the goal of quantitative analysis of landforms so strongly argued for by Strahler (1964).

But there is still something of an illusion at play here, and that, too, will be a recurrent theme in this chapter. The illusion is that digital elevation data can accurately capture the topography of natural landscapes. This illusion is most haunting, perhaps, to those of us concerned with connecting landscape processes and form using field data and at least quasi-mechanistic laws. In the United States, digital elevation data are commonly derived from digitized U.S. Geological Survey (USGS) maps or are supplied directly as gridded digital data by the USGS. Anyone who has walked hillsides in mountainous areas knows that the USGS maps, though extremely useful, typically portray the landscape as much smoother than it actually is. Among other effects, the fine-scale valley density is absent or only partially captured. This presents an important problem, especially when attempting to apply site-specific models for land-use decisions. None of these sets of digital elevation data can capture the topography of river channels, which therefore become operational grid cells without banks, bars, or other fluvial features. Until the grid scale of the information drops well below the scale of influence of a particular process, we shall be wrestling with the compromise between using physical laws and analyzing large landscapes. Once this resolution is reached, we may still need new analytical techniques (and, of course, ever-faster computers) to deal with the vast quantity of information.

We attempt here to illuminate this issue of scale dependence and landscape analysis by asking four questions provoked largely by the advent of digital topographic analysis: How big is a hillslope? How big is a river? What are useful measures of landscape morphology? What process laws are appropriate for modeling landscape evolution? We ask these questions having in mind the goal of understanding real landscapes, both from a long-term evolutionary perspective and from one of practical application.

## 2.2 How Big Is a Hillslope?

A review of introductory texts in geomorphology will reveal many chapters devoted to hillslopes, but little discussion of what defines one (Summerfield, 1991; Bloom, 1991; Ritter, Kochel, and Miller, 1995). Although the question of what constitutes a hillslope sounds like something that would have been answered long ago, it remains central to the issue of landscape scale in geomorphologic process models. The essential property of most landscapes is their division into planform convergent and divergent areas; the convergent areas we call valleys, and the divergent and planar areas we call hillslopes. This morphologic definition implies that valleys are areas of topographically driven focusing of erosional materials and that hillslopes are areas of sediment production and delivery that tend to disperse erosional materials. A hillslope, then, is the elevated land between valley bottoms; hillslopes extend until cut off by valleys. Hence the spacing of valleys governs hillslope size, and the issue of landscape scale is dominated by the definition and recognition of the extent of dissection by valleys (Montgomery and Dietrich, 1992, 1994b).

Although central to the issue of landscape scale, it can be difficult to establish the extent of landscape dissection, as attested to by the variety of ways to define and measure drainage density (Morisawa, 1957; Coffman, Keller, and Melhorn, 1972; Montgomery and Dietrich, 1989, 1994b; Tarboton, Bras, and Rodríguez-Iturbe, 1991; Montgomery and Foufoula-Georgiou,1993; Howard, 1994). Furthermore, drainage density (the total length of the channels in a drainage basin divided by its drainage area), when defined using topographic maps, depends on the resolution of the topographic data. Typically, for a given location, the finer the resolution of the topography, the greater the density of channels (Figure 2.1).

Examination at a fine enough scale, however, reveals that there is a limit to valley dissection (Horton, 1945; Montgomery and Dietrich, 1992). At such a scale, the hillslope surface may exhibit roughness of varying scale, but no organized, persistent convergent areas dissect it. For the purposes of distinction here, we can call this hillslope the "fundamental" hillslope. At any other scale of depiction or analysis, the topography represents a composite of these fundamental hillslopes and the varyingly developed valleys that bound them. At a coarse-scale depiction of a large region, then, the ridge-and-valley topography neither reveals the full drainage density nor defines the fundamental hillslopes. Instead, divergent and planar planform areas are smoothed portrayals of average valley side slopes. So at each scale of analysis there are elevated features that can be called hillslopes, but which represent a different averaging of fundamental hillslopes and valleys.

From the perspective of defining scale invariance, this analysis primarily argues for a limit of dissection and a finite scale, therefore, of fundamental hillslopes (Montgomery and Dietrich, 1992). But it also suggests the idea that different linkages between erosional processes and form occur at different scales. At the finest scale, the hillslope shape (including height) is dictated by the incision rate in the valley, by the hillslope sediment-transport processes, and by the production rate and availability of erodable material. At progressively coarser scales, the hillslope form (the average valley side slope) owes its shape to the incision rate in the valley and to some mixture of the processes operating on the fundamental-hillslope scale and the erosion by smaller channels not depicted at the coarser scale.

In recent years our image of landscapes has become increasingly filtered through digital representations of topography, but the coarse resolution of most digital elevation models (DEMs) emphasizes valley side slopes and major ridges over fundamental hillslopes. The appearance of what constitutes a hillslope depends on the topographic resolution; the hillslope size identifiable in the field can bear little resemblance to that rendered by topographic representations. The effect of DEM resolution on apparent hillslope size is well illustrated by the USGS shaded relief maps of the Allegany, Oregon, 7.5′ topographic quadrangle (Figure 2.2). Although one could define hillslopes and valleys on each of the different-resolution DEMs, there is only one set of fundamental hillslopes – and even these may not be fully captured on the finest-scale DEM. An exception to this argument may be the case of deep-seated landsliding, which can remove entire portions of well-developed drainage basins.

The linkage between process and form (and, by implication, size) is commonly explored through analysis of the longitudinal profiles of rivers and hillslopes. The simple, one-dimensional models that have been developed take into account the

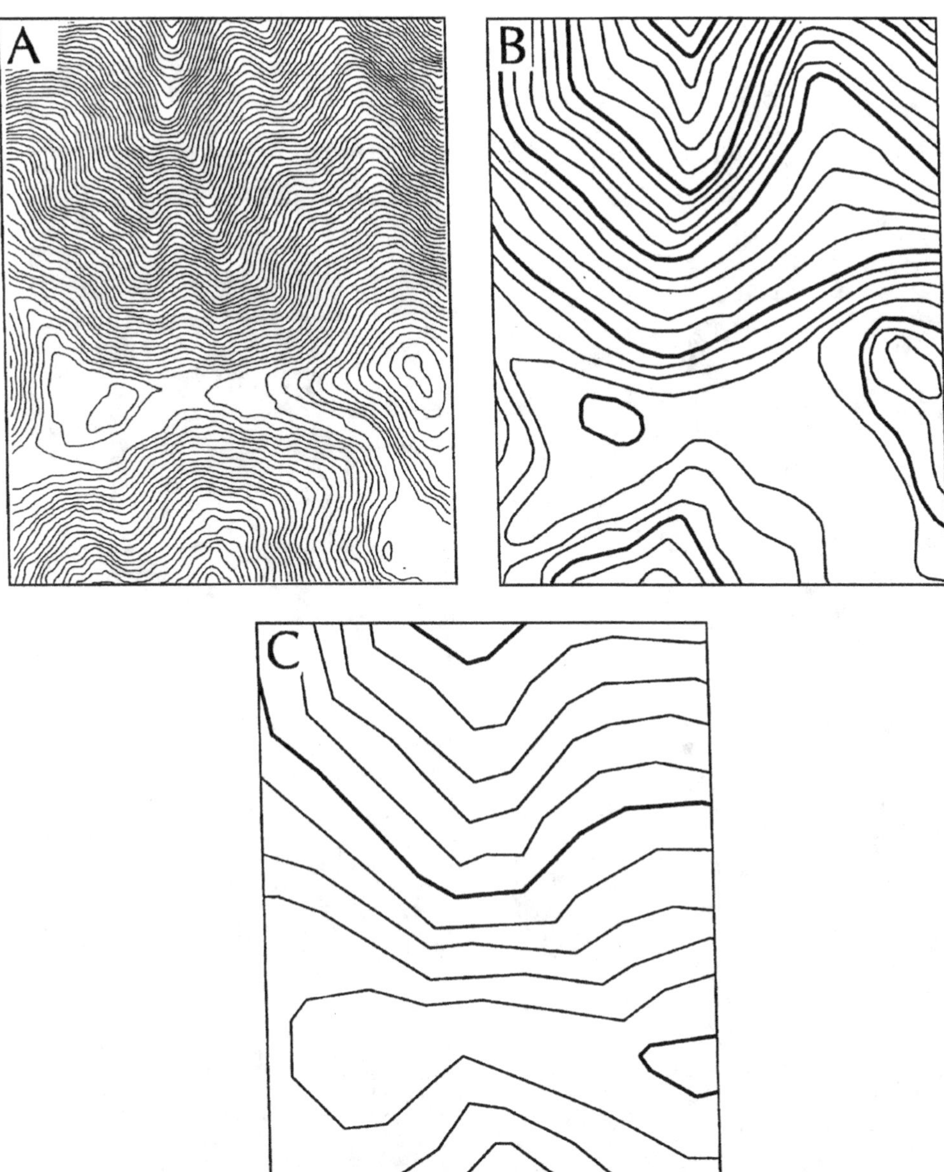


Figure 2.1. Topographic contours for a small valley along Mettman Ridge, near Coos Bay, Oregon, from maps of different scales that illustrate the effect of topographic resolution on the portrayal of hillslope length, and thus drainage density. For each map, north is toward the top of the figure, and the width of the panel is 200 m. (A) Map created from laser altimetry data obtained from a low flying plane (individual survey points were, on average, 2.6 meters apart); contour interval, 20 ft (6.1 m). (B) Map derived from aerial photographs and originally plotted at 1 : 4,800 scale by Weyerhaeuser Company; contour interval 20 ft (6.1 m). (C) Map derived from aerial photographs and originally plotted at 1 : 24,000 as part of the USGS 7.5′ Allegany quadrangle; contour interval 40 ft (12.1 m). Note that the five fine-scale valleys apparent on the field-survey map are reduced to one or two distinct valleys on the coarser-scale maps and that the apparent zone of hillslope convexity increases with map scale. Gallant (1998) discusses map scale effects on morphometric interpretation and proposes a wavelet analysis to quantify the three-dimensional structure of landscapes.

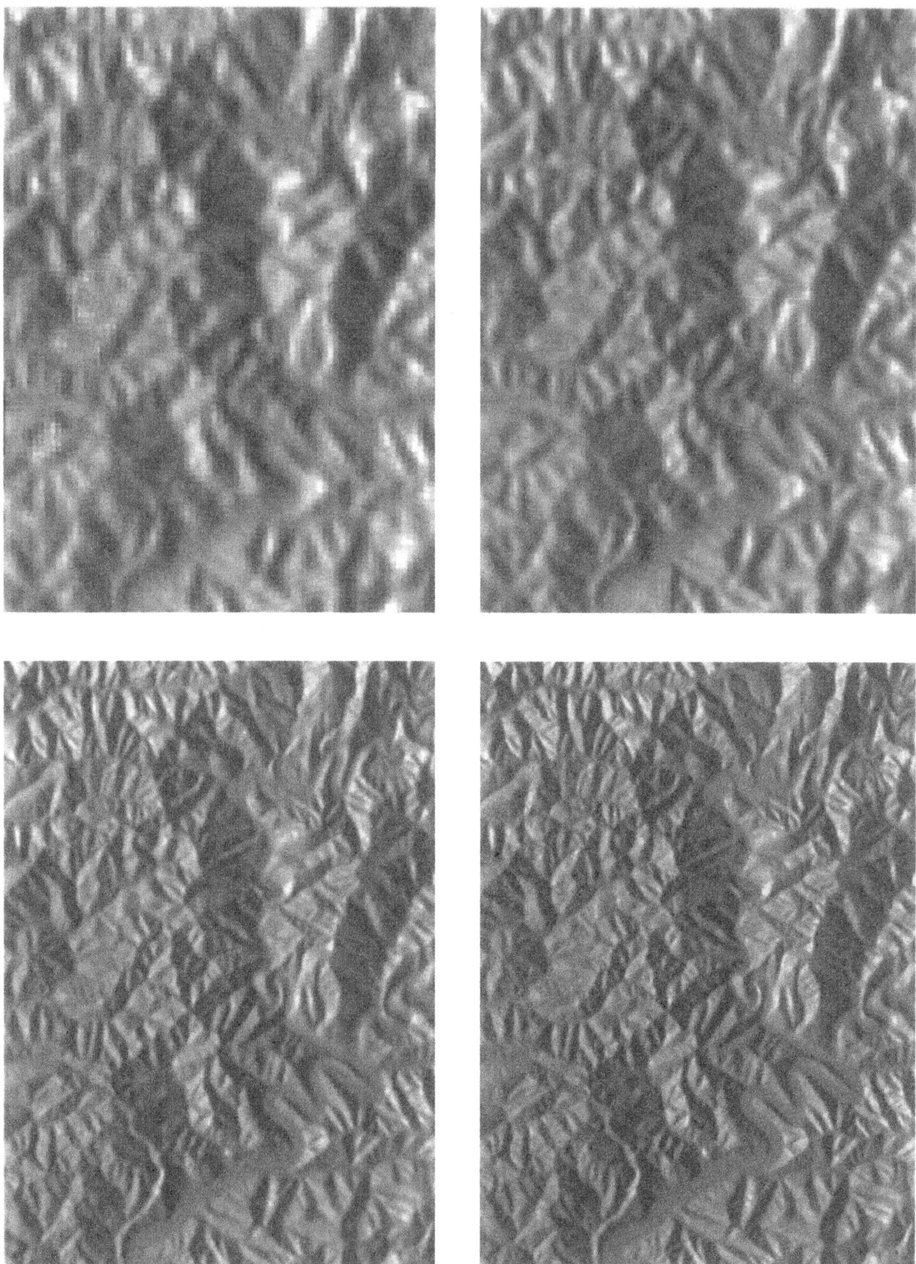

Figure 2.2. Shaded relief image of Allegany, Oregon, created from a 10-m DEM derived from the digital line graph of the contours portrayed on the USGS 7.5′ quadrangle; upper left, 120-m grid; upper right, 90-m grid; lower left, 30-m grid; lower right,10-m grid. Note the progressive loss of resolution, particularly apparent in the portrayal of the fine-scale valleys. Area of image is about 120 km$^2$.

behavior of boundary conditions, the effects of initial conditions, and the roles of different sediment-transport "laws" (e.g., Culling, 1963; Ahnert, 1970; Kirkby, 1971; Howard, Dietrich, and Seidl, 1994). Whereas channels are essentially line transects through the land, hillslopes are not as readily depicted by line transects or profiles because of the divergent topography they may occupy. The most extreme case is a longitudinal profile of a ridge – a feature that coarse-scale DEMs are most likely to capture. Although a continuous line can be drawn along a ridge, the chances that any sediment will follow the ridge line are vanishingly small.

Ridges are the "positive" topographic equivalents of the "negative" valleys (Figure 2.3). As such, they form a network in which the length of a ridge depends on the size (or order) of the valley it parallels (Tokunaga, 1984). But, unlike valleys, concentrated sediment transport does not follow the ridge. The longitudinal shape of a ridge is controlled by the initial surface, by the concavity of the river network, by the bedrock resistance, by the positions and sizes of major valleys bounded by the ridge (affecting the locations of local topographic low points known as saddles), and, perhaps to a lesser extent, by the erosional processes. Hence it is difficult to relate erosional processes to the longitudinal profiles of hillslopes derived from coarse-scale topography that emphasizes the shapes of ridges. Even more difficult to interpret are profiles taken across coarse-scale topography at arbitrary orientations with respect to the maximum fall line on the topography. Process and form (and hence size) are more readily examined at the fundamental-hillslope scale.

The size of a hillslope is measured not only by how long it is but also by how high it is. Whereas hillslope length is determined by valley spacing, the height (for a given hillslope length) is controlled by erosional processes, uplift rates, and material properties. Here we focus on the fundamental hillslope and point out that just as hillslope length is a basic, process-controlled scale imposed on landscapes, so too is the local relief.

Hillslope erosion occurs primarily by slope-dependent or mass-wasting processes (creep, biogenic activity, and landsliding) or by runoff-driven erosion (sheetwash and seepage erosion) (e.g., Selby, 1993). Additional processes, of course, become significant in landscapes where freeze–thaw activity is intense or where glaciers abound. Mass-wasting processes consist of (1) slow creep (shallow continuous or episodic movements typically involving just the soil), (2) shallow landsliding (episodic rapid movements, typically of only the mechanically weakened soil or regolith and usually of limited size, often less than 100 $m^2$), and (3) deep-seated landsliding (movement that occurs into the bedrock, or parent material, and can be hundreds of meters to kilometers long and may move only occasionally, with movement periods that can be separated by millennia). Creep is usually referred to as a diffusional process, and it can occur because of biogenic activity, freeze–thaw, wetting–drying, slow creep, rain splash, or even short bursts of overland flow on long hillslopes (Dunne, 1991). Shallow landslides have been treated as nonlinear diffusional processes over

Figure 2.3. Shaded relief image of a composite of USGS 30-m DEMs from the northeastern corner of the Olympic Mountains in Washington. The upper image shows normal topography, whereas the lower image depicts the same area with inverted topography derived by reversing the sign of the elevation field. Casual inspection of the images suggests similar topography, but close inspection reveals that in this area the ridge tops are narrower than the valley bottoms.

geomorphic time scales (e.g., Anderson and Humphrey, 1989; Howard, 1994; Tucker and Slingerland, 1994). Deep-seated landsliding has received little attention in quantitative geomorphic modeling.

For a longitudinal profile of a fundamental hillslope in which sediment transport is proportional to local slope $(-\partial z/\partial x)$, erosion $[-(\partial z/\partial t)]$ results from downslope changes in surface gradient (e.g., Culling, 1963; Anderson and Humphrey, 1989); that is,

$$-(\partial z/\partial t) = -(\rho_s/\rho_b)K(\partial^2 z/\partial x^2) \tag{2.1}$$

where $z$ is the height above some arbitrary datum, $\rho_s/\rho_b$ is the ratio of soil bulk density to rock density, $x$ is the distance from the drainage divide, $t$ is time, and $K$ is the proportionality constant between sediment transport and local slope [assumed in equation (2.1) to be independent of $x$]. The constant $K$ has units of length squared per time, the same as the diffusivity term in Fick's law; hence $K$ is often called the diffusivity (e.g., Koons, 1989). If we solve equation (2.1) for elevation $z$ for the simplest steady-state case in which uplift equals erosion, $C$ (length per time), and the drainage divide is fixed horizontally and flat, we get (Carson and Kirkby, 1972; Koons, 1989)

$$z = -(\rho_b/\rho_s)(C/2K)(L^2 - x^2) - Ct \tag{2.2}$$

where $L$ is the length of the hillslope. At the drainage divide, $x$ equals zero, and the height difference $H$ between the divide and the base of the hillslope (the local relief) is given by

$$H = -(\rho_b/\rho_s)(C/2K)(L^2) \tag{2.3}$$

As Koons (1989) illustrated in his pioneering study of the evolution of the New Zealand Alps, the height of the hillslope is thus proportional to the uplift or erosion rate and the square of the hillslope length. Where we differ, and this is the most relevant point about scale here, is on the question whether or not equation (2.3) can be mechanistically applied to large mountain slopes. Once slopes begin to approach the frictional strength of the loose weathered mantle of a landscape, landsliding should begin to predominate, leading to a very different set of relationships. On mechanically strong rock, which is mantled with a coarse-textured soil, landsliding probably begins to become important on slopes above 20° and perhaps dominates hillslope sediment transport on slopes above 27°. Equation (2.1) can be solved for the critical length $L_c$, at which landsliding will start to dominate for a given critical slope $S_c$, using the same assumptions that led to equations (2.2) and (2.3):

$$L_c = (\rho_s/\rho_b)(K/C)S_c \tag{2.4}$$

For reasonable values of these parameters, $L_c$ is relatively short compared with mountain-valley side slopes, and the greater the uplift rate, the shorter this length. Field-determined values for the diffusivity of fundamental hillslopes are rare, and for the sites investigated they have ranged from about 10 to 300 $cm^2$/yr (e.g., McKean et al., 1993; Dietrich et al., 1988, 1995). The bulk density ratio is about 0.5; so if $S_c$ is about 0.5, and a typical value for diffusivity is 50 $cm^2$/yr, then for an uplift rate of 0.1 cm/yr (fast), $L_c$ is just over 1 m! Increasing the diffusivity to 300 and lowering the uplift to 0.01 cm/yr will increase the critical length to 75 m. For more typical values of diffusivity (e.g., 50 $cm^2$/yr) and low uplift rate (0.001 cm/yr), the length is still only 125 m. Continuously convex hillslopes of more than 100 m that are not ridges have been rare in our experience. In order to apply a diffusive law to the New Zealand Alps, Koons (1989) found it necessary to use diffusivities of 1,500 to 150,000 $cm^2$/yr. We suggest that a landslide transport law would have been more appropriate.

As yet, there are no field-verified landslide transport laws. The law used by Howard (1994), which would be more applicable to shallow landsliding than to deep-seated landsliding, has the effect of causing sediment transport to approach an infinite rate as the slope approaches a maximum stable slope angle, thus enforcing fairly linear slopes. His equation just for landsliding is

$$q_s = K_f\{[1/(1 - K_x|S|^\alpha)] - 1\} \tag{2.5}$$

where $q_s$ is the sediment flux (volume/time-width), $K_f$ is equivalent to the diffusivity term (i.e., length squared per time) in equation (2.1), $K_x$ is a dimensionless constant set by the condition that the maximum stable hillslope angle must equal $(1/K_x)^{1/\alpha}$. Howard assumed the dimensionless exponent $\alpha$ to be equal to 3. It is important to note that in the original publication the subtraction of the 1 was incorrectly missing from the equation (even though it was properly used in the model, A. D. Howard, 1996, personal communication). Here, if we make the same assumptions used to derive equations (2.3) and (2.4) for a hillslope eroding according to the Howard landslide law, we get

$$z = -\left(\frac{1}{K_x}\right)^{\frac{1}{\alpha}} \int_0^L \left(1 - \frac{1}{(C/K_f)x + 1}\right)^{\frac{1}{\alpha}} dx \tag{2.6}$$

where $C$ is the lowering rate at the base of the hillslope due to channel incision. We could find no analytical solution, but numerical integration using specified values for parameters and hillslope length was accomplished using *Mathematica*™. This maximum relief (the value of $z$ at $x = 0$ at the divide) is equal to the maximum stable hillslope angle $(1/K_x)^{1/\alpha}$ multiplied by the horizontal hillslope length. The relief approaches this maximum value on longer hillslopes with faster lowering rates, lower

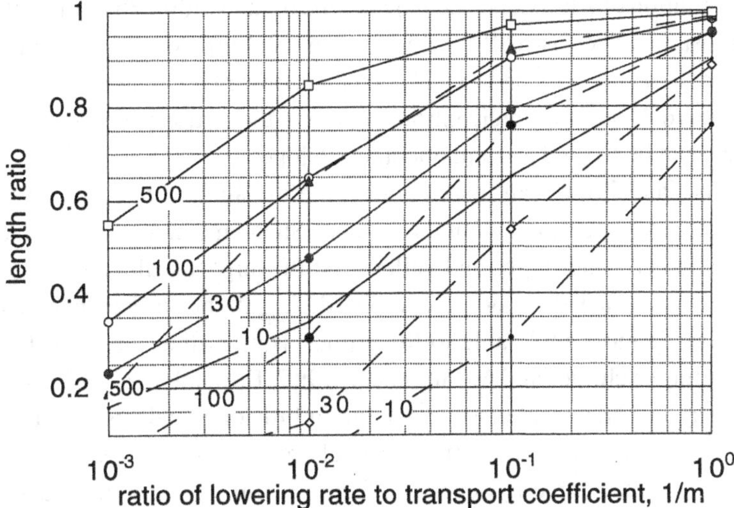


Figure 2.4. Solution to equation (2.6) for estimating relief from the Howard (1994) landslide transport law. The length ratio is the value of the integral in equation (2.6) normalized by the total hillslope length. This value times the maximum stable hillslope angle times the total hillslope length gives the relief or height of the hillslope as controlled just by landsliding processes. Hence the length ratio is the ratio of height of hillslope to maximum possible height of hillslope for a given hillslope length, maximum stable angle, lowering rate, and transport intensity. This length ratio is plotted as a function of the ratio of lowering rate $C$ to the transport coefficient $K_f$ for various hillslope lengths (labeled on lines, in meters) and for two values of the exponent, $\alpha = 3$ (solid lines) and $\alpha = 1$ (dashed lines).

values of the transport coefficient $K_f$, and higher values of the transport exponent $\alpha$. Figure 2.4 shows the value of the integral normalized by the total hillslope length as a function of the ratio of lowering rate $C$ to the transport coefficient $K_f$ for various hillslope lengths (labeled on lines) and for two values of the exponent, $\alpha = 3$ and $\alpha = 1$. Unlike the diffusive-transport case, however, in which the slope relief is directly proportional to the lowering rate and inversely proportional to the diffusion coefficient, the importance of the lowering rate and the sediment-transport coefficient in the landslide case strongly depends on hillslope length, and for longer hillslopes the relief is a weak function of the lowering rate and the sediment-transport coefficient (a variation of three orders of magnitude in the lowering rate may cause a change in the relief of less than a factor of 2).

It has been argued that landsliding imposes a limit to hillslope height, and therefore relief (Schmidt and Montgomery, 1995; Burbank et al., 1996). An expression of the kind used by Howard (1994) would predict such a result, but it would also say that for a given strength property of rock the key control on relief would be channel spacing (and therefore hillslope length), as also implied by the analysis of Schmidt and Montgomery (1995). According to this model, areas with widely differing uplift rates can have essentially the same relief as long as the channel spacing is the same (as Burbank et al., 1996, propose for the Himalaya) and the lowering rate is relatively high. Two problems in comparing this result with field data from mountainous regions are (1) the definition of a hillslope and (2) the type of landsliding that limits the relief. The model just described is really for a "fundamental hillslope," as defined earlier. For landscapes that are rocky and rough, such that convergent areas are discontinuous because of bedrock-outcrop interruptions, perhaps the entire valley side slopes for major valleys can be treated as fundamental hillslopes.

Deep-seated landsliding can result from force imbalances that arise over scales much larger than a single fundamental hillslope. Such landslides can displace entire networks of smaller-scale ridges and valleys. At the largest scale, then, deep-seated landsliding can act on valley side slopes bordering higher-order channels. Such conditions may be most applicable to the proposals regarding the relief limitations set by landsliding (Schmidt and Montgomery, 1995; Burbank et al., 1996).

Another way in which scale issues emerge in hillslope analysis results from the problem that most hillslopes, even fundamental ones, in hilly or mountainous landscapes require so much time to develop a form in equilibrium with boundary conditions and process laws that they rarely develop under a single climatic regime. Fernandes (1994) and Fernandes and Dietrich (1997) applied numerical modeling to diffusive-profile evolution and found that for typical values of diffusion and hillslope lengths, the morphologic response time to temporal variation in boundary conditions or diffusivity could greatly exceed the time period of climatic oscillation, indicating that hillslopes, particularly larger ones, may be continually somewhat out of equilibrium with current conditions. Rinaldo et al. (1995) even suggested that the characteristic form relationships between hillslope profile shapes and the dominant transport processes proposed by Kirkby (1971) may be reversed in landscapes undergoing periodic climatic variations: Instead of profiles being convex-up under diffusion-dominated conditions, hillslopes will tend to have concave-up profiles, because diffusion domination may be associated with reduced river incision, allowing diffusive infilling to create wide concave slopes.

Hence, real landscapes most likely are some integrated form emerging from variable incision rates and variable intensities of erosion (and even changes in erosion processes). So here the issues of time scale and spatial scale merge in any effort to relate process and form. It is essential to ask over what spatial scale and over what period of time one is attempting to relate hillslope shape and size to process controls.

## 2.3 How Big Is a River?

Rivers start at channel heads, drain upland areas, intersect other channels, and create networks that appear scale-invariant; river networks are perhaps the most frequently cited examples of natural systems with fractal properties (Rodríguez-Iturbe and Rinaldo, 1996). But rivers also introduce specific length scales into the landscape in at least four distinct ways. Each of these ways has to do with the question of how big a river is.

First, rivers usually start at distinct channel heads, and commonly in an unchanneled valley, well downslope from the drainage divide; see the review by Dietrich and Dunne (1993). Channels do not branch infinitely to smaller and smaller sizes and eventually cover the entire landscape. Instead, there is a finite channel network that, though prone to expansion and contraction due to climatic, land-use, and tectonic influences, generally occupies a valley. As Horton (1945) noted, simple geometry leads to the realization that the average hillslope length is inversely proportional to half of the drainage density. Hence river networks, while introducing scale independence to landscapes, at the same time define a distinct scale dependence, which is the average length of a hillside. Montgomery and Dietrich (1989, 1994b) and Dietrich and Dunne (1993) reviewed various mechanisms for channel initiation. They concluded from field observations and simple modeling that the positions of channel heads appear to be controlled by erosion thresholds (due to critical shear stress or slope cohesion and frictional strength). Howard (1994) has shown that in a numerical model that has no explicit threshold of erosion, ridges and valleys of finite scale will form. As Smith and Bretherton (1972) originally proposed, valleys formed where the advective transport associated with area-dependent water transport exceeded the diffusive, slope-dependent transport that dominates hillslopes. In either case, a specific scale emerges in the landscape at the transition from the river to the hillslope domain.

A second scale that rivers introduce into landscapes is that associated with channel width. In a river basin where the river discharge of water $Q$ (volume per time) increases with drainage area $A$ (i.e., $Q = cA^e$), the river width $w$ increases downstream (i.e., $w = aQ^b$). Many have argued that stream power per unit area of the riverbed tends either toward a minimum or toward a uniform rate of change through a network (e.g., Leopold, Wolman, and Miller, 1964; Rodríguez-Iturbe et al., 1992). Stream power per unit area, $\Omega$ (watts/area), is

$$\Omega = \rho g Q S / w \tag{2.7}$$

$$= \rho g (c^{1-b} a^{-1}) A^{e(1-b)} S \tag{2.8}$$

If a river network does become structured according to some stream-power expenditure, then the width of the channel will specifically influence the downstream variation in channel slope.

Channel width influences landscape scale in another way. The river-meander wavelength (the distance between alternate bends in which the banks curve in the same direction) typically is about 11 times the channel bankfull width in alluvial floodplains (Leopold and Wolman, 1960). Meanders in resistant banks, such as bedrock, however, more commonly have wavelength-to-width ratios of 20 or more. Hence, as rivers incise their way through bedrock and meander, they introduce a sinuosity to the valleys that is scaled by channel width (Figure 2.5).

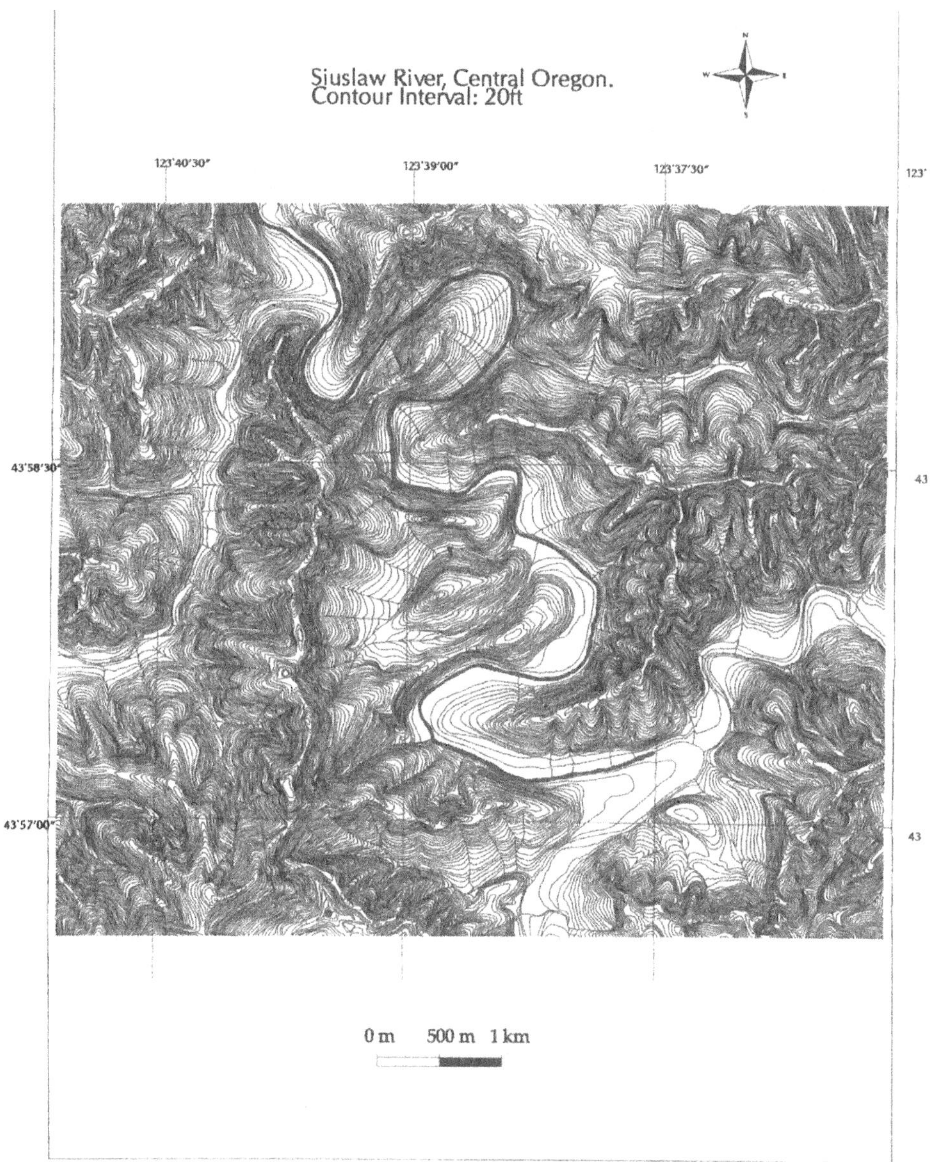


Figure 2.5. Topographic map of the Siuslaw River area in the Coast Ranges of Oregon showing the effects of progressive downcutting, channel meandering, and cutoff on landscape morphology. Ridge sinuosity of the mainstem reflects the meandering main channel. Note the different scales of meanders on tributaries and the mainstem, which runs diagonally across the map. A large abandoned bend of the main channel lies in the center of the map. The approximate channel width is shown on the mainstem. The channels are incised into bedrock and consequently have large wavelength-to-width ratios.

A third scale introduced to landscapes by rivers results from downstream changes in the dominant transport processes that occur along river longitudinal profiles. The rate of river incision (length/time) into bedrock, $E$, has been treated as being proportional to stream power, as reviewed by Howard et al. (1994). That is, from equation (2.8),

$$E = k[\rho g(c^{1-b}a^{-1})A^{e(1-b)}S]^n \tag{2.9}$$

Numerical models show that given adequate time, there is a strong tendency for a landscape to reach an equilibrium in which erosion balances uplift (assumed to be constant) and the erosion rate is spatially uniform (e.g., Howard, 1994; Willgoose, 1994). In that case, $E$ is a constant, and equation (2.9) can be rewritten as

$$S = k_s A^{-m/n} \tag{2.10}$$

where $m = ne(1-b)$ and $k_s = (EK^{-1})^n[\rho g(c^{1-b}a^{-1})]^{-1}$. A river increases its drainage area with distance downstream from the drainage divide (i.e., $A = dx^f$). Hence equation (2.10) can be written as $S = k_s d^{-m/n}x^{-fm/n}$, in which case $fm/n$ corresponds to the curvature of the longitudinal profile in which elevation is plotted as a function of $x$. The curvature of the longitudinal profile will have a large influence on the three-dimensional shape of the basin, as well as on the junction angles for tributaries (Howard, 1971, 1994). Note that $m/n$ equals $e(1-b)$; that is, it is determined by the downstream rate of increase in discharge with area ($e$) and by the rate of increase of channel width downstream with increasing discharge ($b$). If discharge is proportional to drainage area, and $b$ is equal to 0.5 (Leopold et al., 1964), then $m/n$ is 0.5.

A scale-invariant landscape would have a constant $m/n$ throughout the channel network. Observations of longitudinal profiles and inferences about erosion processes suggest, however, that $m/n$ should not be constant. As Seidl and Dietrich (1992) have pointed out, several distinctly different processes act to carve river-long profiles. On steep slopes, periodic debris flows may play a dominant role in channel incision. On gentler slopes, incision may be stream-power-dependent, but because of differences in erosion rates that may occur at tributary junctions, across bedrocks of varying resistance, or because of base-level changes (among other effects), knickpoints may form and propagate upstream. Probably neither the debris-flow erosion mechanism nor the knickpoint-propagation process is well described by a simple $m/n$ value in equation (2.10). Sklar and Dietrich (1996) have argued that sediment supply affects $m/n$. Sediments act as tools for abrasion, but also, when concentrated into bars, sediments can shield the bed from erosion. Channels that are predominantly alluvial or bedrock may have very different $m/n$ values.

Another scale issue that this transport-law analysis raises is the issue of time, particularly with regard to attempting to relate field data to theory. In uplands areas, sediment delivery to channels is stochastically driven, leading to waves of sediment that attenuate through storage and particle breakdown downstream. In many cases, reaches may periodically shift from being sediment-covered to being exposed bedrock. In such a river, can one expect that a single $m/n$ value will capture these effects? Virtually all models make the assumption that the continuous transport laws driving them are representative of the effects of stochastic processes operating over long time periods. Though such an assumption is defensible, it remains possible that there are distinct geomorphic signatures of stochastic processes. The paper by Rinaldo et al. (1995) sheds some light on this issue for the case of long-time-scale climatic oscillations by showing that the effects of previous climatic states are most likely to be preserved in the landscape when uplift is negligible. Furthermore, they show that topography developed under cyclic climate changes may be quite different from that under steady-state conditions. Here another scale enters: the time for completion of a cycle of variation. At a larger spatial scale, Kooi and Beaumont (1996) have used landscape-scale simulations to revisit the classic issue of geomorphic responses to orogenic cycles.

Figure 2.6 shows slope–area relationships for several channel profiles in the Noyo watershed in northern California. These data are from hand-digitized USGS 40-ft contour maps in which the channels were separately digitized onto the contours. No hillslope elements are included. Channel gradients steeper than about 20% (and in this case draining areas smaller than about 1 km$^2$) show no areal dependence (i.e., they are straight). Seidl and Dietrich (1992) and Montgomery and Foufoula-Georgiou (1993)

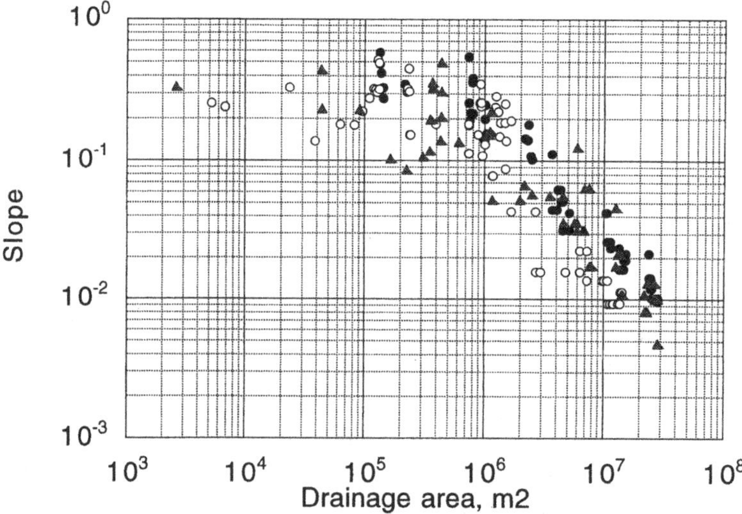


Figure 2.6. Channel-slope–drainage-area relationships for three tributaries of the Noyo River in the Coast Ranges of northern California; data obtained from digitized 40-ft USGS topographic maps. Channels were digitized and referenced to the same datum as the digitized topographic map. Slopes and areas were calculated for each contour crossing of the digitized streamlines. Note the distinct change in the relationship at about 20%.

have proposed that this straight slope is due to periodic debris-flow scour, although the specific reason why the slope tends toward a constant value has not been established. Nonetheless, $m/n$ here is zero. Downstream of the debris-flow-dominated reaches, it is possible that the $m/n$ still should not be a constant. This is because there is a transition reach in which slopes are sufficiently high that cascade and step-pool bed configurations prevail. Here dominant flows tend to be supercritical (e.g., Grant, Swanson, and Wolman, 1990), and it is possible that $m/n$ differs from that found farther downstream where subcritical flows prevail. Hence, although we can assign a single value to $m/n$ for the larger, lower-gradient channels, there may be systematic differences associated with changes in the dominant channel processes that will cause the three-dimensional structure of the river network to differ downstream.

As implied earlier in the discussion of stream power, the value of $m/n$ (and hence the concavity of the longitudinal profile) depends on the relationship between channel width and discharge. Whereas empirically for alluvial channels there is a strong central tendency for the exponent $b$ to equal 0.5 (e.g., Leopold et al., 1964), there is a paucity of data on bedrock channels (which predominate in steep landscapes). Furthermore, there is no general theory for the width–discharge relationship; it is always treated empirically in landscape models, which means that to some degree the answer that is sought, say the value of $m/n$, is put into the model rather than discovered by its workings.

The fourth scale introduced by rivers into a landscape results from their role as conveyors of sediment away from uplifted areas. Elevated landmasses are held up by the rigidity of the crust, by isostatic compensation, or by tectonic forces (e.g., Turcotte and Schubert, 1982). Erosional unloading of surface mass by rivers and reloading where rivers lose their sediment will change these balances and contribute significantly to uplift behavior. In the simplest case, river incision will remove mass, creating a vertical force imbalance that either will be sustained by crustal rigidity or will induce compensating vertical uplift, regenerating an isostatic balance. The most widely used approximation to this geophysical process is the plate-bending theory for various loads (Turcotte and Schubert, 1982). For the simple case of a periodic topographic load and insignificant horizontal forces (which might represent the weight of a mountain on the deformable lithosphere), the degree of topographic compensation $T$ (varying from 0 to 1.0), which is the ratio of topographic deflection relative to deflection with no rigidity, is

$$T = \frac{\rho_m - \rho_c}{\rho_m - \rho_c + (R/g)(2\pi/\lambda)^4} \tag{2.11}$$

where $\rho_m$ and $\rho_c$ are the densities of the mantle and crust, respectively, $g$ is gravitational acceleration, $\lambda$ is the wavelength of the topographic feature, and $R$ (units of newton-meters) is the flexural rigidity of the lithospheric plate.

Figure 2.7 shows the relationships among the amount of compensation, the wavelength of the feature, and the flexural rigidity (which ranges from about $10^{20}$ to

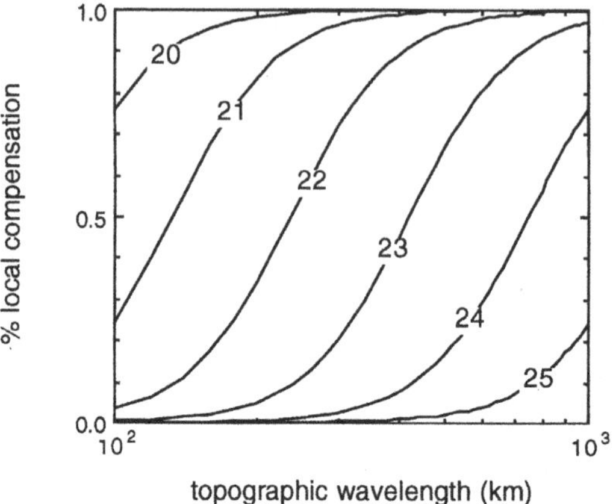


Figure 2.7. Solutions of equation (2.11) showing the predicted normalized percentage of local isostatic compensation as a function of topographic wavelength and the flexural rigidity of the lithosphere. Contours are the exponents for flexural rigidity across a range of $10^{20}$–$10^{25}$ N · m. (From Montgomery, 1994, with permission.)

$10^{25}$ N·m, with typical values around $10^{22}$–$10^{24}$). This suggests that the effect of erosional unloading must operate over length scales greater than 100 km to overcome crustal rigidity and induce a fully compensated uplift of the landmass. This is a one-dimensional analysis of something that is clearly two-dimensional, particularly when considering the competing effects of adjacent river basins cutting into a landmass. As Montgomery (1994) pointed out from this analysis, there may be a tendency for incisions by major rivers to induce uplift that can elevate the bordering mountain peaks. Smaller rivers will have no local effect of this kind, but the average effect of mass losses over a large area will induce compensating uplift. If there were no length-scale effects introduced by crustal rigidity, then very local high rates of erosion, say into mechanically weaker rocks, would induce local high uplifts, causing a kind of runaway effect in erosion and uplift. Instead, regional uplift leads to topographic adjustments that reduce the rapid local erosion rates toward the mean uplift rate. In other words, crustal rigidity enables the Gilbert (1877) law of equal action to act by, in effect, damping out large local gradients in erosion rates, leading to approximately steady-state morphology. See the two-part special publication on tectonics and topography in the 1994 *Journal of Geophysical Research* for further discussion of these issues and for a more realistic analysis of tectonic loading effects.

## 2.4 What Are Useful Measures of Landscape Morphology?

Quantitative description of landscape morphology serves two primary purposes: to provide data with which to test the predictions of landscape models and to provide a topographic framework for hydrologic, ecologic, or other landscape-based models. There have been few high-resolution field surveys of hillslope topography in which the controlling processes have also been quantified. Until recently, topographic analyses of landscapes had relied on tedious measurements obtained by hand from topographic maps. Now, with digital elevation data widely available, rapid analysis of large areas can be easily accomplished. As discussed throughout this chapter, however, digital elevation models (DEMs) usually provide a coarse, inaccurate portrayal of the fine-scale features where linkages between process and form are perhaps most strongly expressed.

Here we shall briefly discuss some useful measures that can be obtained from DEMs, measures that are relevant to the debate about scale effects and landscape development. First, however, let us point out some things that either cannot be done or have proved difficult to do using DEMs. As mentioned earlier, channels do not exist in DEMs. Hence, quantitative analysis of channels, other than of their longitudinal profiles, is not possible. The degree of confinement of a channel – how much floodplain there is (as compared with, say, a terrace surface) – is virtually impossible to identify given the current resolution of DEMs. It would be very useful for ecological modeling, as well as sediment-routing modeling, if such features could be identified. Without digitizing channel positions from either topographic maps or aerial photographs, it is not possible with DEMs to determine accurate channel slopes for sinuous rivers with floodplains. TOPOGRID, the subroutine often used with ARC/INFO (Environmental Systems Research Institute, Redlands, CA) to improve the gridding of digitized contours by using information on channel location, does not overcome this problem. Perhaps surprisingly, the drainage area to each grid cell is not easily determined in some cases. The most trying is the case in which the river lies in a relatively wide valley. Grid cells often will be much smaller than the valley, and because of the low gradient there, the drainage-routing routines will mostly follow the "noise" created from grid generation. That can lead to inaccurate determination of drainage areas along channels.

There is considerable literature on fractal or power-function analysis of topographic properties, and we direct the reader to the work of Rodríguez-Iturbe and Rinaldo (1996) for a thorough review of the topic. The focus in fractal analysis has tended to be on identifying universal power-law relationships for river basins (mostly associated with channel-network structure). There is considerable controversy over whether fractal analysis can provide deeper insight into the controls on landscape morphology or whether the power-law relationships emerge inevitably from the hierarchical river network (channels join but do not redivide downstream), with their generality therefore conveying no special meaning. Here we focus on what features of real landscapes distinguish them not so much from model landscapes but from other real landscapes that have experienced different rates of tectonic and erosional processes. These are the attributes that identify distinct landscape scale and form.

Three attributes that quantitatively distinguish landscapes are drainage density $D$, slope $S$, and relief $H$. Strahler (1958, 1964) combined these attributes to form what he called the "geometry number":

$$HD/S = G \tag{2.12}$$

in which $G$ was found to vary from about 0.4 to 1.0 (a perfect triangular-shaped ridge-and-valley topography would give $G = 0.5$). Drainage density, as determined from maps or DEMs, rather than from field observations, is scale-dependent; the coarser the topographic resolution, the lower the drainage density and the longer the hills. Although DEM-based comparisons of drainage densities between landscapes can use the same grid size, relationships between process and form need to be identified at the scale at which the processes operate. As mentioned already, drainage density sets the average hillslope length, which in turn influences hillslope height, or local relief. But relief is also measured as the height difference between the elevation of the highest peak in the watershed and the elevation of the watershed outlet, or as a function of stream order. These latter measurements of relief should be fairly accurate from DEMs.

Slope is dimensionless, but its mean value or its probability density function for a landscape, however, may reflect dominant transport processes. For example, as Strahler (1950) demonstrated in the Ventura Hills in southern California, Burbank et al. (1996) have recently proposed that the median slope in parts of the Himalayas reflects erosion by landslide processes. Perhaps more informative, and clearly related to scale, are data on the spatial variation in slope. We have used hillslope convexity and the known landscape-lowering rates to estimate diffusivity (Dietrich et al., 1988). Those calculations were based on field measurements of hillslope form; DEMs with resolution of 10 m or more may not be useful for this purpose, because convexities commonly are quite short. We have, however, used a 10-m-grid DEM to predict the variation in soil depth arising from hillslope curvature and have found favorable comparisons with field observations (Dietrich et al., 1995). Howard (1994) proposed the use of a critical value of topographic divergence (convex upslope topographic contours), normalized by the average hillslope gradient, as a means for defining the valley network. Willgoose, Bras, and Rodríguez-Iturbe (1991) argued that for landscapes dominated by overland-flow erosion and channel incision, a nondimensionalization of the process laws leads to the conclusion that tectonic uplift, erodability, and runoff dictate the vertical dimension of a catchment (the relief) and thereby control the slope. In this sense, the vertical and horizontal length scales in mean slope $S$ can be treated separately.

As already described, slope–area relationships should reflect dominant erosion processes. Willgoose (1994) and Howard (1994) have shown how slope–area relationships can be used to assess model behavior. As Montgomery and Foufoula-Georgiou (1993) have pointed out, process-based interpretations of slope–area relationships from DEMs must give consideration to the scale at which processes operate.

They suggest that the inflection in a plot of slope versus area interpreted by Tarboton et al. (1991) to have been due to a change in erosion processes from diffusion-dominated processes to fluvial transport may actually have arisen from changes in erosion processes from debris-flow-dominated processes to fluvial-dominated processes. Slope is expected to increase with drainage area per unit contour length on diffusion-dominated terrain (Willgoose, 1989), but only high-resolution topography can capture this relationship. Whereas the typically available digital data may be good enough to assess the slope–area relationship for channels, rarely are such data of sufficiently fine scale for hillslope analysis. Also, given the fairly narrow range of likely $m/n$ values for channels ($\sim$0.5–1.0), it may not be a very sensitive measure of model performance.

One approach that focuses on process–form relationships is to classify the landscape into divergent, planar, and convergent topographies (Dietrich et al., 1992, 1993). Such a classification, coupled with information on slope and drainage area, may allow one to determine where different processes dominate the landscape forms. Prosser and Abernethy (1996), using the same approach as Dietrich and associates, found that they could also identify areas dominated by different erosion processes. Importantly, their work in Australia led them to identify dominant erosional processes (notably Horton overland-flow erosion) different from those found in the California case reported by Dietrich and associates. Whereas that approach appears to have been instructive, it would be difficult to use in a grid-based analysis in watersheds with valley floors wider than the grid spacing (both Dietrich and associates Prosser and Abernethy used contour-based element construction). If the grid size is small compared with the valley-floor width, then such floors will be seen as flat and as draining small areas, the same as those found on ridge tops. Whereas over the short term this comparison is correct, over geomorphic time scales the valley flats are sometimes occupied by the channels; hence the grid cell in the valley flat owes its geomorphic position to occasional river processes, and a large drainage area is more appropriate for this analysis. One approach that could be tried to solve this problem of time and space scales would be to use variable grid sizes, small on the hillslopes and large in the valley bottoms.

Strahler (1957) proposed that the hypsometric integral (plot of the proportion of the drainage area above a given elevation above the basin outlet) could be used to distinguish the evolutionary state of a landscape, and recent modeling efforts have explored this idea (e.g., Willgoose, 1989; Howard, 1994). Although the hypsometric integral provides some information about basin structure, it is scale-independent. Howard has shown that the shape of the hypsometric integral depends on the erosional history, driven by outlet boundary conditions, but it is not obvious how one could extract such information from measurements of real landscapes.

Drainage density, slope, relief, topographic divergence and convergence, and slope–area relationships provide tools to distinguish real landscapes and identify scale-dependent relationships emerging from dominant erosion processes. It is hoped that the recent availability of digital elevation data will spawn other measures that can serve to differentiate landscapes on the basis of the morphologic signatures of the processes that formed them. Such measures, preferably derived from geomorphic theory, are greatly needed to guide and test landscape-evolution models.

## 2.5 What Process Laws Are Appropriate for Modeling Landscapes?

The question of what process laws are appropriate for modeling landscapes is fundamentally an issue of scale. Can fine-scale mechanistic transport laws be scaled up to explain entire landscapes simply by adjusting the parameters? If a model is driven by rules for processes that cannot be calibrated or, in some cases, even observed in the field, are the model outcomes applicable to real landscapes? Although these and related issues have received considerable attention in hydrologic modeling (e.g., Kalma and Sivapalan, 1995), there has been comparatively little discussion in the geomorphic literature.

Three very different goals have been pursued in the modeling of landscapes. Depending on the goal, the answers to the foregoing questions may differ. One goal has been to explain the apparent similarities of landscapes across widely varying erosional, geologic, and climatic environments, as reviewed by Rodríguez-Iturbe and Rinaldo (1996). Another has been to explore the linkages among erosional processes, boundary conditions (including tectonics), and form (e.g., Howard, 1994; Tucker and Slingerland, 1994; Anderson, 1994; Kooi and Beaumont, 1996). In contrast to the first goal, this effort is directed toward identifying the differences between landscapes. The third goal has been to provide tools for predicting and explaining landscape-scale responses to land-use changes (e.g., Dietrich et al., 1995). Here the goal is to try to provide site-specific insights within a larger context, typically a watershed-scale context. Each of these goals implies different relationships to process laws.

To explain the general tendency for landscapes to have characteristic channel-and-network morphologies, Leopold and his colleagues (e.g., Leopold and Langbein, 1962; Leopold et al., 1964) postulated that the distinctive power-law relationships observed in channel hydraulic geometry are emergent properties from a system adjusting to accomplish, simultaneously, constant stream power per unit area of bed and minimum total stream-power expenditure (i.e., constant power per unit length). They explored these concepts to explain river longitudinal profiles through simple random-walk approaches. They also used random-walk models (Leopold and Langbein, 1962) to derive geometric relationships associated with channel networks, as quantified earlier by Horton (1945). Leopold has argued that landscapes are affected by a complex interplay of processes and are influenced by significant spatial heterogeneities, such that there is an indeterminacy in landscape form at specific local sites, but that overall landforms tend toward a most probable state (Leopold, 1995). In that approach, detailed process laws really are not used.

Although others have pursued various similar ideas since the 1960s, the idea of local and global adjustments in energy expenditure in channel networks has recently received significant attention because of the work of Howard (1990) and Rodríguez-Iturbe et al. (1992). Rodríguez-Iturbe and associates, in particular, have focused on explaining the quantitative aspects of network structure that are common to all natural river networks (the power-law relationships among measures of length, area, and slope). Hence they are concerned with distinguishing between models and real network structures and with finding the general principles that can explain the ubiquitous qualities of network organization. They argue that to explain the common features of all landscapes, only simple parameter-free models are appropriate (Rodríguez-Iturbe and Rinaldo, 1996). In essence, they show that simple assumptions about stream-power expenditure (optimal channel networks), or the use of formulations of the critical-shear-stress type for channel erosion, coupled with diffusive transport, can adequately describe natural network structures.

Two aspects of their modeling approach (and the approaches of many others pursuing similar questions) warrant consideration in our discussion of scale dependence. First, they generate their networks by a computational procedure that does not simulate the actual simultaneous routing of water and sediment off the landscape. Second, their erosion laws cannot be calibrated to particular natural landscapes, in part because of the way the computations are done and in part because the erosion laws are such general descriptions of the phenomena that what few parameters there are cannot be related to local processes. This does not detract from the purpose of their models; rather, it indicates that their models lack specific scale and cannot be used for investigating the evolution of specific landscapes. Because the physical underpinning cannot be tested with field or experimental observations, success with these models depends heavily upon having reliable quantitative measures of real landscapes that are neither inevitable (Kirchner, 1993) nor artifactual – and these pitfalls are difficult to avoid when looking for similarities among landscapes. This is because networks bearing similarities (as measured by such things as Horton's laws) to natural ones are readily generated. In fact, it is fair to say that virtually any kind of rule that allows lines originating from different locations to have some preferred orientation of extension and to join and not separate again will generate things that look like channel networks. Rodríguez-Iturbe and Rinaldo (1996) have summarized why just any rule will not do.

In contrast to those approaches, there is a long tradition in geomorphology of seeking to define quantitative mechanistic "laws" responsible for shaping landscapes, as briefly reviewed by Howard (1994). Even as early as the turn of the century, Davis (1892) and Gilbert (1909) came to realize that the fundamental distinction between hilltops and valley bottoms concerned the relative roles of diffusive processes versus concentrative processes (or advective processes, as proposed by Loewenherz, 1991). Since the seminal works of Hirano (1968), Kirkby (1971), Smith and Bretherton (1972), and Carson and Kirkby (1972), the shorthand for transport laws has become (e.g., Howard, 1994)

$$\text{fluvial:} \quad Q_s = aA^m S^n - (A^m S^n)_c \tag{2.13}$$

$$\text{diffusive:} \quad Q_s = KS \tag{2.14}$$

$$\text{landsliding:} \quad Q_s = F(S, S_m) \tag{2.15}$$

where $Q_s$ is the sediment volumetric transport rate (volume per time), the subscript $c$ refers to a critical threshold value for erosion to occur (this term is often dropped), $K$ is the diffusivity, and $S_m$ is the maximum stable hillslope angle. Though mechanistic-based, these transport laws are still great simplifications of the actual processes, and the challenge is to parameterize these models from field data. Their simplicity, however, points to the overall goal to derive generalizable rules for landscape evolution that will still be capable of being calibrated to specific places such that real landscapes, and particularly their scale-dependent attributes, can be modeled.

An additional transport law, of sorts, is the rate of production of erodable material, and Gilbert (1877) first recognized that it should play a central role in landscape evolution. Following the efforts by Ahnert (1970, 1988) and Carson and Kirkby (1972), the production function is generally expressed as a function of the thickness of the weathered mantle or soil mantle:

$$P = F(h) \tag{2.16}$$

where $h$ is soil depth, and $P$ is the production rate (length/time). The production function is treated as either a negative exponential of soil depth or a more complicated function in which the production rate peaks at some intermediate value of soil depth; see the discussion by Dietrich et al. (1995). Howard (1994) has argued that not only the availability of material for transport but also the erodability of that material (leading to what he termed "detachment-limited transport") can have profound effects on landscape evolution.

All of these "laws" have scale dependences. The critical term in equation (2.13) says that fluvial erosion cannot begin at the drainage divide. As discussed in the earlier section on rivers, the ratio $m/n$ defines the curvature of the river profile and can vary downstream, producing distinct slope breaks. The $K$ term in equation (2.14) has units of length squared per time, and short-wavelength features are selectively removed by this process, destroying scaling relationships (Rodríguez-Iturbe and Rinaldo, 1996). Given steady-state conditions for a landform evolving by the diffusive transport law, the curvature of the topography, $\nabla^2 z$, is equal to the lowering rate divided by the diffusivity (Culling, 1963; Dietrich et al., 1995). As discussed earlier, where diffusion-dominated topography gives way to landslide domination, the curvature should tend to zero. The landslide law, as described earlier, could have scale dependences. Especially in the Howard (1994) model, the ratio between the lowering rate and the transport-rate coefficient determines the local relief.

As mentioned earlier, Howard (1994) has shown that even without a critical area–slope value for fluvial processes, a distinct valley-and-ridge topography will develop in response to the shift in transport dominance from diffusive to fluvial. Hence the parameters controlling the strengths of these two terms introduce a fundamental length scale. Finally, as Anderson and Humphrey (1989) have illustrated, the overall appearance of the landscape, including how rough it is and the shapes of the slopes, is strongly influenced by whether soil production keeps up with erosion or the land surface is stripped to bedrock.

Of these four laws, probably only diffusion has been well calibrated with field data, as discussed by Dietrich et al. (1995). A form of equation (2.13) in which erosion (not transport) is assumed proportional to stream power [as defined in equation (2.8)] has been proposed as an appropriate model for channel incision into bedrock (e.g., Howard and Kerby, 1983; Seidl and Dietrich, 1992). Seidl, Dietrich, and Kirchner (1994) have shown that analysis of Hawaiian river profiles supports the stream-power hypothesis, and J. D. Stock and D. R. Montgomery (unpublished data) have performed analyses of the value of $m$ and the erosion coefficient (assuming $n$ to be equal to 1.0) and have found most values of $m$ to lie between 0.3 and 0.6 for many channels, with $m \approx 0$ in some channels, and the erosion coefficient to vary by five orders of magnitude. We know of no published evidence for a landslide transport law. Heimsath et al. (1997) have reported field data that support the assumption of depth dependence in the soil-production law. There is still much to be done in calibrating and improving upon these transport laws.

Different processes dominate over particular length scales; hence parameterization of the transport laws in landscape-evolution models is most appropriately accomplished at the scale at which the process is significant. This has been done only rarely. Koons (1989) and Anderson and Humphrey (1989) have pointed out that the effective diffusivity needed to produce desired sediment-yield values systematically increases with landscape size. This tuning of the diffusivity is similar to what is done with effective hydraulic conductivity in runoff modeling. In large-landscape models, where individual cells can be measured in square kilometers, we can write equations (2.13) and (2.14) (ignoring the effects of landslides) as

$$\begin{aligned} Q_s &= K_1 S + K_2 A^m S^n \\ &= K^* S \end{aligned} \tag{2.17}$$

where

$$K^* = K_1 + K_2 A^m S^{n-1}$$

Here, then, the use of a diffusion equation implicitly or explicitly relies on the assumption that the combined effects at the fine scale of fluvial transport and diffusive hillslope transport can be represented by an effective diffusivity. One can then tune the $K^*$ value in a model to provide the desired sediment flux. Although equation (2.17) can be used to simulate landscape evolution, it is not clear that such a model really has much to do with process. Koons (1989) has shown, for example, that for a given channel spacing, the elevation of the divide varies as $1/K^*$. But this relationship sheds no light on what processes actually control local relief; rather, $K^*$ is just a knob to tune the model to the desired result. Beven (1995) has argued that in hydrologic modeling, the recent trend of using "effective parameters," such as effective conductivity, to aggregate soil hydrologic properties is "doomed to failure" (p. 265), as it ignores inherent nonlinearities and ignores the scale dependences in the effective parameters.

The problem here, then, is that the transport laws operate over finite scales and when related to real landscapes have specific and often limited ranges in real parameters. These transport laws can be used in coarse-scale simulations of landscapes, but it is not clear that this warrants a process-based interpretation of the model results. For that to happen, an intermediate step would need to be performed in which it would be shown that the scaled-up versions of the transport laws could capture the essential behavior and morphology of the real landscape; that is, it would have to be shown that the process laws really did capture the essential physics at the large scale. Anderson (1994) suggested one approach to this problem in his effort to model the Santa Cruz mountains in California. Howard et al. (1994) have reviewed the problem of modeling river processes at the subgrid scale. Beven (1995) has discussed the problem of subgrid parameterization in hydrologic models and has proposed a specific set of objectives and principles to follow.

Figure 2.8 illustrates this problem by showing a hypothetical channel network predicted from a grid-based model of a large landscape. In the lower box the grid cells are 1 $km^2$, and in the expanded view above, the cells are $1,275$ $m^2$. In neither case is the actual channel network fully represented by the coarse grid topography, and consequently the hillslope length over which diffusion and landsliding occur is not correctly captured. In the large-landscape case, all the processes that actually shape the land are of subgrid scale, but the mathematical expressions representing these processes are applied at a scale at which they do not physically occur. Because a channel responds to water and sediment derived from its entire catchment, these large-landscape models may more accurately capture the effects of river-network development (but see the later discussion of the problem of channel width).

Anyone seeking to attribute accuracy to large-landscape models cannot simply rely on the use of mechanistic laws, because at the scale at which the laws are being applied there may no longer be a corresponding relationship between topography and process. Hence, innovative large-landscape modeling of the kind reported by Tucker and Slingerland (1994), Kooi and Beaumont (1996), Anderson (1994), and others

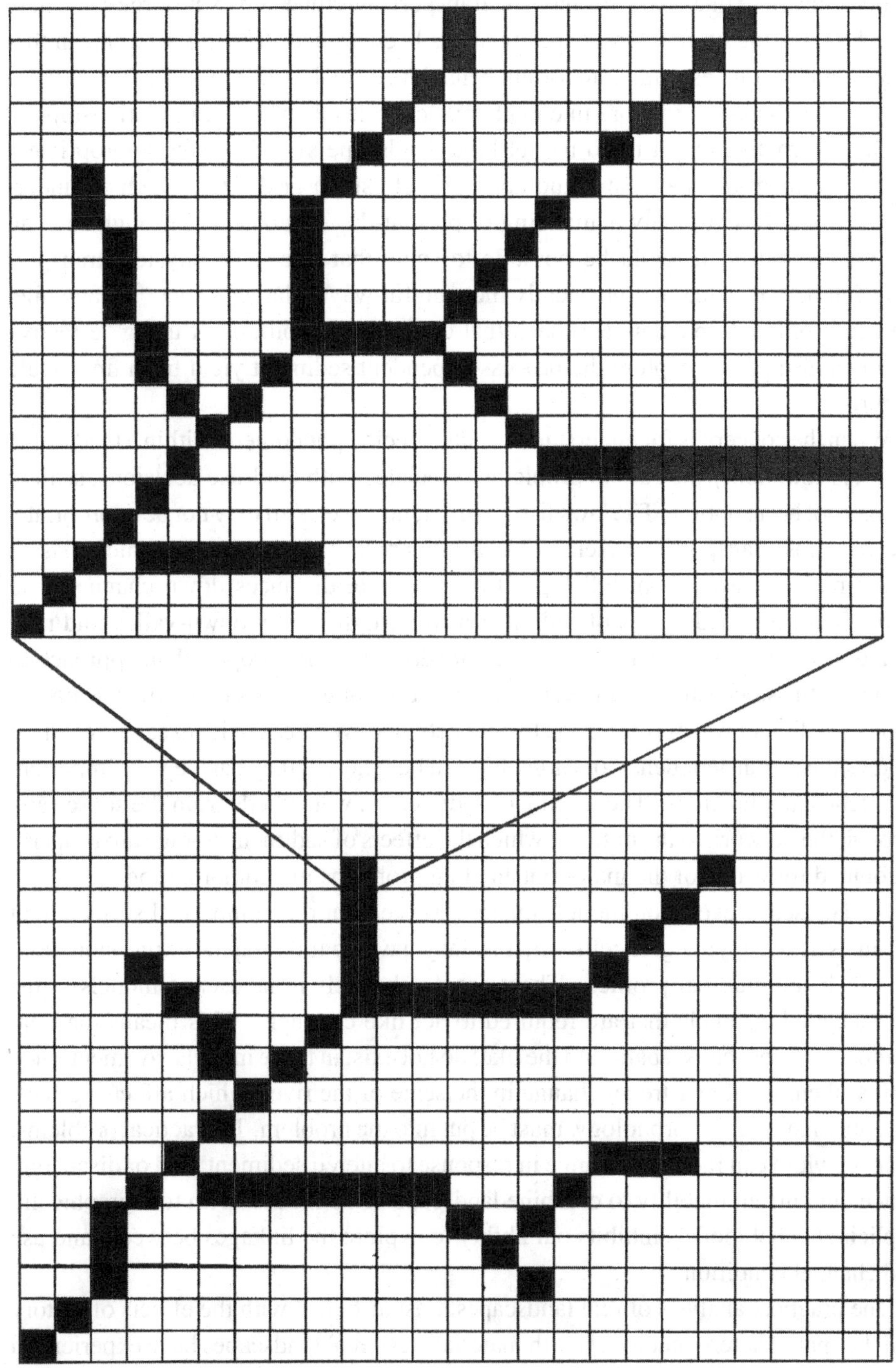

Figure 2.8. A cartoon to illustrate that at coarse grid scales the actual channel networks are poorly represented. In some simulations, every grid cell has a channel and lacks a hillslope, in which case in the lower map all cells would be black.

who rely on these fine-scale mechanistic transport laws tuned up to the large scale may have less bearing on real landscapes than has been argued. This does not mean that they do not capture essential, important behaviors of real landscapes. Rather, at these large scales the laws are more like mathematical rules than physical relationships.

One alternative might be to model large-landscape systems using appropriately scaled sediment-transport laws and appropriately small grid sizes. Such an undertaking would be extremely demanding of present-day computers. Through such an exercise, however, it might be possible to show that fine-scale physical transport laws can be translated in a physically meaningful way to larger scales. Meanwhile, we need some intermediate approach that can avoid the pitfalls of using "effective parameters" and can capture the process-dependent sediment yield from finer-scale features.

When the concern is for understanding site-specific phenomena within a landscape, which is commonly the case in problems associated with land-use decisions, scaling up may not be an option. The laws listed earlier, however, still may not be appropriate, because most transport is driven by stochastic precipitation events in which masses of sediment released from hillslopes travel discrete distances down channels and stop. These released pulses of sediment travel progressively down-valley and may temporarily transform the river channel. Benda (1994) has proposed an approach to modeling the stochastically driven release of sediment by debris flows and the attenuation of sediment waves through a channel network. Whereas in landscape-evolution models these transient behaviors arguably can be ignored, they can be very important in practical applications. The routing of sediment in watersheds from the source and through the network to the outlet at which the effects of sediment storage are properly accounted for is one of the major outstanding problems in geomorphology.

All landscape-evolution models that are worked out on a numerical surface treat channels as occupying grid cells – that is, they have no topographic expression other than their longitudinal profiles. There are no channels in landscape models – just places called channels that are required to act like channels. This means that real channel features (banks, bars, and the like) do not exist in these models. As mentioned earlier, then, the downstream change in the scale of the river, which affects perhaps the entire landscape morphology, must be put into the problem. In practical problems, channel width can radically change in response to altered sediment load or discharge; hence our current inability to combine landscape-scale models with topographically explicit channel models inhibits our ability to explore the linkages between land use and channel condition.

The practical analysis of real landscapes must also deal with the effects of history and the spatial heterogeneity of earth materials. All real landscapes have experienced climate changes, and the effects may be simply obvious, such as scoured glacial topography, or subtle, such as small changes in soil properties. Virtually all landscapes have also experienced land-use effects, and the alterations in some features, such as river channels, can be so complete as to render it nearly impossible to reconstruct the prior condition. Many of these climatic and land-use effects alter landscapes at scales well below the resolution of the typically available digital elevation data. Here we face the same problems with which hydrologists (e.g., Kalma and Sivapalan, 1995) have been wrestling: The spatial variations in the controlling parameters are virtually unknowable, even for small-hillslope studies. The challenge before us is to define appropriate scales for landscape modeling in which subgrid, historical, and earth-material effects will not dominate the phenomena of interest. In other words, we are seeking to ask questions about landscapes in which the modeling can be parameterized with field data and verified with field observations. This has proved extraordinarily difficult to do in hydrology, but it is our hope that the long-term averaging effects arising from the slowly changing morphology of a landscape relative to the stochastic driving forces and the local heterogeneities of earth materials will permit quantitative linkages between process and form.

## 2.6 Conclusion

Although the repeated pattern of ridge-and-valley topography over a wide range of scales leads to the appearance of scale invariance in landscapes, process dominance over distinct spatial domains imparts clear scale dependence to landscape morphology. These scales are most clearly evident at the finest resolution, where closer inspection reveals no further subdivision of landscapes into ridge-and-valley topography. Hillslope length, height, and curvature, drainage density, and channel slope–area relationships all express scale-specific attributes of landscapes and provide clues about dominant erosion processes. The wide availability of digital elevation data is radically changing how geomorphic studies are being conducted. These digital data, however, tend to underestimate slopes and drainage density and therefore may not provide the fine-scale topography needed to test process–form analysis. Furthermore, the grid cells of 90 m, 1 km, or larger used in regional scale simulations will not capture the scale of the fundamental hillslopes at which process and form are linked (except in the case, perhaps, of deep-seated landsliding). Large-scale simulations have provided important insights into the coupling of tectonic and erosional processes. The use of scale-dependent process laws in large-scale models, however, may produce interesting topography, but may have little to do with real landscapes.

## References

Ahnert, F. 1970. A comparison of theoretical slope models with slopes in the field. *Z. Geomorphologie* [*Suppl.*] 9:88–101.

Ahnert, F. 1988. Modelling landform change. In: *Modelling Geomorphological Systems*, ed. M. G. Anderson, pp. 375–400. New York: Wiley.

Anderson, R. S. 1994. Evolution of the Santa Cruz Mountains, California, through tectonic growth and geomorphic decay. *J. Geophys. Res.* 99:20161–79.
Anderson, R. S., and Humphrey, N. F. 1989. Interaction of weathering and transport processes in the evolution of arid landscapes. In: *Quantitative Dynamic Stratigraphy*, ed. T. A. Cross, pp. 349–61. Englewood Cliffs, NJ: Prentice-Hall.
Benda, L. E. 1994. Stochastic geomorphology in a humid mountain landscape. Unpublished Ph.D. dissertation, University of Washington.
Beven, K. 1995. Linking parameters across scales: subgrid parameterizations and scale dependent hydrologic models. In: *Scale Issues in Hydrologic Modelling*, ed. J. D. Kalma and M. Sivapalan, pp. 263–82. New York: Wiley.
Bloom, A. L. 1991. *Geomorphology*. Englewood Cliffs, NJ: Prentice-Hall.
Burbank, D. W., Leland, J., Fielding, E., Anderson, R. S., Brozovic, N., Reid, M. R., and Duncan, C. 1996. Bedrock incision, rock uplift and threshold hillslopes in the northwest Himalayas. *Nature* 379:505–10.
Carson, M. A., and Kirkby, M. J. 1972. *Hillslope Form and Process*. Cambridge University Press.
Coffman, D. M., Keller, E. A., and Melhorn, W. N. 1972. New topologic relationship as an indicator of drainage network evolution. *Water Resour. Res.* 8:1497–505.
Culling, W. E. H. 1963. Soil creep and the development of hillside slopes. *J. Geology* 71:127–61.
Davis, W. M. 1892. The convex profile of bad-land divides. *Science* 20:245.
Dietrich, W. E., and Dunne, T. 1993. The channel head. In: *Channel Network Hydrology*, ed. K. Beven and M. J. Kirkby, pp. 175–219. New York: Wiley.
Dietrich, W. E., Montgomery, D. R., Reneau, S. L., and Jordan, P. D. 1988. The use of hillslope convexity to calculate diffusion coefficients for a slope dependent transport law. *EOS, Trans. AGU* 69:346.
Dietrich, W. E., Reiss, R., Hsu, M.-L., and Montgomery, D. R. 1995. A process-based model for colluvial soil depth and shallow landsliding using digital elevation data. *Hydrological Processes* 9:383–400.
Dietrich, W. E., Wilson, C. J., Montgomery, D. R., and McKean, J. 1993. Analysis of erosion thresholds, channel networks and landscape morphology using a digital terrain model. *J. Geology* 101:259–78.
Dietrich, W. E., Wilson, C. J., Montgomery, D. R., McKean, J., and Bauer, R. 1992. Channelization thresholds and land surface morphology. *Geology* 20:675–9.
Dunne, T. 1991. Stochastic aspects of the relations between climate, hydrology and landform evolution. *Trans. Japanese Geomorph. Union* 12:1–24.
Fernandes, N. F. 1994. Hillslope evolution by diffusive processes: the problem of equilibrium and the effects of climatic and tectonic changes. Ph.D. dissertation, University of California, Berkeley.
Fernandes, N. F., and Dietrich, W. E. 1997. Hillslope evolution by diffusive processes: the timescale for equilibrium adjustments. *Water Resour. Res.* 33:1307–18.
Gallant, J. C. 1998. Scale and Structure in Landscapes. Ph.D. dissertation, Australian National University.
Gilbert, G. K. 1877. *Geology of the Henry Mountains*. Washington, DC: U.S. Geographical and Geological Survey.
Gilbert, G. K. 1909. The convexity of hill tops. *J. Geology* 17:344–50.
Grant, G. E., Swanson, F. J., and Wolman, M. G. 1990. Pattern and origin of stepped-bed morphology in high-gradient streams, Western Cascades, Oregon. *Geol. Soc. Am. Bull.* 102:340–52.
Heimsath, A. M., Dietrich, W. E., Nishiizumi, K., and Finkel, R. C. 1997. The soil production function and landscape equilibrium. *Nature* 388:358–61.

Hirano, M. 1968. A mathematical model of slope development: an approach to the analytical theory of erosional topography. *J. Geosciences Osaka City University* 11:13–52.

Horton, R. E. 1945. Erosional development of streams and their drainage basins; hydrophysical approach to quantitative morphology. *Geol. Soc. Am. Bull.* 56:275–370.

Howard, A. D. 1971. Simulation model of stream capture. *Geol. Soc. Am. Bull.* 82:1355–76.

Howard, A. D. 1990. Theoretical model of optimal drainage networks. *Water Resour. Res.* 26:2107–17.

Howard, A. D. 1994. A detachment-limited model of drainage basin evolution. *Water Resour. Res.* 30:2261–85.

Howard, A. D., Dietrich, W. E., and Seidl, M. A. 1994. Modeling fluvial erosion on regional to continental scales. *J. Geophys. Res.* 99:13971–86.

Howard, A. D., and Kerby, G. 1983. Channel changes in badlands. *Geol. Soc. Am. Bull.* 94:739–52.

Kalma, J. D., and Sivapalan, M. (eds.). 1995. *Scale Issues in Hydrologic Modelling.* New York: Wiley.

Kirchner, J. W. 1993. Statistical inevitability of Horton's laws and the apparent randomness of stream channel networks. *Geology* 21:591–4.

Kirkby, M. J. 1971. Hillslope process–response models based on the continuity equation. *Spec. Publ. Inst. Br. Geogr.* 3:15–30.

Kooi, H., and Beaumont, C. 1996. Large-scale geomorphology: classical concepts reconciled and integrated with contemporary ideas via a surface processes model. *J. Geophys. Res.* 101:3361–86.

Koons, P. O. 1989. The topographic evolution of collisional mountain belts: a numerical look at the Southern Alps, New Zealand. *Am. J. Sci.* 289:1041–69.

Leopold, L. B., 1995. *A View of the River.* Harvard University Press.

Leopold, L. B., and Langbein, W. B. 1962. *The Concept of Entropy in Landscape Evolution.* U.S. Geological Survey professional paper 282-D.

Leopold, L. B., and Wolman, M. G. 1960. River meanders. *Geol. Soc. Am. Bull.* 71:769–94.

Leopold, L. B., Wolman, M. G., and Miller, J. P. 1964. *Fluvial Processes in Geomorphology.* San Francisco: Freeman.

Loewenherz, D. S. 1991. Stability and the initiation of channelized surface drainages: a reassessment of short wavelength limit. *J. Geophys. Res.* 96:8453–64.

McKean, J. A., Dietrich, W. E., Finkel, R. C., Southon, J. R., and Caffee, M. W. 1993. Quantification of soil production and downslope creep rates from cosmogenic $^{10}$Be accumulations on a hillslope profile. *Geology* 21:343–6.

Montgomery, D. R. 1994. Valley incision and the uplift of mountain peaks. *J. Geophys. Res.* 99:13913–21.

Montgomery, D. R., and Dietrich, W. E. 1989. Source areas, drainage density, and channel initiation. *Water Resour. Res.* 25:1907–18.

Montgomery, D. R., and Dietrich, W. E. 1992. Channel initiation and the problem of landscape scale. *Science* 255:826–30.

Montgomery, D. R., and Dietrich, W. E. 1994a. A physically-based model for the topographic control on shallow landsliding. *Water Resour. Res.* 30: 1153–71.

Montgomery, D. R., and Dietrich, W. E. 1994b. Landscape dissection and drainage area–slope thresholds. In: *Process Models and Theoretical Geomorphology*, ed. M. J. Kirkby, pp. 221–46. New York: Wiley.

Montgomery, D. R., and Foufoula-Georgiou, E. 1993. Channel network source representation using digital elevation models. *Water Resour. Res.* 29:3925–34.

Morisawa, M. E. 1957. Accuracy of determination of stream lengths from topographic maps. *EOS, Trans. AGU* 38:86–8.

Prosser, I. P., and Abernethy, B. 1996. Predicting the topographic limits to a gully network using a digital terrain model and process thresholds. *Water Resour. Res.* 32:2289–98.

Rinaldo, A, Dietrich, W. E., Rigon, R., Vogel, G. K., and Rodríguez-Iturbe, I. 1995. Geomorphological signatures of varying climate. *Nature* 374:632–5.

Ritter, D. F., Kochel, R. C., and Miller, J. R. 1995. *Process Geomorphology*. Dubuque: Wm. C. Brown.

Rodríguez-Iturbe, I., and Rinaldo, A. 1996. *Fractal River Basins: Chance and Self-organization.* Cambridge University Press.

Rodríguez-Iturbe, I., Rinaldo, A., Rigon, R. J., Bras, R. L., and Iijasz-Vasquez, E. 1992. Energy dissipation, runoff production and the three dimensional structure of channel networks. *Water Resour. Res.* 28:1095–103.

Schmidt, K. M., and Montgomery, D. R. 1995. Limits to relief. *Science* 270:617–20.

Seidl, M., and Dietrich, W. E. 1992. The problem of channel incision into bedrock. In: *Functional Geomorphology*, ed. K.-H. Schmidt and J. de Ploey, pp. 101–24. Catena supplement 23. Cremlingen-Destedt: Catena Verlag.

Seidl, M. A., Dietrich, W. E., and Kirchner, J. W. 1994. Longitudinal profile development into bedrock: an analysis of Hawaiian channels. *J. Geology* 102:457–74.

Selby, M. J. 1993. *Hillslope Materials and Processes*. Oxford University Press.

Sklar, L., and Dietrich, W. E. 1996. The influence of sediment supply on river incision into bedrock. *EOS, Trans. AGU* 77:F251.

Smith, T. R., and Bretherton, F. P. 1972. Stability and the conservation of mass in drainage basin evolution. *Water Resour. Res.* 8:1506–29.

Strahler, A. N. 1950. Equilibrium theory of erosional slopes approached by frequency distribution analysis. *Am. J. Sci.* 248:673–96, 800–14.

Strahler, A. N. 1957. A quantitative analysis of watershed geomorphology. *EOS, Trans. AGU* 38:913–20.

Strahler, A. N. 1958. Dimensional analysis applied to fluvially eroded landforms. *Geol. Soc. Am. Bull.* 69:279–300.

Strahler, A. N. 1964. Quantitative geomorphology of drainage basins and channel networks. In: *Handbook of Applied Hydrology*, ed. V. T. Chow, pp. 4-39–4-76. New York: McGraw-Hill.

Summerfield, M. A. 1991. *Global Geomorphology*. London: Longman.

Tarboton, D. G., Bras, R. L., and Rodríguez-Iturbe, I. 1991. On the extraction of channel networks from digital elevation data. *Hydrological Processes* 5:81–100.

Tokunaga, E. 1984. Ordering of divide segments and law of divide segment numbers. *Trans. Japanese Geomorph. Union* 7:125–35.

Tucker, G. E., and Slingerland, R. L. 1994. Erosional dynamics, flexural isostasy, and long-lived escarpments: a numerical modeling study. *J. Geophys. Res.* 99:12229–43.

Turcotte, D. L., and Schubert, G. 1982. *Geodynamics*. New York: Wiley.

Willgoose, G. R. 1989. A physically based channel network and catchment evolution model. Ph.D. dissertation, Massachusetts Institute of Technology.

Willgoose, G. 1994. A statistic for testing the elevation characteristics of landscape simulation models. *J. Geophys. Res.* 99:13987–96.

Willgoose, G. R., Bras, R., and Rodríguez-Iturbe, I. 1991. Results from a new model of river basin evolution. *Earth Surf. Proc. Landforms* 16:237–54.

# 3

# Scaling in River Networks

ANDREA RINALDO and IGNACIO RODRÍGUEZ-ITURBE

## 3.1 Introduction

What do biological evolution, the physics of glassy materials, a superfluid near its phase transition, the brain, a slowly driven sandpile, and a fluvial network have in common? Little, it might seem. Nevertheless, they give rise to similar theoretical notions and statistical features (Bak, 1996). Central to their similitude is the power-law nature of the probability distribution that describes key geometric and topologic quantities, or the descriptors of time activity that yield signals (usually termed "$1/f$ noise," from the algebraic decay with exponent $-f$ of the power spectrum) having components of all durations. The algebraic decay (the so-called fat tail) of the distribution ensures that extreme events will be much more likely than in any "regular" case, where the exponential decay of the probability of large events makes catastrophic events (the extremes) prohibitively unlikely. It is this feature that is responsible for the enhanced probability of individual, anomalously large events.

Power laws are also the essential ingredients of scaling arguments; that is, they describe a scale-free arrangement of the parts and the whole such that no characteristic scale is present in the growth of the structure and such a scale can arise only through the size of the system. The foregoing description postulates that in infinite domains, no embedded scale typical of the process is found. In perhaps simpler terms, this means that "events" (or values for the quantity under study) range from infinitely small to infinitely large without preference, and thus they scale with the system size.

A fundamental question among scientists in diverse disciplines is related to the dynamic reasons behind scale-free growth. The linkage between the ubiquitous observations of fractal forms in nature (a geometric, descriptive problem) and the common occurrence of $1/f$ noise in time signals arising in nature from a variety of sources has recently been established through theories of critical self-organization (Bak, Tang, and Wiesenfeld, 1987, 1988). According to such theories, open dissipative systems with many degrees of freedom tend to display dynamic patterns that are robustly attracted toward dynamically recursive states, where the system is free of characteristic spatial or temporal scales. It has also been suggested that this feature is characteristic of energy expenditure in locally optimal niches in the complex, rugged "fitness landscapes" in nature (Rodríguez-Iturbe et al., 1992b; Kauffman, 1993; Rinaldo et al., 1996). In essence, as Bak (1996) has put it rather nicely, we wonder if nature as we see it now has any preferential state from an evolutionary viewpoint.

The dynamics of river networks fits perfectly into such schemes. On one hand, scaling properties have been shown to exist over several logarithmic decades by some of the most accurate data sets available in science for the study of complex dissipative systems observed in nature. On the other hand, dynamic theories and their validations (e.g., Sinclair and Ball, 1996) have become well established. Thus the very landscape whose gradients identify drainage directions and network structure is to be seen as the fitness landscape itself that provided a working image for scientists active in the most diverse disciplines. In fact, the lowest mean elevation compatible with constraints of dynamic origin and with the complete drainage of an assigned region yields the desired state (i.e., the most stable and the most likely to be attained by the dynamics). Nevertheless, the dynamic accessibility of optimal niches in the fitness landscape yields other implications, quite outside the scope of this chapter.

River networks provide a benchmark for the testing of the scaling hypothesis concerning many quantities of geomorphologic and hydrologic interest and, because of the massive amount of data provided by digital terrain models (DTMs), allow a thorough testing of the theoretical notions. Not that scaling properties were not already being empirically observed by geomorphologists. On the contrary, some "laws" of geomorphology are manifest consequences of a deeper regularity of landforms that can be organically interpreted only within a scaling framework. One such example is Hack's law, which relates the area of a basin to the length of its longest stream through a power law characterized by a nontrivial scaling exponent. This long-studied macroscopic regularity (e.g., Hack, 1957; Gray, 1961; Muller, 1973) has been reexamined in the light of theoretical schemes, as described in the following sections, and it is now clear that Hack's law is merely a reflection of the drive toward fractality shown by the dynamics of landform evolution. This probably holds for a wide class of empirical relationships. The search for regularity in the variety of landforms in nature thus finds in this context a validation through scaling theories.

Moreover, recent research suggests that the emergence of scaling properties for many features in river-basin landscapes corresponds to a stationary state resulting from an erosion dynamics that eventually settles into a stable state. This stationary state is "frustrated" by initial and boundary conditions, as well as by geologic constraints; that is, it cannot completely escape their imprinting during its evolution. This is a heuristic definition of a dynamically accessible niche in a fitness landscape, because a stationary, stable state can be shown to correspond to a state featuring locally minimum energy (or maximum fitness) of the system. Around this niche, the landscape is defined by the energy levels in different configurations of the system, where one system configuration is a point displaced by its evolutionary process. Within our context, the network configurations corresponding to stable, dynamically accessible niches in their fitness landscape show recursive scaling properties (Rinaldo et al., 1996) and display fractal arrangements that make the parts statistically indistinguishable from the whole.

In this chapter we examine the experimental evidence from several river basins in very different geologic, climatic, and vegetational environments. The analysis identifies the basic facts to which theoretical tools must refer and singles out what we believe to be the most important scaling features found in fluvial geomorphology.

## 3.2 Hack's Law and Scaling Processes

A river basin is an anisotropic system defined by a typical longitudinal length $L_{\parallel}$ (which we shall identify with the linear size of the system) and a typical transverse length $L_{\perp} \leq L_{\parallel}$ (Figure 3.1a). Interchangeably, we shall call $L_{\parallel}$ the diameter of the basin, and $L_{\perp}$ its width. We shall assume that the latter scales as

$$L_{\perp} \propto L_{\parallel}^{H} \qquad (0 \leq H \leq 1) \tag{3.1}$$

and call basins self-affine if $H < 1$ and self-similar if $H = 1$. Here $H$ is still called the Hurst exponent because of its similarity to the analogue in the fractional Brownian-motion context ( e.g., Mandelbrot, 1983; Feder, 1988).

Equation (3.1) postulates that basin boundaries are self-affine *curves* for which $L_{\parallel}$ and $L_{\perp}$ can be seen respectively as diameter and width. It is a fairly general property of self-affine boundaries that their embedded area, say $A$, is related to the diameter and width via

$$A \propto L_{\parallel}^{1+H} \tag{3.2}$$

which is valid whatever the anisotropic (i.e., self-affine) scalings of the boundaries. A summary of the experimental values obtained for $L$ and $L_{\parallel}$ for a number of basins is shown in Table 3.1.

Hack's law (Hack, 1957) relates the length of the longest stream $L$ in the drainage region, measured to the divide, with the drainage area of the basin, say $A$, as $L \sim A^{h}$, where extensive experimental analyses, though constrained by rudimentary techniques and data (e.g., Gray, 1961; Muller, 1973), have shown that the values for the exponent are in the range $h = 0.56$–$0.6$.

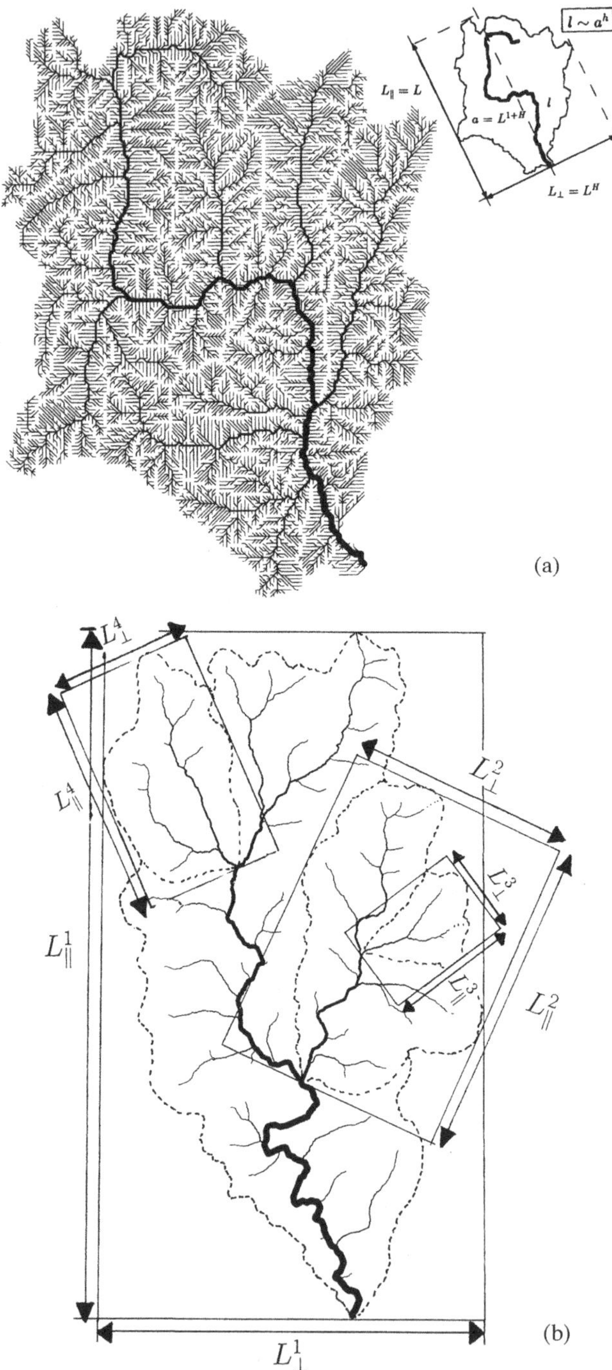


Figure 3.1. (a) Sketch of a river basin, its diameter $L_\parallel$, and its width $L_\perp$. (b) Some subbasins are also drawn. For any subbasin, the longest sides of the rectangle enclosing the network are parallel to the diameter $L_\parallel$ defined as the straight line from the outlet to the farthest point in the basin. The shortest sides are $L_\perp$. (Adapted from Maritan et al., 1996, and Rigon et al., 1996.)

Table 3.1. *Geometric data for the basins analyzed*

| Basin | Location | Area ($km^2$) | $L$ (km) | $L_{\parallel}$ (km) |
|---|---|---|---|---|
| Salt Creek | ID | 2091 | 97.8 | 66.8 |
| Guyandotte | WV | 2088 | 145.1 | 75.8 |
| Tug Fork | WV | 1442 | 97.8 | 66.9 |
| Salmon River | CA | 1936 | 94.9 | 64.5 |
| Boise River | ID | 1847 | 94.8 | 63.3 |
| Little Coal | WV | 984 | 90.5 | 57.5 |
| North Fork River | VA | 825 | 70.1 | 37.0 |
| South Fork, Smith River | CA | 606 | 52.3 | 28.3 |
| Dry Fork | WV | 586 | 63.7 | 41.6 |
| Johns Creek | KY | 484 | 68.8 | 45.7 |
| Big Coal | WV | 449 | 56.2 | 40.7 |
| Racoon Creek | PA | 448 | 53.6 | 35.2 |
| Pingeon Creek | WV | 405 | 49.1 | 35.3 |
| Moshannon Creek | PA | 393 | 49.7 | 33.8 |
| Wooley Creek | CA | 384 | 38.75 | 30.41 |
| Clear Fork | WV | 333 | 59.0 | 34.9 |
| Brushy Creek | AL | 322 | 52.4 | 29.9 |
| Rockcastle Creek | KY | 310 | 45.9 | 33.5 |
| Sturgeon Creek | KY | 295 | 46.3 | 27.1 |
| Edinburgh Creek | VA | 294 | 43.8 | 27.9 |
| Reynolds | ID | 266 | 36.0 | 30.4 |
| Island Creek | WV | 260 | 30.5 | 23.0 |
| Wolf Creek | KY | 212 | 30.3 | 21.7 |
| Sexton Creek | KY | 186 | 33.3 | 23.4 |
| Elkhorn Creek | WV | 190 | 36.5 | 24.6 |
| Big Creek | ID | 147 | 25.6 | 16.9 |
| Indian Creek | WV | 148 | 41.0 | 28.3 |
| Big Creek (Cald well) | ID | 146 | 21.5 | 16.9 |
| Muddy Run | PA | 141 | 28.8 | 21.0 |
| Schoharie Creek headwater | NY | 108 | 18.5 | 15.8 |
| Pound Creek | KY | 105 | 23.5 | 17.8 |
| Tipton River | PA | 64 | 18.8 | 12.5 |
| Blackberry Fork | KY | 53 | 18.0 | 13.3 |
| Catskill Creek | NY | 33 | 15.4 | 10.9 |

*Note:* Data were obtained by processing USGS digital terrain maps (DTMs). In natural basins, many features, including stream lengths, geometric characteristics, and total contributing areas, are amenable to detailed and objective experimental investigation through analyses of DTMs (e.g., Tarboton et al., 1991; Dietrich et al., 1992), which can provide extensive evidence covering several length and area scales, say from the order of 10 m to $10^5$ m. Most important, data analysis from DTMs can be handled objectively, rather than on a case-by-case basis – a major pitfall for earlier geomorphologic analyses. It is only recently, with the availability of digital elevation maps, that detailed and extensive study of networks has been made routinely possible. DTMs generally consist of elevations in a rectangular grid; in the United States, grids with 30 m to a side are common. Each grid block is termed a pixel.
*Source:* Adapted from Rigon et al. (1996).

Hack's equation can be rewritten as

$$\frac{A}{L^2} \propto A^{-0.15\pm0.05} \tag{3.3}$$

thus implying that catchments of all sizes are not entirely similar in shape. Rather, as the area increases, $A/L^2$ decreases, leading to the conclusion that there is a tendency toward elongation of the larger catchments.

Interestingly, Hack's law has long been recognized as an outgrowth of some scaling invariance. For instance, Mandelbrot (1983) suggested that an exponent larger than 0.5 in $L \propto A^h$ could arise from the fractal character of river channels that causes the measured length to vary with the spatial scale of the object. The argument is that the length of a stream channel is expected to increase as the power $D/2$ of its drainage area, where $D$ is the fractal dimension (e.g., Feder, 1988) of the stream. In a nutshell, the length of a smooth curve can be measured by stepping along it with dividers, or rulers, of length $\delta$. The length is thus $L(\delta) = N\delta$, where $N$ is the number of ruler steps. When measuring natural forms, where finer detail ($\delta \to 0$) always reveals new details, a finite value of the measure $L(\delta)$ in the magnifying limit ($N\delta^D$) exists only for a nontrivial exponent $D$, which, following Mandelbrot, we call fractal dimension. Thus equation (3.3) will be a reflection of a fractal dimension of river channels close to $D = 2 \times 0.6 = 1.2$. However, as explained in detail by Feder (1988), to derive the fractal dimension of a river from the foregoing argument we need to make specific invariance assumptions about the shape of the basin. Other interpretations have also been proposed (e.g., Mesa and Gupta, 1987; Peckham, 1995).

Figure 3.2a shows the result of Hack's analysis carried out for the DTMs of the river basins derived from reasonably homogeneous geologic and climatic conditions among the basins described in Table 3.1. Although the best estimate for the slope is close to the commonly found value of 0.56, we notice that a closer look at the morphology of the fluvial basins allowed by DTMs suggests statistical fluctuations that were not considered in the original work by Hack and by the following experimental work. These fluctuations are enhanced if we consider all the basins in Table 3.1.

Figure 3.2b shows the result obtained by plotting a modified Hack relationship, as obtained by plotting $L_\parallel$ versus $A$ for the same basins analyzed in Figure 3.2a. We obtain a relationship of the type $L_\parallel \propto A^{h'}$, with $h' \sim 0.52$. This suggests a statistical relationship of the type $L \propto L_\parallel^{h/h'}$, with $h/h' \sim 1.1$, which will be discussed later. It is also clearly observed that the scatter that arises in the foregoing relationships, neglected in previous experimental observations, calls for some statistical interpretation, which we shall provide – following Maritan et al. (1996) and Rigon et al. (1996) – in the framework of scaling structures.

In any given simply connected planar figure we have $A \propto L_\parallel^2$, where $A$ is the area, and $L_\parallel$ is a characteristic length. The constant of proportionality depends on the

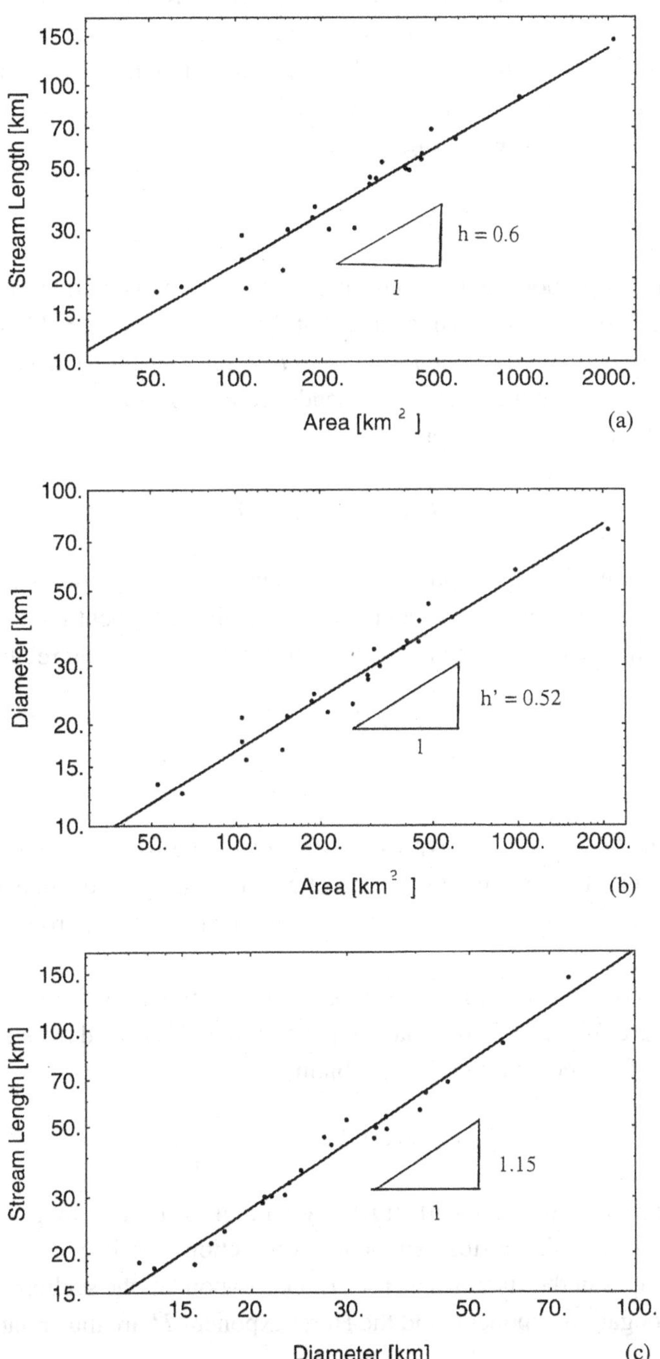


Figure 3.2. (a) Hack's law for geologically and climatically similar basins described in Table 3.1. (b) Elongation for the same set of data. An exponent greater than 0.5 means that basins elongate. (c) Stream length versus diameter for the basins in Table 3.1. (Adapted from Rigon et al., 1996.)

shape of the figure. We define elongation as a property related to the magnification of the basin describing the effect that when $L_\parallel$ increases, $A/L_\parallel^2$ decreases. The shape remains geometrically similar when $A/L_\parallel^2$ is constant, regardless of changes in $L_\parallel$. Consider now a set of river basins, and express the area of the $i$th basin as a function of the characteristic distance by (Figure 3.1b)

$$A(L_\parallel^i) = a(L_\parallel^i)\ (L_\parallel^i)^2 \tag{3.4}$$

where $A(\cdot)$ and the proportionality constant $a(\cdot)$ have been explicitly written as functions of $L_\parallel$. If we order the basins such that $L_\parallel^1 < L_\parallel^2 < \cdots < L_\parallel^i < \ldots$, then elongation will occur if $a(L_\parallel^1), a(L_\parallel^2), \ldots, a(L_\parallel^i), \ldots$ constitutes a decreasing sequence. Thus, assuming a power-law dependence of $a(L_\parallel)$ on $L_\parallel$, we say that the series of river basins shows elongation when

$$a(L_\parallel) \propto L_\parallel^{-q} \qquad (q > 0) \tag{3.5}$$

and $q$ will be defined as the elongation exponent. We shall now explicitly relate Hack's law to the elongation exponent $q$ and a scaling exponent $\phi_L$ related to the divider fractal dimension of stream channels through the following relationships:

$$L \propto L_\parallel^{\phi_L} \tag{3.6}$$

$$A = a(L_\parallel)L_\parallel^2 \tag{3.7}$$

where equations (3.6) and (3.7) express, respectively, the fractal character of mainstream lengths and the general dependence of area on the squared value of a characteristic length that in this case is the longest straight-line distance from the outlet to the divide.

Figure 3.2c shows an experimental validation of equation (3.6) for the basins listed in Table 3.1. Here $\phi_L \sim 1.15$, thus matching the ratio $h/h'$ derived from Figures 3.2a and 3.2b. Also, from equation (3.7), we obtain

$$a(L_\parallel) \propto L_\parallel^{\phi_L/h-2} \tag{3.8}$$

Thus Hack's law is consistent with fractality and elongation as long as $q = 2 - \phi_L/h > 0$. If $q < 0$, the basins experience contraction, and if $q = 0$ they remain with similar shapes in the statistical sense. Hack's exponent, the scaling exponent of lengths, the elongation exponent, and the Hurst exponent $H$ are thus related through

$$\begin{aligned} q &= 2 - \frac{\phi_L}{h} \\ H &= \frac{\phi_L}{h} - 1 \end{aligned} \tag{3.9}$$

We observe that equation (3.9) imposes constraints between $\phi_L$ and $h$. Thus Hack's exponents between 0.57 and 0.6 require values of $\phi_L$ smaller than 1.14 and 1.2, respectively, for elongation to occur. Higher values of $\phi_L$ would imply contraction of river basins when increasing with size. Our experimental evidence regarding values for $\phi_L$ and $h$ fulfills the condition that $q > 0$ without exceptions, implying that indeed there seems to be an elongation of river basins as their size increases.

## 3.3 Scaling of Areas

Our starting point is the study of the total contributing area, say $A$, at a site. This is a key variable because of its intrinsic capability to describe the nested aggregation structure embedded in the fluvial landforms and its important physical implications (e.g., Rodríguez-Iturbe and Rinaldo, 1997).

In a river basin, the total contributing area of a site $i$ (i.e., $A_i$) is identified through drainage directions and is measured by the number of sites upstream of the site connected by the network. The integral equation yielding $A_i$ pointwise is

$$A_i = \sum_{j \in nn(i)} A_j + 1 \tag{3.10}$$

where $nn(i)$ are the nearest neighbors to the $i$th site actually draining into it. Note that equation (3.10) gives an oversimplified notation for a technical problem (i.e., that of the definition of which of the eight neighbors actually drains into $i$), requiring careful attention somewhat removed from the scope of this chapter (e.g., Tarboton, Bras, and Rodríguez-Iturbe, 1991; Jenson and Domingue, 1988).

From the computational viewpoint, $A$ can also be regarded as a measure of the flow rate if a unit rainfall (mimicking a constant rate of mass injection) is applied uniformly over the basin. Thus we shall assume for convenience that a significant measure of discharge, say $Q$, be it the mean annual value or some landscape-forming value, can be surrogated by the total contributing area $A$ (i.e., $Q \sim A$). This holds correctly as long as the assumption of a spatially uniform rainfall rate is meaningful. One could also show that the influence of random forcings (e.g., spatially varying rainfall) described by an equation like $A_i = \sum_j A_j + R_i$, where $R_i$ is described by a spatial random-noise field, does not change our main conclusions (Maritan et al., 1996).

Channelized pixels from DTMs can be defined in a number of manners. The easiest way, which was implemented in the early studies on this subject, defines channels by their contributing areas automatically found from the DTM (e.g., Tarboton et al., 1991). Streams are identified as those pixels with cumulative contributing areas greater than a support threshold, say $A_t$. Although this can become a rather crude approximation, especially if the range of scales investigated is not exceedingly larger than the lower cutoff for scaling processes ruled by hillslope processes, the method is still valued for its viability. Issues surrounding the correct identification of the threshold area and other related topics have been dealt with, for instance, by Dietrich et al. (1988), Howard (1994), and Rinaldo et al. (1995).

This section specifically addresses the issue of scaling for the total contributing areas at an arbitrary point within the basin. Scaling postulates that the distribution function obeys, within an idealized boundless domain, a power law of the type

$$P[A \geq a] \propto a^{-\beta} \tag{3.11}$$

where $a$ is the arbitrary value of the random variable $A$. The coefficient $\beta$ characterizes the entire aggregation pattern. We shall first examine this key feature without reference to domain-size issues (i.e., the power-law character exhibited by the observational evidence provided by the total contributing area at an arbitrary site in a real river basin). Later on, we shall focus on a more comprehensive view (i.e., that of finite-size scaling).

In the early study that recognized this important character, five basins of very different characteristics throughout North America were used by Rodríguez-Iturbe et al. (1992b) for a study of the scaling properties of discharge and energy in river basins. All of them had areas greater than 100 km$^2$ – a fact that bears some relevance in view of a lower cutoff for the scaling behaviors clearly indentified by crossovers with hillslope processes (Montgomery and Dietrich, 1992). The basins studied by Rodríguez-Iturbe et al. (1992b) are included among those listed in Table 3.1.

Figure 3.3 shows the results for the five basins. The distributions show good agreement with a power-law form for nearly three logarithmic scales. The result found by Rodríguez-Iturbe et al. (1992b) is

$$P[A \geq a] \propto a^{-0.43 \pm 0.02} \tag{3.12}$$

with the important characteristic of the exponent being relatively robust with respect to size, vegetation, geology, climate, or orientation of the basin. The deviations that are observed at very large values of areas are finite-size effects, analyzed later in this section.

It has also been observed that the value of the scaling exponent $\beta$ is relatively unaffected by the size of the support threshold used in identification of the network or by the method used for screening where channels begin (Rodríguez-Iturbe and Rinaldo, 1997). It is important to note that the estimates for the exponent $\beta$ are similar in many basins, approximately equal to 0.43, though there are fluctuations. The recurring similar values for the exponent suggest a common "average" scaling behavior whereby a spatially extended dissipative dynamic system naturally evolves into similar states with no characteristic time or length scales. The dynamic background for this phenomenon has been discussed by, for instance, Rinaldo et al. (1993).

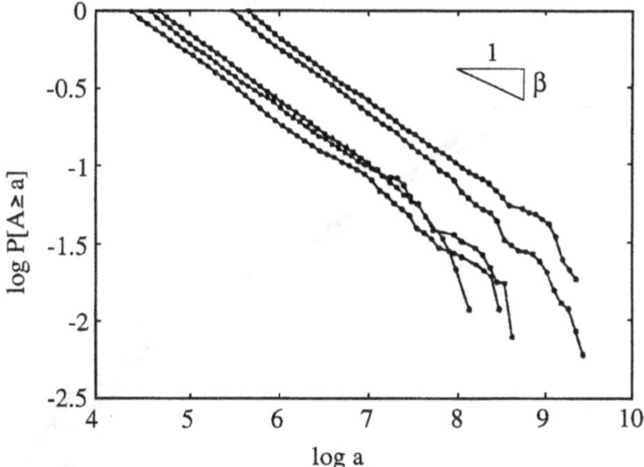


Figure 3.3. Log-log plot for the probability of exceedance of the total contributing area, $P[A \geq a]$, versus $a$ for five basins (Brushy, 311 km$^2$, $A_t = 25$ pixels; Caldwell, 147 km$^2$, $A_t = 40$ pixels; Raccoon, 448 km$^2$, $A_t = 50$ pixels; Schoharie, 2,408 km$^2$, $A_t = 70$ pixels; St. Joe, 2,834 km$^2$, $A_t = 50$ pixels). This choice of basins and threshold areas reproduces the original experimental analysis by Rodríguez-Iturbe et al. (1992b). The estimate of the slope (i.e., the scaling exponent $\beta$) is $\beta = 0.43 \pm 0.02$.

Now we shall show, following Maritan et al. (1996), how a simple finite-size scaling *Ansatz* leads to natural explanations for many empirical and experimental facts, as well as for the scaling properties exhibited by real basins and their suitable numerical simulations. As seen in the preceding section, real river basins exhibit a power-law probability of exceeding the total contributing area, say $P[A \geq a] = P(a)$ (notice the slight change in notation, chosen for convenience), that scales like $P(a) \propto a^{-\beta}$, with $\beta = 0.43 \pm 0.02$ [equation (3.12)]. We shall now explore the inference of the finite-size effect yielding $P[A \geq a, L_{\parallel}] = P(a, L_{\parallel})$, where $L_{\parallel}$ is a *given* linear size of the system. Figure 3.4 shows $P(a, L_{\parallel})$ for four subbasins with different $L_{\parallel}$ values within the Fella River region in Italy (Maritan et al., 1996). The departure from the straight line for small areas indicates the existence of a reasonable threshold for channelized areas (i.e., the suitable minimum area supporting a channel head). Notice also that the largest basin covers four $\log_{10}$ scales of $a$.

Let us define the probability density distribution $p(a, L_{\parallel}) = -\partial P(a, L_{\parallel})/\partial a$ for a basin to have area $a$ for a given linear size $L_{\parallel}$. As in many other scaling systems, the key idea is that the constraint imposed by the size of the system, $L_{\parallel}$, selectively affects the areas such that by rescaling a system through a factor $A/L_{\parallel}^{\phi}$ (where $\phi$ is a suitable exponent), an equivalent system is obtained. Thus, what matters is a size "relative" to a cutoff (e.g., Fisher, 1971; Meakin, Feder, and Jossang, 1991). Thus one expects that the finite-size distribution $p(a, L_{\parallel})$ will obey a scaling form of the type

$$p(a, L_{\parallel}) \propto a^{-\tau} f\left(\frac{a}{L_{\parallel}^{\phi}}\right) \qquad (1 \leq \tau \leq 2) \qquad (3.13)$$

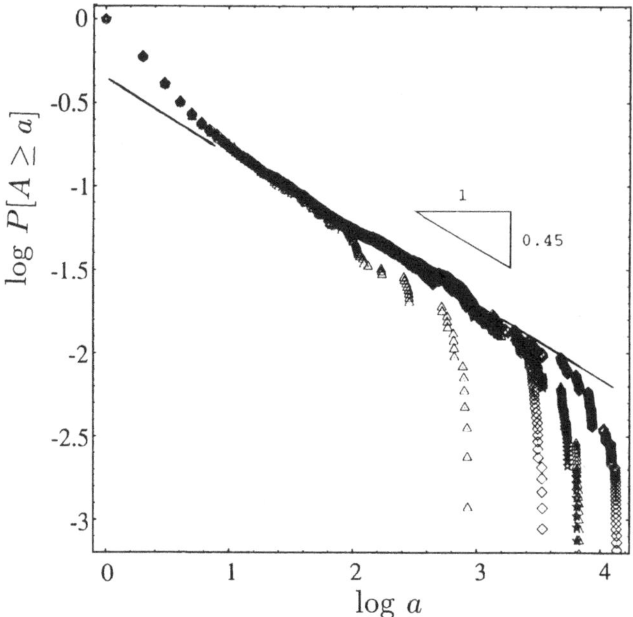


Figure 3.4. Log-log plot for the probability of exceeding the total contributing area, $P[A \geq a] = P(a, L_{\parallel})$, for four subbasins of the Fella River characterized by different $L_{\parallel}$ values.

and

$$\tau = 1 + \beta \tag{3.14}$$

where $L_{\parallel\phi} \Rightarrow L_{\parallel}^{\phi}$ is a characteristic area, and $f(x)$ is a scaling function satisfying the following properties:

$$\lim_{x \to \infty} f(x) = 0 \qquad \text{(sufficiently fast)} \tag{3.15}$$

$$\lim_{x \to 0} f(x) = c \tag{3.16}$$

where $c$ is a suitable constant (Maritan et al., 1996). The first equation ensures the correct behavior at infinity, and the second equation gives a power-law behavior in the infinite size limit [i.e., $p(a, \infty) = ca^{-1-\beta}$]. Because of equation (3.2), we know that the characteristic area $L_{\parallel}^{\phi}$ yields

$$\phi = 1 + H \tag{3.17}$$

Because we require normalization,

$$1 = \int_1^{\infty} da\, a^{-\tau} f\left(\frac{a}{L_{\parallel}^{\phi}}\right) = L_{\parallel}^{\phi(1-\tau)} \int_{L_{\parallel}^{-\phi}}^{\infty} dx\, x^{-\tau} f(x) = \frac{c}{\tau - 1} \tag{3.18}$$

Notice that because $1 \leq \tau \leq 2$, we cannot allow the lower cutoff to go to zero. Therefore we find

$$p(1, L_{\parallel} \to \infty) = c = \tau - 1 \tag{3.19}$$

that is, the proportion of the number of sources is independent of the size of the system.

The probability of exceeding the total contributing area at any site is then

$$P(a, L_{\parallel}) = \int_a^{+\infty} dA\, p(A, L_{\parallel}) \tag{3.20}$$

Because $\phi = 1 + H$, and using equations (3.13) and (3.20) jointly with the fact that $a_c(L_{\parallel}) \sim L_{\parallel}^{\phi}$, we get

$$P(a, L_{\parallel}) = a^{-\beta} \mathcal{F}\left(\frac{a}{L_{\parallel}^{\phi}}\right) \tag{3.21}$$

with

$$\beta = \tau - 1 \tag{3.22}$$

where we have defined

$$\mathcal{F}(x) = x^{\beta} \int_x^{+\infty} d\chi\, \chi^{-(1+\beta)} f(\chi) \tag{3.23}$$

Many network models that have been studied in different contexts have had a *directed* character because of the fact that they typically have been applied on slopes that have given preferred directions to the flows. Under this condition, we expect (Takayasu, 1990) that

$$\langle a \rangle = \int_1^{+\infty} da\, ap(a, L_{\parallel}) \propto L_{\parallel} \tag{3.24}$$

Equation (3.24) can be obtained assuming that the spanning tree defining the river network is strictly direct, as, for example, in the Scheidegger (1967) model. This means that along the path from any site to the outlet, the tangent always has a positive projection along a fixed direction. If the river network is not direct, a correction is needed, as will be seen in what follows. The lower cutoff in equation (3.24) can be allowed to go to zero, because the integral is convergent for $L_{\parallel} \to \infty$, and thus we easily find

$$\phi = \frac{1}{2 - \tau} \tag{3.25}$$

which means that there is only *one* independent exponent in equation (3.21) (Meakin et al., 1991; Maritan et al., 1996). Notice, however, that equation (3.24) holds only if statistically relevant river configurations are directed. Using equations (3.17), (3.22), and (3.25), we get

$$\beta = \frac{H}{H+1} \tag{3.26}$$

in substantial agreement with the theoretical results to be derived in the next section and with the experimental data, from which we expect $H$ in the range 0.75–0.8 (Ijjasz-Vasquez, Bras, and Rodríguez-Iturbe, 1994) and $\beta = 0.43 \pm 0.02$. Interestingly, the elongation exponent $q$ turns out to be $q = 1 - H$ [see equation (3.5)], thereby showing elongation whenever $H < 1$.

We shall now show that a non-direct character in the network yields a minor modification to equation (3.26). In the general case (non-directed networks), from equation (3.24) we obtain (Maritan et al., 1996)

$$\langle a \rangle = L_{\parallel}^{\phi(2-\tau)} \int_{1/L_{\parallel}^{\phi}}^{\infty} dx\, x^{1-\tau} f(x) \tag{3.27}$$

where the lower cutoff is irrelevant because the integral converges for $L_{\parallel} \to \infty$. It thus follows that

$$\langle a \rangle \propto L_{\parallel}^{\phi(2-\tau)} \tag{3.28}$$

The foregoing results also allow us to relate the exponents $\tau$ and $\phi$, thus linking the planar aggregation structure to the elongation, because

$$\langle a \rangle = \ell \propto L_{\parallel}^{\varphi} \tag{3.29}$$

where $\varphi$ is a new scaling exponent, derived as follows. If we characterize the sum of the total lengths to the outlet $\sum_i x_i$ (where $i$ spans all sites from 1 to $N$, where $N$ is the total number of sites and $x_i$ is the length to the outlet from site $i$ measured along the network), we have $\sum_i x_i = \sum_i \sum_{j \in \gamma(i)} \Delta x_{ij} = \sum_i a_i$, where $j$ indexes the intermediate steps in the path $\gamma(i)$ from $i$ to the outlet, and $\Delta x_{ij}$ is the lattice size (i.e., it is unity in isotropic lattices). Thus, if $M$ is the maximum distance from source to outlet, we can reverse the order of summation and write, in general, for loopless structures, $\langle a \rangle = \sum_{x=1}^{M} x W(x) = \ell$, where $W(x)$ counts the relative proportion of sites at distance $x$ from the outlet (i.e., the geomorphologic width function) (Gupta and Mesa, 1988), and $\ell$ is the mean distance to the outlet of the network. The length $\ell$ is characteristic of the entire aggregation pattern of the area distribution. The relationship with a Euclidean length $L_{\parallel}$ has been experimentally

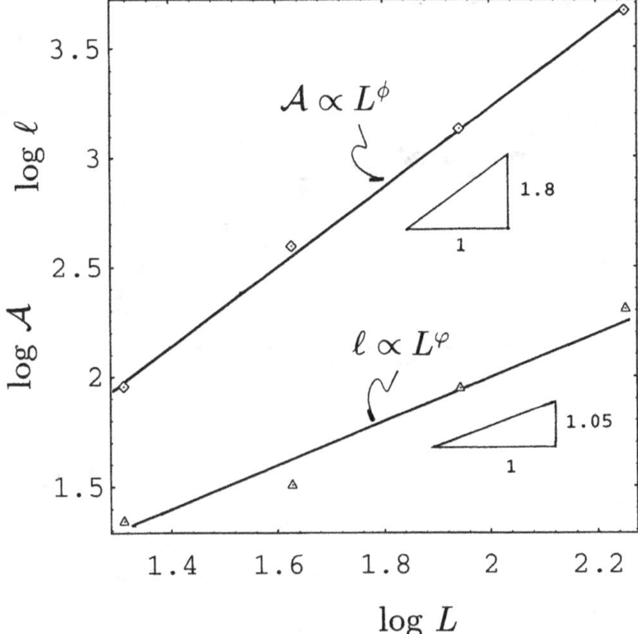


Figure 3.5. Experimental assessment of scaling assumptions from the DTM for the Fella River: $\log \mathcal{A}$ versus $\log L_\parallel$ and $\log \ell$ versus $\log L_\parallel$ for the Fella River ($\phi = 1.8 \pm 0.01, \varphi = 1.05 \pm 0.02$). (Adapted from Maritan et al., 1996.)

established as $\ell \propto L_\parallel^\varphi$, with $\varphi = 1.05$ (Maritan et al., 1996). Figure 3.5 shows the original experimental evaluation of $\varphi$. Notice that $\varphi$ does not coincide, in principle, with $\phi_L$ [see equation (3.6)] because $\ell$ is the mean distance to the outlet computed along the network for *all* sites [i.e., the centroid of the width function $W(x)$], whereas $L$ is just the *mainstream* length.

Thus we find the relationship linking the scaling coefficients:

$$\phi = \frac{\varphi}{2 - \tau} \tag{3.30}$$

where $\varphi \approx 1.05$. Notice that in the case $\varphi = 1$, we recover equation (3.24). The foregoing relationship is analytically verified, for instance, in the simple Scheidegger (1967) model of network development, where $\tau = \frac{4}{3}$ and $\varphi = 1$ yield exactly $\phi = \frac{3}{2}$. We also note that in the general case we obtain

$$\beta = \frac{1 + H - \varphi}{H + 1} \tag{3.31}$$

We note, finally, that from equations (3.13) and (3.17) we obtain

$$\langle a^n \rangle \propto L_\parallel^{(n - \tau - 1)\phi} \tag{3.32}$$

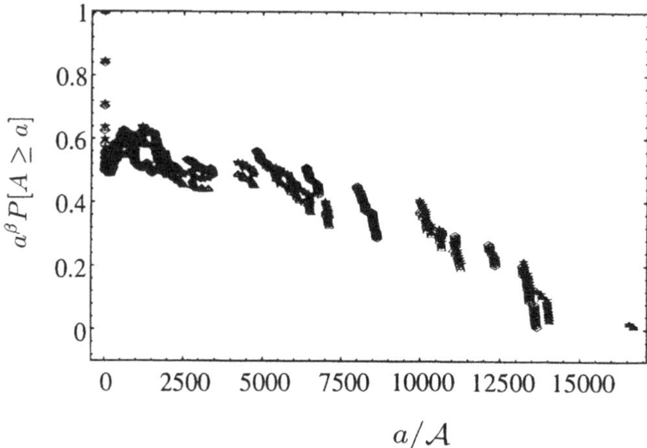


Figure 3.6. Collapse of $a^\beta P(a, \mathcal{A})$ versus $a/\mathcal{A}$ for four subbasins of the Fella River basin ($\beta = 0.45$). (Adapted from Maritan et al., 1996.)

and thus we can provide an alternative definition of the characteristic *size* of contributing area $\mathcal{A}$, as

$$\mathcal{A} = \frac{\langle a^2 \rangle}{\langle a \rangle} \propto L_{\parallel}^{\phi} \tag{3.33}$$

which, once substituted into equation (3.13), yields the final scaling relationship in the form

$$p(a, \mathcal{A}) = a^{-\tau} f\left(\frac{a}{\mathcal{A}}\right) \tag{3.34}$$

From the data shown in Figure 3.4 we obtain the collapse of the various curves shown in Figure 3.6 for the experimental value $\tau = 1.45$, obtained for the Fella River basin, in domains ranging from 2,200 km$^2$ to 140 km$^2$. The collapse of the various curves is quite good, and thus we can conclude that the generalized scaling law (3.13) is supported by experimental and theoretical data.

## 3.4 Scaling of Lengths

Several characterizations of the stream lengths that make up a drainage network have been proposed (Rodríguez-Iturbe and Rinaldo, 1997). A common one is the probability distribution of Horton stream lengths. Thus a population of stream lengths can be defined by using the Horton-Strahler stream-ordering procedure. The lengths can be defined as straight-line distances between the end points of a stream or as distances obtained following the stream course. In both cases the stream itself is defined by the ordering procedure. As discussed by Tarboton et al. (1991), in both cases the probability distributions of stream lengths follow a power-law distribution with exponents 2 and 1.9, respectively. It is appealing to define stream-length properties without linking the analysis to any particular ordering scheme. One way to do this is to analyze the random variable defined as the longest distance $\mathcal{L}$, measured through the network from a randomly chosen point to the boundary of the basin. The probability distribution of this distance can be derived by making an assumption about Hack's law and the distribution of contributing areas. We shall assume that Hack's law is valid when considering not only different basins but also subbasins defined by the choice of any point inside a basin. This is not the same type of analysis on which Hack's law is based, and therefore we should not necessarily expect either the same value of the exponent or the same excellent fit observed when dealing with different basins. Nevertheless we would hope for the exponent to be larger than 0.5, with a good fit in the relationship. We shall also provide an interpretation for this type of analysis based on scaling properties.

Defining as $\mathcal{L}$, the random stream length to the divide, we have

$$P[\mathcal{L} > L] \propto P[A^h > L] = P[A > L^{1/h}] \propto L^{-\beta/h} \tag{3.35}$$

where $\beta$ is defined in equation (3.11), and $h$ is Hack's exponent. The power law for lengths, defined in equation (3.35), has been verified for numerous basins, with excellent fits throughout several logarithmic scales (Table 3.2). It links the aggregation pattern for an area with the stream-length structure of the network and Hack's law. The exponent in equation (3.35) will fall between 0.70 and 0.80 for values of $\beta$ between 0.43 and 0.45 and $h$ between 0.57 and 0.60.

Figure 3.7 shows the experimental values of $P[\mathcal{L} \geq L]$ for a choice of river basins whose DTM features are described in Table 3.1. In particular, Figure 3.8 shows the detailed results obtained for the Brushy Creek basin, where we used a support area $A_t$ of 50 pixels to identify the network. The lengths were measured in kilometers. The exponent in the power law for lengths in this case is 0.74, the value of $\beta$ is 0.43, and the value of $h$ resulting from equation (3.35) is approximately 0.58.

Figure 3.9 shows the same results for the Raccoon Creek basin. In this case we observe a slightly higher value for the slope of the probability of exceedance of lengths, and a worse adaptation of the theoretical prediction in equation (3.35) to the common values of Hack's exponent. Nevertheless, as we shall see in the following, we expect statistical departures from Hack's values.

Table 3.2 provides a synthesis of the experimental analyses conducted on the DTMs for the basins in Table 3.1, all of them analyzed with the same support area of 50 pixels. The average exponents of the probability distributions indeed are close to the standard values of 0.80 for the lengths and 0.43 for total contributing areas.

Recall that the stream length at any point, $\mathcal{L}$, is defined as the main distance measured through the network from the point to the boundary of the basin. Technically one defines the mainstream pattern upstream of any junction following the site having maximum area (in case of equal contributions, one chooses at random) until a source

Table 3.2. *Values for the sinuosity exponent $\phi_L$, Hack's exponent $h$, the Hurst exponent $H$, and the power-law exponents for areas ($\beta$) and lengths ($\gamma$) for the rivers in Table 3.1 whose areas are larger than 200 km²*

| Basin | $\phi_L$ | $h$ | $H$ | $\phi_L/h-1$ | $\beta$ | $\gamma$ | $\beta/h$ |
|---|---|---|---|---|---|---|---|
| Guyandotte | 1.06 | 0.56 | 0.92 | 0.89 | 0.43 | 0.74 | 0.77 |
| Little Coal | 1.06 | 0.56 | 0.92 | 0.89 | 0.41 | 0.74 | 0.73 |
| Tug Dry Fork | 1.07 | 0.54 | 1.00 | 0.98 | 0.41 | 0.77 | 0.76 |
| Johns Creek | 1.06 | 0.60 | 0.75 | 0.77 | 0.40 | 0.67 | 0.67 |
| Big Coal | 1.05 | 0.56 | 0.89 | 0.88 | 0.42 | 0.75 | 0.75 |
| Raccoon | 1.02 | 0.52 | 1.00 | 0.96 | 0.45 | 0.89 | 0.86 |
| Pingeon | 1.06 | 0.55 | 0.92 | 0.93 | 0.45 | 0.75 | 0.82 |
| Moshannon | 1.02 | 0.52 | 0.96 | 0.96 | 0.45 | 0.86 | 0.85 |
| Brushy Creek | 1.04 | 0.54 | 0.96 | 0.93 | 0.43 | 0.82 | 0.80 |
| Rock Castle | 1.06 | 0.55 | 0.92 | 0.93 | 0.43 | 0.79 | 0.78 |
| Sturgeon | 1.03 | 0.52 | 1.01 | 0.98 | 0.46 | 0.92 | 0.88 |
| Island | 1.07 | 0.54 | 0.96 | 0.98 | 0.43 | 0.78 | 0.80 |
| Wolf | 1.07 | 0.55 | 0.92 | 0.96 | 0.41 | 0.75 | 0.75 |

*Note:* Data were obtained after extracting the drainage networks from DTMs using a threshold $A_t = 50$ pixels as the maximum contributing area necessary to maintain a channel. The exponent $h$ was estimated by fitting $\langle \mathcal{L} \rangle_a$ versus the contributing area in a log-log plot (in fact, $\langle \mathcal{L} \rangle_a \propto a^{h(2-\xi)}$). The value for $H$ was obtained from $\langle \mathcal{L}_\parallel \rangle_a \sim a^{1/H+1}$, where $\langle \mathcal{L}_\parallel \rangle_a$ is the expected value of diameters with the contributing area $a$. In column 4, the relationship $H = \phi_L/h - 1$ is used to estimate $H$ from the values of $\phi_L$ and $h$ whose values are reported in columns 2 and 3, respectively. In column 7, $\gamma$ is estimated directly from the individual plots and the collapse of the distribtions, to be compared with the theoretical value $\beta/h$ of column 8.

is reached. To relate to the notation adopted throughout this chapter, we notice that for subbasins of area $a$ we indicate the mainstream length as $L$, and obviously at the closure of the basin, where $a = A$, we have $\mathcal{L} = L$. Let us now define the probability distribution $\pi(\mathcal{L}, a)$ for the lengths defined in this way *for points with a given area a.* The constraint of choosing sites with a given area $a$ plays an important role in the revisited interpretation of Hack's law proposed by Rigon et al. (1996). The length distribution is then given by

$$\hat{\pi}(\mathcal{L}, L_\parallel) = \int_1^{+\infty} da\, \pi(\mathcal{L}, a) p(a, L_\parallel) \tag{3.36}$$

We shall further denote by $\Pi(\mathcal{L}, a)$ and $\hat{\Pi}(\mathcal{L}, L_\parallel)$ the corresponding probabilities of exceedence [e.g., $\Pi(\mathcal{L}, a) = \int_{\mathcal{L}}^{\infty} \pi(x, a)\, dx$]. In general, nothing can be said about the distribution $\hat{\pi}(\mathcal{L}, L_\parallel)$ or $\hat{\Pi}(\mathcal{L}, L_\parallel)$ unless the distribution $\pi(\mathcal{L}, a)$ is known. A strict deterministic assumption of the validity of Hack's law [equation (3.3)] will require that the function $\pi(\mathcal{L}, a)$ be a sharply peaked function of one variable with

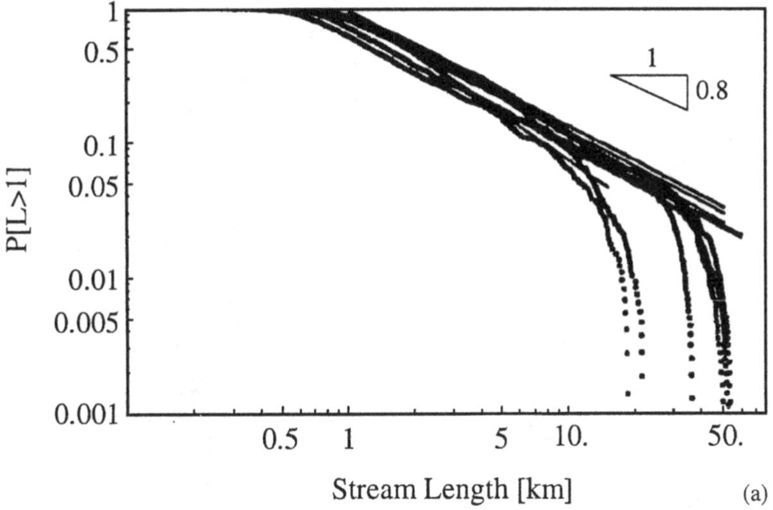


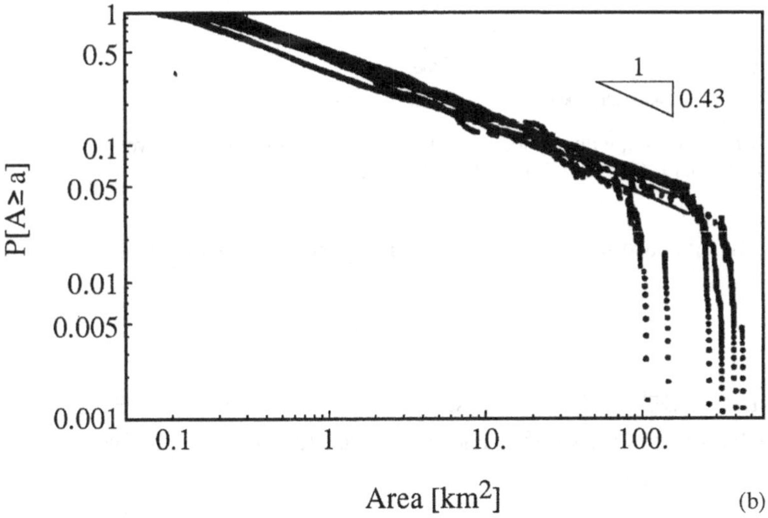


Figure 3.7. A synthesis of the experimental analyses. Plots of (a) $P[\mathcal{L} \geq L]$ versus $L$ for several basins in Table 3.1 and (b) $P[A \geq a]$ versus $a$ for the same basins. The detailed data for the scaling exponents are shown in Table 3.2.

respect to the other, thus leading to an effective constraint between areas and lengths. Thus we can assume that

$$\pi(\mathcal{L}, a) = \delta(\mathcal{L} - a^h) \tag{3.37}$$

which is a mathematical statement of Hack's law when the law is assumed to hold without dispersion and, moreover, it is assumed to hold for all subbasins embedded

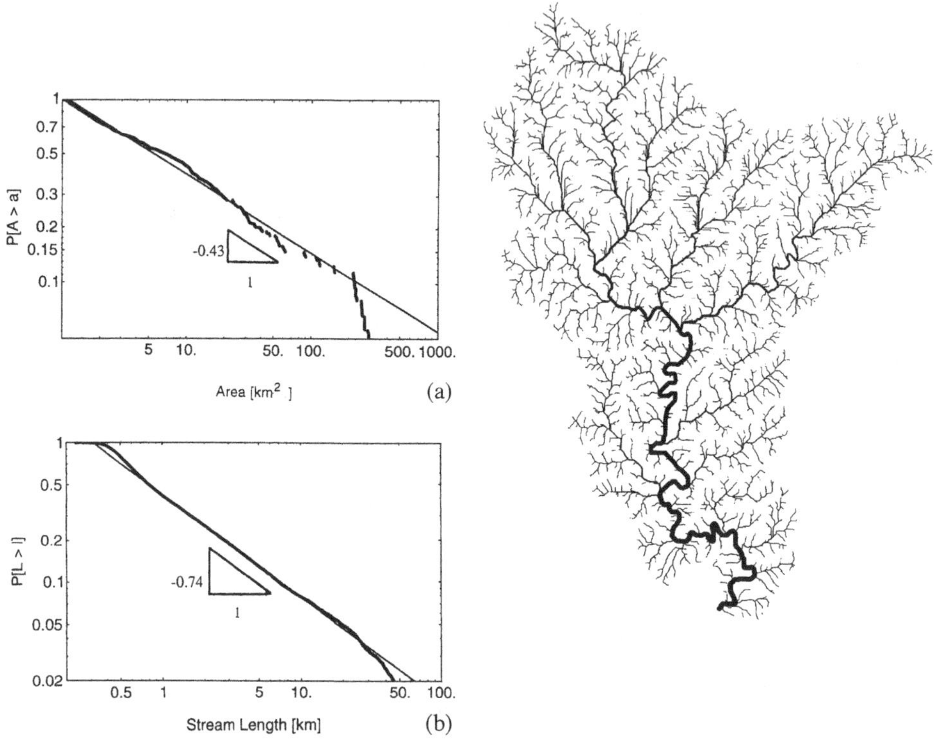


Figure 3.8. Brushy Creek, with support area $A_t = 50$ pixels: (a) $P[A > a]$ versus $a$ ($\beta \sim -0.43$); (b) $P[\mathcal{L} \geq L]$ versus $L$ ($\gamma \sim -0.74$). (Adapted from Rigon et al., 1996.)

in a basin. In this case it is easily derived that in the $L_\parallel \to \infty$ limit,

$$\hat{\Pi}(\mathcal{L}, L_\parallel \to \infty) \propto \mathcal{L}^{-\beta/h} \tag{3.38}$$

and $\hat{\pi}(\mathcal{L}, L_\parallel) \propto \mathcal{L}^{-(1+\beta/h)}$, where use has been made of equation (3.13). This result is equal to that in equation (3.35). The same result (3.38) can be derived if we assume for $\pi(\mathcal{L}, a)$ a scaling form like

$$\pi(\mathcal{L}, a) = \frac{1}{\mathcal{L}} g\left(\frac{\mathcal{L}}{a^h}\right) \tag{3.39}$$

which is a generalization of equation (3.37). We notice that equation (3.39) does not strictly presume Hack's law, because the original assumptions do not consider statistical fluctuations about the mean value. Our interpretation suggests that fluctuations are indeed scaling, and Hack's law holds for the mean.

Finally, we observe that a powerful scaling argument is obtained when describing the statistical variability of the mainstream length for any point with a given area through a finite-size probability distribution. Following Rigon et al. (1996), we

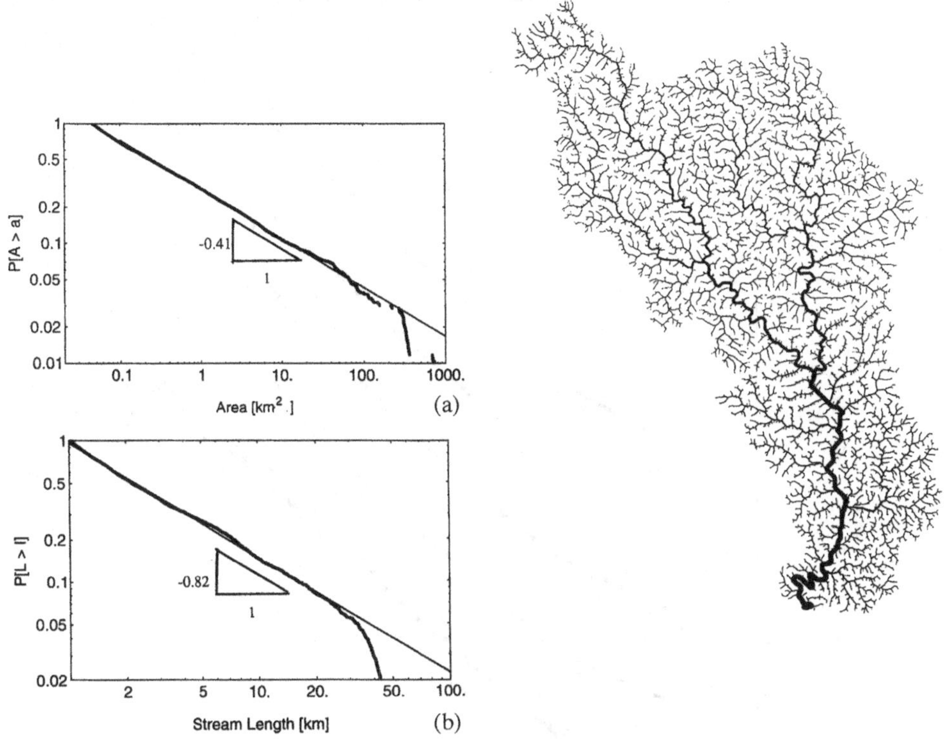


Figure 3.9. Little Coal River, with support area $A_t = 50$ pixels: (a) $P[A > a]$ versus $a$ ($\beta \sim -0.41$); (b) $P[\mathcal{L} \geq L]$ versus $L$ ($\gamma \sim -0.82$). (Adapted from Rigon et al., 1996.)

generalize equation (3.39) to the usual type of scaling relationship:

$$\pi(\mathcal{L}, a) = \mathcal{L}^{-\xi} g\left(\frac{\mathcal{L}}{a^h}\right) \tag{3.40}$$

where, as usual, $\pi(\mathcal{L}, a)$ is the probability density of length $\mathcal{L}$ given a drainage area $a$, and $\xi$ is a scaling exponent. Equation (3.40) is another mathematical statement of Hack's law. It postulates fractality because it embeds a basic similarity in the distribution of lengths when rescaled by a factor $a^h$. Notice that, as shown in equation (3.32), we have

$$\langle \mathcal{L}^n \rangle \propto a^{h(n-\xi-1)} \tag{3.41}$$

where $\langle \mathcal{L}^n \rangle \equiv \langle \mathcal{L}^n \rangle_a$ (i.e., it is the $n$th moment of the distribution of mainstream lengths for a given area $a$). Straightforward manipulation yields

$$\frac{\langle \mathcal{L}^n \rangle}{\langle \mathcal{L}^{n-1} \rangle} \propto a^h \tag{3.42}$$

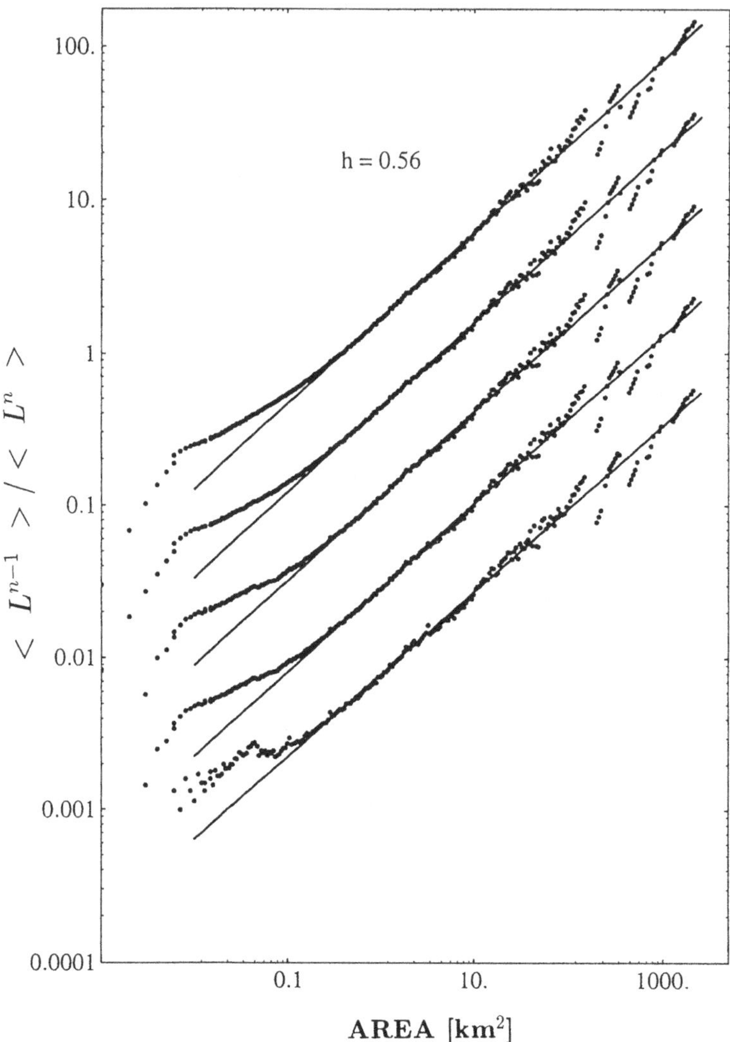


Figure 3.10. Analysis of the scaling of the moments of stream lengths with area for Guyandotte River. From top to bottom: the first moment, followed by the ratio $\langle \mathcal{L}^n \rangle / \langle \mathcal{L}^{n-1} \rangle \propto a^h$ for $n = 2, 3, 4,$ and 5. The plot shows a well-defined simple scaling for at least two decades. At the very small scales, the scaling is broken because of the transition from fluvial to hillslope geometry.

Rigon et al. (1996) have tested the assumption behind equations (3.40) and (3.42) using double logarithmic plots of the ratio of consecutive moments versus area from the large DTMs for different basins described in Table 3.1.

Figure 3.10 shows the ratios of five consecutive moments for the Guyandotte River basin (Table 3.1) defining a different subbasin with its contributing area at any link of the network. We observe that a clear common straight line emerges whose slopes are $0.56 \pm 0.01$ for at least three decades. The case $n = 1$ also suggests that $\xi \sim 1$; that is, equation (3.40) implies the validity of Hack's law for the mean, as seen in equations (3.37) and (3.39). The deviations appearing for small contributing areas reflect the presence of unchannelized areas. For large areas (say $a > 500$ km$^2$), the statistics progressively lose significance because of the scarcity of the data available.

Although technicalities arise, particularly in regard to the required binning of lengths in adequate intervals of area $\Delta a$, as described by Rigon et al. (1996), it is clearly seen that the data support the assumption of equation (3.40). We emphasize that this internal consistency suggests that Hack's relationship is to be viewed within a statistical framework and not necessarily in connection with arbitrary definitions of suitable subbasins such as, for example, at predefined outlets.

Scaling arguments provide an appealing framework because they bypass the tedious and misleading techniques for estimating the goodness of fit for the regressions of past analyses on Hack's law. Our assessment of a value for Hack's exponent defined through scaling arguments proves close to the original one in which suitable nonoverlapping basins were plotted. In our framework, such an occurrence is viewed as likely whenever each realization is close to the mean value of the distribution.

It is of interest to analyze results analogous to those in Figure 3.10 for two basins developed in different geologic contexts. Figures 3.11 and 3.12 illustrate the finite-size scaling assumption, equation (3.40), for the Tug Fork and Johns River basins described in Tables 3.1 and 3.2. The quality of the fit is also excellent over at least three decades, and the slope (i.e., the coefficient $h$) is respectively $h = 0.54 \pm 0.01$ and $h = 0.60 \pm 0.01$. From the foregoing results we suggest that Hack's law is indeed an outgrowth of fractality. In fact, the result in equation (3.26) shows that the aggregation characteristics cannot be isolated from their geologic context because the features of the boundaries are related to the internal organization of the basin. As such, Hack's exponent cannot be universal. Nevertheless, because commonly a free competition for drainage occurs for the migration of divides, average characters appear consistent – and seemingly universal – for certain choices of basins (e.g., nonoverlapping, closed at a confluence, control-free).

The preceding scaling analysis essentially shows that Hack's law is a reflection of the basic similarity of the part and the whole in river-basin landforms. Mandelbrot's idea of the inference of Hack's law from the fractal characters of the mainstream was therefore incomplete but visionary, because many fractal forms in a river basin are related. The structural similarity in shape carries throughout the length scales of a fluvial basin, as clearly illustrated by the scaling plots of the ratios of moments of mainstream lengths versus area (Figures 3.10–3.12). This similarity is conditional on the constraints (or the absence thereof) imposed on the developing basin. The consistency of the scaling structure with the results that we found from our data is remarkable. We also suggest that the extensions of Hack's original result to basins of completely different regions may be misleading. Although the "average" characters point toward similar behaviors, the claimed universality is untenable unless clearly referred to a common dynamic and geologic framework.

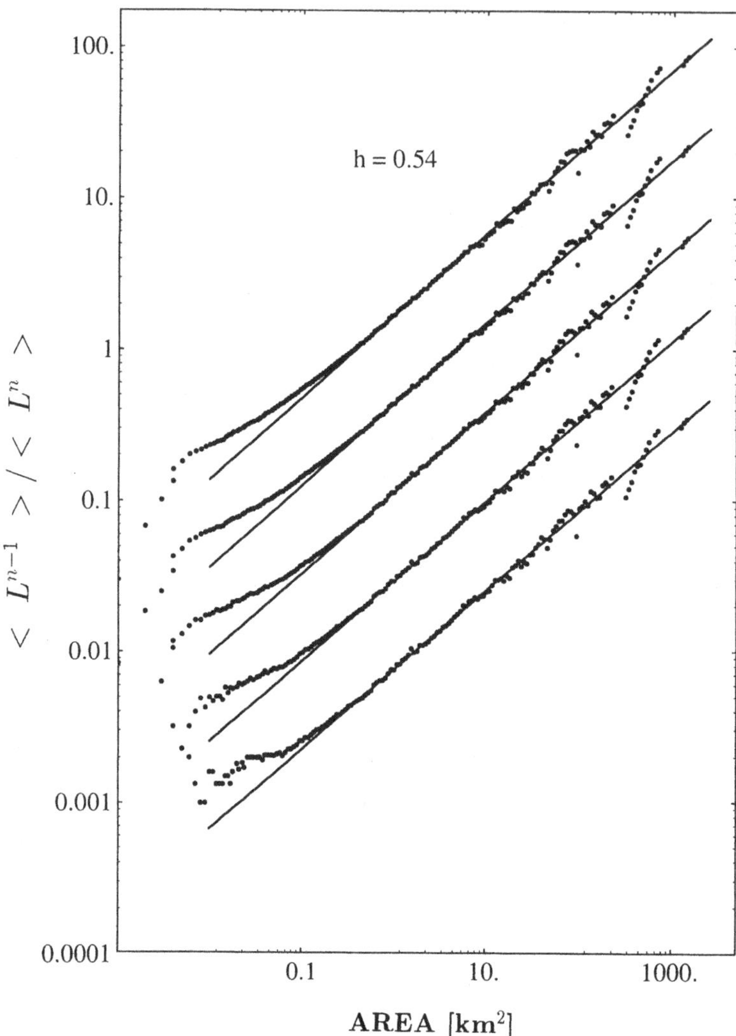


Figure 3.11. Analysis of the scaling of the moments of stream lengths with area for Tug Fork. Notation as in Figure 3.10.

## 3.5 Conclusions

We have shown, on the basis of detailed observational evidence, that seemingly unrelated hydrologic laws of empirical origin describing river-network morphology and several experimental facts find a natural explanation through a finite-size scaling argument. This basic *Ansatz* is characterized by the power-law nature of the distribution of key geomorphic variables describing the morphology of a river basin (e.g., drainage areas, stream lengths, elongations). In this chapter we have observed the emergence and the implications of scaling in river-basin morphology and have

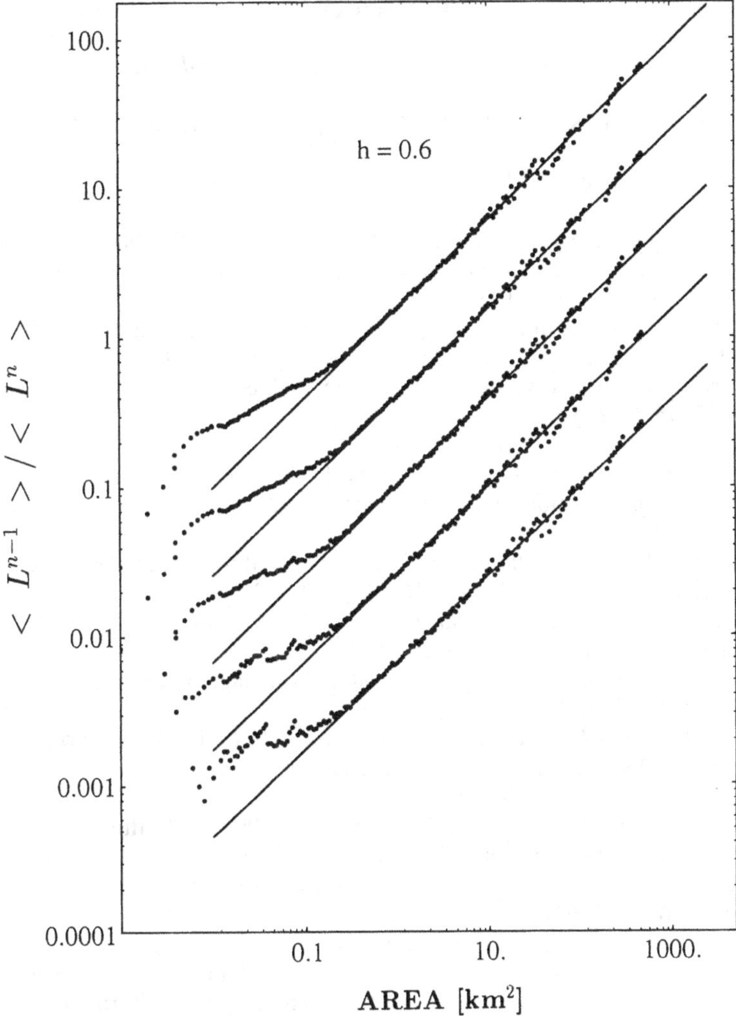


Figure 3.12. Analysis of the scaling of the moments of stream lengths with area for Johns River. Notation as in Figure 3.10.

recalled the analytical derivation of a set of consistently linked scaling exponents. Such emergence suggests nature's drive toward fractality in the making of the fluvial landforms – though confined by lower and upper cutoffs of clear physical nature. We therefore suggest that fractal geometric arrangements in a river basin are diverse because they are by necessity imperfect (i.e., constrained by random forcings and quenched disorder). Nevertheless, they are remarkably coherent in their major structural characters, which are invariably scaling. Thus the scaling properties observed in a river basin suggest further elements supporting a particular view, nonreductionist and self-organizing, of how nature works.

## References

Bak, P. 1996. *How Nature Works: The Science of Self-organized Criticality*. New York: Copernicus.

Bak, P., Tang, C., and Wiesenfeld, K. 1987. Self-organized criticality: an explanation of $1/f$ noise. *Phys. Rev. Lett.* 59:381–5.

Bak, P., Tang, C., and Wiesenfeld, K. 1988. Self-organized criticality. *Phys. Rev.* A38:364–74.

Dietrich, W. E., Montgomery, D. R., Reneau, S. L., and Jordan, P. 1988. The use of hillslope convexity to calculate diffusion coefficients for a slope dependent transport law. *EOS, Trans. AGU* 69:1123–4.

Dietrich, W. E., Wilson, C., Montgomery, D. R., McKean, J., and Bauer, R. 1992. Erosion threshold and land surface morphology. *J. Geology* 20:675–9.

Durret, R., Kesten, H., and Waymire, E. 1991. On weighted heights of random trees. *J. Theor. Prob.* 18:223–37.

Feder, J. 1988. *Fractals*. New York: Plenum.

Fisher, M. E. 1971. *Critical Phenomena*. New York: Academic Press.

Gray, D. M., 1961. Interrelationships of watershed characteristics. *J. Geophys. Res.* 66:1215–33.

Gupta, V. K., and Mesa, O. J. 1988. Runoff generation and hydrologic response via channel network geomorphology. *J. Hydrol.* 102:3–28.

Hack, J. T. 1957. *Studies of Longitudinal Profiles in Virginia and Maryland*. U.S. Geological Survey professional paper 294-B.

Howard, A. D. 1994. A detachment-limited model of drainage basin evolution. *Water Resour. Res.* 30:2261–85.

Ijjasz-Vasquez, E., Bras, R. L., and Rodríguez-Iturbe, I. 1993. Hack's relation and optimal channel networks: the elongation of river basins as a consequence of energy minimization. *Geophys. Res. Lett.* 20:1583–6.

Ijjasz-Vasquez, E., Bras, R. L., and Rodríguez-Iturbe, I. 1994. Self-affine scaling of fractal river courses and basin boundaries. *Physica* A209:288–300.

Jenson, S. K., and Domingue, J. O. 1988. Extracting topographic structures from digital elevation models. *Hydrol. Proc.* 5:31–44.

Kauffman, S. 1993. *The Origins of Order*. Oxford University Press.

Mandelbrot, B. B. 1983. *The Fractal Geometry of Nature*. San Francisco: Freeman.

Maritan, A., Rinaldo, A., Rigon, R., Giacometti, A., and Rodríguez-Iturbe, I. 1996. Scaling laws for river networks. *Phys. Rev.* E53:1510–22.

Meakin, P., Feder, J., and Jossang, T. 1991. Simple statistical models of river networks. *Physica* A176:409–29.

Mesa, O. J., and Gupta, V. K. 1987. On the main channel length–area relationships for channel networks. *Water Resour. Res.* 23:2119–22.

Montgomery, D. R., and Dietrich, W. E. 1992. Channel initiation and the problem of landscape scale. *Science* 255:826–30.

Muller, J. E. 1973. Re-evaluation of the relationship of master streams and drainage basins: reply. *Geol. Soc. Am. Bull.* 84:3127–30.

Peckham, S. 1995. New results for self-similar tree with application to river networks. *Water Resour. Res.* 31:1023–9.

Rigon, R., Rinaldo, A. Maritan, A., Giacometti, A., Tarboton, D. R., and Rodríguez-Iturbe, I. 1996. On Hack's law. *Water Resour. Res.* 32:3374–8.

Rinaldo, A., Dietrich, W. E., Rigon, R., Vogel, G. K., and Rodríguez-Iturbe, I. 1995. Geomorphological signatures of varying climate. *Nature* 374:632–6.

Rinaldo, A., Maritan, A., Colaiori, F., Flammini, A., Rodríguez-Iturbe, I., and Banavar, J. R. 1996. Thermodynamics of fractal networks. *Phys. Rev. Lett.* 76:3364–8.

Rinaldo A., Rodríguez-Iturbe, I., Rigon, R., Bras, R. L., and Ijjasz-Vasquez, E. J. 1993. Self-organized fractal river networks. *Phys. Rev. Lett.* 70:822–6.

Rodríguez-Iturbe, I., Ijjasz-Vasquez, E., Bras, R. L., and Tarboton, D. G. 1992a. Power-law distributions of mass and energy in river basins. *Water Resour. Res.* 28:988–93.

Rodríguez-Iturbe, I., and Rinaldo, A. 1997. *Fractal River Basins: Chance and Self-organization.* Cambridge University Press.

Rodríguez-Iturbe, I., Rinaldo, A., Rigon, R., Bras, R. L., Marani A., and Ijjasz-Vasquez, E. J. 1992b. Energy dissipation, runoff production, and the 3-dimensional structure of river basins. *Water Resour. Res.* 28:1095–103.

Scheidegger, A. E. 1967. A stochastic model for drainage patterns into a intramontane trench. *IAHS Bull.* 12:15–20.

Sinclair, K., and Ball, R. C. 1996. A mechanism for global optimization of river networks from local erosion rules. *Phys. Rev. Lett.* 76:3359–63.

Takayasu, H. 1990. *Fractals in the Physical Sciences.* Manchester University Press.

Tarboton, D. G., Bras, R. L., and Rodríguez-Iturbe, I. 1991. On the extraction of channel networks from digital elevation data. *Hydrol. Proc.* 5:81–100.

# 4

# Spatial Variability and Scale Invariance in Hydrologic Regionalization

VIJAY K. GUPTA and EDWARD C. WAYMIRE

## 4.1 Problems of Scale and Regionalization in River Basins

The occurrence of the hydrologic cycle covers a very wide range of space and time scales, and involves physical, chemical, and ecological processes. Therefore, in modeling and making predictions, one is required to understand how various properties and measurements behave under a change of scale. At the spatial and temporal scales of interest in river basins, space–time variability and fluctuations are displayed in input, output, and storage elements of the components of the hydrologic cycle and in their interactions. This variability is part of the physics and in this sense is different from the measurement noise. We can use the term *physical-statistical* or *statistical-dynamical* to describe such systems. An understanding of the physics of these systems in the presence of variability and fluctuations through mathematical notions of randomness has been and continues to be one of the central challenges of hydrology and constitutes the main theme of this chapter.

A river basin contains a channel network, as shown in Figure 4.1, and systems of hills on both sides of the channels in the network. Rainfall and/or snowmelt are transformed into runoff, and sediments are eroded over hills, and these in turn are fed into a channel network for their journey toward an ocean. The hydrologic cycle on a hillside involves transformation of rainfall to surface runoff, infiltration through the near-surface unsaturated soils, and evapotranspiration from the soil surface into the atmosphere. The infiltrated water goes into recharging the soil moisture in the unsaturated zone, the aquifers, and some of it also appears as subsurface runoff in a channel network. All of these physical processes are highly variable in space and time.

Much of the spatial variability can be ignored at "small" spatial scales on the order of 0.1–1.0 m. Indeed, the scientific understanding of individual hydrologic processes at laboratory scales, such as flow through saturated and unsaturated columns of porous media, is fairly well advanced. This understanding has been used to investigate the long-term water balance as governed by the temporal interactions among the

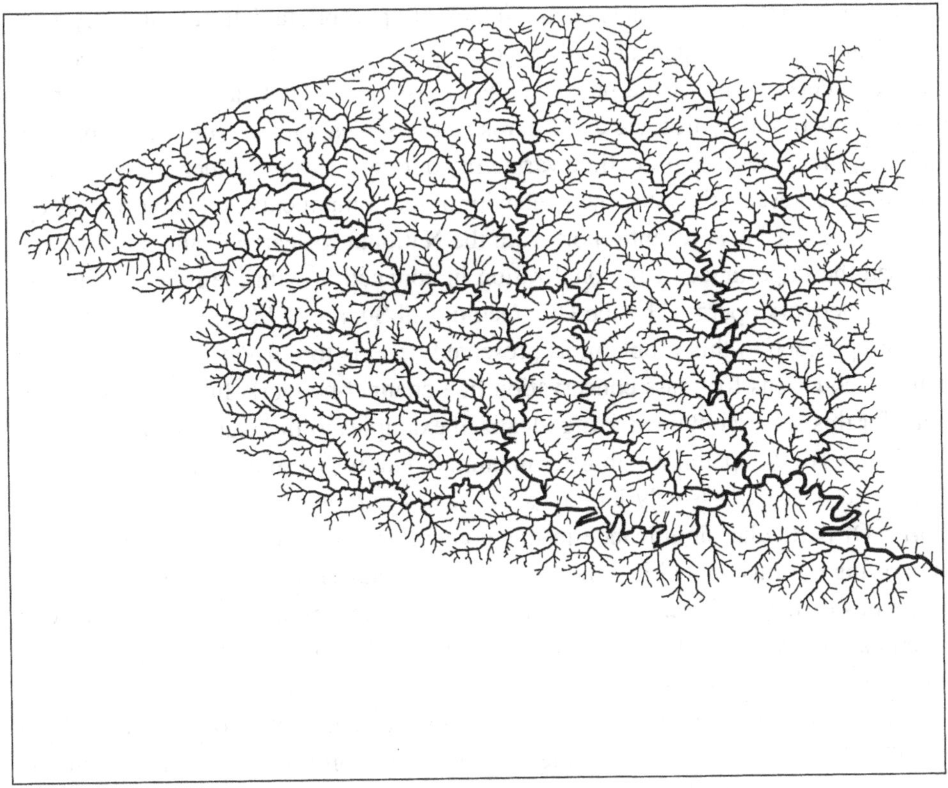

Figure 4.1. Drainage network for the Kentucky River basin.

elements of the hydrologic cycle through the processes of rainfall, evapotranspiration, infiltration, and runoff as a statistical-dynamical system (Eagleson, 1978).

As the spatial scale under consideration increases to that of a single hillside, on the order of 10–$10^2$ m, spatial variability becomes important, and new elements begin to influence the hydrologic mass balance, such as the topography of the hillside. In particular, one wants to know how the laboratory-scale equations can be spatially integrated so as to describe the hydrologic cycle over a hillside. Some recent work to extend Eagleson's theory of water balance to a hillside has been reported by Salvucci and Entekhabi (1995). Nonetheless, a comprehensive theory for the elements of the hydrologic cycle and their interactions over a single hillside remains a major challenge.

Beyond a single hillside, a river basin can be viewed as a channel-network–hills system. Typical spatial scales for river basins span about four decades, from $10^2$ to $10^6$ m. The hydrologic cycle for larger subbasins involves the spatially integrated behaviors of several hills along a channel network. An understanding of the spatial variability among hillsides and their interactions through a channel network is necessary for this integration. A major obstacle to understanding this integration, or *hydrologic regionalization*, has been the lack of appropriate data sets and physical-statistical theories for such a broad range of spatial scales. Regionalization requires a basic understanding of the physical and statistical relationships that connect the components of the hydrologic cycle across these scales within some geographic region. This includes the space–time variability of runoff as a fundamental component, because it represents the integrated behavior of the components of the hydrologic cycle in a river basin. A recent report from the National Research Council (1992) has stressed the importance of regionalization in connection with the role of a stream-gauging network operated by the U.S. Geological Survey (USGS).

Our approach to this problem, as presented here, is to introduce a general set of coupled equations for mass conservation in a channel-network–hills system. Solutions for this set are obtained using idealized examples to illustrate that many of the key features of the spatial variabilities in rainfall, landforms, and runoff can be unified and tested over successively larger spatial scales within a broad theoretical framework. Of particular significance in this context are the roles of two "boundaries" in carrying out this spatial integration. These boundaries are defined by the spatial properties of the landforms and channel networks and by the spatial variability in precipitation. Notions of spatial-scale invariance and scale dependence provide the foundations for this theory.

The time scales in the illustrations used are kept small, on the order of a "rainfall event" or less. Over longer time scales, temporal variabilities in rainfall, evapotranspiration, and infiltration become very important in the dynamics of water balance, as Eagleson (1978) has illustrated for very small spatial scales. Such temporal variability is also influenced by the climatic variability over interseasonal, interannual, interdecadal, and longer time scales. However, their precise connections remain unclear. By restricting attention to time scales on the order of a rainfall event, the effects of the temporal variabilities on the integrated spatial variability of runoff are treated here in a very restricted sense. A generalization of this theory to longer time scales has identified a very important area for hydrologic research.

Development of a general theoretical framework that can couple various components of the hydrologic cycle in river basins has been a long-standing challenge to the science of river-basin hydrology. A theoretical formulation will allow us to abstract the basic properties of real systems and develop new intuitions and insights into their connections by a combination of computations for simple examples and attention to the wider range of data. As theory advances, simple examples give way to the development and testing of more complex models that capture more of the essential characteristics of real systems. Careful tests of models can provide a basis for new experiments and observations. In due course these continual interactions among theory, data, and modeling should lead to new physical understanding and an improved basis for scientific predictions.

We begin this discussion with a general physical formulation of the problem around a set of coupled equations for mass conservation in a channel-network–hills system. An explanation of the scientific scope of this chapter and its organization is deferred until the next section, where that can best be explained in the context of this basic set of physical equations of hydrology. Several major problems will be identified throughout this chapter. This chapter is twin to another paper that is similar in scope, but in which we have emphasized precise mathematical formulations and results (Gupta and Waymire, 1998).

## 4.2 A General Physical Approach to Regionalization

### *4.2.1 A General Set of Coupled Equations of Continuity for a Channel-Network–Hills System*

Streamflows are measured at point locations within a drainage network. Each such location is called a gauging station. At a gauging station, the graph of river discharge, or the volumetric flow per unit time, as a function of time is called a *hydrograph*. To represent this behavior analytically in general physical terms, we shall use the *equation of mass conservation*, or the *continuity equation*. It is one of the three fundamental physical equations of hydrology. The second and third are the *equation of momentum* and the *equation of energy conservation*. However, the continuity equation can be regarded as the most basic equation of hydrology from the point of view of first principles.

In standard texts on fluid mechanics the continuity equation and other conservation equations are commonly written as differential equations. This formulation rests on the key assumption that space–time can be approximated as a continuum, so that derivatives with respect to space and time can be defined. However, in the present context, the presence of a drainage network as a discrete, binary-tree graph subject to certain space-filling constraints renders a continuum model inappropriate; see Figure 4.1 for a depiction of a drainage network. The "space-filling properties" of drainage networks have been and continue to be of great interest in both fluvial geomorphology and river-basin hydrology (Jarvis and Woldenberg, 1984). Therefore, it is most natural to formulate the continuity equation as a "local-difference equation" at an appropriate space–time scale for a channel-network–hills system. In order to write this equation, some basic terminology is needed regarding the *topology* or the branching structure of a network. We shall first introduce this terminology and then formulate the continuity equation. Following that we shall have several comments regarding this equation in an effort to orient the reader to the overall scope of this chapter. Several major problems that remain unsolved will be discussed throughout this and the following sections.

Assume that at an appropriate scale of resolution a *drainage network* $\tau$ can be represented as a *finite, binary-rooted-tree graph*, consisting of links (edges), junctions and sources (vertices), and the outlet (root). A *junction* (interior vertex) is defined as a point where three channels meet. A *source* (exterior vertex) denotes the starting point of an unbranched channel, and a *link* (edge) is a channel segment between two junctions, a source and a junction, or the outlet and a junction. The *outlet* is a privileged junction $\phi$ that "directs" or drains water from any location within the entire channel network, and no other junction has this property. Each link $e$ directs water flow from its upper vertex $\bar{e}$ to its lower vertex $\underline{e}$. We identify links and vertices according to the convention that to each link $e$ there is a unique vertex $v$ such that, $\bar{e} = v$. A ghost link may be added to $\phi$ when necessary to complete the convention. The notations $e \in \tau$ and $v \in \tau$ indicate that a link $e$ and a vertex $v$ respectively belong to the network $\tau$. The symbol $\tau(e)$ represents the subtree above $e$ for which $\bar{e} = \phi$. The *size* of a tree $\tau$ is denoted by $\|\tau\|$. A natural way to assign a size to a tree is by equating it to the total number of sources, called *magnitude*, and the junctions belonging to it. For example, for a binary tree with magnitude $m$, we can readily see that $\|\tau\| = 2m - 1$ (Shreve, 1967).

In the present context, we can think of a link as a *control volume* for a network–hills system for which a general equation of continuity needs to be formulated; see Figure 4.2 for a schematic depiction of this control volume. This is analogous to the differential control volume widely used in fluid mechanics. However, identification

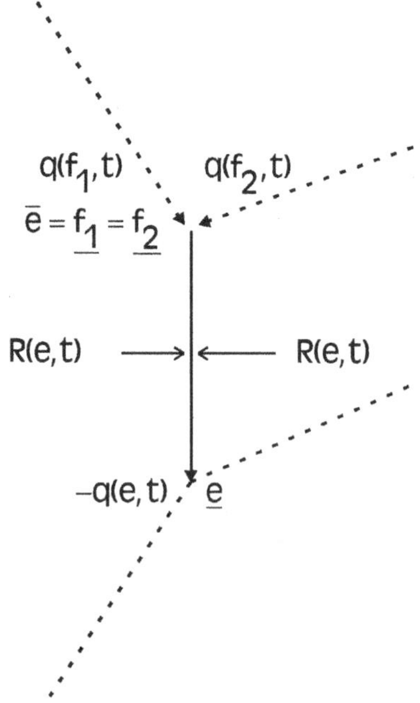


Figure 4.2. A schematic plot of a control volume for a channel-network–hills system.

of a link depends on the scale of resolution of a channel network, because more links appear as the resolution increases (see fig. 12 of Leopold and Miller, 1984). Moreover, empirical investigations have shown that the spatial variabilities in the network topology and in the link geometry (link lengths, slopes, etc.) exhibit both a systematic component and a random component. For example, systematic variability in network topology has been observed as a scale-invariant property in many, but not all, large river networks (Peckham, 1995a). Likewise, the mean link length in networks is approximately a constant, but the mean link slope decreases systematically from the drainage divide to the basin outlet. These systematic features essentially determine the theoretical structures, even in the presence of random variability in link geometry (Gupta and Waymire, 1989; Gupta, Mesa, and Waymire, 1990). An overview of these and other mathematical results, as reported earlier (Gupta and Waymire, 1998), suggests that the systematic and random variabilities in network topology and the systematic variability in link geometry are of primary importance. The random variability in the link geometry is of secondary importance provided the probability distributions describing this variability have finite statistical moments. Therefore, in order to focus on key physical and mathematical issues in the development of a theory, we ignore the random variability in the link geometry and assume that at some fine scale of resolution all the links have a common length, say $l$.

Assume that flow in a network is observed in multiples of a time scale $\Delta t$ at times $t = n\Delta t,\ n = 0, 1, 2, \ldots$. As an example, one may think of $\Delta t$ as the time it takes for water to flow through a link with some "mean velocity" $v$ (i.e, $\Delta t = l/v$). From here on, we shall suppress $\Delta t$ and simply write $t = 0, 1, 2, \ldots$. Let $q(e, t),\ e \in \tau,\ t \geq 1$, be a space–time field representing river discharge, or the volume of flow per unit time, across a link $e$ in the time interval $(t-1, t)$ of unit length $\Delta t$. The flow out of a link from its bottom vertex $\underline{e}$ in the time interval $(t-1, t)$ is defined as $q(\underline{e}, t) = -q(e, t)$. Let $R(e, t)a(e)$ denote the volume of runoff into a link $e$ from the two adjacent hills in the time interval $(t-1, t)$, where $a(e)$ denotes the area of the hillsides, and $R(e, t)$ is the runoff intensity measured in units of length per unit time. This term can also include the volume of runoff depletion per unit time due to channel-infiltration losses or evaporation from the channel surface in link $e$, but we shall not analyze the problem in such physical generality in this chapter. Let $S(e, t)$ denote the total volume of runoff stored in link $e$ in the time interval $(0, t)$, and the change in the total volume of runoff stored in the time interval $(t-1, t)$ is denoted as $\Delta S(e, t)$. Then the equation of continuity for a network–hills system can be written as

$$\frac{\Delta S(e, t)}{\Delta t} = -q(e, t) + \sum_{f:\,\underline{f}=\bar{e}} q(f, t-1) + R(e, t)a(e) \qquad (e \in \tau,\ t \geq 1) \quad (4.1)$$

The sum on the right-hand side of equation (4.1) consists of discharges from all those links that are connected with the upper vertex of link $e$, as shown in Figure 4.2. The topology of a network enters into the solution of equation (4.1) through this term. The formulation of the continuity equation does not assume that the channel network is binary. However, it does assume that no loops are present in the network.

The runoff intensity $R(e,t)$ in equation (4.1) can be determined from a water-balance equation for hillslopes draining into link $e$. For this purpose, let $\theta(e,t)$ denote the volumetric storage per unit area on hillsides, consisting of overland storage, soil moisture, and aquifer storage, in the time interval $(0,t)$. Further, assume that the space–time dependence of runoff comes from its dependence on storage only; that is, $R(e,t) \equiv \tilde{R}[\theta(e,t)]$, where $\tilde{R}$ is an arbitrary function. Then the water balance on hillsides can be written as

$$\frac{\Delta\theta(e,t)}{\Delta t} = I_r(e,t) - E_T[\theta(e,t)] - \tilde{R}[\theta(e,t)] \qquad (e \in \tau,\ t \geq 1) \tag{4.2}$$

where $I_r(e,t)$ is a space–time precipitation intensity field on the drainage basin, and $E_T[\theta(e,t)]$ is the evapotranspiration rate in the time interval $(t-1,t)$, and both are measured in units of length per unit time. Ignoring overland storage, the dependence of $E_T[\theta(e,t)]$ on storage primarily comes through soil moisture on a hillside. It should be noted that every link in a network will have a version of equation (4.2). These equations in general are coupled together. The effect of moisture recycling on the precipitation term is not included in equation (4.2), nor have we said anything about the storage term $S(e,t)$ in equation (4.1). A specification of this term from first principles is a topic of current research, and some further comments regarding this issue are made in Section 4.5.

A key physical problem in river-basin hydrology is to solve the system of coupled equations (4.1) and (4.2) to predict the space–time components of the water balance, such as $E_T[\theta(e,t)]$, $\tilde{R}[\theta(e,t)]$, and the discharge field $q(e,t)$, for every link $e \in \tau$ within some large river network and time $0 \leq t \leq t_1$, where $(0,t_1)$ can range from a few hours to several days, months, decades, and so forth. For this purpose, equations (4.1) and (4.2) can be solved iteratively over the entire network. Spatial variability enters into their solution through the fields $S(e,t)$, $E_T[\theta(e,t)]$, $R(e,t)$ and through the topology of a drainage network. The precipitation-intensity field $I_r(e,t)$ is highly variable in space and time, and this feature alone requires that a statistical-dynamic framework be adopted for solving this set of coupled equations over a wide range of space and time scales. In addition, there are other major sources of spatial variability (soil hydraulic conductivity, vegetation among hillsides, network topology, etc.) that can be included reasonably well only within a statistical-dynamic framework. Because this line of investigation within a space–time context is new to river-basin hydrology, the scope of this chapter is restricted to small time scales of the order of several hours, which are typical of "rainfall events." Over longer time scales, the development of a statistical-dynamic space–time framework for solutions of equations (4.1) and (4.2) has identified a major topic for future research. For example, as a starting point, it would require a generalization of the temporal theory of climate-soil-vegetation by Eagleson (1978) to include space and time, and some comments in this respect are made in Section 4.7.

### 4.2.2 Scientific Scope of the Chapter

Theories are normally developed to explain the empirical features exhibited by sets of data. This raises a key question regarding what major empirical features of the spatial fields of hydrographs over short time scales are sufficiently well known that they might be explained by a theory. Unfortunately, we are unaware of any comprehensive data sets that exist or have been analyzed over this range of space–time scales in river basins. Perhaps such data will be compiled in the future. A major reason for this data gap is that hydrologic investigations regarding rainfall–runoff relationships have focused primarily on the temporal behavior of runoff at the outlet of a basin, rather than on its space–time structure.

By contrast, the USGS has carried out empirical regional analyses of annual floods that have been used to identify "signatures of spatial variability" in the snowmelt- and rainfall-generated floods and to formulate appropriate physical hypotheses (Gupta and Dawdy, 1995). These empirical regional analyses for the state of New Mexico are described in Section 4.3. These analyses provide a natural starting point for the development of a physical-statistical theory to explain these signatures of spatial variability in network runoff. In particular, we shall study the regional statistics of peak flows over a short time interval $(0, t_1)$ defined as

$$Q(e) = \max_{0 \leq t \leq t_1} q(e, t) \qquad (e \in \tau) \tag{4.3}$$

on the basis of simplified solutions for equation (4.1). Another major problem, twin to the flood problem, is to study the spatial field of low flows as a solution to the set of equations (4.1) and (4.2) within a statistical framework; see Furey and Gupta (1998) for some preliminary investigations. However, regionalization of low flows is not discussed here.

In Section 4.4, the notion of a particular form of statistical scale invariance, called *simple scaling*, will be introduced to illustrate how some of the empirical signatures in snowmelt-generated floods in New Mexico can be interpreted mathematically. This interpretation takes the very first steps toward the development of a general statistical framework. This signature is also observed in regional floods generated by stratiform rainfall systems in the state of Washington (Dawdy and Gupta, 1995). Simple scaling is related to the well-known index-flood assumption in regional-flood-frequency analyses, and this connection is explained here.

The objective of Section 4.5 is to solve equations (4.1) and (4.2) under idealized conditions in order to illustrate the physics of statistical simple scaling for peak flows. We set $\Delta S = 0$, $\Delta\theta = 0$, $E_T = 0$ in solving these equations. A physical specification of $\Delta S$ requires a basic understanding of the *downstream hydraulic-geometric relations* in river networks, involving specification of channel widths, depths, velocities, slopes, and friction coefficients throughout a network. Some past empirical research and some current theoretical work on hydraulic geometry will be briefly reviewed. We shall study equation (4.1) assuming that the rainfall $I_r = \tilde{R} = R$ is spatially uniform: $a(e)R(e,t) = aR(t), t \geq 0, e \in \tau$, where $a$ is the common area of hillsides. Under these idealizations, equation (4.1) shows that the spatial variability in runoff hydrographs in different links of a drainage network can be represented as a convolution between the so-called geomorphologic width function (Lee and Delleur, 1976; Kirkby,1984) and $R(t)$. An idealized example, with $R(t) = R$ when $t = 0$, and $R(t) = 0$ otherwise, will be worked out to interpret a statistical scaling parameter for floods as a topologic-geometric dimension of a drainage network with a nonrandom self-similar topology (Gupta, Castro, and Over, 1996). Some of the physical ideas regarding runoff-generation mechanisms on hillslopes will be explained in the context of computing a probability distribution for a fraction of $R$ that contributes to the peak flows. Finally, the effect of a finite duration of $R(t) = R$, $0 \leq t \leq T$, on the flood scaling exponent will be illustrated using a general result for the mean width function in the presence of randomness in network topology (Troutman and Karlinger, 1984). These results will be used to suggest a physical interpretation for two sets of regional-flood parameters computed for the New Mexico data.

Simple scaling cannot accommodate some of the empirical signatures of spatial variability in rainfall-generated floods in the New Mexico data (Gupta and Dawdy, 1995) and other data sets (Smith, 1992; Gupta, Mesa, and Dawdy, 1994). Another form of statistical scale invariance, called *multiscaling*, is introduced in Section 4.6 to interpret those signatures. A multiscaling framework is being developed within the context of the theory of random cascades in an effort to understand some of the key signatures of spatial variability in rainfall. Some elementary but important features of this theory will be explained in Section 4.6. A key result that illustrates how spatial variability in rainfall enters into the spatial statistics for floods will be explained through the same idealized-network model introduced in Section 4.5. This example will solve equation (4.1) when $R(e,t) = R(e), t = 0, e \in \tau$, is given by a random cascade, and it will give an overview of the effect of spatial variability in rainfall on flood statistics. This chapter will conclude with our perspective on future research directions in this newly emerging theory of hydrologic regionalization.

### *4.2.3 Some Brief Comments Contrasting This Chapter with Past Approaches*

Traditionally, rainfall–runoff models have considered deterministic predictions for a runoff hydrograph at a fixed spatial location, namely, the outlet of a drainage network

(Dooge, 1973), or statistical predictions for peak flows at the outlet (Eagleson, 1972; Sivapalan, Wood, and Beven, 1990). Even though Eagleson (1972) discussed some implications of his analysis for regionalization of flood frequencies, for the most part the common practice of analyzing the temporal variability of runoff only at the outlet, and nowhere else within a basin, is perhaps one of the main reasons why little attention has been paid to hydrologic regionalization. Even relatively recent approaches to *distributed rainfall–runoff modeling* in river basins have been based on solving some version of equation (4.1) under different specifications of $S(e, t)$ and $R(e, t)$. The term "distributed" is somewhat of a misnomer, for though inputs and flow routing are treated in some spatially distributed sense, these approaches do not seek a spatially distributed prediction of runoff in a network. This problem of outflow-hydrograph prediction only at the outlet, and nowhere else within a basin, is deeply rooted in the hydrology literature. Its origins lie in engineering applications of rainfall–runoff relations dating back almost to the turn of the twentieth century; see Dooge (1973) for some historical perspective. Even relatively recent efforts to introduce statistical elements into the dynamics of rainfall–runoff transformations have maintained the mismatch between the input and output fields mentioned earlier (Sivapalan et al., 1990). In this sense, these existing physical approaches, either deterministic or statistical, are not regional. The framework to be described here represents a major departure from those traditional approaches, as it emphasizes regionalization and thereby offers new insights and raises new questions.

As a contrast to rainfall–runoff modeling, most of the literature on spatial or regional analysis of runoff has dealt with annual flood peaks in a purely statistical sense. This literature has investigated the spatial relationships of annual flood frequencies within a geographic region. Kinnison and Colby (1945) introduced a regional-flood-frequency hypothesis to the USGS known as the "index-flood assumption." During the last half of the twentieth century this assumption has become one of the cornerstones of regional-flood-frequency analysis. However, despite the wide recognition in the literature of the need to develop a sound hydrologic basis for regional-flood-frequency approaches (Potter, 1987; National Research Council, 1988), not much progress has been made on this front. Consequently, these approaches have remained largely statistical; see Bobee and Rasmussen (1995) for a recent review. By contrast, a major objective of this chapter is to illustrate how the spatial variability in the hydrology of runoff generation and transport can be unified with the spatial statistics.

Rainfall–runoff models generally lack testability, because they calibrate parameters derived from historical runoff data. By contrast, physically based predictions of regional statistics for peak flows, as developed here, can be tested against empirical observations without requiring calibration. This feature distinguishes the theory presented here from previous approaches to rainfall–runoff modeling or regional-flood-frequency analysis, because testability is a necessary condition for further developments of this theory and for understanding the key interactions among the elements of the hydrologic cycle over a wide range of space and time scales. Moreover, a testable theory also lends itself to a new approach to the classic unsolved problem of "prediction (of floods in real time or flood frequencies) from ungauged basins" (PUB), which has numerous practical applications.

## 4.3 Empirical Evidence of Scale Invariance and Scale Dependence in Regional-Flood Frequencies

Our purpose here is to explain how statistical analyses of the annual maximum flow rates within a basin or a region can be used to formulate physical hypotheses about their spatial variability. Specifically, frequencies are computed at several gauging stations within a region from the annual time series of the peak flow rates. The two main approaches to regional analysis have been the index-flood method mentioned earlier and the quantile-regression method of the USGS. We shall illustrate the key features of both of these analyses using the regional-flood-frequency data from New Mexico. These features will provide a basis for development of a physical-statistical theory of regional flows in the subsequent sections.

The USGS performs regional-flood-frequency analyses on a state-by-state basis throughout the United States. These data and their analyses contain a wealth of regional information on flood responses to precipitation input and can serve as major components in developing a physically based understanding of regional-flood-frequency behaviors (Gupta and Dawdy, 1995). We consider New Mexico, which is divided into eight regions with an average area of 15,200 square miles (39,000 km$^2$) (Waltemeyer, 1986). The quantile-regression method of the USGS consists in fitting a logarithmic Pearson III probability distribution to flood data at each station in a region and then using those distributions to compute the quantile discharges corresponding to return periods of 2, 5, 10, 25, 50, 100, and 500 years. The logarithms of these computed quantile discharges are then regressed against logarithms of several physical variables, such as drainage area, mean elevation, mean rainfall, basin shape, and many others, in a stepwise multiple-regression analysis. The equations derived for regional-flood-frequency analysis show that often, but not always, the drainage area serves as the key basin descriptor for many regions in the United States, because the reduction in the standard error is usually observed to be small by regressing on more variables.

Let $Q(A)$ be a random variable denoting the annual peak floods from subbasins $A$ with drainage areas $|A|$ within a basin or a region $D$. Then the $p$th quantile $q_p(A)$ is defined as

$$P\{Q(A) > q_p(A)\} = p \tag{4.4}$$

where $1/p$ is defined as the *return period*. Regressions of $\log q_p(A)$ against $\log |A|$

Table 4.1. *Exponents and standard errors of estimate in flood-quantile–drainage-area relations for four New Mexico regions*

| | Region 1 (northeast plains) | | Region 5 (northern mountains) | | Region 2 (northwest plateau) | | Region 4 (southeast plains) | |
|---|---|---|---|---|---|---|---|---|
| Quantile | EXP | SE | EXP | SE | EXP | SE | EXP | SE |
| Q2 | .56 | .35 | .912 | .34 | .52 | .38 | .671 | .45 |
| Q5 | .55 | .31 | .920 | .32 | .47 | .33 | .591 | .34 |
| Q10 | .55 | .30 | .924 | .32 | .44 | .31 | .546 | .28 |
| Q25 | .55 | .30 | .929 | .34 | .41 | .30 | .498 | .23 |
| Q50 | .55 | .31 | .933 | .35 | .39 | .30 | .465 | .21 |
| Q100 | .56 | .32 | .936 | .37 | .37 | .30 | .436 | .19 |
| Q500 | .581 | .32 | .940 | .39 | .365 | .32 | .408 | .19 |

*Source:* Adapted from Gupta and Dawdy (1995).

for different $p$ values generally show log-log linearity:

$$q_p(A) = q_p(1)|A|^{\theta(p)} \tag{4.5}$$

where both the coefficient $q_p(1)$ and the exponent $\theta(p)$ are functions of the probability of exceedance $p$. In the analysis of the New Mexico data, only the drainage area is used as the physical descriptor for theoretical reasons that are explained later. This allows one to examine how the exponent $\theta(p)$ behaves as a function of $p$. Table 4.1 gives the computed values for the exponents and the corresponding standard errors for four regions in New Mexico.

Region 5 is mountainous and contains the headwaters of the Canadian River. Not only are the exponents relatively constant, but also the runoff is almost directly related to the drainage area, because all the exponents are greater then 0.9. Waltemeyer stated that "in the northern mountain region, floods generally are produced from snowmelt runoff." Region 1 is downstream from region 5, and its mainstream stations receive flows from region 5. It also exhibits an approximate constancy of the exponents, but with a considerably smaller value than 0.9. By contrast, regions 2 and 4 have generally decreasing exponents, with increasing return periods for floods. These two regions, as well as region 6 between them, which is not shown here, constitute a wide band across the middle of the state and exhibit similar decreases in exponents. Waltemeyer further stated that "those [floods] in the plains, plateaus, valleys, and deserts generally are produced by rainfall." The decreasing exponents suggest that rainfall-generated flood peaks exhibit regional relationships different from those of snowmelt-generated floods. The reader should refer to Gupta and Dawdy (1995), Gupta et al. (1994), and Smith (1992) for further details of the New Mexico study and similar analyses for three other regions in the United States.

We can now explain the index-flood method in the context of the foregoing analyses. It consists of two assumptions. The first is that the mean annual peak flow, $E[Q(A)]$, can be computed by regressing on "various descriptors" of a drainage basin, which is taken to be drainage area. It is the most important and sometimes the only descriptor used. The second assumption states that the ratio of floods of any given return period to the mean annual flood does not depend on the drainage area. To express this mathematically, let $g(|A|)$ be some function of the drainage area. Then the *index-flood assumption* says that

$$q_p(A) = q_p(1)g(|A|) \tag{4.6}$$

where, $q_p(1)$ is called the *regional quantile*. Equivalently, equation (4.6) can be written in terms of the equality of the probability distributions of the random variables $Q(A)$ and $g(|A|)Q(1)$ as

$$Q(A) \stackrel{d}{=} g(|A|)Q(1) \tag{4.7}$$

For regions 1 and 5 in New Mexico, data analysis shows that $g(|A|) = |A|^{\theta}$, where the exponent $\theta$ does not vary with respect to the return period. Substituting this expression for $g(|A|)$ into equation (4.6) gives

$$q_p(A) = q_p(1)|A|^{\theta} \tag{4.8}$$

To understand how regions 1 and 5 differ from regions 2 and 4, let $q_m(A)$ denote the mean annual flood; that is, $q_m(A) = E[Q(A)]$, and $m = P\{Q(A) > E[Q(A)]\}$. It follows from equation (4.6) that for regions 1 and 5,

$$q_p(A)/q_m(A) = q_p(1)/q_m(1) \tag{4.9}$$

which shows that the same ratio $q_p/q_m$ can be used for different-size basins, as it does not depend on the drainage area. By contrast, for regions 2 and 4, equation (4.5) shows that

$$\frac{q_p(A)}{q_m(A)} = \frac{q_p(1)}{q_m(1)}|A|^{\theta(p)-\theta(m)} \tag{4.10}$$

According to Table 4.1, if $p < m$, then $\theta(p) < \theta(m)$. This implies that the ratio $q_p/q_m$ decreases as the drainage area increases. This feature was observed by Dawdy (1961), who found that the ratio of peak-flow quantiles with return periods of 100–199 years to the median flood for streams in New England and the ratio of the 100-year flood to the mean flood for streams in the Missouri–Mississippi basin were not constant, but decreased as drainage area increased. That observation was one of the main reasons that ultimately led the USGS to abandon the index-flood approach in favor of the quantile-regression approach to regional-flood-frequency analysis (Benson, 1962).

The basic difference between the results from these two approaches is not as widely recognized as it should be (National Research Council, 1988, p. 38). For example, the differences in the behaviors of the exponents for the four New Mexico regions distinguish between snowmelt-generated and rainfall-generated floods. Only the former corresponds to the index-flood approach (Gupta et al., 1994; Gupta and Dawdy, 1995). However, Dawdy and Gupta (1995) have noted that in four subregions in western Washington State, the index-flood approach seems to describe the regional-frequency behavior of rainfall-generated floods as well.

Another example is given in Figure 4.3, which shows how the coefficient of variation (CV) for regional floods varies with respect to drainage area. This plot can be viewed as a depiction of the spatial variability of floods in a homogeneous region (defined in Section 4.4). This qualitative feature was suggested by an analysis of Appalachian flood data (Smith, 1992). If the index-flood assumption were applied over the entire Appalachian region, then the CV would be a constant. Scale-invariant analyses of spatial variability in rainfall, as explained in Section 4.6, first suggested the hypothesis that rainfall is the dominant factor underlying the decrease in CV for floods in basins larger than some critical size $A_c$ (Gupta et al., 1994). That led to speculation that perhaps the geometry of channel networks influences the structure of the CV in basins smaller than $A_c$ (Gupta and Dawdy, 1995). Recent work by Robinson and Sivapalan (1997) supports this speculation. Moreover, the physical and mathematical developments in the next three sections provide some basic underpinning for the behavior of flood CV as a function of scale within a homogeneous region $D$.

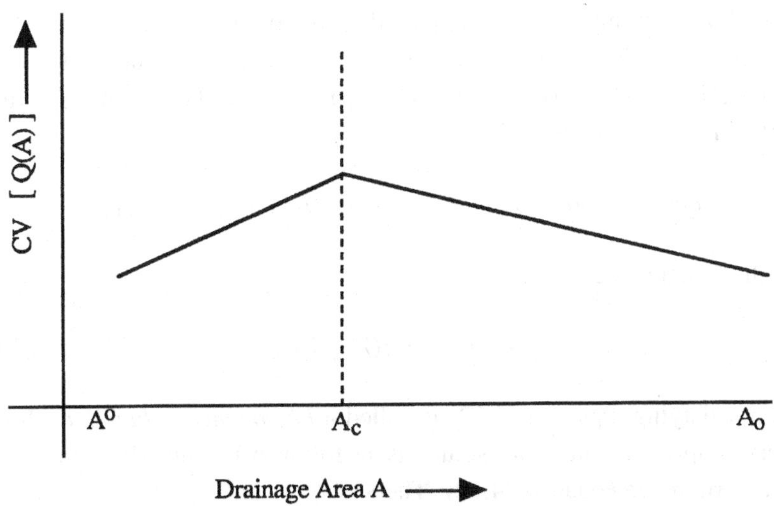


Figure 4.3. A schematic plot of the coefficient of variation for peak flow with respect to drainage area. (From Gupta and Dawdy, 1994, with permission.)

## 4.4 Statistical Simple Scaling and Regional Homogeneity

Consider a large drainage network $\tau$ embedded in a region $D$, and assume that river discharges are being measured in different links $e \in \tau$ in this network. We use the notation that each subtree $\tau(e)$, defined in Section 4.2, is embedded in the subbasin $A(e) \subset D$, with drainage area $|A(e)|$, that drains into the link $e$. Moreover, we assume that the size of a subtree is equal to the drainage area up to a proportionality constant that we henceforth take to be unity: $\|\tau(e)\| = |A(e)|$ for each $e \in \tau$. This assumption is quite reasonable and has been widely used in the literature (Shreve, 1967). The variable of interest here is the maximum instantaneous flow at a guage within some time interval of interest, as defined by equation (4.3). The collection of peak flows or floods at each junction $\{Q(e)\}$ can be viewed as a random field indexed by $e \in \tau$. The discharge at each junction necessarily depends on the size of the upstream subbasin contributing flows to this junction. We define *spatial homogeneity* of the random field $\{Q(e); e \in \tau\}$ to mean that, except for their dependence on the drainage areas $|A(e)|$ of the subregions $A(e) \subset D$, the probability distributions for the peak-flow random field are translation-invariant on the drainage network $\tau$.

### 4.4.1 Statistical Simple Scaling

Let $\{Q(A)\}$ be a random field indexed by subregions $A$ of $D$, and let $\lambda > 0$ be an arbitrary scalar. The random field is defined to exhibit *simple scaling* if for a positive *scaling function* $g(\cdot)$ the following equality holds:

$$Q(\lambda A) \stackrel{d}{=} g(\lambda) Q(A) \tag{4.11}$$

where $\lambda A$ is the set $A$ rescaled by the factor $\lambda > 0$. This equality is understood to mean that all of the finite-dimensional joint distribution functions of the two random fields are the same. To evaluate the scaling function $g(\cdot)$, we take $|A| = 1$ as a *unit reference basin* and consider two arbitrary positive scalars, $\lambda$ and $\mu$. We apply equation (4.11) iteratively as

$$Q(\lambda\mu) \stackrel{d}{=} g(\lambda)Q(\mu) \stackrel{d}{=} g(\lambda)g(\mu)Q(1) \stackrel{d}{=} g(\lambda\mu)Q(1)$$

to obtain the functional equation

$$g(\lambda)g(\mu) = g(\lambda\mu) \tag{4.12}$$

A function satisfying equation (4.12) is called a *homogeneous function*. It can be solved very simply for continuous solutions, as follows: Define $z(\log \lambda) = \log g(\lambda)$, and take logarithms in equation (4.12). Then

$$z[\log(\lambda) + \log(\mu)] = z[\log(\lambda)] + z[\log(\mu)]$$

A solution of this functional equation is $z(\log \lambda) = \theta \log \lambda$ for some constant $\theta$, which says that $\log g(\lambda)$ is a linear function of $\log \lambda$ (Parzen, 1967, p. 121). Therefore, the solution of equation (4.12) is given by a *power law*:

$$g(\lambda) = \lambda^{\theta} \tag{4.13}$$

The fundamental statistical parameter $\theta$ is called a *scaling exponent*. It can take either positive or negative values. This is an important result, because it shows how the power laws are connected with statistical simple scaling. Some well-known examples of simple scaling include the Brownian-motion (BM) and fractional-Brownian-motion (FBM) processes, which belong to the class of "additive" stochastic processes (Feder, 1988).

Some of the empirical results for flood quantiles described in Section 4.3 have implications for simple scaling. To see this connection precisely, consider the $p$th quantile defined in equation (4.4). Take $\lambda = |A|$, where $|A|$ denotes drainage area. It then follows from the definition of a quantile and from equations (4.11) and (4.13) that

$$q_p(A) = q_p(1)|A|^{\theta} \tag{4.14}$$

where $q_p(1)$ is the $p$th quantile of the discharge in a region of unit area. The reader can see that equation (4.14) is identical with the empirical equation (4.8) for flood quantiles. This identification raises the question of the basic physical significance of this correspondence. The physical interpretation of the scaling function in equation (4.14), based on the movement of water through a channel network, is explained in Section 4.5. Likewise, how the probability distribution of peak discharge in a region of unit area can be obtained from mechanisms of runoff generation on hillslopes is explained in Section 4.5.

Let us briefly turn attention to the implications of simple scaling for the statistical moments of floods. First we note that if two random variables $X$ and $Y$ have identical probability distributions, then all of their statistical moments are the same. This can be easily checked from the definition of moments. Therefore, for a simple-scaling stochastic process, it follows from equations (4.11) and (4.13) that

$$E[Q^n(A)] = |A|^{\theta n} E[Q^n(1)] \tag{4.15a}$$

or

$$\log E[Q^n(A)] = n\theta \log|A| + c_n \tag{4.15b}$$

where $c_n = \log E[Q^n(A)]$ when $|A| = 1$. This equation shows two very important features, namely, (1) log-log linearity between statistical moments and scale parameter and (2) a linear increase in slope, $s(n) = \theta n$. These two features play important roles in data analysis, but care is needed in addressing the issue of bias in the estimation of moments from data and how it might affect the interpretation of slopes (Kumar, Guttorp, and Foufoula-Georgiou, 1994; Gupta and Dawdy, 1995).

### *4.4.2 Regional Homogeneity*

Before closing this section, it is pertinent to briefly discuss the issue of regional statistical homogeneity of floods. In the current literature (National Research Council, 1988, p. 38), a region is defined to be "homogeneous" if the CV for floods is constant in that region. Because most of the current research in regional-flood-frequency analysis is based on the validity of the index-flood assumption, this definition of homogeneity has been instigated and is implied by the index-flood assumption. For example, the reader can see from equation (4.15a) that $\mathrm{CV}[Q(A)] = \mathrm{CV}[Q(1)]$. In the flood data, observed deviations from the index-flood assumption are generally attributed to a lack of regional homogeneity.

The preceding notion of homogeneity has been criticized by Gupta et al. (1994) as being "ad hoc and not very useful" for developing a physically based statistical theory of regional floods. To appreciate the reasons for this criticism, first note that a simple-scaling stochastic process defined by equations (4.11) and (4.13) implies a constant CV. The reader can easily check the constancy of CV on the basis of the moment expression given in equation (4.15a). By contrast, a multiscaling stochastic process defined in Section 4.6 does not exhibit a constant CV. Because simple scaling and multiscaling appear to be related to differences in the spatial variability of flood-generating mechanisms (e.g., snowmelt versus rainfall), it is desirable to define homogeneity in such a way as to place these two physical situations on the same theoretical footing. This new definition of homogeneity provides the basis for developing a regional statistical-hydrologic theory of floods in Sections 4.5 and 4.6.

Certainly there are regions or basins where the foregoing assumption of homogeneity will not hold. A simple physical example is seen in the systematic changes in precipitation with changes in elevation in a large mountainous basin. This feature is likely to violate the criterion of homogeneity of peak flows. Theoretically, this additional feature can be incorporated by constructing a stochastic flow field on a network whose probability distributions depend on the sizes of subbasins and on elevation (i.e., on a multivariate-parameter set). However, the development of a general theory that can incorporate heterogeneity or mixing within a region or among regions with multivariate parameters will come at some time in the future. In summary, this discussion demonstrates that the constancy of CV for peak flows as a definition of homogeneity, which was motivated by purely statistical considerations in the past, needs to be revised in developing a physically based understanding of the statistics of regional floods.

### 4.5 Toward a Physical Basis for Simple Scaling in Regional Floods

Our main objective in this section is to illustrate how the abstract notion of statistical simple scaling in regional floods can be understood physically in terms of solutions for the set of equations (4.1) and (4.2). Let us assume that simple scaling is applicable to floods within a large homogeneous region. Taking $\lambda = |A_1|/|A|$ in equation (4.11) as the ratio of the areas of two arbitrary subbasins and combining it with equation (4.13) gives

$$Q(A_1) \stackrel{d}{=} \left(\frac{|A_1|}{|A|}\right)^{\theta} Q(A) \tag{4.16}$$

This equation can be used to scale up the statistics of peak flows from some reference basin $A$ to other drainage basins $A_1, A_2, \ldots,$ provided the exponent $\theta$ can be estimated physically and the probability distribution for peak flows can be predicted for a reference basin $A$. This program would also provide a scientific basis for testing flood frequency predictions in gauged basins and for making predictions in ungauged river basins.

The main focus of this section is to solve equation (4.1) under idealized conditions and obtain a physical interpretation of the scaling exponent $\theta$ so that it can be estimated independently of the runoff data. Specifically, we assume that $\Delta S(e,t) = 0$ and that the hillside runoff field $a(e)R(e,t) = aR(t),\ t \geq 0,\ e \in \tau$, is *spatially uniform.* Recall from Section 4.2 that $a$ denotes the combined area of hills on either side of a channel link. It now follows from equation (4.1) that the river runoff in each link is given by a *convolution,*

$$q(e,t) = a \sum_{s=0}^{t-1} W_e(t-1-s)R(s) \qquad (t \geq 1,\ e \in \tau) \tag{4.17}$$

where $W_e(x),\ x \geq 0$, is called the *local geomorphologic width function* of a network with outlet $\phi = e$, for each $e \in \tau$. It is defined as the number of links at a distance $x$ from the outlet. In order to introduce the notion of distance, we define the lengths of vertices as $|\bar{e}| = |\underline{e}| + l$, and $|\phi| = 0$. This definition can be used iteratively to assign a distance to each link from the outlet and thereby compute the width function.

To further understand the importance of the width function in the present context, assume that runoff is applied instantaneously on a network; that is, $R(s) = R$ if $s = 0$, and $R(s) = 0$ otherwise. This simplifies equation (4.17) as

$$q(e,t) = W_e(t-1)Ra \qquad (t \geq 1,\ e \in \tau) \tag{4.18}$$

This means that the width function, up to a scale transformation from space to time via $x = t/v$, determines the response of the channel network to an instantaneous, spatially uniform runoff $R$. Known as an *instantaneous unit hydrograph* (IUH), this key concept has been widely used in engineering hydrology since the 1950s. An IUH is defined as "the response of a basin at the outlet due to a unit volume of net rainfall applied instantaneously and spatially uniformly over it." Extensions to finite-duration runoff $R$ (also called effective rainfall) are widely used on the basis of a convolution operation shown in equation (4.17); see Dooge (1973) for an excellent review of the state of the art in rainfall–runoff modeling up to the early 1970s.

Determination of an IUH from the topology and the geometry of channel networks introduced the important concept of the *geomorphologic instantaneous unit hydrograph* (GIUH) into the literature. Two different methods of incorporating channel-network geometry into a GIUH were introduced. One of these was based on the width function (Lee and Delleur, 1976; Kirkby, 1984), and the other used Horton-Strahler ordering of a network, as explained in Section 4.5.1 (Rodríguez-Iturbe and Valdez, 1979; Gupta, Waymire, and Wang, 1980). The GIUH was investigated extensively in the 1980s on both empirical and theoretical grounds; see Rinaldo and Rodríguez-Iturbe (1996) for a recent review, and Gupta and Mesa (1988) for a somewhat dated review. Even though a detailed explanation of these developments regarding GIUH is outside the scope of this chapter, one key theoretical result from the GIUH theory is described in Section 4.5.3.

In view of the definition of peak flows given in equation (4.3), it follows from equation (4.18) that

$$Q(e) = \max_t q(e,t) = \max_t W_e(t-1)Ra \qquad (e \in \tau) \tag{4.19}$$

The physical basis of simple scaling can be illustrated through idealized examples, and this objective is the main focus of this section. Section 4.5.1 explains the concept of a network with a deterministic self-similar topology (SST). As before, we use "self-similarity" and "scale invariance" synonymously. The SST structure is observed in many real drainage networks (Peckham, 1995a). We take an idealized example of an SST network, the so-called Peano network (Mandelbrot, 1983, ch. 7; Marani, Rigon, and Rinaldo, 1991), and show that with respect to the size of the subnetworks $\|\tau(e)\|$, $\max_t W_e(t-1)$ scales in a power-law manner, with an exponent of log 3/ log 4. If $R$, the spatially uniform runoff from hillsides, is assumed to be a random variable, then the peak flow field in equation (4.19) becomes identical with the simple-scaling representation given in equation (4.16), and $\theta = \log 3/\log 4$. To complete the physical specification of simple scaling, the probability distribution for $R$ must be obtained from the mechanisms for runoff generation from hillsides given by equation (4.2), and this is sketched in Section 4.5.2.

Section 4.5.3 considers the finite-duration runoff forcing from equation (4.17) [i.e., $R(t) = R,\ 0 \le t \le T$] and illustrates the effect of duration $T$ on the scaling exponent within the context of the Peano example (Castro, 1998). Next, the effect of finite duration on the scaling exponent is investigated when randomness is present in the network topology. The presence of randomness in the network topology, on the average, does not alter the qualitative behavior of the scaling exponent. Therefore we apply it to suggest a possible physical explanation for two of the scaling exponents in the New Mexico flood data from Section 4.3.

Throughout this section, and in Section 4.6, the water-flow dynamics in natural river networks will be reflected in the constant-velocity assumption through the time scale $\Delta t$ in equation (4.1), even though the term $\Delta S(e, t)$ has been set to zero there. Let us discuss this and various other issues surrounding the specification of the storage term. Water-flow dynamics in open channels is governed by the forces of gravity and channel friction. It causes attenuation of water waves, and this results in channel storage. The hydraulics of water flow in open channels has been widely described by a nonlinear momentum-balance equation in one spatial coordinate and time, known as the Saint Venant equation. A commonly and widely used approximation of this full dynamic equation that retains its nonlinearity is known as the kinematic-wave equation (Eagleson, 1970, ch. 15). In many investigations regarding flow routing in open channels it has been common to linearize the dynamic equation. Flow-routing methods in river basins have used either some version of the one-dimensional dynamic equation or ad hoc storage–discharge relationships, such as that given by the Muskingum method (Dooge, 1973).

In a channel network, the channel slopes, widths, depths, velocities, and friction coefficients, known as the *hydraulic-geometric variables*, will vary spatially throughout in response to spatially varying discharge. These variations can be quantified by functionally relating the variables to bankfull discharge through power laws. These systematic variations in a channel network are studied as *downstream hydraulic-geometric relations* (Leopold, Wolman, and Miller, 1964). Early empirical studies showed that velocity changes little with discharge in the downstream direction and can be treated approximately as a constant. Current research by Peckham, Gupta, and Smith (1998) shows that the downstream hydraulic-geometric relations can be predicted from the steady-state equations of mass and momentum balance for water in two spatial coordinates, coupled with a postulate of *dynamic self-similarity* or scale invariance. Moreover, previously published predictions of the hydraulic-geometric exponents by Leopold et al. (1964) and Rodríguez-Iturbe et al. (1992) have turned out to be special cases of this general theory. This set of empirical and theoretical results not only provides a physical basis for the (approximate) constant-velocity assumption made here but also suggests that a correct, physically based specification of the change in the storage field $\Delta S(e, t)$ in a natural river network would necessarily entail downstream hydraulic-geometric considerations. An example of some recent attempts to introduce hydraulic-geometric ideas into flow routing through a network is provided by the work of Snell and Sivapalan (1995). However, the development of a rigorous theory of runoff routing in natural channel networks is still to be achieved in the future.

### *4.5.1 A Physical Determination of the Scaling Exponent in Networks with Deterministic Self-similar Topologies*

Investigations initiated by Robert Horton and first reported in 1945, based on a parameterization of network topologic structure according to a stream-ordering system that was later refined by Strahler, led to the widely known *Horton laws* of drainage composition (Horton, 1984). Briefly, the Horton-Strahler (H-S) stream ordering is defined by assigning an order $\omega = 1$ to the source links. When two links of order $\omega$ meet, the downstream link is assigned an order $\omega + 1$. When two links of different orders meet, the downstream link carries the higher of the two orders. A Horton-Strahler stream of order $\omega$ consists of a chain of links of order $\omega$ that is joined above by two links of order $\omega - 1$ and is joined below by a link of order $\omega + 1$ or by the outlet. A very insightful derivation of the H-S stream ordering has been given by Melton (1959).

Systematic theoretical inquiry into the nature of Horton laws did not begin until the middle of the 1960s. One of the most widely studied models in this regard is the random model of Shreve (1967). Within the past decade, research has begun to show that river networks exhibit a high degree of self-similarity or scale invariance across scales in their branching structures, unless geologic controls break this invariance by imposing characteristic scales. For example, theoretical ideas of deterministic *self-similar topology* (SST) by Tokunaga (1966, 1978), as explained later, have been used to illustrate that the Horton laws of drainage composition arise from this invariance property in the topologic structure of river networks (Peckham, 1995a; Tarboton, 1996). Empirical tests of these new findings have been possible because of the widespread availability of three-dimensional topographic data sets called digital elevation models (DEMs).

One way to define SST is to stipulate that the number of side tributaries $T_{\omega,\omega-k}$ of H-S order $\omega - k$ joining a stream of order $\omega$ does not depend on $\omega$; that is,

$$T_{\omega,\omega-k} = T_k \qquad (k = 1, 2, \ldots, \omega - 1) \tag{4.20}$$

Tokunaga (1966) postulated that $T_{k+1}/T_k = c$, which leads to the expression that $T_k = ac^{k-1}$, $k = 1, 2, \ldots$, where $T_1 = a$. Recent DEM-based analyses of river networks have shown that this expression of the network generators, referred to as a *Tokunaga tree*, describes the average branching structure for natural river networks. The well-known *random model* of Shreve (1967) obeys this SST property in an average sense over a statistical ensemble of networks, and the generators of the *average Shreve tree* are given by $T_k = 2^{k-1}$ (Tokunaga, 1978).

Recent work by Burd and Waymire (1997) has considered network models with a *stochastic self-similar topologic structure*. Without going into the mathematical details of this definition here, it is sufficient to mention two key results. First, the deterministic SST property appears as a special case of this class of random self-similar networks. Second, the random model (Shreve, 1967) is the only branching stochastic process that exhibits this stochastic self-similar topology. Various other results have been obtained regarding statistical tests of simple scaling in channel-network geometry using both H-S-ordering-based enumeration (Peckham, 1995b) and link-based enumeration (Gupta and Waymire, 1989; Tarboton, Bras, and Rodríguez-Iturbe, 1989). However, a detailed discussion of these results is outside the scope of this chapter. Many of the classic papers on channel networks, including the random model, have been reprinted in the volume edited by Jarvis and Woldenberg (1984), and several of the articles referenced in this chapter are contained in that book. The reader is urged to read those articles for some basic background in quantitative channel-network geomorphology.

We consider a Peano tree as an idealized model of a channel network, as previously studied by Marani et al. (1991), and recently by Gupta et al. (1996), to illustrate scale invariance in peak flows. Unlike natural river networks, a Peano tree is not binary. However, a Peano network has the SST property needed here, and it is easily embedded in a two-dimensional space, a property that is needed in the example given in Section 4.6. To understand the construction of a Peano tree in a two-dimensional unit square, place a channel across the diagonal of a unit square at the initial stage of construction. At the first stage, this channel is subdivided into halves, and two side tributaries are added at the node in the middle. At the second stage of construction, each link of length $\sqrt{2}/2$ is subdivided into halves, and two side tributaries are added at each node in the middle. This construction continues ad infinitum, leading to a Peano tree. This idea can be used to construct a variety of other networks with the SST property (Peckham, 1995a). A three-stage Peano network is shown in Figure 4.4.

In an $n$-stage Peano basin, each link is of length $l_n = \sqrt{2}(\frac{1}{2})^n$. In fact, the Peano basin is made up of a discrete set of length scales, $\lambda_k = (\frac{1}{2})^k$, $k = 0, 2, \ldots, n$, that represent the side lengths of subbasins at different scales of spatial resolution. The H-S stream order, $\omega = n-k+1$, $k = 0, 1, \ldots, n$, also changes at each of the tributary junctions, as shown in Figure 4.4. Here $\Omega = n + 1$ is the basin order. The number of H-S streams $N_\omega$ of order $\omega$ in a Peano tree obeys the recursion equation

$$N_{\Omega-j} = 4N_{\Omega-j+1} - 1 \qquad (N_\Omega = 1,\ j = 1, 2, \ldots, \Omega - 1) \tag{4.21}$$

One can easily see from equation (4.21) that as the order becomes large, *Horton's bifurcation ratio*, defined as $R_b = N_\omega / N_{\omega+1}$, approaches 4. The relation $l_n = \sqrt{2}(\frac{1}{2})^n$ predicts that *Horton's length ratio*, defined as $R_l = l_{n+1}/l_n$, is 2. These values can be used to compute the *topologic fractal dimension* of a network, as given

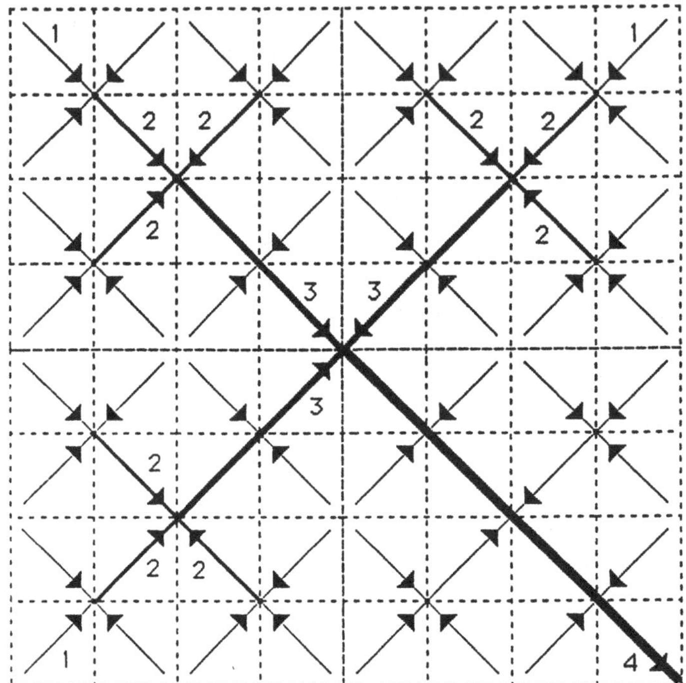


Figure 4.4. A Peano network after three stages of construction. (From Gupta et al., 1996, with permission.)

by the expression $D_p = R_b/R_l$ (La Barbera and Rosso, 1989; Peckham, 1995a; Tarboton, 1996). In particular, the value $D_p = 2$ for a Peano network exhibits its space-filling property. It is interesting to note that whereas these three topologic parameters are the same for the average Shreve tree, its topologic-network structure is quite different from that of a Peano network.

For the problem at hand, assume that a finite network $\tau$ is given by an $n$-stage Peano basin. The main problem is to investigate how the maximum-width function, $\max_x W_e(x)$, $e \in \tau$, in equation (4.19) scales with respect to the size of subtrees $\|\tau(e)\|$ for a Peano basin. First, a natural length scale $\lambda_\omega$ can be assigned to each subtree $\tau(e_\omega)$ draining into a link $e_\omega$ at the bottom of a stream of order $\omega$ as

$$\lambda_\omega = \left(\frac{1}{2}\right)^{n+1-\omega} \qquad (\omega = 1, 2, \ldots, n+1) \tag{4.22a}$$

Then

$$\|\tau(e_\omega)\| = \lambda_\omega^2 \qquad (\omega = 1, 2, \ldots, n+1) \tag{4.22b}$$

This suggests that peak flows $Q_n(e_\omega)$, $\omega = 1, 2, \ldots, n+1$, need be measured only at the bottom of each stream of a given H-S order, rather than at the bottom of each link in a network. In view of the constant-velocity assumption, it is easy to see from Figure 4.4 that $3^{\omega-1}$ links make up the maximum-width function at the bottom of each H-S stream of order $\omega = 1, 2, \ldots, n+1$ and thereby contribute to the peak flows. Therefore,

$$\max_x W_{e_\omega}(x) = 3^{\omega-1} \qquad (\omega = 1, 2, \ldots, n+1) \tag{4.23}$$

Substituting the expression for $\omega$ in terms of $\lambda_\omega$ from equation (4.22a) into this equation, and making simple algebraic manipulations, we get

$$\max_x W_{e_\omega}(x) = 3^n (\lambda_\omega)^{\log 3/\log 2} \qquad (\omega = 1, 2, \ldots, n+1) \tag{4.24}$$

Finally, substituting equation (4.24) into equation (4.19), noting that the volume of runoff in a first-order subbasin of the Peano basin is given by $R(\frac{1}{4})^n$, because $\|\tau(e_1)\| = (\frac{1}{4})^n$, and assuming that $R$ is a random variable, we get a simple scaling expression for peak flows in a Peano network:

$$\begin{aligned} Q(e_\omega) &\stackrel{d}{=} R_n(\lambda_\omega)^{\log 3/\log 2} \qquad (\omega = 1, 2, \ldots, n+1) \\ &= R_n\left(\lambda_\omega^2\right)^{\log 3/\log 4} = R_n(|A_\omega|)^{\log 3/\log 4} \end{aligned} \tag{4.25}$$

where $|A_\omega|$ is the cumulative area of a subtree draining into a stream of order $\omega$, and $R_n = R(\frac{3}{4})^n$, is the peak flow rate from the entire basin.

The simple-scaling exponent log 3/log 2 admits a natural geometric interpretation as a box-counting dimension (Feder, 1988) of the *maximum contributing set* $C_s$. It is defined as the region occupied by the number of pixels $N_{\lambda_k} = 3^k$ that make up the maximum-width function at each length scale $\lambda_k = (\frac{1}{2})^k$, $k = 0, 1, 2, \ldots, n$. Therefore, the *box-counting dimension* or *fractal dimension* of $C_s$ is given by (Marani et al., 1991, p. 3046)

$$D(C_s) = \lim_{n\to\infty} \frac{\log N_{\lambda_n}}{\log(1/\lambda_n)} = \frac{\log 3}{\log 2} = 1.585 \tag{4.26}$$

A network-based geometric interpretation of the statistical parameter $\theta = \log 3/\log 2$ governing spatial variability in peak flows brings an exciting new theoretical perspective to the problem of hydrologic regionalization. It parallels many other known examples from physics, such as the BM model of molecular diffusion ($\theta = \frac{1}{2}$), Kolmogorov's spectrum of turbulent-velocity distributions ($\theta = \frac{1}{3}$), and the Holtzmark distribution of force ($\theta = \frac{2}{3}$) (Gupta and Waymire, 1993). The geometric interpretation of the scaling exponent for floods suggests that this type of theoretical result can be empirically tested by carrying out precise topographic and peak-flow data analyses in those drainage basins in which peak flows are generated by snowmelt runoff or stratiform rainfall. However, general theoretical results similar to equation (4.25) will be required for networks that are physically more realistic than a Peano basin.

### 4.5.2 Probability Distribution for the Runoff Ratio from a Hillside

Equation (4.25) clearly illustrates how randomness in the peak flows over the entire Peano network arises from the term $R$ in $R_n = R(\frac{3}{4})^n$. Because $R$ denotes the flow rate for a first-order basin, or a hillside, a specification of the probability distribution for $R$ should be based on a physical understanding of runoff-generation mechanisms on hillsides. This section illustrates how qualitative insights about runoff-generation mechanisms from hillsides can be used in computing the probability distribution for a fraction of $R$ that contributes to the peak flows.

Within the past two decades there have been many efforts to model runoff-generation from hillsides as a complex interaction among soil-moisture storage, infiltration through the unsaturated-soil zone, subsurface runoff through the saturated zone, and surface runoff due to "excess" rainfall; see the studies by Duffy (1996) and Robinson and Sivapalan (1995) and the references therein. Our intent here is only to illustrate, in a conceptually simple and self-contained manner, how a physical understanding of runoff-generation mechanisms from a single hillside can be used to begin to formulate and compute a probability distribution function for runoff that will apply to an ensemble of hillsides in a basin.

Three distinct physical runoff-generation mechanisms on hillsides have been identified in the literature: (1) Hortonian infiltration-excess runoff, (2) saturated overland flow, and (3) subsurface storm flow. Different runoff-generation mechanisms have been reported to lead to distinct travel velocities for hillside flows. For instance, Dunne (1982) reported that flow velocities due to Hortonian infiltration-excess runoff can vary between 10 and 500 $\mathrm{m \cdot h^{-1}}$. Saturated-overland-flow velocities on hillsides near channels typically can range from 0.3 to 100 $\mathrm{m \cdot h^{-1}}$. These values are several orders of magnitude larger than the flow velocities associated with the subsurface storm flow from hillsides to a network, which are on the order of $10^{-4}$ $\mathrm{m \cdot h^{-1}}$. Even in highly permeable soils with steep slopes the storm-flow velocity can reach only as high as 0.2 $\mathrm{m \cdot h^{-1}}$, and that occurs only rarely. Because of this time-scale separation of flow velocities, the runoff hydrograph from a hillside can be conceived as the sum of a "fast" overland-flow component and a "slow" base-flow component (Mesa and Mifflin, 1986). Recent detailed investigations have supported this qualitative physical view of runoff generation from hillsides (Duffy, 1996).

Because of time-scale separation between the fast and slow runoff components, the entire runoff $R$ does not contribute to the peak flow. Assume that the fast runoff component, which is only a fraction of the term $R$, contributes to the peak flow.

The remainder of $R$, corresponding to the subsurface storm flow or the slow runoff component, generates low stream flows. Let $\pi_1$ and $\pi_2$ denote the *runoff ratios* corresponding to the fast and slow runoff components, respectively, such that $\pi_1 + \pi_2 = 1$. For illustration purposes, we ignore the time variability of a hillslope hydrograph. Moreover, we assume that the runoff rate is in a steady state [i.e., $\Delta\theta(e, t) = 0$] and that the time scales are small, so that $E_T[\theta(e, t)] = 0$ in equation (4.2). This gives $I_r = R$. Such simplifications are also necessary for obtaining qualitative theoretical insights that can guide more detailed theoretical, field and modeling studies in the future. The problem before us is to compute $\pi_1 R$.

Let $I_f$ denote a spatially uniform infiltration rate in millimeters per hour, and let $C$ be the fraction of a hill region saturated with water from below. Then the fast component of the runoff rate, $\pi_1 R$, due to Hortonian infiltration excess and saturated overland flow can be expressed as

$$\begin{aligned} \pi_1 R &= (I_r - I_f)(1 - C) + I_r C \quad && \text{if } I_r > I_f \\ &= I_r C && \text{if } I_r < I_f \end{aligned} \tag{4.27}$$

The infiltration rate in unsaturated soils is governed by the initial soil moisture, the forces of soil suction and gravity, and a fundamental soil parameter called the unsaturated hydraulic conductivity. Hydraulic conductivity varies with soil moisture. This soil parameter depends on the properties of the porous medium as well as those of the fluid and air phases in it. The infiltration rate decreases with time from some initial rate and reaches a limiting value. This behavior can be described by Horton's empirical infiltration equation or by other equations based on the physics of soil-water movement that apply to small spatial scales. Extension of these equations to larger spatial scales remains an important open problem.

For this discussion, we assume that the infiltration rate is equal to the saturated hydraulic conductivity, $K_s$ in millimeters per hour and that $K_s$ is spatially uniform across and below the hillslope surface. However, the infiltration rate always lies between the rainfall intensity $I_r$ and $K_s$, in the sense that $0 \leq 1 - (I_f/I_r) \leq 1 - (K_s/I_r)$. Dividing both sides of equation (4.27) by $I_r$ gives an expression for the surface runoff ratio $\pi_1$ as

$$\begin{aligned} \pi_1 &= \left(1 - \frac{K_s}{I_r}\right)(1 - C) + C \quad && \text{if } \frac{K_s}{I_r} < 1 \\ &= C && \text{if } \frac{K_s}{I_r} \geq 1 \end{aligned} \tag{4.28}$$

The key problem is to find an expression for $C$ in terms of the ratio $K_s/I_r$ and the topographic slope $S$. In the past, detailed and complex model simulations were carried out to illustrate, among other things, the effects of soil hydraulic conductivity,

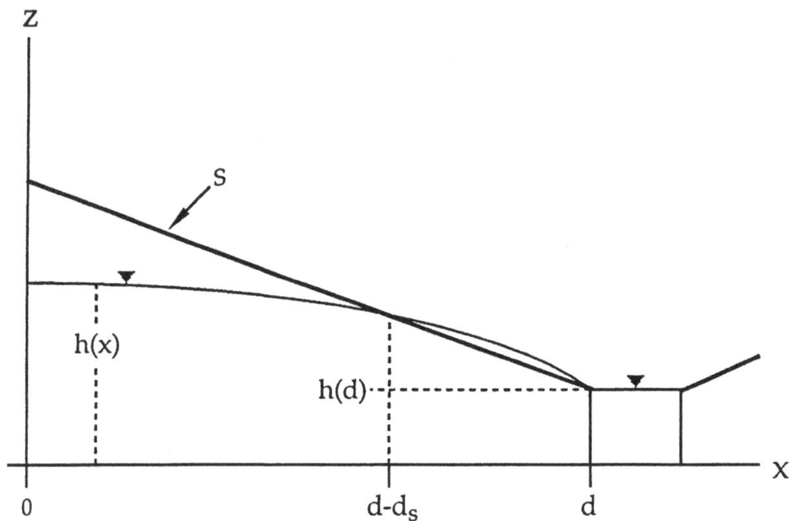


Figure 4.5. Schematic depiction of how topographic and groundwater profiles intersect on a hillside.

topography, and climate on the saturated overland flow from a hillslope (Freeze, 1980). Such models, because of their complexity, are unsuitable to provide the qualitative physical insights that we are after. By contrast, the simplified formulation of saturated overland flow and subsurface storm flow given by Dietrich and Dunne (1993) is similar in spirit to the presentation that follows.

Consider a hillslope, as shown in Figure 4.5. Let the left-hand point at the drainage divide denote the origin of an $x$–$z$ coordinate system, with $d$ the width of the hillslope. We consider a simple one-dimensional problem of the lateral inflow into a stream per unit length of the hillslope along the stream from the aquifer. Let $h(x)$ be the height of the water table or the *hydraulic head* at a distance $x$ from the origin. In general, $h(x)$ also varies with time. However, if attention is restricted to a steady-state situation, the time dependence of the head is ignored. In a steady state, the discharge per unit length along the stream is balanced by the infiltration rate per unit length. This, in turn, is equal to the rainfall rate times the unit length $I_r x$, provided that $K_s \geq I_r$. If $K_s \leq I_r$, then the excess rainfall $I_r - K_s$ becomes Hortonian runoff, and the infiltration occurs at the rate $K_s$. Assume that the Dupit-Forchheimer assumptions hold, namely, (1) the hydraulic gradient is equal to the slope of the water table, and (2) the water-table gradient is small, so that the flow is parallel to the horizontal direction. In a steady state, the discharge per unit length of the hillslope is given by a well-known relation, Darcy's law:

$$-K_s h \frac{d}{dx} h(x) = I_r x \tag{4.29}$$

where

$$h(x)|_{x=d} = h_1$$

denotes the depth of water in the stream as a boundary condition. Equation (4.29) can be easily solved to obtain

$$h^2(x) - h_1^2 = \frac{I_r}{K_s}(d^2 - x^2) \tag{4.30}$$

The physical idea regarding saturated overland flow is tied to the observation that the groundwater profile $h(x)$ can intersect with the land-surface profile and thereby create saturated conditions in valley bottoms adjacent to a stream. For analytical simplicity, assume that the land-surface profile is a straight line with slope $S$, and it meets the water surface at $h_1$ in the stream. Let $d_s$ denote the length of the saturated surface. Then the horizontal distance of the intersection of these two surfaces from the origin is $d - d_s$. By definition, at the point of intersection,

$$\frac{h(d - d_s) - h_1}{-d_s} = S \tag{4.31}$$

Substitution of $x = d - d_s$ into equation (4.30) gives another equation. These two equations can be solved for the two unknowns $h(d - d_s)$ and $d_s$. Solving for $d_s$ gives the fractional saturated-hillslope area per unit length of stream as

$$C = \frac{d_s}{d} = \frac{2(I_r/K_s - h_1/d)}{S^2 + I_r/K_s} \approx \frac{2I_r/K_s}{S^2 + I_r/K_s} \qquad (h_1/d \ll 1) \tag{4.32}$$

This simple expression shows how $C$ depends on the ratio $I_r/K_s$ and the topographic slope $S$. Notice that as the rainfall rate $I_r \to 0$, so does $C \to 0$, as one expects. When $I_r = K_s$, then the two surfaces do not intersect in the hillslope region $0 < x < d$, unless $S > 1$. This simply means that the entire hillslope becomes saturated if $S \leq 1$. Finally, note from equation (4.28) that when $K_s < I_r$, the overland-flow rate is $(I_r - K_s)(1 - C) + I_r C$, and the subsurface storm-flow rate is $K_s(1 - C)$, and the two add to $I_r$ by mass conservation. When $K_s \geq I_r$, then the overland-flow rate $I_r C$ is due to only saturated overland flow, and the subsurface storm-flow rate is $I_r(1 - C)$.

Equations (4.28) and (4.32) completely specify the runoff ratio. Its probability distribution can be computed from the (joint) probability distributions of the ratio $I_r/K_s$ and $S$. We do not intend to pursue this computation, as it is not difficult conceptually. The computation of the runoff ratio shows that even when the rainfall field in a drainage network is spatially uniform, the term $\pi_1(e)$, $e \in \tau$, exhibits spatial variability because of the effects of variabilities in soil hydraulic properties and topographic slopes on it. In its simplest form, this variability can be quantified by assuming that $R(e,t) = \pi_1(e)R$ for $t = 0$, and $R(e,t) = 0$ otherwise. Suppose that the values of $\pi_1(e)$, $e \in \tau$, can be regarded as samples from independent and identically distributed (IID) random variables having a common probability distribution. This distribution applies to an ensemble of hills in a basin, rather than to a single hill, and violates the assumption of spatial uniformity of runoff from different hills made in Section 4.5.1. Computation of spatial-scaling statistics for peak flows via spatial aggregation of runoff in a river network in the presence of this spatial variability remains an unsolved problem. However, some results have been obtained in a similar context for low streamflows (Furey and Gupta, 1998).

We shall end this section with a potentially important observation in regard to the probability distribution for the runoff ratio. This observation concerns the distributional forms of the rainfall intensity and the saturated hydraulic conductivity. There is a fairly large body of empirical evidence to suggest that $I_r$ exhibits a lognormal type of distribution (Bell, 1987). Likewise, $K_s$ has been widely assumed to be lognormal in the literature (Rogowski, 1972; Loague and Gander, 1990). Because a lognormal distribution has a multiplicative structure, it follows that the ratio $I_r/K_s$ is also lognormal, provided that $\log I_r$ and $\log K_s$ are jointly normal. Lognormality may seem like a technical convenience, but its multiplicative structure is potentially important in generalizing this theory to accommodate spatial variabilities in rainfall and soil hydraulic conductivity among hillsides in subbasins. As discussed in Section 4.6, the statistical theory of spatially variable rainfall that exhibits multiscaling in floods has a multiplicative structure.

### *4.5.3 Effect of Rainfall Duration on the Scaling Exponent*

Equation (4.25) showed that the peak flows from an instantaneous or very short duration, uniformly distributed, spatial runoff (rainfall) in a Peano basin at different scales are determined by the fractions of the total number of pixels, which are proportional to $(\frac{3}{4})^{w-1}$, $\omega = 1, 2, \ldots, n+1$. These fractions clearly decrease as the order or the size of the subbasins increases. When the runoff term is not instantaneous, but persists for a longer time [i.e., $R(t) \geq 0,\ 0 \leq t \leq T$], then smaller-order subbasins begin to "saturate." This means that the entire drainage basin contributes to the peak flow, rather than only a fraction of it, and the peak flow becomes proportional to the drainage area, with an exponent of unity. This time to saturation is called the *concentration time*, at which the river discharge reaches its maximum value. The discharge remains in a steady state after the concentration time as long as $R(t) > 0$. If $j(T)$ denotes the scale index for the largest subbasin in the Peano example that has become saturated at a fixed duration $T$, then the peak-flow scaling in larger basins with areas $\lambda_k^2$ is still given by $(\lambda_k^2)^{\log 3/\log 4}$, $k > j(T)$. The saturation extends to larger subbasins as duration $T$ increases. This argument illustrates that the rainfall duration has a very significant effect on the scaling exponent for peak flows (Castro, 1998).

To further illustrate the effect of duration on the scaling exponent, as explained earlier, consider the random model. It has been studied extensively in previous investigations regarding the structure of the width function, and it exhibits a *stochastic self-similar topologic* structure, as mentioned earlier. Because of randomness in the topologic structure of the network, the runoff in equation (4.17) is modeled as a random field. Taking the means, conditioned on the size of the network, on both sides of equation (4.17), we have

$$E[q(e,t)|\|\tau(e)\|] = a\sum_{s=0}^{t-1} E[W_e(t-1-s)|\|\tau(e)\|]R(s) \qquad (t \geq 1,\ e \in \tau) \tag{4.33}$$

We shall investigate the effect of duration on $\max_t E[q(e,t)|\|\tau(e)\|]$ via its scaling exponent.

Because $\|\tau(e)\| = 2m - 1$, let us consider the asymptotic expression (for large $m$) for the mean-width function conditioned on magnitude, $\mu_m(x) = E[W_m(x)|m]$. Under the conditions that the link lengths are IID random variables with mean $\alpha$ and finite moments of all higher orders, and the random model specifies the branching structure of the network, Troutman and Karlinger (1984) showed that

$$\mu_m(y)\,dy \sim \frac{y}{\alpha}\exp\{-y^2/4\alpha^2 m\}\,dy \qquad (y > 0 \quad \text{for large } m) \tag{4.34}$$

The expression on the right-hand side is the so-called Rayleigh probability density function. An overview of a precise mathematical formulation and various generalizations of this result has been given by Gupta and Waymire (1998). Substitution of $t\nu = y$ into equation (4.34), where $\nu$ is the constant velocity, gives an expression for the mean instantaneous hydrograph:

$$\nu\mu_m(t)\,dt \sim \nu\frac{2t}{\beta}\exp\{-t^2/\beta^2 m\}\,dt \qquad (t > 0 \quad \text{for large } m) \tag{4.35}$$

Where $\beta = 2\alpha/\nu$, and $\mu_m(t) \equiv \mu_m(\nu t)$.

We replace summation by integration in equation (4.33), for analytical convenience, and substitute $E[W_e(t-1-s)|\|\tau(e)\|] \sim \mu_m(t-s)$. Note that the spatial dependence of the mean-width function on each link $e$ is parameterized by the dependence of $\mu_m(t)$ on magnitude $m$, since $\|\tau(e)\| = 2m - 1$. Because the scaling exponent depends only on the width function, we take $R(s) = 1$ as the runoff of unit intensity and duration $T$, with $a = 1$ in equation (4.33), to obtain

$$Q_{m,T}(t) = \nu \int_0^T \mu_m(t-s)\, ds = S_m(t) - S_m(t-T) \qquad (t > T) \tag{4.36a}$$

where

$$S_m(t) = \nu \int_0^t \mu_m(s)\, ds = \nu \int_0^t \frac{2s}{\beta} \exp\{-s^2/\beta^2 m\}\, ds \tag{4.36b}$$

is called an *S-hydrograph*. The relationship between an S-hydrograph and a *unit hydrograph of duration T* is exhibited by equation (4.36) up to a constant. It is widely used in engineering hydrology (Dooge, 1973, p. 86).

To calculate $\max_t Q_{m,T}(t)$, let $t_p$ denote the time to peak. Differentiating equation (4.36), it follows that $t_p$ is given by the implicit equation

$$t_p \exp\left\{-t_p^2/\beta^2 m\right\} = (t_p - T)\exp\{-(t_p - T)^2/\beta^2 m\} \tag{4.37}$$

The solution to this equation depends parametrically on duration and magnitude, that is, $t_p \equiv t_p(m, T)$. An expression for the $\max_t Q_{m,T}(t)$ is obtained by evaluating equation (4.36) at $t = t_p$:

$$\begin{aligned} Q_{m,T} &\equiv Q_{m,T}(t_p) = S_m(t_p) - S_m(t_p - T) \\ &= m\nu\beta\left[\exp\{-(t_p - T)^2/\beta^2 m\} - \exp\{-t_p^2/\beta^2 m\}\right] \end{aligned} \tag{4.38}$$

Substituting equation (4.37) into equation (4.38) simplifies it to

$$Q_{m,T} = \frac{m\nu\beta T}{t_p(m,T)} \exp\left\{-[t_p(m,T) - T]^2/\beta^2 m\right\} \tag{4.39}$$

To investigate the influence of duration $T$ on the scaling behavior of $Q_{m,T}$ with respect to magnitude, first consider the behavior of $t_p$ in two limiting cases: as $T \to 0$ and as $T \to \infty$. In the first case, it can be checked from equation (4.39) that $t_p \sim \beta\sqrt{m}$. Substituting this expression into equation (4.39) gives $Q_{m,T} \sim \sqrt{m}$. By contrast, as $T \to \infty$, so does $t_p$. Moreover, $t_p - T \to 0$ and $t_p/T \to 1$. Therefore, it follows from equation (4.39) that $Q_{m,T} \sim m$. The latter behavior has a simple physical interpretation, as explained earlier. It corresponds to the situation in which rainfall duration exceeds the time of concentration for a basin, so that the entire basin contributes to the peak flow. The appearance of the exponent $\frac{1}{2}$ for a short time, or a large magnitude, comes from the random-model assumption for the network topology. In the Peano-basin example, this exponent is $\log 3/\log 4$ instead of $\frac{1}{2}$.

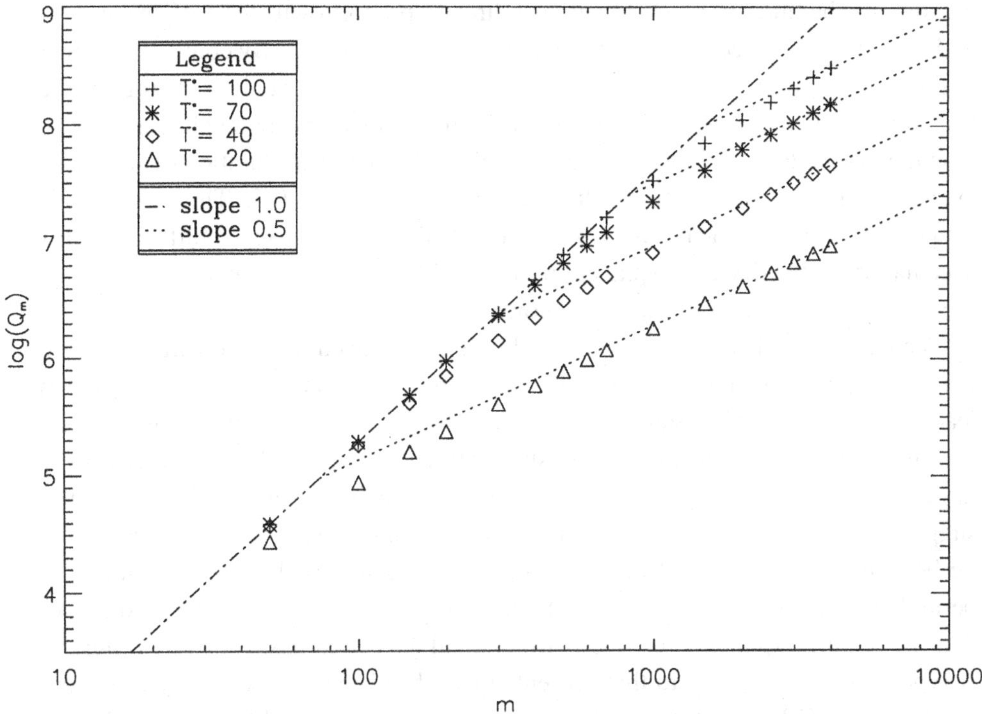


Figure 4.6. Schematic depiction of the effects of rainfall duration and basin size on the scaling exponents for floods.

The foregoing results can be applied to demonstrate how rainfall duration affects the scaling exponent for maximum mean flows. Let us fix a duration $T$ and consider the behavior of the scaling exponent as magnitude changes from $m_1$ to $m_2$, $m_1 \ll m_2$. Numerically solving equation (4.37) and then substituting it into equation (4.39) gives $Q_{m,T}$. Figure 4.6 shows the graph of $\log(Q_{m,T})$ versus $\log(m)$ for values of $m$ between 50 and 4,000, with $\nu\beta = 1$, for $T = 20$, 50, 75, and 100. Interestingly, one can see from this figure that small-magnitude basins up to some threshold magnitude $m(T)$ exhibit the long-duration behavior with a scaling exponent of 1. In subbasins with magnitudes $m > m(T)$, the scaling exponent is approximately $\frac{1}{2}$, which corresponds to the short-duration behavior. Figure 4.6 also suggests that as a first-order approximation, the small transition zone where the exponents lie between $\frac{1}{2}$ and 1 can be ignored.

As an application of the effect of duration on the flood exponent depicted in Figure 4.6, it is very tempting to speculate about a physical explanation for the computed exponent values for the New Mexico regions 1 and 5 described in Section 4.3. Recall from our discussion there that region 5 is mountainous, where floods are generally produced from snowmelt runoff. A part of region 5 constitutes the headwaters of the Canadian River, and it is physically reasonable to suppose that these headwater basins approximately satisfy the small-magnitude, long-duration condition. Therefore the runoff is almost directly proportional to the drainage area, and indeed this is reflected in the exponents being around 0.92. By contrast, region 1 is downstream from region 5, and many of its mainstream stations, being on the Canadian River, receive flows from region 5. These subbasins are much larger in size than the headwater basins in region 5. Therefore, for the same duration of snowmelt events as region 5, they should exhibit an exponent close to 0.5. This argument is supported by the empirically observed values of 0.56. Of course, to further test this speculation and verify our explanation, it would be necessary to undertake a much more detailed data analysis than that shown here.

The main point that we wish to make through the preceding demonstration is that a rigorous, physically based statistical theory of regional-flood frequency is certain to lead to a much better understanding of the hydrology than are pure statistical analyses, in which the statistical parameters are simply fitted from data. For example, a fairly large body of engineering literature, as cited by Robinson and Sivapalan (1997), suggests that the empirical values for the flood scaling exponent $\theta$ are observed to lie between 0.6 and 0.9, but that literature does not address how these values can be understood and predicted physically, as illustrated here. Figure 4.6 also shows that the scaling exponents for floods from small and large basins can be different. A type of analysis similar to that presented here has been carried out by Robinson and Sivapalan (1997) to illustrate that the empirically observed differences in the CV values for peak flows between large and small basins in Figure 4.3 have a physical basis. However, that demonstration assumes a multiscaling structure in the spatial variability of rainfall, which is the topic of the next section.

## 4.6 Spatial Variability and Multiscaling in Peak Flows

Recall from our discussion in Section 4.3 that the rainfall-generated regional peak flows for regions 2 and 4 in New Mexico did not exhibit a constancy of slope with respect to the return period. This feature cannot be accommodated within the framework of statistical simple scaling. More precisely, preliminary analysis of a variety of data sets suggests that although equation (4.15b) governing the log-log linearity between the statistical moments and a scale parameter holds, the linear growth of slope for simple scaling in it does not (Gupta and Waymire, 1990; Gupta et al., 1994). Smith (1992) carried out a maximum-likelihood test of simple scaling versus multiscaling using the regional-flood data from the Appalachian region in the United States and rejected the hypothesis of simple scaling. Our main objective in this section is to give the reader an orientation regarding some of the ideas currently being explored in connection with developing a physical basis for the spatial variability in regional floods within the multiscaling framework.

We introduce the statistical notion of multiscaling in Section 4.6.1 and explain how it fits with the empirically observed features in the regional-flood data. This is followed in Section 4.6.2 by a brief overview of the structure of the space–time variability in rainfall, past efforts to model this variability, and an application of the newly developing theory of random cascades to understand this variability. Section 4.6.3 briefly describes a study illustrating how multiscaling applies to the peak flows in a Peano basin via the spatial variability in rainfall. It solves equation (4.1) under the assumption that the runoff field for a hillslope is given by $R(e, t) = R(e)$, for $t = 0$ with $e \in \tau$, and $R(e, t) = 0$ otherwise. The statistical structure of the spatial field $R(e)$ is specified by a random cascade. Some of the unsolved problems that arise in this connection are also explained.

### *4.6.1 Statistical Multiscaling*

A formal way to construct a multiscaling random field, that exhibits (1) log-log linearity between the statistical moments of order $n$ and a scale parameter such that (2) the slope $s(n)$ is a nonlinear function of $n$ is to assume that the following equality holds:

$$Q(\lambda A) \overset{d}{=} G(\lambda) Q(A) \tag{4.40}$$

Here $G(\lambda)$ is a positive random function that is statistically independent of $\{Q(A)\}$. Unlike the case for simple scaling, we get two representations of $G(\lambda)$ corresponding to whether $\lambda < 1$ or $\lambda > 1$. Here we consider only the case $\lambda < 1$, because it is most relevant to understanding the spatial variability in rainfall and floods. However, the argument for the case $\lambda > 1$ is almost identical with the case $\lambda < 1$, and we leave it as an exercise for the reader; see Gupta and Waymire (1990) for further discussion regarding these two representations.

Equation (4.40) can be formally iterated with respect to the two positive scalars $\lambda_1 < 1$ and $\lambda_2 < 1$, in a manner similar to that discussed for simple scaling in Section 4.4. This gives a functional equation:

$$G(\lambda_1 \lambda_2) \overset{d}{=} G(\lambda_1) G(\lambda_2) \tag{4.41}$$

Because $G(\lambda)$ is a random function of $\lambda$, the equality in equation (4.41) holds in the sense of probability distributions. Taking logs,

$$\log\{G(\lambda_1 \lambda_2)\} \overset{d}{=} \log\{G(\lambda_1)\} + \log\{G(\lambda_2)\} \tag{4.42}$$

Let $\{Z(t) : t > 0\}$ be a stochastic process defined by

$$Z(t) = \log\{G(e^{-t})\}$$

or, equivalently,

$$Z[\log(1/t)] = \log\{G(t)\} \qquad (0 < t < 1) \tag{4.43}$$

Then equation (4.42) can be expressed as

$$Z\{\log(1/\lambda_1) + \log(1/\lambda_2)\} \stackrel{d}{=} Z\{\log(1/\lambda_1)\} + Z\{\log(1/\lambda_2)\}$$

which shows that $\{Z(t) : t > 0\}$ has "additive growth," in the sense of

$$Z(t+s) \stackrel{d}{=} Z(t) + Z(s) \tag{4.44}$$

Equivalently, $\{Z(t) \pm \mu t : t > 0\}$ also has additive growth, where $\mu$ is an arbitrary scalar. It now follows from equation (4.43) that the random scaling function $G(\lambda)$ can be represented as

$$G(\lambda) = \exp\{\pm\mu \log\lambda + Z[\log(1/\lambda)] \qquad (0 < \lambda < 1) \tag{4.45}$$

Any stochastic process with stationary increments, such as a BM process, has additive growth; however, the converse need not hold.

To illustrate the behavior of the moments for equation (4.45), let us take $\{Z(t) : t > 0\}$ to be a BM process starting at zero [i.e., $Z(0) = 0$]. This means that $Z(t)$ has independent increments and has a normal distribution, with mean zero and variance $\sigma^2 t$, where $\sigma^2 > 0$. Using the well-known expression for the moment-generating function of a normal random variable, it follows that

$$E[G^n(\lambda)] = \exp\left\{\mu n \log\lambda - \frac{1}{2}\sigma^2 n^2 \log\lambda\right\} \tag{4.46}$$

Equation (4.46) exhibits (1) log-log linearity between the moments and the scale parameter $\lambda$ and (2) a nonlinear slope function $s(n) = \mu n - \frac{1}{2}\sigma^2 n^2$; see Gupta and Waymire (1990) for other examples of this kind. These two features of a stochastic process or a random field are being called *multiscaling*, to distinguish them from the simple-scaling stochastic process described in Section 4.4. Multiscaling is also a scale-invariant property.

It is not difficult to determine how the quantiles of a multiscaling stochastic process behave with respect to a scale parameter. The derivation given by Gupta et al. (1994) shows that the log-log linearity between quantiles and the scale parameter holds only approximately, and the slopes in these plots decrease as the probability of exceedance decreases or as the return period increases. This feature was noted to be present in the flood data for regions 2 and 4 in New Mexico. However, as it is theoretically more convenient to work with moments instead of quantiles, in the subsequent discussion we shall primarily use moment expressions similar to that in equation (4.46). The reader can refer to Gupta et al. (1994) for a derivation of the foregoing result for quantiles and for a demonstration of how well equation (4.46) exhibits the scaling of moments in the Appalachian regional-flood data.

A physical basis for the multiscaling theory of regional floods involves computing the random scaling function $G(\lambda)$ in equation (4.40) in terms of the space–time variability of rainfall, the channel-network geometry (e.g., the width function), and the runoff routing parameters in river networks. In addition, a computation of the distribution of the runoff field from hillsides, $R(e, t),\ t \geq 0, e \in \tau$, in terms of spatially variable runoff-generation processes among hillsides is also required. In this sense, the research program for developing a physical basis for multiscaling in peak flows follows a path that is in general similar to that illustrated for simple scaling in Section 4.5. However, the mathematical framework for multiscaling is more technical than that for simple scaling, and the theory is still under development (Waymire and Williams, 1995). For these reasons, as well as because of space limitation, here we give only a brief nontechnical overview of this theory.

### *4.6.2 Spatial Variability in Rainfall and Random Cascades: A Brief Overview*

Rainfall intensity can vary substantially over space and time. A widely observed hierarchical spatial structure of rainfall-intensity fields within synoptic-scale regions can be described as follows. The smallest spatial structures consist of clusters of regions of high-intensity rainfall, called "cells," that are embedded within regions of low rainfall intensity, called small mesoscale areas (SMSAs). These, in turn, are clustered within still larger regions of even lower rainfall intensity, called large mesoscale areas (LMSAs). The LMSAs are contained within synoptic regions that typically are larger than $10^4$ km$^2$. This organization is shown schematically in Figure 4.7.

Some of the early empirical evidence for this kind of structure came from Austin and Houze (1972), who observed hierarchical structure within storms of a variety of synoptic types occurring in New England. In the Global Atmospheric Research Program (GARP) Atlantic Tropical Experiment (GATE), carried out in the tropical Atlantic in the mid-1970s, precipitation was observed to follow a life cycle, that at its peak included clusters of convective cells embedded within stratiform rainfall. It has been found that groups of thunderstorms embedded within a mesoscale system, termed *mesoscale convective complexes*, are major contributors of summertime rainfall over the central United States. The regions of distinct rainfall intensities shown in Figure 4.4 are observed to have distinct life cycles. For example, the rainfall cells typically have a life cycle of about half an hour to an hour. During that period, initially a cell experiences an increase in rainfall intensity, reaching a maximum, and finally

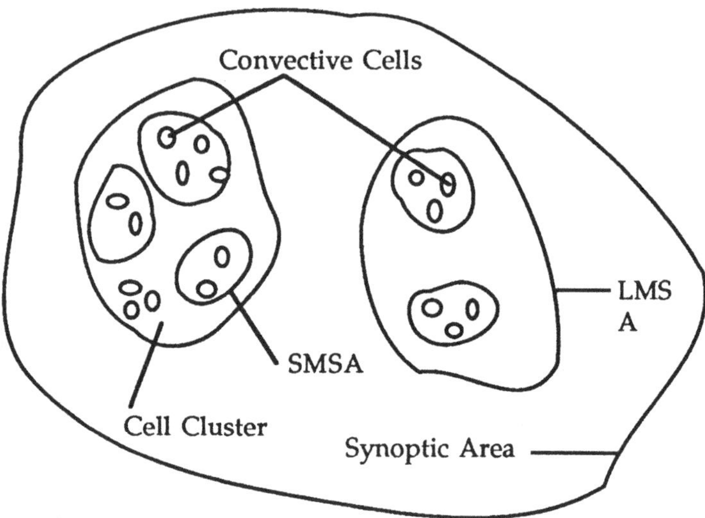


Figure 4.7. Schematic display of hierarchical organization in the spatial field of a rainfall event. (From Gupta and Waymire, 1979, with permission.)

it dissipates. Similarly, the SMSAs have life cycles of about 2–4 hours, and LMSAs last 6–12 hours. A review of the observational evidence for such systems and their effects has been published by Houze (1989).

This somewhat idealized geometric structure for rainfall has guided the development of statistical models for the space–time variabilities of rainfall over the past 20 years. However, some of the first steps in that direction were taken more than three decades ago by LeCam (1961). Two distinct sets of theories began to be developed during the late 1970s, continuing into the 1980s and 1990s. The approaches in the first set were based on the assumption of distinct spatial scales, as depicted in Figure 4.7, and corresponding time scales for rainfall (Waymire, Gupta and Rodríguez-Iturbe, 1984). By contrast, theories in the second set assumed that rainfall followed no characteristic spatial scale and that scale invariance in the form of statistical simple scaling held (Lovejoy and Mandelbrot, 1985). It turned out that neither of those sets of theories could adequately describe the spatial variability in rainfall fields (Schertzer and Lovejoy, 1987; Gupta and Waymire, 1993).

Schertzer and Lovejoy (1987) first began to explore the theoretical framework of random cascades and multifractal measures to describe the spatial variability of rainfall. This framework has continued to be developed and tested on rainfall data sets. It also eliminates some of the difficulties that were encountered with the two previous sets of approaches. There are two types of cascade theories, continuous and discrete, in the literature. Our focus here is on discrete cascades, and the reasons for this focus will become clear from the subsequent discussion. For an introduction to the theory of continuous cascades, readers should consult the work of Lovejoy and Schertzer (1990) and Tessier, Lovejoy, and Schertzer (1993). A recent review article on progress in modeling and data analysis for the space–time variabilities in rainfall discusses many other developments that are not covered in this chapter (Foufoula-Georgiou, 1998).

The theory of cascades and multifractal measures arose from two distinct sources. The theory of deterministic multifractal measures had its origin in nonlinear dynamical systems, where as the theory of random multifractal measures was derived from the statistical theory of fully developed turbulent flows governing transfer of energy from some large spatial scale to smaller scales (Kolmogorov, 1941, 1962). The multifractal formalism arose naturally in connection with Kolmogorov's lognormal model (Kolmogorov, 1962); see also Mandelbrot (1974) and She and Waymire (1995). Applications of multifractal theory to a variety of other physical systems (viscous fingering in porous media, Rayleigh-Benard convection, diffusion-limited aggregation etc.) have been described by Feder (1988).

A discrete random cascade is constructed on a $d$-dimensional bounded region, which can be taken to be the unit square $J = [0, 1]^2$ in the context of rainfall distribution. It redistributes some initial mass, say $\mu_0$, over $J$ by successive subdivisions of space into smaller regions and multiplication by IID nonnegative random variables $W$ called *cascade generators*. The generators must satisfy some type of mass-conservation condition through this redistribution process as a necessary condition for the mathematics of random cascades to apply. For example, one of the simplest conditions is to assume that mass is conserved on the average, that is, $E[W] = 1$. A simple random-cascade model that is proving to be very valuable in modeling the spatial variability of rainfall, particularly its spatial intermittence, is the *beta model* (Over and Gupta, 1994). It was introduced by Frisch, Sulem, and Nelkin (1978) in connection with the statistical theory of turbulence. The probability distribution for the generators in the beta model can be defined as

$$P(W = 0) = 1 - b^{-\beta}, \qquad P(W = b^{\beta}) = b^{-\beta} \tag{4.47}$$

which implies that $E[W] = 1$ holds for the beta model. The parameter $b$ is called the *branching number*, because each square at each level is subdivided into $b$ subsquares. In our context, $b = 4$. The length scale for a subsquare at the $n$th generation is simply $\lambda_n = b^{-n/2}$. In the limit as $n \to \infty$, the redistributed random mass $\mu_\infty$ exhibits a very spiky and intermittent spatial structure. A schematic for a cascade construction on a unit interval with $b = 2$ is shown in Figure 4.8.

Let us illustrate how IID cascade generators give rise to multiscaling. Let $\Delta_n^i$ denote a typical pixel in a unit square at the spatial scale $\lambda_n = b^{-n/2}$ [i.e., $\Delta_n^i = \lambda_n^2 = b^{-n}$, $i = 1, 2, \ldots, b^n$]. The mass on this pixel in the limit as $n \to \infty$ is given by a *basic recursion equation* for cascades (Holley and Waymire, 1992):

$$\mu_\infty(\Delta_n^i) \stackrel{d}{=} Z_\infty(i)\mu_n \stackrel{d}{=} Z_\infty(i) W_1 \ldots W_n b^{-n} \qquad (i = 1, 2, \ldots, b^n) \tag{4.48}$$

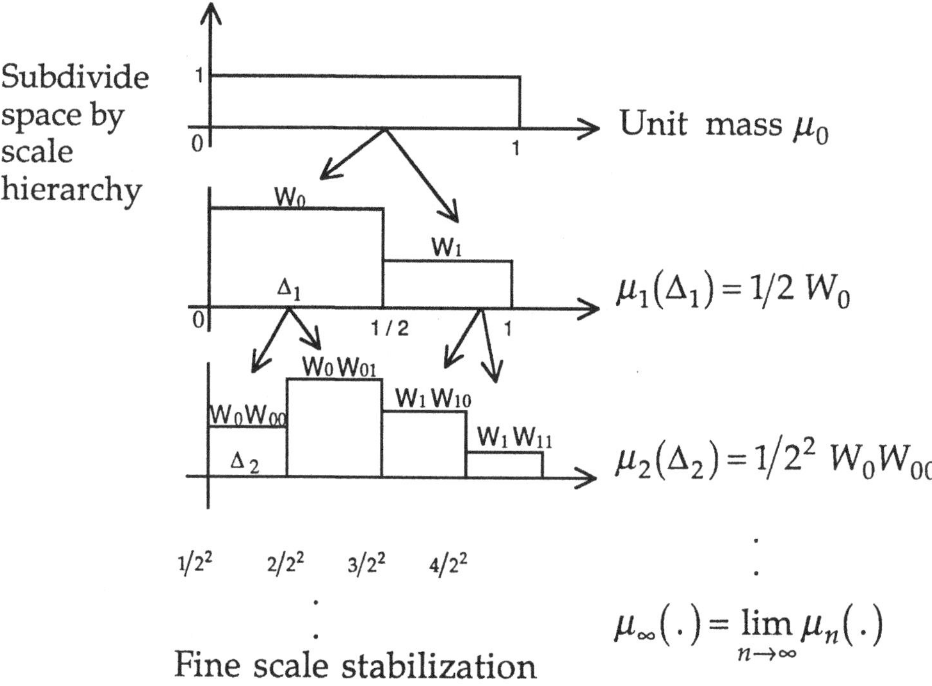


Figure 4.8. Schematic of a cascade construction on a unit interval.

where, the random variable $Z_\infty$ is distributed as the total limiting mass on $J$ and is statistically independent of the mass $\mu_n$. Equation (4.48) can be understood as follows. The effect of spatial variability below the spatial scale $\lambda_n$ is reflected in the high-frequency term $Z_\infty$, whereas $\mu_n$ represents the low-frequency effect above the scale $\lambda_n$. Equation (4.48) depicts the multiplicative structure of the cascade mass $\mu_\infty$.

Let us define rainfall intensity as, $R_\lambda = \mu_\lambda/\lambda^2$, which corresponds to the mass density up to division by a time constant. It follows from equation (4.48) that the $h$th moment of $R_{\lambda_n}$ at scale $\lambda_n$ can be computed as

$$E\left[R^h_{\lambda_n}\right] = \frac{E\left[\mu^h_\infty(\Delta^i_n)\right]}{b^{-nh}} = E\left[Z^h_\infty\right]E^n[W^h] \tag{4.49}$$

Taking $\log_b$ on both sides of equation (4.49), we get

$$\log_b E\left[R^h_{\lambda_n}\right] = \log_b E\left[Z^h_\infty\right] - \kappa(h)\log_b \lambda^2_n \tag{4.50}$$

This is a log-log linear relationship between the statistical moments and the scale parameter, such that the slope function

$$\kappa(h) = \log_b E[W^h] = \chi_b(h) + (h-1) \tag{4.51}$$

is nonlinear in $h$. This nonlinearity in $\kappa(h)$ distinguishes multiscaling in a process, as illustrated in Section 4.6.1, from simple scaling. The slope function for the beta model can be easily computed from equation (4.47) as

$$\kappa(h) = \chi_b(h) + (h - 1) = (h - 1)\beta \tag{4.52}$$

This equation is nonlinear because of a nonzero intercept, $-\beta$. Gupta and Waymire (1993) give many examples of $\chi_b(h)$ for different models of cascade generators.

The function, $\chi_b(h)$ plays a basic role in a variety of calculations involving random cascades. For example, Kahane and Peyriere (1976) showed that a necessary and sufficient condition for the limiting mass $\mu_\infty$ to be nondegenerate is given by $\chi_b'(1) < 0$. Applying this condition to equation (4.52) gives $\beta < 1$ for the beta model. Kahane and Peyriere (1976) also showed that the subset of the unit square over which the limiting mass $\mu_\infty$ is distributed, called the *support*, is a fractal set with a *Hausdorff dimension*: $D = -2\chi_b'(1)$. The notion of the Hausdorff dimension is somewhat technical (Feder, 1988), and the reader can take it to be the box-counting dimension in this section. It follows from equation (4.52) that the dimension of the support in the beta model is

$$D_\beta = 2 - 2\beta \qquad (0 < \beta < 1) \tag{4.53}$$

The parameter $2\beta$ is called the codimension of the support and is the difference between the dimension of the unit square and the dimension of the support for the cascade mass (Feder, 1988, p. 206). Readers can refer to Gupta and Waymire (1993) and Over (1995, ch. 2) for further mathematical details of random cascades.

### *4.6.3 A Physical Determination of the Multiscaling Peak-Flow Exponents in a Network with a Deterministic Self-similar Topology*

Figure 4.8 shows how a random cascade redistributes mass along a branching topology. This suggests that perhaps the naturally occurring scale-invariant branching structure of a river network can be used within a cascade theoretical framework for distributing rainfall intensities over a basin and all of its subbasins. Our objective in this section is to illustrate how the topology of a Peano network combines naturally with a cascade structure for rainfall and how the geometry of network and rainfall appears in the multiscaling exponents for peak flows (Gupta et al., 1996). We also highlight some important issues that arise in extending this idealized example to networks that are physically more realistic than a Peano basin.

The example is based on solving equation (4.1) for a finite Peano-basin channel network under the previously stated assumption that $S(e, t) = 0$, and $R(e, t) = R(e)$ for $t = 0$, with $e \in \tau$, and $R(e, t) = 0$ otherwise, is specified by a random cascade. Assume that the runoff coefficient $\pi_1 = 1$; so $R(e)$, with $e \in \tau$, denotes an instantaneously applied runoff-intensity field over the network. Because a random cascade generates a spatial random field, its specification on a drainage basin mandates that the network be embedded in a two-dimensional (or three-dimensional) space. The Peano network considered here (Figure 4.4) (Mandelbrot, 1983, ch. 7; Marani et al., 1991) is embedded in a two-dimensional space. Nonetheless, the general problem of spatial embedding remains unsolved, and that adds a major new dimension to the overall problem of hydrologic regionalization. Only recently has the stochastic modeling of networks begun to address the problem of spatial embedding; see the review article by Troutman and Karlinger (1998) on spatial-network models.

The problem of the distribution of rainfall over a Peano basin is simple, because we can easily use the network partitioning to construct a rainfall cascade. A key question that arises is whether or not that choice makes any theoretical sense. Surprisingly, a general result reported by Over (1995) essentially says that for a broad class of cascade generators, including the beta and the lognormal, the choice of the branching number is arbitrary, and it can be selected according to one's convenience. That result, stated precisely by Gupta and Waymire (1998), provides an important avenue for analysis of spatially distributed rainfall fields in channel networks which are more realistic than a Peano basin.

Even under the simplified conditions stated earlier, a precise theoretical approach to computing statistics for peak flows in a Peano network, as a solution for equation (4.1), remains an unsolved problem. Extensive numerical simulations for beta-cascade rainfall in a Peano basin were carried out by Gupta et al. (1996). Their findings suggested the key approximation that up to moderately large values of $\beta$, the peak-flow statistics for different scales in a Peano network can be approximated by the statistics of peak flows coming from the maximum contributing set. Recall from Section 4.5.1 that $Q_n(e_\omega)$ denotes peak flow at the bottom of a stream of order $\omega = 1, 2, \ldots, n+1$ in a level-$n$ Peano network. Let $\tilde{Q}_n(\lambda_\omega)$, $\omega = 1, 2, \ldots, n+1$, denote the peak flow generated by the maximum contributing set at the scale $\lambda_\omega$. On the basis of numerical simulations reported by Gupta et al. (1996) we can replace $Q_n(e_\omega)$ by $\tilde{Q}_n(\lambda_\omega)$, $\omega = 1, 2, \ldots, n+1$, for small and moderately large values of $\beta$. Moreover, for the spatially uniform inputs treated in Section 4.5.1, $\tilde{Q}_n(\lambda_\omega) = Q_n(e_\omega)$.

Let $E[\tilde{Q}_n^h(\lambda_\omega)]$ and $C_h$ represent the $h$th statistical moments of the peak flow in subbasins at the $\omega$th and the $n$th scales, respectively. Then

$$\log_4 E\left[\tilde{Q}_n^h(\lambda_\omega)\right] = C_h + \left\{\left(\frac{\log 3}{\log 2} - 2\beta\right)h + 2\beta\right\}\log_4(\lambda_\omega) \tag{4.54}$$

This shows a multiscaling property for peak flows via (1) log-log linearity between the statistical moments and the scale parameter and (2) a nonlinear slope function. Moreover, the scaling exponent $(\log 3/\log 2) - 2\beta$ combines the spatial geometry of rainfall and that of a Peano network, and this result makes it possible to test the theory. The peak-flow scaling exponent can be geometrically interpreted as the box-counting (Hausdorff) dimension of the intersection of the maximum contributing set of the Peano basin with the fractal set supporting the rainfall distribution from a beta-cascade model. The result in equation (4.54) easily generalizes to other cascade generators for rainfall (e.g., the lognormal). However, the analytical calculations involving the multiscaling exponents for peak flows under spatial aggregation, leading to equation (4.54), are somewhat lengthy and technical, and we refer the reader to Gupta et al. (1996) for details.

A generalization of this example to spatially variable runoff from hillslopes due to spatial variabilities in rainfall, topographic slopes, and saturated-soil hydraulic conductivity has not been carried out. Robinson and Sivapalan (1997) assumed a multiscaling spatial structure for rainfall and then advanced some physical arguments to illustrate that the interaction between rainfall duration and time to peak flow is a major factor in producing the empirically observed differences in CV values for peak flows between large and small basins, as shown in Figure 4.3. Many other interesting problems regarding further development of this theory have been discussed by Gupta and Waymire (1996, 1998) and Gupta et al. (1996).

## 4.7 A Perspective on Future Research

Our discussion and illustrations show that the currently popular theme of spatial variability arises naturally within the context of hydrologic regionalization. Notions of scale invariance and scale dependence in spatial variability provide a natural foundation on which disjoint theories regarding the components of the hydrologic cycle in river basins can be unified within a precisely defined theoretical framework. Unification is a common goal in all of science, and it is a rich scientific theme in river-basin hydrology. It is our view that further research on regionalization can be expected to lead to exciting new developments in the science of hydrology in the next decade and beyond.

For example, a generalization from the short time scales of a single rainfall event to longer time scales involving multiple events requires that the effect of hillside storage on runoff generation be analyzed; see equation (4.2). A key contribution to the storage term in equation (4.2) comes from soil moisture, which governs land-climate-vegetation interactions. As we mentioned in the introduction, this issue was investigated by Eagleson (1978) in the context of annual water balance at small spatial scales. The set of coupled equations (4.1) and (4.2) provides a framework to significantly generalize Eagleson's theory from small spatial scales to a wide range of spatial scales for drainage basins. In this respect, the first and foremost problem is to develop a space–time stochastic theory of rainfall intensity $I_r(e, t)$ on a channel network. Efforts have already begun to generalize the spatial-cascade theory from space to space–time (Over and Gupta, 1996). Such a theory of rainfall would enable us to investigate under what conditions on the terms $E_T$ and $\tilde{R}$ the space–time stochastic differential equation (4.2) admits a steady-state probability density. Answers to these types of questions would serve as the building blocks to understand the annual water balance on drainage basins via *network-land-climate-vegetation interactions*.

Space–time variability in rainfall and evapotranspiration over larger spatial scales can be also influenced by large-scale atmospheric forcing. For example, analysis of tropical oceanic rainfall data by Over and Gupta (1994) shows that the $\beta$ cascade parameter in a spatial rainfall model depends on large-scale spatial-average rainfall, which is a signature of large-scale atmospheric forcing. This dependence makes the time evolution of space–time rainfall a nonstationarity random field, and this feature influences the runoff field via equation (4.2). Indeed, some empirical analyses have independently shown that the El Niño Southern Oscillation (ENSO) influences flood frequencies (Cayan and Webb, 1992; Waylen and Caviedes, 1986) in certain regions of the world. Equations (4.1) and (4.2) furnish a concrete context for investigating how seasonal, interannual, interdecadal, and other longer-time-scale climatic variabilities influence the hydrologic extremes on drainage basins.

However, no theory can long stand without careful tests against field data. Two major points must be made regarding this important issue. First, coordinated field observations are critically needed over time scales on the order of a rainfall event. These must include space–time intensity measurements of rainfall from radar and rain gauges, spatially distributed streamflow hydrographs, spatial measurements of surface topography, channel-network geometry, and hydraulic geometry, and soil characteristics (e.g., infiltration rates and water-table positions) on several hillslopes so that the spatial and ensemble characteristics of runoff generation from individual hillslopes can be developed and tested. Such data sets have not yet been compiled. The second issue concerns the testability of the physical-statistical theories. Until appropriate data sets become available, preliminary tests of physical-statistical regional theories can be carried out using the available peak-flow and low-flow data set. This was illustrated for peak flows in Section 4.5. A similar analysis for regional low flows has been carried out by Furey and Gupta (1998). Such a program not only would help with the design of new field experiments and monitoring programs but also would require that models and simulations be used to formulate and test new hypotheses. Moreover, such a field program would not be limited to only physical aspects, but would include ecological, geochemical, and water-quality aspects.

## Acknowledgments

We gratefully acknowledge many insightful discussions over the past 20 years with Dave Dawdy, Ignacio Rodríguez-Iturbe, Brent Troutman, Shaun Lovejoy, and Daniel Schertzer. Their ideas and contributions have been major influences in our thinking.

The late Steven Mock of the Army Research Office supported our vision in this line of research in the early 1980s, when our ideas were at best sketchy and the path was unclear. Without the innovative energy of our former and current graduate students this body of research would not have been possible. We gratefully acknowledge the comments of Murugesu Sivapalan, Bryson Bates, an anonymous referee, and the book editor, Gary Sposito, on this chapter. V.K.G. also acknowledges support and hospitality, during sabbatical, from LTHE, Grenoble, France, and the University of Western Australia, Nedlands, Australia, where parts of this chapter were written. This research was supported by grants from the National Aeronautics and Space Administration and the National Science Foundation.

## References

Austin, P. M., and Houze, R. A. 1972. Analysis of the structure of precipitation patterns in New England. *J. Appl. Meteorol.* 11:926–35.

Bell, T. L. 1987. A space–time stochastic model of rainfall for satellite remote sensing studies. *J. Geophys. Res.* 92:9631–43.

Benson, M. A. 1962. *Factors Influencing the Occurrence of Floods in a Humid Region of Diverse Terrain.* USGS water supply professional paper 1580B.

Bobee, B., and Rasmussen, P. 1995. Recent advances in flood frequency analysis. *Rev. Geophys.* [*Suppl.*] 33:1111–6.

Burd, G., and Waymire, E. 1997. On stochatic network self-similarity. Preprint.

Castro, S. 1998. Scaling exponents of floods from scale-invariant rainfall and river networks. M.S. thesis, University of Colorado.

Cayan, D. R., and Webb, R. H. 1992. El-Niño/Southern Oscillation and streamflow in the western United States. In: *El Niño*, ed. H. Diaz and V. Markgraf, pp. 29–68. Cambridge University Press.

Dawdy, D. R. 1961. *Variation of Flood Ratios with Size of Drainage Area.* U.S. Geological Survey Research, 424-C, paper C36. Reston, VA: USGS.

Dawdy, D., and Gupta, V. K. 1995. Multiscaling and skew separation in regional floods. *Water Resour. Res.* 31:2761–7.

Dietrich, W. E., and Dunne, T. 1993. The channel head. In: *Channel Network Hydrology*, ed. K. Beven and M. J. Kirkby, pp. 190–219. New York: Wiley.

Dooge, J. C. I. 1973. *Linear Theory of Hydrologic Systems.* Technical bulletin 1468. Wasington, DC: Agricultural Research Service, U.S. Department of Agriculture.

Duffy, C. 1996. A two state integral balance model for soil moisture and groundwater dynamics in complex terrain. *Water Resour. Res.* 32:2421–34.

Dunne, T. 1982. Models of runoff processes and their significance. In: *Scientific Basis of Water Resource Management*, pp. 17–31. Washington, DC: National Academy Press.

Eagleson, P. 1970. *Dynamic Hydrology.* New York: McGraw-Hill.

Eagleson, P. 1972. Dynamic of food frequency. *Water Resour. Res.* 8:878–98.

Eagleson, P. 1978. Climate, soil and vegetation. 1. Introduction to water balance dynamics. *Water Resour. Res.* 14:705–12.

Feder, J. 1988. *Fractals.* New York: Plenum.

Foufoula-Georgiou, E. 1998. On stochastic theories of space–time rainfall; recent results and open problems. In: *Advanced Series on Statistical Sciences and Applied*

*Probability*, Vol. 7: *Stochastic Methods in Hydrology: Rainfall, Landforms and Floods*, ed. O. E. Barndorff-Nielsen, V. K. Gupta, V. Perez-Abreu, and E. C. Waymire. Singapore: World Scientific.

Freeze, R. A. 1980. A stochastic conceptual model of rainfall-runoff processes on a hillslope. *Water Resour. Res.* 16:391–408.

Frisch, U., Sulem, P. L., and Nelkin, M. 1978. A simple dynamical model of fully developed turbulence. *J. Fluid Mech.* 87:719–36.

Furey, P., and Gupta, V. K. 1998. A physical-statistical approach to regionalizing low flows (preprint).

Gupta, V. K., Castro, S., and Over, T. M. 1996. On scaling exponents of spatial peak flows from rainfall and river network geometry. *J. Hydrol.* 187:81–104.

Gupta, V. K., and Dawdy, D. R. 1994. Regional analysis of flood peaks: multiscaling theory and its physical basis. In: *Advances in Distributed Hydrology*, ed. R. Rosso, A. Peano, I. Becchi, and G. A. Bemporad, pp. 149–68. Highlands Ranch, CO: Water Resources Publications.

Gupta, V. K., and Dawdy, D. R. 1995. Physical interpretations of regional variations in the scaling exponents of flood quantiles. *Hydrological Processes* 9:347–61.

Gupta, V. K., and Mesa, O. 1988. Runoff generation and hydrologic response via channel network geomorphology: recent progress and open problems. *J. Hydrol.* 102:3–28.

Gupta, V. K., Mesa, O., and Dawdy, D. R. 1994. Multiscaling theory of flood peaks: regional quantile analysis. *Water Resour. Res.* 30:3405–21.

Gupta, V. K., Mesa, O., and Waymire, E. 1990. Tree dependent extreme values: the exponential case. *J. Appl. Prob.* 27:124–33.

Gupta, V. K., and Waymire, E. 1979. A stochastic kinematic study of subsynoptic space–time rainfall. *Water Resour. Res.* 15:637–44.

Gupta, V. K., and Waymire, E. 1989. Statistical self-similarity in river networks parameterized by elevation. *Water Resour. Res.* 25:463–76.

Gupta, V. K., and Waymire, E. C. 1990. Multiscaling properties of spatial rainfall and river flow distributions. *J. Geophys. Res.* 95:1999–2009.

Gupta, V. K., and Waymire, E. C. 1993. A statistical analysis of mesoscale rainfall as a random cascade. *J. Appl. Meteorol.* 12:251–67.

Gupta, V. K., and Waymire, E. C. 1996. Multiplicative cascades and spatial variability in rainfall, river networks and floods. In: *Reduction and Predictability of Natural Disasters*, ed. J. Rundle, W. Klein, and D. Turcotte, pp. 71–95. Reading, MA: Addison-Wesley.

Gupta, V. K., and Waymire, E. C. 1998. Some mathematical aspects of rainfall, landforms and floods. In: *Advanced Series on Statistical Sciences and Applied Probability*, Vol. 7: *Stochastic Methods in Hydrology: Rainfall, Landforms and Floods*, ed. O. E. Barndorff-Nielsen, V. K. Gupta, V. Perez-Abreu, and E. C. Waymire. Singapore: World Scientific.

Gupta, V. K., Waymire, E., and Wang, C. T. 1980. A representation of an instantaneous unit hydrograph from geomorphology. *Water Resour. Res.* 16:855–62.

Holley, R., and Waymire, E. 1992. Multifractal dimensions and scaling exponents for strongly bounded random cascades. *Ann. Appl. Prob.* 2:819–45.

Horton, R. 1984. Erosional development of streams and their drainage basins; hydrophysical approach to quantitative morphology. In: *Benchmark Papers in Geology*, Vol. 80: *River Networks*, ed. R. S. Jarvis and M. J. Woldenberg, pp. 15–36. Stroudsburg, PA: Hutchinson Ross. (Originally published 1945.)

Houze, R. A. 1989. Observed structure of mesoscale convective systems and implications for large scale forcing. *Q. J. Roy. Meteor. Soc.* 115:525–61.

Jarvis, R. S., and Woldenberg, M. J. (eds.). 1984. *Benchmark Papers in Geology*, Vol. 80: *River Networks*. Stroudsburg, PA: Hutchinson Ross.

Kahane, J. P., and Peyriere, J. 1976. Sur certaines martingales de Benoit Mandelbrot. *Adv. Math.* 22:131–45.

Kinnison, H. B., and Colby, B. R. 1945. Flood formulas based on drainage-basin characteristics. *ASCE Trans.* 110:849–904.

Kirkby, M. 1984. Tests of random network model, and its application to basin hydrology. In: *Benchmark Papers in Geology*, Vol. 80: *River Networks*, ed. R. S. Jarvis and M. J. Woldenberg, pp. 353–68. Stroudsburg, PA: Hutchinson Ross. (Originally published 1976.)

Kolmogorov, A. N. 1941. Local structure of turbulence in an incompressible liquid for very large Reynolds numbers. *Comptes Rendus (Doklady) de l'academie des sciences de l'URSS* 30:301–5.

Kolmogorov, A. N. 1962. A refinement of previous hypotheses concerning the local structure of turbulence in a viscous inhomogeneous fluid at high Reynolds number. *J. Fluid Mech.* 13:82–5.

Kumar, P., Guttorp, P., and Foufoula-Georgiou, E. 1994. A probability-weighted moment test to assess simple scaling. *Stoch. Hydrol. Hydraul.* 8:173–83.

La Barbera, P., and Rosso, R. 1989. On the fractal dimension of stream networks. *Water Resour. Res.* 25:735–41.

Lee, M. T., and Delleur, J. W. 1976. A variable source area model of the rainfall-runoff process based on the watershed stream network. *Water Resour. Res.* 12:1029–35.

LeCam, L. 1961. A stochastic description of precipitation. In: *Fourth Berkeley Symposium on Mathematical Statistics and Probability*, vol. 3, pp. 165–86. Berkeley: University of California Press.

Leopold, L. B., and Miller, J. P. 1984. Ephemeral streams–hydraulic factors and their relation to the drainage net. In: *Benchmark Papers in Geology*, Vol. 80: *River Networks*, ed. R. S. Jarvis and W. J. Woldenburg, pp. 302–11. Stroudsburg, PA: Hutchinson Ross. (Originally published 1956.)

Leopold, L. B., Wolman, M. G., and Miller, J. P. 1964. *Fluvial Processes in Geomorphology*. San Francisco: Freeman.

Loague, K. M., and Gander, G. A. 1990. R-5 revisited: spatial variability of infiltration on a small rangeland watershed. *Water Resour. Res.* 26:957–71.

Lovejoy, S., and Mandelbrot, B. B. 1985. Fractal properties of rain and a fractal model. *Tellus* 37A:209–32.

Lovejoy, S., and Schertzer, D. 1990. Multifractals, universality classes, and satellite and radar measurements of cloud and rain fields. *J. Geophys. Res.* 95:2021–31.

Mandelbrot, B. B. 1974. Intermittent turbulence in self-similar cascades: divergence of high moments and dimension of the carrier. *J. Fluid Mech.* 62:331–58.

Mandelbrot, B. B. 1983. *The Fractal Geometry of Nature*. San Francisco: Freeman.

Marani, A., Rigon, R., and Rinaldo, A. 1991. A note on fractal channel networks. *Water Resour. Res.* 27:3041–9.

Melton, M. A. 1959. A derivation of Strahler's channel-ordering system. *J. Geology* 67:345–6.

Mesa, O. J., and Mifflin, E. 1986. On the relative role of hillslope and network geometry on basin response. In: *Scale Problems in Hydrology*, ed. V. K. Gupta, I. Rodríguez-Iturbe, and E. Wood, pp. 1–17. Dordrecht: Reidel.

National Research Council. 1988. *Estimating Probabilities of Extreme Floods: Methods and Recommended Research*. Washington, DC: National Academy Press.

National Research Council. 1992. *Regional Hydrology and the USGS Stream Gauging Network*. Washington, DC: National Academy Press.

Over, T. 1995. Modeling space–time rainfall at the mesoscale using random cascades. Ph.D. dissertation, Program in Geophysics, University of Colorado.
Over, T., and Gupta, V. K. 1994. Statistical analysis of mesoscale rainfall: dependence of a random cascade generator on large-scale forcing. *J. Appl. Meteorol.* 33:1526–42.
Over, T., and Gupta, V. K. 1996. A space–time theory of mesoscale rainfall using random cascades. *J. Geophys. Res.* 101:26,319–31.
Parzen, E. 1967. *Stochastic Processes.* New York: Wiley.
Peckham, S. 1995a. New results for self-similar trees with applications to river networks. *Water Resour. Res.* 31:1023–9.
Peckham, S. 1995b. Self-similarity in the three-dimensional geometry and dynamics of large river basins. Ph.D. dissertation, Program in Geophysics, University of Colorado.
Peckham, S., Gupta, V. K., and Smith, J. D. 1998. A dynamic self-similarity hypothesis and predictions for downstream hydraulic geometry, longitudinal river profiles, and Hack's law (preprint).
Potter, K. 1987. Research on flood frequency analysis: 1983–1986. *Rev. Geophys.* 25:113–18.
Rinaldo, A., and Rodríguez-Iturbe, I. 1996. Geomorphological theory of the hydrological response. *Hydrological Processes* 10:803–29.
Robinson, J. S., and Sivapalan, M. 1995. Catchment scale runoff generation model by aggregation and similarity analysis. In: *Scale Issues in Hydrologic Modeling*, ed. J. Kalma and M. Sivapalan, pp. 311–30. New York: Wiley.
Robinson, J. S., and Sivapalan, M. 1997. An investigation into the physical causes of scaling and heterogeneity in regional flood frequency. *Water Resour. Res.* 33:1045–59.
Rodríguez-Iturbe, I., Rinaldo, A. I., Rigon, R., Bras, R. L., Marani, A., and Ijjasz-Vasquez, E. 1992. Energy dissipation, runoff production, and the 3-dimensional structure of river basins. *Water Resour. Res.* 28:1095–103.
Rodríguez-Iturbe, I., and Valdez, J. B. 1979. The geomorphologic structure of hydrologic response. *Water Resour. Res.* 15:1409–20.
Rogowski, A. S. 1972. Watershed physics: soil variability criteria. *Water Resour. Res.* 8:1015–23.
Salvucci, G. D., and Entekhabi, D. 1995. Hillslope and climate control on hydrologic fluxes. *Water Resour. Res.* 31:1725–39.
Schertzer, D., and Lovejoy, S. 1987. Physical modeling and analysis of rain and clouds by anisotropic scaling multiplicative processes. *J. Geophys. Res.* 92:9693–714.
She, Z. S., and Waymire, E. C. 1995. Quantized energy cascade and log-Poisson statistics in fully developed turbulence. *Phys. Rev. Lett.* 74:262–5.
Shreve, R. L. 1967. Infinite topologically random channel networks. *J. Geology* 75:178–86.
Sivapalan, M., Wood, E. F., and Beven, K. J. 1990. On hydrologic similarity. 3. A dimensionless flood frequency model using a generalized geomorphic unit hydrograph and partial area runoff generation. *Water Resour. Res.* 26:43–58.
Smith, J. 1992. Representation of basin scale in flood peak distributions. *Water Resour. Res.* 28:2993–9.
Snell, J., and Sivapalan, M. 1995. Application of the meta-channel concept: construction of the meta-channel hydraulic geometry for a natural catchment. In: *Scale Issues in Hydrologic Modeling*, ed. J. Kalma and M. Sivapalan, pp. 241–61. New York: Wiley.
Tarboton, D. 1996. Fractal river networks, Horton's laws and Tokunaga cyclicity. *J. Hydrol.* 187:105–17.

Tarboton, D., Bras, R. L., and Rodríguez-Iturbe, I. 1989. Elevation and scale in drainage networks. *Water Resour. Res.* 25:2037–51.

Tessier, Y., Lovejoy, S., and Schertzer, D. 1993. Universal multifractals: theory and observations for rain and clouds. *J. Appl. Meteorol.* 32:223–50.

Tokunaga, E. 1966. The composition of drainage networks in Toyohira river basin and valuation of Horton's first law (in Japanese, with English summary). *Geophys. Bull. Hokkaido Univ.* 15:1–19.

Tokunaga, E. 1978. Consideration on the composition of drainage networks and their evolution. *Geogr. Rep., Tokyo Metrop. Univ.* 13:1–27.

Troutman, B., and Karlinger, M. 1984. On the expected width function of topologically random channel networks. *J. Appl. Prob.* 22:836–49.

Troutman, B., and Karlinger, M. 1998. Spatial channel network models in hydrology. In: *Advanced Series on Statistical Sciences and Applied Probability*, Vol. 7: *Stochastic Methods in Hydrology: Rainfall, Landforms and Floods*, ed. O. E. Barndorff-Nielsen, V. K. Gupta, V. Perez-Abreu, and E. C. Waymire. Singapore: World Scientific.

Waltemeyer, S. D. 1986. *Techniques for Estimating Flood-Flow Frequency for Unregulated Streams in New Mexico.* USGS Water Resources Investigations, report 86-4104.

Waylen, P. R., and Caviedes, C. N. 1986. El Niño and annual floods on the Peruvian littoral. *J. Hydrol.* 89:141–56.

Waymire, E. C., Gupta, V. K., and Rodríguez-Iturbe, I. 1984. A spectral theory of rainfall intensity at the meso-$\beta$ scale. *Water Resour. Res.* 20:1453–65.

Waymire, E., and Williams, S. 1995. Multiplicative cascades: dimension spectra and dependence. *J. Fourier Analysis Appl.* (special issue in honor of Jean-Pierre Kahane), pp. 589–609.

# 5

# An Emerging Technology for Scaling Field Soil-Water Behavior

DONALD R. NIELSEN, JAN W. HOPMANS, and KLAUS REICHARDT

## 5.1 Introduction

In 1955, Miller and Miller created a new avenue for research in soil hydrology when they presented their pioneering concepts for scaling capillary-flow phenomena. Their description of the self-similar microscopic soil-particle structure and its implications for the retention and transport of soil water stimulated many studies to test how well laboratory-measured soil-water retention curves could be coalesced into a single-scale mean function (e.g., Klute and Wilkinson, 1958). Because the results from ensuing tests were not particularly encouraging, except for soils composed of graded sands, their scaling concepts lay idle for several years. At that time, when the pressure-outflow method (Gardner, 1956) and other transient methods for estimating the value of the hydraulic conductivity in the laboratory were still in their infancy, few attempts were being made to scale the hydraulic-conductivity function, owing to the paucity of data from its quantitative observation. It was during that same period that the classic works of Philip (1955, 1957), describing a solution for the Richards equation, shifted attention to infiltration. With field measurements of soil-water properties only beginning to be reported (e.g., Richards, Gardner, and Ogata, 1956), little information was available on the reliability of their measurements or on the variations in their magnitudes to be expected within a particular field or soil mapping unit.

During the 1960s, the development and acceptance of the portable neutron meter to measure soil-water content spurred research into field-measured soil-water properties. Because of its availability, combined with the well-known technology of tensiometry, field studies of soil-water behavior accelerated in the 1970s. However, soil physicists were soon faced with a dilemma – how to deal with the naturally occurring variability in field soils (Nielsen, Biggar, and Erh, 1973) and concomitantly measure, within reliably prescribed fiducial limits, the much-needed soil-water functions associated with the Darcy-Buckingham equation and the Richards equation. Extension of the concepts of Miller scaling was thought to be a promising approach. The first purpose of this chapter is to provide a historical and pedagogic summary of the efforts to scale field-measured unsaturated soil-water regimes. The second purpose is to suggest additional incentives to continue research and development toward a reliable field technology for ascertaining soil-water functions and parameters based upon scaling concepts. With many different kinds of invasive and noninvasive techniques available today to measure soil water and related soil properties (Hopmans, Hendricks, and Selker, in press), the scaling approach continues to appear both promising and provoking.

## 5.2 The Principle of Miller Scaling

Scale-invariant relationships for water properties in homogeneous soils based upon the microscopic arrangements of their soil particles and the viscous flow of water within their pores were proposed by Miller and Miller (1955a,b). Each soil was assumed to be characterized by a water-retention curve $\theta(h)$, where $\theta$ is the volumetric soil-water content, and $h$ the soil-water pressure head. Through the law of capillarity, the value of $h$ for a particular $\theta$ is related to a function of $r^{-1}$, where $r$ is the effective radius of the largest soil pores remaining filled with water. According to Miller and Miller, two soils or porous media are similar when scale factors exist that will transform the behavior of one of the porous media to that of the other. Figure 5.1 illustrates their concept of self-similar microscopic soil-particle structure for two soils. The relative size for each of the geometrically identical particles is defined by the particular value of the microscopic scale length $\lambda_i$. This kind of similarity leads to the constant relation $r_1/\lambda_1 = r_2/\lambda_2 = r_3/\lambda_3 = \cdots = r_i/\lambda_i$ and to the formulation

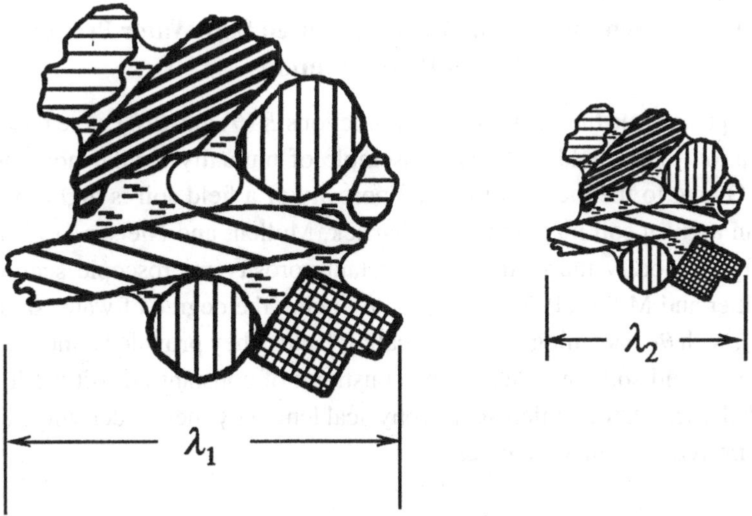


Figure 5.1. Geometrically similar soil-particle arrangements at the microscopic level provide the basis for Miller scaling. (From Miller and Miller, 1955a,b, with permission.)

of a scaled, invariant soil-water pressure head $h_*$ such that

$$\lambda_1 h_1 = \lambda_2 h_2 = \cdots = \lambda_i h_i = \lambda_* h_* \tag{5.1}$$

where $h_*$ is the scale mean pressure head, and $\lambda_*$ is the mean scale length. Dividing each scale length by the mean scale length reduces (5.1) to

$$\alpha_1 h_1 = \alpha_2 h_2 = \cdots = \alpha_i h_i = h_* \tag{5.2}$$

where $\alpha_i$ are the scale factors having a mean value of unity. The hydraulic-conductivity function $K(\theta)$ that relates the soil-water flux density to the force acting on the soil water is analogously scaled

$$\frac{K_1(\theta)}{\lambda_1^2} = \frac{K_2(\theta)}{\lambda_2^2} = \cdots = \frac{K_i(\theta)}{\lambda_i^2} = \frac{K_*(\theta)}{\lambda_*^2} \tag{5.3}$$

where $K_*$ is the scale mean hydraulic-conductivity function. Written in terms of scale factors $\alpha_i$, equation (5.3) becomes

$$\frac{K_1(\theta)}{\alpha_1^2} = \frac{K_2(\theta)}{\alpha_2^2} = \cdots = \frac{K_i(\theta)}{\alpha_i^2} = K_*(\theta) \tag{5.4}$$

Note that the scale length $\lambda_i$ has a physical interpretation and that the porosities of the soils are assumed identical. A constant porosity across "similar" soils is an important assumption made in this approach.

## 5.3 Initial Attempts to Scale Field-measured Soil-Water Properties during Redistribution

Initial attempts to scale the field-measured functions $K(\theta)$ and $\theta(h)$ were based upon the assumption that a field soil is an ensemble of mutually similar homogeneous domains. Owing to the fact that the total porosity of a field soil is highly variable, even within a given soil mapping unit, Warrick, Mullen, and Nielsen (1977) found it necessary to modify the restrictive, constant-porosity microscopic scaling concept of Miller and Miller (1955a,b). By introducing the degree of water saturation $s(= \theta\theta_S^{-1})$, with $\theta_S$ becoming a second scaling factor, they provided a more realistic description of field soils by relaxing the constraint of constant porosity. Moreover, they avoided a search for a microscopic physical length by merely deriving values of $\alpha$ that minimized the sum of squares

$$\mathrm{SS} = \sum_{r=1}^{N}\sum_{i=1}^{M}(h_* - \alpha_r h_i)^2 \tag{5.5}$$

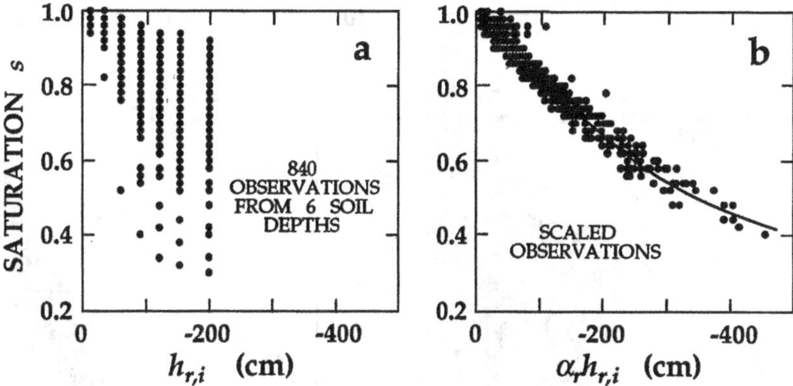


Figure 5.2. (a) Unscaled observations of $s(h)$ and (b) scaled observations $s(h_*)$ for panoche soil. (From Warrick et al., 1977, with permission.)

for $N$ macroscopic locations within a field soil and $M$ observations of $h$. For example, with this minimization, the 840 measurements of $(\theta, h)$ [samples taken at six soil depths and 20 sites ($N = 120$) within an agricultural field and analyzed in the laboratory with seven values of $h$ ($M = 7$)] shown in Figure 5.2a as $h(s)$ were coalesced into the single curve

$$h_* = -6{,}020s^{-1}[(1 - s) - 2.14(1 - s^2) + 2.04(1 - s^3) - 0.69(1 - s^4)] \tag{5.6}$$

in Figure 5.2b (Warrick et al., 1977). The 2,640 values of $(K, \theta)$ stemming from field measurements analyzed by the instantaneous-profile method for six soil depths and 20 locations shown in Figure 5.3a were coalesced and described by the regression expression

$$\ln K_* = -23.3 + 75.0s - 103s^2 + 55.7s^3 \tag{5.7}$$

as shown in Figure 5.3b. Although Warrick et al. (1977) abandoned the concept of microscopic geometric similarity (Miller and Miller, 1955a,b) and based their scaling method on the similarity between soil hydraulic functions, they noted that values of $\alpha_r$ required for scaling $h$ in equation (5.6) were not equal to those for scaling $K$ in equation (5.7). Here it should also be noted that the values of $h(\theta)$ scaled in (5.6) were those measured in the laboratory on soil cores removed from the field, and the values of $K(\theta)$ scaled in (5.7) relied on the laboratory-measured values of $h(\theta)$ to obtain estimates of $\theta(t)$ based upon tensiometric measurements taken in the field.

During the next decade, several others attempted to scale field-measured hydraulic properties (e.g., Ahuja, Naney, and Nielsen, 1984b; Ahuja et al., 1989b; Hills, Hudson, and Wierenga, 1989). Rao et al. (1983) and others found that the scale

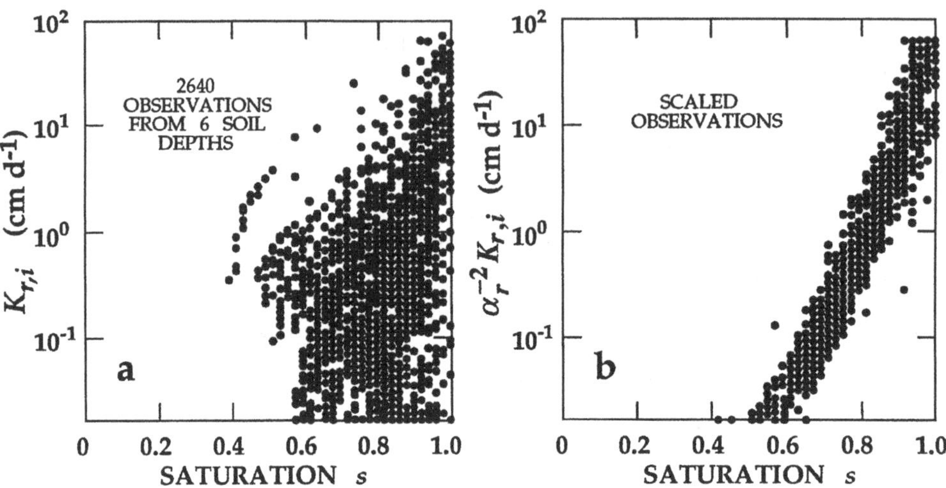


Figure 5.3. (a) Unscaled determinations of $K(s)$ and (b) scaled determinations $K_*(s)$ for Panoche soil. (From Warrick et al., 1977, with permission.)

factors that coalesced the field-measured functions of $K(\theta)$ differed from those that coalesced the field-measured functions of $\theta(h)$.

Encouraged by the results of Warrick et al. (1977) and the findings being discussed with a group of soil physicists from 11 countries, eventually published by the International Atomic Energy Agency (IAEA, 1984), Simmons, Nielsen, and Biggar (1979) suggested scaling the redistribution of soil water in the instantaneous-profile method (Watson, 1966). They assumed a unit hydraulic gradient at the lower boundary of the soil profile and used a common value of $\beta$ in

$$K(\theta) = K_0 \exp[\beta(\theta - \theta_0)] \tag{5.8}$$

for all locations within a field. That assumption was based on the finding that in a given field, the slope $\beta$ of the graph of $\ln K$ versus $(\theta - \theta_0)$ was normally distributed and was characterized by a reasonably small coefficient of variation. Hence, for redistribution in the absence of evaporation, the flux density for water at the lower boundary of the soil profile becomes (Libardi et al., 1980)

$$az\frac{d\theta}{dt} = -K_0 \exp[\beta(\theta - \theta_0)] \tag{5.9}$$

with $\theta_0$ being the soil-water content at the beginning of the redistribution, and $a$ being defined by

$$\bar{\theta} = a\theta + b \tag{5.10}$$

where $b$ is a constant, and $\bar{\theta}$ is the mean soil-water content from the soil surface to depth $z$, from field data. Integration of equation (5.9) yields

$$\theta = \theta_0 - \frac{1}{\beta}\ln\left(1 + \frac{\beta K_0 t}{az}\right) \tag{5.11}$$

With a common value of $\beta$ and a common initial value for $\bar{\theta}_0$, observations of $\theta$ measured at different locations and depths as functions of time were scaled with

$$\theta = \bar{\theta}_0 - \frac{1}{\beta}\ln\left(1 + \frac{\beta K_* \tau}{z_*}\right) \tag{5.12}$$

where $z_*$ is a reference depth, $\tau = \omega^2 z_* t(az)^{-1}$, and $\omega$ is a scale factor defined by $K_0 = \omega^2 K_*$, where $K_*$ is the scale mean for all $K_0$. Simmons et al. (1979) attempted to use equation (5.12) to coalesce 608 neutron-probe measurements of $\theta$ at 128 locations within four small field plots during redistribution for 1 month following steady-state infiltration. With their assumptions, the necessity of installing tensiometers was eliminated, thereby allowing much greater numbers of locations and depths to be sampled with only a neutron probe. With that simplified method it was envisioned that a large number of scaling factors could adequately quantify the spatial variability of $K_0$ and its scale mean within the experimental site containing the four plots. The 608 values for $\theta$ during redistribution were reasonably coalesced about the scale mean curve, equation (5.12), only if the value $\theta_0$ for each depth was adjusted to that of $\bar{\theta}_0$.

## 5.4 Initial Attempts to Scale Soil-Water Properties During Infiltration

The technology to accurately measure water behavior in homogeneous soil columns improved significantly during the second decade after the pioneering concepts of Miller and Miller were published. For example, the gamma-attenuation method for measuring soil-water content and soil bulk density, miniature pressure transducers to quickly and accurately measure soil-water pressure, highly permeable porous plates and cups for improved measurements and control of soil-water pressure, and improved theoretical methods for ascertaining $K(\theta)$ and $D(\theta)$ from laboratory observations all became available over a relatively short period. The improved technology soon led to attempts to scale transient soil-water conditions in both the laboratory and the field.

### *5.4.1 Initial Laboratory Experiments*

Using inspectional analysis (Ruark, 1935), Reichardt, Nielsen, and Biggar (1972) extended the microscopic scaling concepts of Miller and Miller (1955a,b) by attempting to scale macroscopic observations of horizontal infiltrations in different kinds of initially dry, homogeneous soils. For an arbitrary macroscopic length $L$, Reichardt et al. (1972) used the scale mean values

$$K_* = \frac{\mu K_r}{\rho g \lambda_r^2}, \qquad D_* = \frac{\mu D_r}{\sigma \lambda_r}, \qquad h_* = \frac{\lambda_r \rho g h_r}{\sigma}, \qquad t_* = \frac{\lambda_r \sigma t_r}{\mu L^2} \tag{5.13}$$

to reduce the Richards equation for horizontal flow in the $x$ direction to

$$\frac{\partial \Theta}{\partial t} = \frac{\partial}{\partial x_*}\left[D(\Theta)\frac{\partial \Theta}{\partial x_*}\right] \tag{5.14}$$

where $x_* = xL^{-1}$, $\Theta = (\theta - \theta_n)(\theta_0 - \theta_n)^{-1}$, $\theta_n$ is the initial soil-water content, $\theta_0$ is the soil-water content at $x = 0$ for $t > 0$, $g$ is acceleration due to gravity, and $\rho, \sigma$, and $\mu$ are the density, surface tension, and viscosity of water, respectively. Reichardt and co-workers chose the definitions (5.13) to achieve a formal resemblance to those for microscopic similarity of Miller and Miller. [For equations (5.13)–(5.22) we retain the original symbolism $\lambda$ for scaling length or factor.] With the solution of equation (5.14) being

$$x_* = \phi_*(\Theta) t_*^{1/2} \tag{5.15}$$

plots of the distance to the wetting front $x_f$ versus the square root of infiltration time for different soils (Fig. 5.4) yielded values of $\lambda_r$ defined by

$$\lambda_r \lambda_*^{-1} = m_r^2 m_*^{-2} \tag{5.16}$$

where $m_r$ is the slope $x_{f_r} t_r^{-1/2}$. Arbitrarily choosing the value of $\lambda$ for Fresno soil to be equal to unity, the distances to the wetting front $x_f$ versus $t^{1/2}$ for eight soils shown

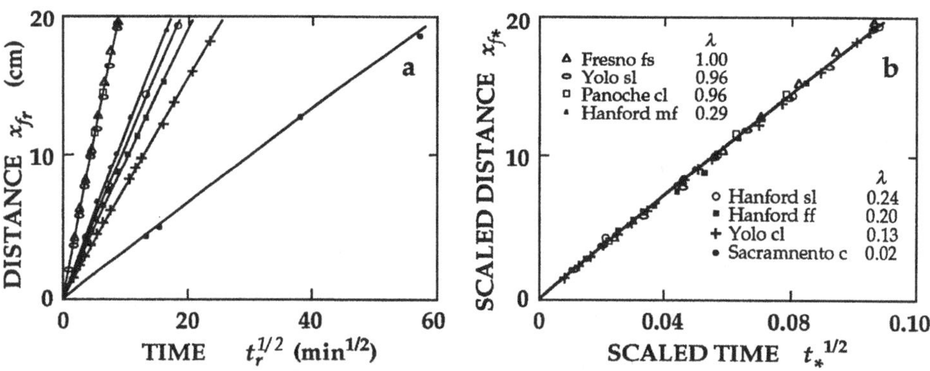


Figure 5.4. (a) Unscaled distance to the wetting front $x_{f_r}$ versus $t_r^{-1/2}$ and (b) scaled distance $x_{f_*}$ versus $t_*^{1/2}$ for horizontal infiltration into 12 air-dry soils. (From Reichardt et al., 1972, with permission.)

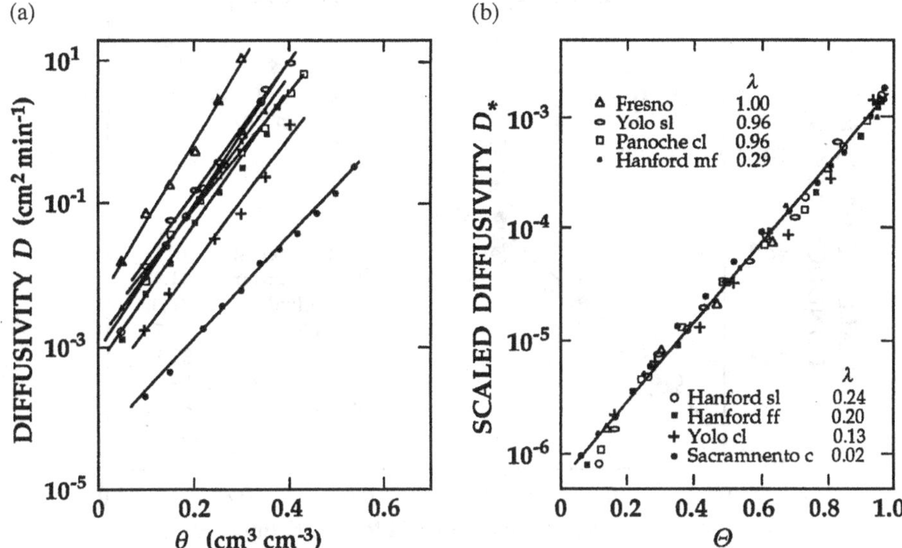


Figure 5.5. (a) Unscaled soil-water diffusivity $D(\theta)$ and (b) scaled soil-water diffusivity $D_*(\Theta)$ for 12 soils. (From Reichardt et al., 1972, with permission.)

in Figure 5.4a were scaled into the single curve of equation (5.15) for $t_*$ defined by equation (5.13) with $L = 1$. See Figure 5.4b. The soil-water diffusivity functions $D(\theta)$ calculated for the soils in Figure 5.5a were successfully scaled (Figure 5.5b) using the scaled diffusivity function

$$D_* = 6 \cdot 10^{-7}\exp(8\Theta) \tag{5.17}$$

Miller and Bresler (1977) subsequently included the slope $m_*$ for the Fresno soil in (5.17) and suggested that a universal equation

$$D(\theta) = 10^{-3}m^2\exp(8\Theta) \tag{5.18}$$

may exist. Additional research has not been conducted to confirm or reject their suggestion. It should also be noted that $h_*$ and $K_*$ defined by equation (5.13) were not able to coalesce independently measured values of $\theta(h)$ and $K(h)$ using the scaling procedures of Reichardt et al. (1972).

Later, Reichardt, Libardi, and Nielsen (1975) extended their testing of equation (5.15) for infiltration into 12 temperate- and tropical-zone soils whose $\lambda$ values ranged over two orders of magnitude. They showed that $K_*$ derived from equation (5.13) for all 12 soils could be described by

$$K_*(\Theta) = 2.65 \cdot 10^{-19}\exp(-12.23\Theta^2 + 29.06\Theta) \tag{5.19}$$

as illustrated in Figure 5.6. Somewhat later, Youngs and Price (1981) observed that different cumulative infiltration curves for a variety of porous materials individually

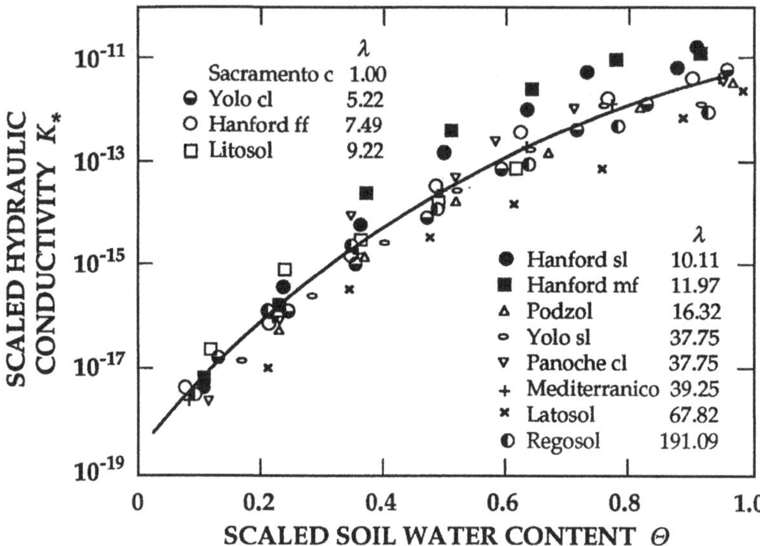


Figure 5.6. Scaled hydraulic conductivity $K_*$ as a function of scaled soil-water content Θ (From Reichardt et al., 1975, with permission.)

packed into laboratory columns could be scaled to coalesce into a unique curve. Similar to the results of Reichardt et al. (1972), the scale factors defined in equation (5.13) for $K$ differed from those for $h$ as well as those for the sorptivity $S$.

### 5.4.2 Initial Field Experiments

Using regression techniques similar to those employed by Reichardt et al. (1972) and Swartzendruber and Youngs (1974), Sharma, Gander, and Hunt (1980) attempted to scale the 26 sets of field-measured cumulative infiltration data from the 9.6-ha watershed shown in Figure 5.7a. Values of $S$ and $A$ for each of the 26 data sets were obtained by regression using the two-term, truncated version of the cumulative infiltration $I$ into a homogeneous soil (Philip, 1957):

$$I = St^{1/2} + At + Bt^{3/2} + \cdots \tag{5.20}$$

where the first term $S$ is the sorptivity, and the remaining terms account for the force of gravity. The solid line in Figure 5.7a is calculated from

$$I = \bar{S}t^{1/2} + \bar{A}t \tag{5.21}$$

where $\bar{S}$ and $\bar{A}$ are the mean values for each of the respective sets of $S_r$ and $A_r$. The solid line in Figure 5.7b is the scaled cumulative infiltration

$$I_* = \bar{S}t_*^{1/2} + \bar{A}t_* \tag{5.22}$$

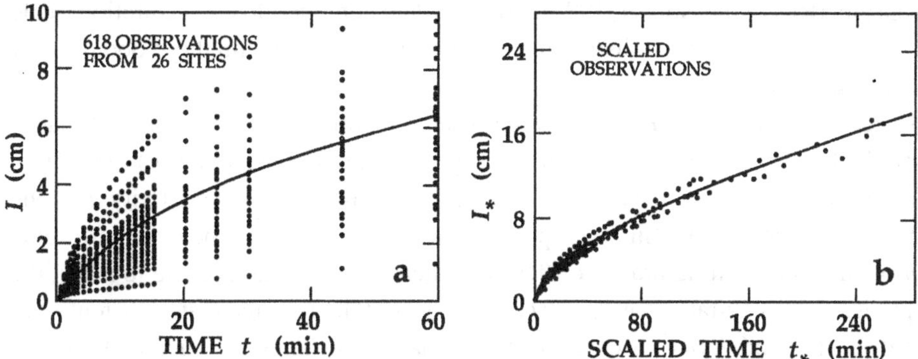


Figure 5.7. (a) Unscaled cumulative infiltration versus time and (b) scaled cumulative infiltration versus scaled time for 26 sites in a 9.6-ha watershed. (From Sharma et al., 1980, with permission.)

where $I_* = \lambda_r I \lambda_*^{-1}$, $t_* = \lambda_r^3 t \lambda_*^{-3}$, $r$ is the site index, and $\lambda_*$ is arbitrarily chosen as unity. Scale factors $\lambda_r$ for each of the $r$ sites were obtained by minimizing the sum of squares,

$$\mathrm{SS} = \sum_{r=1}^{N} \sum_{i=1}^{M} [I_*(t_{*i}) - I_r(t_i)]^2 \tag{5.23}$$

for all $r$ locations and $i$ observations within the watershed. As shown in Figure 5.7b, the 26 sets of cumulative infiltration observations were nicely coalesced into a unique curve with these values of $\lambda_r$. They also calculated two additional sets of scale factors. One set was derived from the observations of $I(t_i)$ throughout the watershed using the scaling relation $\bar{S} = S_r \lambda_r^{-1/2}$ with equation (5.21). Another set was similarly derived using the scaling relation $\bar{A} = A_r \lambda_r^{-2}$. Although each of those sets of scale factors tended to coalesce the original observations $I$ into a single curve, neither curve was as well defined as that given in Figure 5.7b. The three sets of scale factors differed significantly, but were nevertheless correlated. The frequency for each set of scale factors was lognormally distributed, and each manifested no apparent spatial correlation.

Russo and Bresler (1980a) also followed the suggestion of Reichardt et al. (1972) to scale soil-water profiles during infiltration. They assumed that cumulative vertical infiltration into a field soil could be described by equation (5.20) truncated to only the sorptivity term provided that $t$ approached zero. For a Green and Ampt (1911) piston-type wetting profile, the truncated (5.20) becomes

$$x_f(\theta_0 - \theta_f) = St^{1/2} \tag{5.24}$$

where $x_f$ is the distance to the wetting front. Using (5.16) with field-measured values of $S$ from (5.24), they obtained scale factors $\alpha_i$ for each of the 120 sites in the

0.8-ha plot (Russo and Bresler, 1981). At soil-water contents close to saturation, scale factors for $S$ were normally distributed and highly correlated with those obtained for $h(s)$ using (5.21).

In addition, Russo and Bresler (1980a) used a scaling technique similar to that of Warrick et al. (1977) to scale calculated, not measured, values of $\theta(h)$ and $K(h)$. They used field-measured values of the saturated hydraulic conductivity $K_S$, the water-entry value for the soil-water pressure head $h_w$ [defined as the minimum value of $h$ on the main wetting branch of $\theta(h)$ at which $d\theta/dh$ remains equal to zero], the sorptivity $S$, the saturated soil-water content $\theta_S$, and the residual water content $\theta_r$ to calculate $\theta(h)$ and $K(h)$ for each of the 120 data sets. They assumed that the soil hydraulic properties were described by

$$\begin{aligned} \theta(h) &= (\theta_S - \theta_r)(h_w h^{-1})^\beta + \theta_r \qquad && (h < h_w) \\ &= \theta_S && (h \geq h_w) \\ K(h) &= K_S(h_w h^{-1})^\eta && (h < h_w) \\ &= K_S && (h \geq h_w) \end{aligned} \tag{5.25}$$

where $h_w$ is the water-entry value of $h$ (Bouwer, 1966), and $\beta$ and $\eta$ are soil constants (Brooks and Corey, 1964) calculated from

$$\eta = -5K_S h_w(\theta_S - \theta_r)[\pi S^2(\theta_S, \theta_r)]^{-1} + 1.25 \tag{5.26}$$

$$\beta = (\eta - 2)2^{-1} \tag{5.27}$$

Functions (5.25) expressed in terms of the degree of water saturation $s$ at each location $r$ were scaled according to

$$h_r(s) = h_*(s)\alpha_r^{-1} \tag{5.28}$$

$$K_r(s) = K_*(s)\alpha_r^2 \tag{5.29}$$

using a minimization procedure similar to that of Warrick et al. (1977), where the field scale means of $h(s)$ and $K(s)$ for $N$ locations were defined as

$$h_*(s) = N^{-1}\left\{\sum_{r=1}^{N}\left[h_r^{-1}(s)\right]\right\}^{-1} \tag{5.30}$$

$$K_*(s) = N^{-2}\left\{\sum_{r=1}^{N}[K_r(s)]^{1/2}\right\}^2 \tag{5.31}$$

subject to the condition that

$$N^{-1}\sum_{r=1}^{N}\alpha_r = 1 \tag{5.32}$$

Here their results were similar to those of Warrick et al. (1977), inasmuch as values of $\alpha_r$ for $h(s)$ were correlated but not necessarily equal to those for $K(s)$. Nevertheless, they optimistically concluded that the use of a single scaling factor $\alpha$ as a representative of the soil hydraulic properties of a field remained a practical possibility provided that the water content was expressed as the degree of water saturation $s$.

## 5.5 Functional Normalization: An Empirical Attempt to Scale Soil-Water Regimes

Tillotson and Nielsen (1984), describing some of the different kinds of scaling techniques used in the physical sciences and engineering, attempted to clarify the differences between the similar-media concept of Miller and Miller and other scaling approaches. Reviewing dimensional methods and noting that no universal nomenclature exists in the literature for distinguishing between them, they grouped all of the methods into two categories (dimensional analysis and inspectional analysis), provided simple examples of each, and introduced the term "functional normalization," a regression procedure by which scale factors for soil processes are determined from sets of experimental observations.

The scaling of groundwater levels in relation to the rate of water draining from sandy-soil profiles $P$ measured at the outlet of a watershed is an example of functional normalization (Hopmans, 1988). The depth to the groundwater $z_w$ within the watershed is influenced by the topography, the spacing between drainage channels, and the field-measured soil-water properties. Ernst and Feddes (1979) had previously derived the empirical relation

$$P = \chi \exp(-\varepsilon z_w) \tag{5.33}$$

where $\chi$ and $\varepsilon$ are parameters determined from measurements of $P$ and $z_w$. Because $P$ is measured only at the outlet of the watershed, and $z_w$ is measured with observation wells at $r$ locations, the values of $\chi$ differ for each location. Assuming that a common value for $\varepsilon$, derived from its average values for all locations $\bar{\varepsilon}$, defines equation (5.33) for each location, then (5.33) becomes

$$P_r = \chi_r \exp(-\bar{\varepsilon} z_w) \tag{5.34}$$

Assuming that a scale factor $\alpha_r$ exists such that $\chi_* = \alpha_r \chi_r$, the scaled watershed discharge $P_*$ is given by

$$P_* = \chi_* \exp(-\bar{\varepsilon} z_w) \tag{5.35}$$

In Figure 5.8a, the discharge rate measured only at the outlet of the watershed is plotted against groundwater levels measured at 83 locations within the watershed.

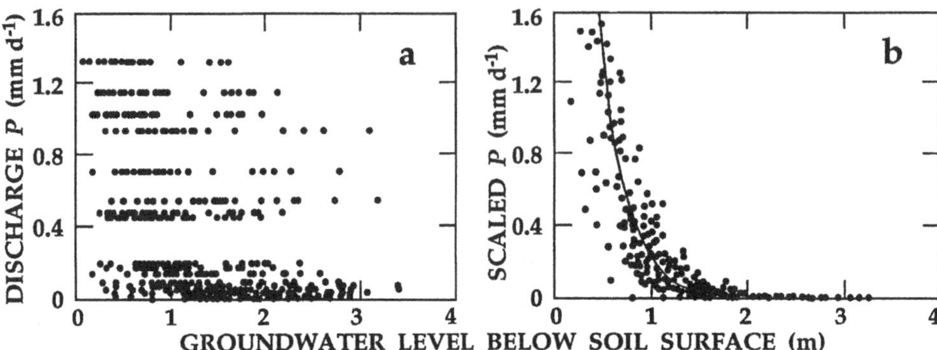


Figure 5.8. (a) Unscaled discharge from a watershed as a function of groundwater level below the soil surface. (b) Scaled discharge, with solid line described by equation (5.35). (From Hopmans, 1988, with permission.)

Equation (5.35) describes the scaled data, where $\chi_*$ is 8.52 mm·d$^{-1}$, and $\bar{\varepsilon}$ is 3.59 m$^{-1}$. Hence, by functional normalization, Hopmans (1988) found a set of $\alpha_r$ values that coalesced a large amount of data satisfying equation (5.34) at each location $r$ to be described by equation (5.35). Although those values of $\alpha_r$ were not directly related to Miller scaling and have no explicit physical meaning, they were potentially useful to express the variability of the water-table levels in a single parameter that could be correlated with other parameters, properties, or processes operating within the watershed.

## 5.6 Analyzing Initial Scaling Attempts

After more than a decade of research in testing the applicability of the concept of Miller similitude and related extended theories to laboratory- and field-measured soil-water properties, the absence of a capacious theory provoked controversy regarding the utility of scaling. The criteria for acceptance or rejection of their application to describe field-measured soil-water behavior were without foundation. Moreover, the situation was exacerbated by the fact that at that time no paradigms for local and regional scales of homogeneity in pedology and soil classification had been developed.

Sposito and Jury (1985) significantly advanced the analysis of scaling concepts and theories for soil-water retention and movement. They improved our understanding of the scaling of soil-water behavior and interpreted the findings from many scaling experiments. At that time, they provided a unified classification scheme for the three most common macroscopic scaling approaches that had been applied to the Richards equation for one spatial dimension. They scaled the Richards equation subject to those initial and boundary conditions for which most scaling experiments had been performed, namely, that the initial water content throughout the profile was $\theta(z, 0) = \theta_n$, and the boundary condition at the soil surface was either $\theta(0, t) = \theta_0$ or the water flux density described in terms of $K[\theta(h)]$ and $D[\theta(h)]$. They showed that many scaling parameters developed in an inspectional analysis depend not only upon the initial and boundary conditions but also upon whatever special or unique physical hypotheses are assumed. From their examination of inspectional analyses, three macroscopic similitude approaches were identified: Miller similitude, Warrick similitude, and Nielsen similitude. Miller similitude, based upon viscous flow and capillary forces, differs from the original Miller and Miller microscopic approach, because a scaling parameter for the volumetric soil-water content is required, and no hypothesis regarding the microscopic geometry of the soil's pore structure is used. Warrick similitude derives scaling factors based on equation (5.4) and arbitrarily selected expressions for $K(s)$ and $h(s)$. Nielsen similitude is based on a flux density of zero at the soil surface during redistribution and on the assumption that the hydraulic conductivity, the soil-water diffusivity, and the soil-water retention are exponential functions of soil-water content.

Jury, Russo, and Sposito (1987) suggested that the description of the spatial variability of any transient-water-flow problem involving the scaling of both $K(s)$ and $h(s)$ requires the use of at least three stochastic variates: $K_S$, $\alpha$ [as in equation (5.28)], and $\eta$, defined by $K = h^{-\eta}$. [Twenty years earlier, Corey and Corey (1967) had successfully scaled the Richards equation solved for the drainage of laboratory columns, each packed with sands of different sizes, provided that the hydraulic conductivity was described by $K = h^{-\eta}$.] Sposito and Jury (1990) showed theoretically that the scale factors $\alpha_h$ and $\alpha_K$, defined from equations of the form (5.28) and (5.29), respectively, were related to a third parameter $\omega$ by

$$\alpha_K = \alpha_h^{\omega} \tag{5.36}$$

Using the theory of Lie groups, Sposito (1990) showed that the Richards equation will be invariant under scaling transformations of the soil hydraulic properties only when both $K$ and $D$ are either exponential or power functions of $\theta$. Earlier, Ahuja, et al. (1989b) had proposed an equation similar to (5.36) relating scaling factors for water-saturated hydraulic conductivity $K_S$ and soil-water pressure head $h_f$ at the wetting front.

Noting the suggestion of Sposito and Jury (1990) that the microscopic length $\lambda$ could be defined as that of the "effective" pores, rather than that depicted in Figure 5.1, Snyder (1996) introduced the possibility of a porosity scale factor $\alpha_P$. With $\alpha_P$, an alternative form of (5.36) becomes

$$\alpha_K = \alpha_h \alpha_P^n \tag{5.37}$$

where $n$ is a constant that depends on the model used to describe the hydraulic-conductivity function. He suggested that the inclusion of $\alpha_P$ would account for the fact that various investigators had obtained values for $\alpha_h$ and $\alpha_K$ that were correlated as well as values that were not correlated (Rao et al., 1983). He also suggested that equation (5.37) could account for the results of Ahuja et al. (1984a, 1989a), where the field-measured values of $K_S$ were proportional to the fourth or fifth power of the effective porosity.

## 5.7 Recent Attempts to Scale Soil-Water Properties

The scaling approach of Miller and Miller (1955a,b) and the approaches of others discussed earlier, based on inspectional analysis and functional normalization, sought scale factors that could simplify problems by expressing them in terms of a small number of reduced variables. New scaling concepts with potential applications to field conditions continue to be developed. For example, Kutílek et al. (1991) theoretically scaled the Richards equation under an invariant-boundary-flux condition. This type of scaling has application when water is ponded on a soil having a very thin, relatively impermeable crust at its surface – a condition commonly observed in many field soils. Warrick and Hussen (1993) developed scaled solutions more general than those of Kutílek and co-workers, in that $h$ specified boundary and initial conditions were considered and were invariant with respect to $K_S$ and $h_w$. Nachabe (1996) developed theoretical relationships for infiltration involving macroscopic capillary length, sorptivity, and the shape factor (a measure of the nonlinearity of the soil-water diffusivity). He showed that the predicted infiltration rate is not greatly sensitive to different values of the shape factor, provided that the macroscopic capillary length remains the same. With shape factors being difficult to ascertain in the field, the macroscopic capillary length serves as a scale factor. These and other ideas await field investigation.

### *5.7.1 Field Approaches to Scaling Soil-Water Properties and Processes*

Recently, however, at least four different approaches have been used to scale soil-water behavior in the field. The first involves a vertically heterogeneous or layered soil profile that is transformed into a uniform profile using scale factors to stretch or shrink the thickness of each of the nonuniform soil layers. The second approach is that of linear-variability scaling in combination with an inverse technique, solving the Richards equation to estimate the in situ hydraulic properties of soils. The third approach uses the slopes of the log-log $h(\theta)$ and $K(\theta)$ relations as scaling factors. The fourth approach is that of fractal scaling.

### *5.7.2 Time-invariant Hydraulic Gradients in Layered Soils*

Virtually every field soil is heterogeneous with depth and possesses more or less distinct layers or genetic horizons. Hence the distribution of $h(z, t)$ observed in a field will depend upon the choice of depths at which $h$ is measured. This scaling approach (Sisson, 1987) begins with an analysis of the redistribution of soil water under a unit hydraulic gradient (Sisson, Ferguson, and van Genuchten, 1980) to estimate $K(\theta)$ without the need for calculating hydraulic gradients and water flux densities from the differences in noisy time- and depth-averaged measurements. With a unit hydraulic gradient, the Richards equation reduces to

$$\frac{\partial \theta}{\partial t} = -\frac{dK}{d\theta}\frac{\partial \theta}{\partial z} \tag{5.38}$$

For an initial condition of

$$\theta(z, 0) = \theta_n(z) \tag{5.39}$$

the solution of (5.38), known as a Cauchy problem (Lax, 1972), is simply

$$\left.\frac{dK}{d\theta}\right|_{\theta_n} = \frac{z}{t} \tag{5.40}$$

where the soil depth $z$ and the time $t$ are those associated with measured values of $\theta$. Assuming that $K(\theta)$ is an exponential or power function of $\theta$, then the function differentiated with respect to $\theta$ is equated to $zt^{-1}$ and solved explicitly for $\theta(z, t)$. For a given soil depth, the parameters defining $K(\theta)$ are obtained by regression of a plot of $\theta$ versus $t$.

Sisson (1987) extended the foregoing analysis to a time-invariant hydraulic gradient occurring in layered field soils. The layered soil profile is transformed into a uniform profile by scaling soil depth $z$ in relation to $\theta$ and $h$ such that mass and energy are conserved. The Richards equation containing the depth-dependent hydraulic properties $h(\theta, z)$ and $K(\theta, z)$ is transformed into a scaled Richards equation having hydraulic properties $h_*(\theta_*)$ and $K_*(\theta_*)$, which are independent of soil depth for a scaled depth $z_*$. Such a transformation is achieved by requiring the following relationships:

$$\frac{\partial K(\theta, z)}{\partial K_*(\theta_*)} = 1 \tag{5.41}$$

$$\frac{\partial h(\theta, z)}{\partial h_*(\theta_*)} = \frac{\partial h}{\partial h_*} \tag{5.42}$$

$$\frac{\partial \theta_*}{\partial \theta} = \frac{\partial z}{\partial z_*} \tag{5.43}$$

Note that (5.41) requires that all $K(\theta, z)$ curves be parallel. Shouse, van Genuchten, and Sisson (1991) assumed a linear relationship between the scaled soil-water profile and the profile measured in the field:

$$\theta_*(z_* t) = a(z) + b(z)\theta(z, t) \tag{5.44}$$

where the two scale factors $a$ and $b$ depend upon $z$. Using (5.44) in (5.42) and (5.43), assuming a unit hydraulic gradient during redistribution, and satisfying (5.41) by selecting a reference depth at which the unscaled hydraulic conductivity $K(\theta, z)$ is chosen to equal the scaled hydraulic conductivity $K_*(\theta_*)$, they obtained the scaled gravity-drainage equation

$$\frac{\partial \theta_*}{\partial t} = -\frac{\partial K_*(\theta_*)}{\partial z_*} \tag{5.45}$$

which has a solution,

$$\frac{dK_*}{d\theta_*} = \frac{z_*}{t} \tag{5.46}$$

similar to (5.40). Using (5.46) to obtain values of $\theta_*$ for each measured value of $\theta$ at each soil depth, scale factors $a(z)$ and $b(z)$ were obtained by regression using (5.44).

Shouse et al. (1992a,b) demonstrated the applicability of the foregoing scaling method for observations of soil water within a 4-m$^2$ plot of a layered soil. Observations of soil-water content were obtained during a 3-week period of redistribution following an initial 30-day ponded condition. $\theta(h)$ was estimated from measurements made in the laboratory on soil cores (5 cm in diameter, 7 cm in length). By using (5.40), the original soil-water profile distributions to a depth of 120 cm for 11 redistribution times in Figure 5.9a led to $dK/d\theta$ versus $\theta$ for the five depths shown in Figure 5.9b. Shown in Figure 5.10a are the scaled soil-water profiles that coalesce the data from all five depths into the unique curve of $dK_*/d\theta_*$ shown in Figure 5.10b using equation (5.41). Hence, this approach requires one set of reference functions for $\theta(h)$ and $K(\theta)$, as well as two scale factors for each additional horizon or depth

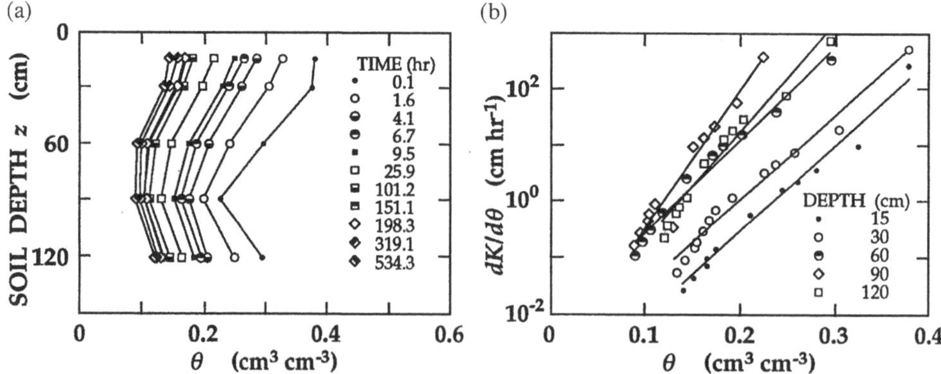


Figure 5.9. (a) Measured soil-water-content profiles for selected times during redistribution and (b) calculated values for $dK/d\theta$ versus $\theta$ for five soil depths. (From Shouse et al., 1992a, with permission.)

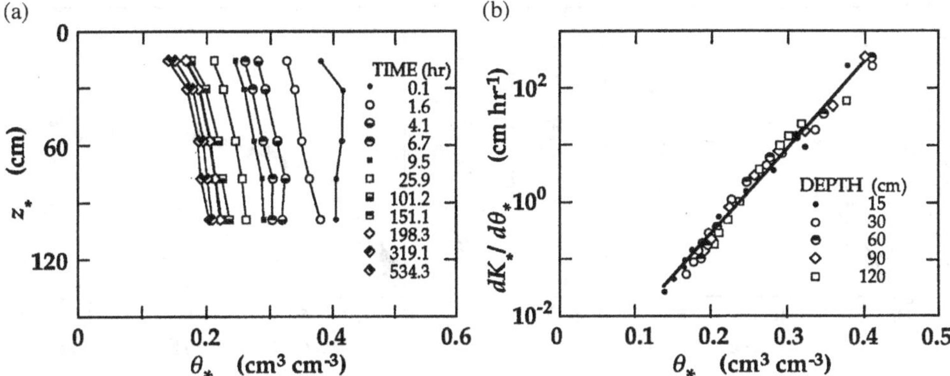


Figure 5.10. (a) Scaled soil-water-content profiles $z_*(\theta_*)$ and (b) scaled values for $dK_*/d\theta_*$ versus $\theta_*$ for the observations in Figure 5.9 (From Shouse et al., 1992a, with permission.)

considered. It should be noted that this kind of linear $\theta(z)$ scaling is independent of the form of $K(\theta)$ and that the same scale factors scale both $\theta(h)$ and $K(\theta)$.

### *5.7.3 Inverse Solution of the Richards Equation*

The second scaling approach combines the linear-variability scaling technique introduced by Vogel, Cislerova, and Hopmans (1991) with an inverse solution of the Richards equation to determine soil hydraulic properties. After irrigating a 32-ha agricultural field with 0.3 m of water, they estimated the amount of water draining from the soil based on observations of soil-water content measured with a neutron meter at 44 locations during a period of 125 days (Eching, Hopmans, and Wallender, 1994). With rainfall equaling the estimates of evapotranspiration during that 125-day period, the net changes in soil-water storage from the soil surface to the 2.1-m soil depth were attributed to drainage only. Hence the Richards equation was solved assuming a condition of zero flux for the soil surface and a known flux condition at the bottom of the 2.1-m profile.

At each location $r$, they assumed that the drainage flux density $q_r$ was described by the exponential function (Belmans, Wesselling, and Feddes, 1983)

$$q_r = a_r \exp(-b_r t_r) \tag{5.47}$$

where $a_r$ and $b_r$ are fitting parameters, and $t_r$ is the drainage time, beginning when the soil-water storage in the profile is a maximum. They used the linear scaling relations proposed by Vogel et al. (1991) for each location $r$:

$$\begin{aligned} K_r(h_r) &= \alpha_{K_r} K_*(h_*) \\ \theta_r(h_r) &= \theta_{\text{res}_r} + \alpha_{\theta_r}[\theta_*(h_*) - \theta_{\text{res}_*}] \\ h_r &= \alpha_{h_r} h_* \end{aligned} \tag{5.48}$$

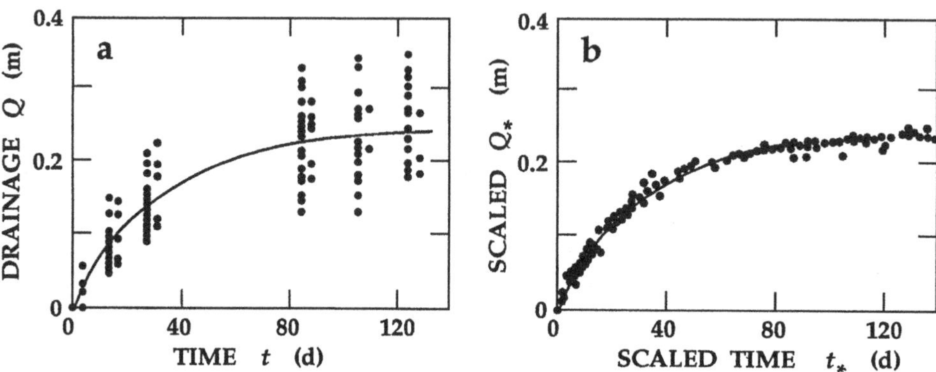


Figure 5.11. (a) Unscaled cumulative drainage versus time and (b) scaled cumulative drainage at the 2.1-m depth in a 32-ha field. (From Eching et al., 1994, with permission.)

where $\theta_{\text{res}}$ is the residual soil-water content $\theta_r$ as in (5.25), and $K_*(h_*)$ and $\theta_*(h_*)$ are the space-invariant scale mean hydraulic functions for the soil. The distributions of $\alpha_K$ and $\alpha_\theta$ were defined such that each would have an arithmetic mean of unity, and $\alpha_{h_r} = 1$ throughout the profile at any location $r$.

Derived from equation (5.48), values of $a_r = \alpha_{K_r} a_*$, $b_r = a_r b_r \alpha_{\theta_r}^{-1} a_*^{-1}$, and $t_r = \alpha_{\theta_r} \alpha_{K_r}^{-1} t_*$ were substituted into (5.47) to obtain the scale mean flux density

$$q_* = a_* \exp(-b_* t_*) \tag{5.49}$$

The cumulative drainage $Q_r$ obtained by integrating (5.47) for each location $r$,

$$Q_r = \frac{a_r}{b_r}[1 - \exp(-b_r t_r)] \tag{5.50}$$

was similarly scaled to provide the scale mean cumulative drainage $Q_*$:

$$Q_* = \frac{a_*}{b_*}[1 - \exp(-b_* t_*)] \tag{5.51}$$

In order to meet the constraint that the arithmetic means $\bar{\alpha}_K$ and $\bar{\alpha}_\theta$ were each unity, each set of scale factors was normalized by dividing the values by their respective arithmetic mean. Use of the scale mean values $a_* = \bar{\alpha}_K \bar{a}$ and $b_* = a_* \bar{b}(\bar{\alpha}_\theta \bar{a})^{-1}$ in (5.49) or (5.51) defined the lower boundary condition for which the Richards equation was solved. Note that $\bar{a}$ and $\bar{b}$ are the arithmetic means of $a_r$ and $b_r$, respectively, in (5.47) for the 44 locations. With the inverse solution yielding scale mean functions $\theta_*(h_*)$ and $K_*(\theta_*)$, the soil hydraulic functions at each of the 44 locations were determined from (5.48).

The cumulative drainage $Q(t)$ measured at each of the 44 locations is shown in Figure 5.11a, and the cumulative drainage $Q_*(t_*)$ scaled using (5.51) is shown in

Figure 5.11b. Obtaining scale factors from the simple exponential equation describing drainage from the lower boundary of the soil profiles, the hydraulic functions were easily estimated with a minimal number of soil data by combining the inverse solution of the Richards equation with the linear-variability scaling concept.

### *5.7.4 One-Parameter Scale Model for $h(\theta)$ and $K(h)$*

In the third approach, Ahuja and Williams (1991) started with the one-parameter model for $h(\theta)$ proposed by Gregson, Hector, and McGowan (1987). Assuming that $\theta_r$ is zero and $h < -50$ cm, Gregson and co-workers used (5.25) in the form

$$\ln[-h_r(\theta)] = u_r + v_r \ln\theta \tag{5.52}$$

to describe $h(\theta)$ at each location $r$ for a large number of soils representing 41 textural classes from Australia and the United Kingdom. They found that the $u_r$-versus-$v_r$ linear relation

$$u_r = \delta + \zeta v_r \tag{5.53}$$

derived from these diverse soils coalesced together into one common relation, with only small scatter. In other words, Gregson and co-workers found that $\delta$ and $\zeta$ each had essentially the same value for all the soils studied. When Ahuja and Williams (1991) found similar results for 10 different U.S. soils, they were encouraged to develop an approach for scaling $h(\theta)$ that would be applicable across soil types and textural classes. Substituting (5.53) into (5.52) and rearranging, they obtained

$$\ln\theta = \{\ln[-h_r(\theta)] - \delta\}v_r^{-1} - \zeta \tag{5.54}$$

Assuming that the values for $\delta$ and $\zeta$ are constant and independent of location, the only parameter in (5.54) that depends upon location is $v_r$. Hence $v_r$ serves as a scaling factor, and for a fixed value of $\theta$ the right-hand side of (5.54) is the same for all $r$ locations.

Ahuja and Williams extended their approach to also scale data on unsaturated hydraulic conductivity $K(h)$ derived from sets of field measurements for six U.S. soils. Assuming that $h < h_w$, they used (5.25) in the form

$$\ln K_r = U_r + V_r \ln(-h) \tag{5.55}$$

to describe $K(h)$ at each location $r$. They found that the $U_r$-versus-$V_r$ linear relation

$$U_r = \xi + \varsigma V_r \tag{5.56}$$

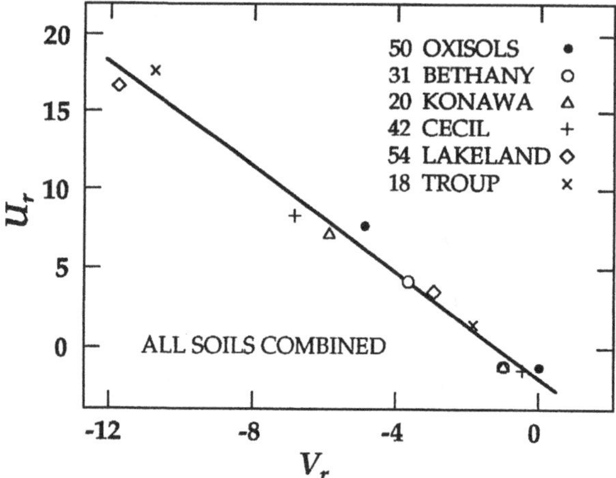


Figure 5.12. Linear $U_r$ and $V_r$ relationships derived by fitting $U_r$ and $V_r$ data for individual sites within each different soil and the site data for all soils combined to equation (5.56) (units for $K$ and $h$ are cm $\cdot\, h^{-1}$ and cm, respectively). The number of sites within each soil is noted next to the soil's name in the figure. (From Ahuja and Williams, 1991, with permission.)

derived from the six data sets shown in Figure 5.12 coalesced together into a unique relation, with $R^2 = 0.94$. For the individual soils, only the end points of the derived relations within the experimental data range are shown in the figure. The site $U_r$ and $V_r$ values were obtained by regressing (5.55) to the experimental $K(h)$ data for the respective sites. Substituting (5.56) into (5.55), they obtained

$$\ln(-h) = (\ln K_r - \xi)V_r^{-1} - \varsigma \tag{5.57}$$

and used $V_r$ as a scaling factor for each $r$ location. Using constant values of $\xi$ and $\varsigma$ obtained from data for all soils combined in (5.57), the scaled values were adequately coalesced into a single curve.

### *5.7.5 Fractal Scaling*

For the fourth approach, Tyler and Wheatcraft (1990) provided fractal-scaling insights into the power-law models of soil-water retention, equation (5.25), developed empirically by Brooks and Corey (1964) and Campbell (1974). They considered the porous structure of a soil to be represented by a simple fractal in two dimensions, known as the Sierpinski carpet, generated by starting with an initial square of side length $a$ and removing one or more subsquares of size $ab_1^{-1}$. Increasing the values of $b$, where $b = b_i$, to the $i$th recursive level yields a carpet everywhere filled with holes, as shown in Figure 5.13, but where only two levels of the recursion are indicated. For $i = 1$, the large square in the center of the carpet having an edge equal to $ab_1^{-1}$

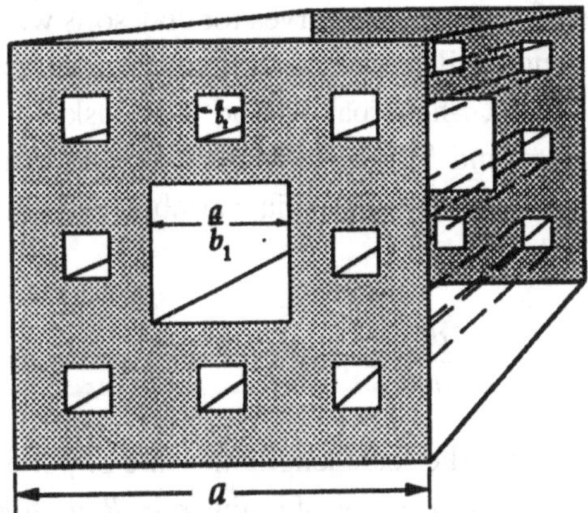


Figure 5.13. Simulated soil using Sierpinski carpet as a conceptual model for pore structure. (From Tyler and Wheatcraft, 1990, with permission.)

is first removed. For the first recursion, $b_1 = a3^{-1}$. Next, for $i = 2$, eight squares having an edge equal to $ab_2^{-1} (b_2 = a3^{-2})$ is removed. Assuming that the open areas represent the cross sections of capillary tubes, and after an arbitrary large number of recursions, we have a distribution of pores in a soil having porosity

$$\phi(b) = 1 - b^{D-2} \tag{5.58}$$

The parameter $b$ is inversely related to the smallest pore size, and $D$ is the fractal dimension of the soil, given by

$$D = \frac{\log(b_1^2 - l_1)}{\log b_1} \tag{5.59}$$

where $b_1^{-1}$ represents the size of the largest pore, and $l_1$ represents the number of pores of size $b_1^{-1}$. The soil-water content $\theta(b)$ associated with water held in pores of size $b^{-1}$ or smaller is, from (5.58),

$$\theta(b) = b^{D-2} \tag{5.60}$$

From the capillary-rise equation, $h$ is proportional to $b$, and hence

$$\theta(b) \propto h^{D-2} \tag{5.61}$$

Normalizing (5.61) in terms of $\theta_S$, the Brooks and Corey (1964) or Campbell (1974) form of the water-retention curve (5.25) is obtained:

$$s = \left(hh_w^{-1}\right)^{D-2} \tag{5.62}$$

They demonstrated that water-retention curves for clay soils would tend to have large values for $D$, whereas sandy soils would have smaller values for $D$.

More recently, Pachepsky, Shcherbakov, and Korsunskaya (1995) extended the fractal concepts of Friesen and Mikula (1987), Tyler and Wheatcraft (1990), and Brakensiek et al. (1992) to quantify the spatial variability of $\theta(h)$ in field soils. They started with the fractal relationship

$$\frac{d\theta}{dR} = AR^{2-D} \tag{5.63}$$

where the effective radius $R$ is the scale length measure of pores, and $A$ is a constant reflecting the geometry of the soil. Pores of radius $r < R$ are filled with water. From (5.63) they derived

$$\theta = \frac{\theta_c}{2}\text{erfc}\left[\frac{1}{\sigma\sqrt{2}}\ln\left(\frac{h}{h_*}\right)\right] \tag{5.64}$$

where $h_*$ is the value of $h$ at $r_*$,

$$\ln r_* = \ln \bar{r} + \sigma^2(3-D) \tag{5.65}$$

$$\ln \theta_c = \ln(A\bar{r}^{3-D}) + \sigma^2(3-D)^2/2 \tag{5.66}$$

and $\bar{r}$ and $\sigma^2$ are the geometric mean radius and variance of the pore radius distribution, respectively.

Water-retention curves measured for 84 samples of a clay loam taken from a 12,000-m$^2$ field were used to determine the parameters $\theta_c$, $h_*$, and $\sigma$ in (5.64). Plots of $\ln r_*$ versus $\sigma^2$ and $\ln \theta_c$ versus $\sigma^2$ shown in Figures 5.14a and 5.14b, respectively, yielded nearly identical values for the fractal dimension $D$. In Figure 5.14a, the

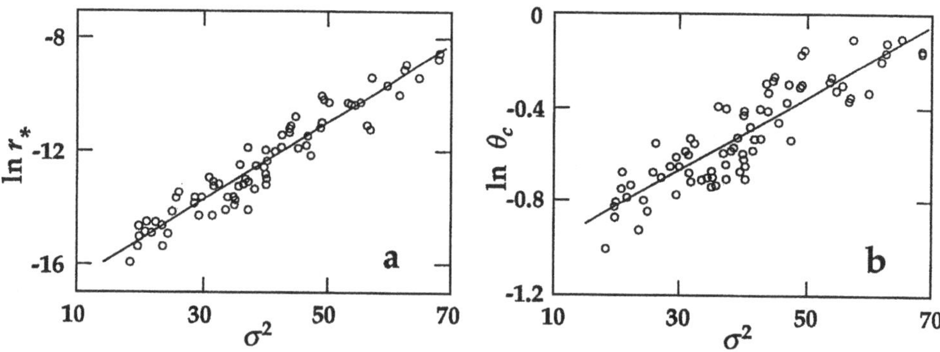


Figure 5.14. (a) Linear plot of $\ln r_*$ versus the variance of the pore radius distribution $\sigma^2$. (b) Linear plot of $\ln \theta_c$ versus the variance of the pore radius distribution $\sigma_2$. (From Pachepsky et al., 1995, with permission.)

value for $D$, according to equation (5.65), is 2.86, whereas in Figure 5.14b its value, according to equation (5.66), is 2.82. Results obtained from 60 samples of a loam from a 1,500-m$^2$ field were also encouraging. Hence, assuming that the geometric mean radius and the fractal dimension are constant for the field soil investigated, Pachepsky and associates suggested that the variance of the pore radius distribution $\sigma^2$ could be used as a single variable to quantify the spatial variability of a soil-water property. They further suggested that when the probability distribution for $\sigma^2$ and the values for parameters $A$ and $\bar{r}$ are known, the proposed scaling allows the generation of spatially noncorrelated random fields of $\theta(h)$.

## 5.8 The Efficacy of Scaling Field Soil-Water Behavior

Although a generalized theory for comprehensively scaling the behavior of field soil-water regimes does not yet exist, though its development may be said to be in the cradle stage, scaling has provided an encouraging degree of success for those coping with the heterogeneity of field soils. For examples indicating the progress achieved, see Warrick (1990), as well as a few additional examples indicated later.

With the potential to characterize the spatial variability of field soil-water properties, captured with one or more scaling factors, several investigators have used scale factors assumed to be spatially independent to simulate hydrologic processes as well as for measurements of the soil-water properties in fields (e.g., Luxmoore and Sharma, 1980; Bresler and Dagan, 1983; Dagan and Bresler, 1983; Tseng and Jury, 1993). Recognizing that soil-water properties should be spatially correlated, Jury et al. (1987) provided a method for obtaining different scale factors that included their spatial-correlation structure and illustrated its use with the data sets of Nielsen et al. (1973) and Russo and Bresler (1980a,b).

From scaling observations of infiltration, values for scaling factors have revealed differences in crop management. After measuring the rates of infiltration at 50 locations in a transect across an agricultural field, Hopmans (1989) scaled the modified Kostiakov equation in a manner similar to equation (5.22). The scale factors, shown in Figure 5.15, manifest different average values across the transect. The first portion of the transect was located in a region planted to sorghum, and the remainder was in fields that had been fallow prior to the infiltration observations. The larger mean value for $\bar{\alpha}$ in the region planted to sorghum probably reflects crop-root-induced differences in the soil's pore structure or in soil-water content at the time infiltration was measured.

Rockhold, Rossi, and Hills (1996) successfully used scale factors to conditionally simulate water flow and tritium transport measured at the Las Cruces trench site. Water-retention data from 448 core samples were scaled according to (5.2) into a scale mean curve using the Brooks and Corey model (5.25) to obtain values for $\alpha_h$. Parameters for soil-water retention were used in the Burdine (1953) relative-permeability

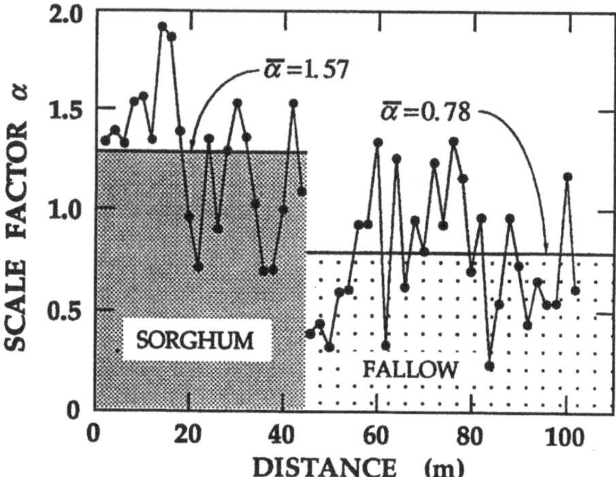


Figure 5.15. Scale factors for infiltration rate measured along a 100-m transect in a partially cropped field. (From Hopmans, 1989, with permission.)

model to estimate $K(\theta)$. Saturated values for the hydraulic conductivity $K_{Sl}$ were measured in the laboratory on the 448 soil cores, and saturated values for the hydraulic conductivity $K_{Sf}$ were measured in the field using a borehole permeameter at nearly 600 locations. The latter two sets of data were scaled according to equation (5.4). The probability distribution for each of the three scale factors was found to be lognormal. The horizontal variograms for the log-transformed scale factors for soil-water retention and field-measured $K_S$ (Figure 5.16 a, b) show remarkably similar spatial structures to about a 5-m lag. Interestingly, the log-transformed scale factors for $K_{Sl}$ measured in the laboratory on the same cores used for measuring $\theta(h)$ manifest (Figure 5.16c) virtually no spatial structure (spatially random behavior).

Although Rockhold et al. (1996) followed the advice of Jury et al. (1987) to quantify the spatial structure of the soil hydraulic properties for the Las Cruces trench site,

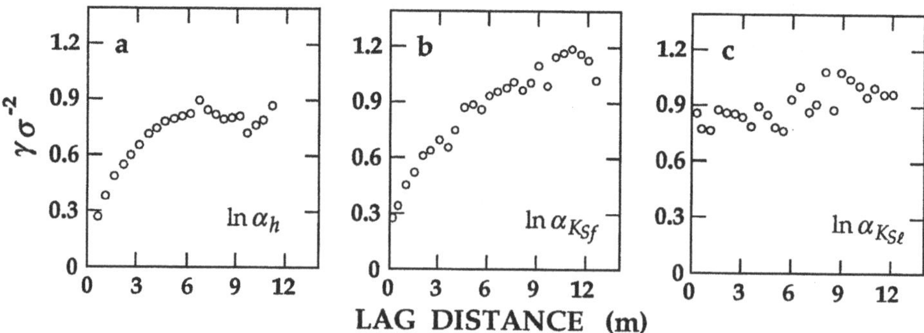


Figure 5.16. (a) Normalized horizontal variograms for ln-transformed scaling factors $\alpha_h$ for $\theta(h)$ determined in the laboratory on soil core samples, (b) $\alpha_{K_{Sf}}$ for field-measured saturated hydraulic conductivity, and (c) $\alpha_{K_{Sl}}$ for saturated hydraulic conductivity determined in the laboratory on soil core samples. (From Rockhold et al., 1996, with permission.)

they found it not necessary to use three stochastic variates to condition their simulations of water and solute transport. Using a constant value for $\theta_S$, constant values for the slopes of $s(h)$ and $K(s)$, and the same distribution of $\alpha$ values in equations (5.2) and (5.4) to condition the hydraulic properties of the field, simulations of water flow adequately agreed with those measured. They were sufficiently encouraged by their results to speculate that in the future the simulation of unsaturated water flow at field scale would evolve from its present-day stochastic analysis to a deterministic analysis.

## 5.9 Expectations

We expect that the intellectual curiosity derived from or supported by observations in the field and the laboratory will continue to kindle investigations into the scaling of soil-water regimes. Because potential avenues for the development of a comprehensive set of different kinds of scaling theories remain largely unexplored, opportunities to quantitatively ascertain the efficacy of scaling soil-water regimes in the field must await additional inquiry and creativity. Without a unified, comprehensive theory, our fragmented theoretical considerations can provide only criteria that are inadequate for success.

We do not anticipate abundant progress until a complete set of field-measured soil-water properties for several locations within at least one field can be simultaneously and directly observed, analyzed, and published. To date, in every study cited, as well as in studies not cited in this chapter, critical field measurements have been lacking. For example, Nielsen et al. (1973) and Shouse et al. (1992a,b) estimated the soil-water contents in fields from laboratory measurements of $\theta(h)$ on soil cores. Russo and Bresler (1980b) estimated the functions $\theta(h)$ and $K(\theta)$ from field observations of sorptivity and other parameters. Ahuja and Williams (1991) and Rockhold et al. (1996) estimated field values for $\theta(h)$ from measurements on soil cores analyzed in the laboratory. Although Eching et al. (1994) measured $\theta(z, t)$ in the field with a neutron meter, they recorded no observations of $h(z, t)$. Moreover, they offered no independent confirmation that the functional relations assumed for the hydraulic properties in the inverse method were descriptive of the field soil studied. On the other hand, both Eching et al. (1994) and Rockhold et al. (1996) explicitly showed that values for $K_S$ measured in the laboratory on soil cores were different from those measured *in situ*. In Figure 5.17, the laboratory-determined values for $K_S$ are an order of magnitude greater than those estimated in the field.

Progress toward improved scaling concepts should be accelerated as investigators take the opportunity to simultaneously study the details of both an experiment and a theory (e.g., Flühler, Ardakani, and Stolzy, 1976). Present-day scaling attempts are being confounded by failure to recognize that most experimental observations are subject to space- and time-dependent instrument responses (Baveye and Sposito, 1985). And more attention should be given to the consequences of selecting simplified

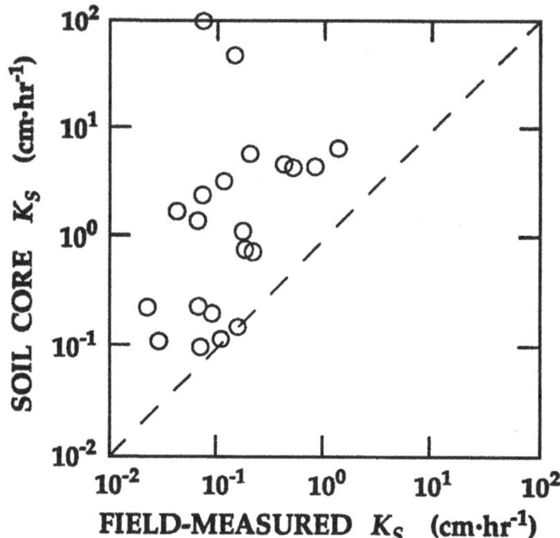


Figure 5.17. Values for water-saturated hydraulic conductivity $K_S$ measured in the laboratory versus those measured in the field. (From Eching et al., 1994, with permission.)

theoretical models to analyze and scale field-measured data (Tseng and Jury, 1993). Presently, no criteria have been established to ascertain the appropriate soil-depth or time intervals at which observations should be taken. The choice of horizontal spacings between observations remains *ad hoc*. Functional forms of soil hydraulic properties remain without theoretical foundation. Indeed, a dilemma persists regarding how to include "preferential" flow near water saturation. If $K_S$ is dominated by preferential flow, should the relative hydraulic-conductivity function $K(s)K_S^{-1}$ be scaled (Jury et al., 1987; Ahuja and Williams, 1991), or should that preferential flow be described by equations other than the Richards equation (e.g., Germann and Beven, 1985), and scaled independently? When and how should laboratory studies complement field investigations? Equipment and methods appropriate for taking the essential observations are readily available to those wishing to make a contribution to the development of scaling technology. Paradigms for scaling steady-state, one-directional Buckingham-Darcy flow are anticipated to be less restrictive than those for the Richards equation describing transient flow in one or more directions.

We believe that information derived from laboratory investigations at the soil-pore scale, obtained with computed microtomography, magnetic-resonance imaging, and other noninvasive techniques will improve the applications of the fractal concepts used by Tyler and Wheatcraft (1990) to describe $\theta(h)$ and by Shepard (1993) to calculate $K(\theta)$. The logical next step based on fractals will be to extend the descriptions and calculations to a field scale as other fractal properties and processes within field soils become better known and understood (Burrough, 1983a,b).

Eventually, appropriate scale factors for field-measured soil-water properties and processes will be measured in sufficient quantity and detail to permit analysis and documentation of their spatial and temporal statistical-variance structures across and within the landscape. With their values being linked to other soil properties through state-space analysis and other regionalized analyses of variables (e.g., Wendroth et al., 1993), we anticipate that new paradigms for local and regional scales of homogeneity in pedology and soil classification will emerge. With soil mapping units embracing magnitudes and distributions of spatial and temporal soil-water scale factors, unlimited opportunities will unfold. We expect the numerous uniquely scaled solutions for the Buckingham-Darcy and Richards equations now only theoretically available (e.g., Kutílek et al., 1991; Warrick and Hussen, 1993; Nachabe, 1996) to be extended to specific landscape and field regions categorized by mapping units described by information containing scale factors for their soil-water properties.

## References

Ahuja, L. R., Cassel, D. K., Bruce, R. R., and Barnes, B. B. 1989a. Evaluation of spatial distribution of hydraulic conductivity using effective porosity data. *Soil Sci.* 148:404–11.

Ahuja, L. R., Naney, J. W., Green, R. E., and Nielsen, D. R. 1984a. Macroporosity to characterize spatial variability of hydraulic conductivity and effects of land management. *Soil Sci. Soc. Am. J.* 48:699–702.

Ahuja, L. R., Naney, J. W., and Nielsen, D. R. 1984b. Scaling to characterize soil water properties and infiltration modeling. *Soil Sci. Soc. Am. J.* 48:970–3.

Ahuja, L. R., Nofziger, D. L., Swartzendruber, D., and Ross, J. D. 1989b. Relationship between Green and Ampt parameters based on scaling concepts and field-measured hydraulic data. *Water Resour. Res.* 25:1766–70.

Ahuja, L. R., and Williams, R. D. 1991. Scaling water characteristics and hydraulic conductivity based on Gregson-Hector-McGown approach. *Soil Sci. Soc. Am. J.* 55:308–19.

Baveye, P., and Sposito, G. 1985. Macroscopic balance equations in soils and aquifers; the case of space- and time-dependent instrumental response. *Water Resour. Res.* 21:1116–20.

Belmans, C., Wesselling, J. G., and Feddes, R. A. 1983. Simulation model of the water balance of cropped soil: SWATRE. *J. Hydrol.* 63:271–81.

Bouwer, H. 1966. Rapid field measurement of air entry value and hydraulic conductivity of a soil as significant parameters in flow system analysis. *Water Resour. Res.* 2:729–38.

Brakensiek, D. L., Rawls, W. J., Logsdon, S. D., and Edwards, W. M. 1992. Fractal description of macroporosity. *Soil Sci. Soc. Am. J.* 56:1721–3.

Bresler, E., and Dagan, G. 1983. Unsaturated flow in spatially variable fields. 2. Application of water flow models to various fields. *Water Resour. Res.* 19:421–8.

Brooks, R. H., and Corey, A. T. 1964. *Hydraulic Properties of Porous Media.* Hydrology Paper no. 3. Fort Collins: Colorado State University.

Burdine, N. T. 1953. Relative permeability calculations from size distribution data. *Am. Inst. Min. Metal. Pet. Eng.* 198:71–7.

Burrough, P. A. 1983a. Multiscale sources of spatial variation in soil. I. The application of fractal concepts to nested levels of soil variation. *J. Soil Sci.* 34:577–97.

Burrough, P. A. 1983b. Multiscale sources of spatial variation in soil. II. A non-Brownian fractal model and its application in soil survey. *J. Soil Sci.* 34:599–620.

Campbell, G. S. 1974. A simple method for determining unsaturated hydraulic conductivity from moisture retention data. *Soil Sci.* 117:311–14.

Corey, G. L., and Corey, A. T. 1967. Similitude for drainage of soils. *J. Irri. Drain. Div., Proc. ASCE.* 93:3–23.

Dagan, G., and Bresler, E. 1983. Unsaturated flow in spatially variable fields. 1. Derivation of models of infiltration and redistribution. *Water Resour. Res.* 19:413–20.

Eching, S. O., Hopmans, J. W., and Wallender, W. W. 1994. Estimation of in situ unsaturated soil hydraulic functions from scaled cumulative drainage data. *Water Resour. Res.* 30:2387–94.

Ernst, L. F., and Feddes, R. A. 1979. *Invloed van grondwateronttrekking voor beregening en drinkwater op de grondwaterstand.* Nota 1116. Wageningen, The Netherlands: ICW.

Flühler, H., Ardakani, M. S., and Stolzy, L. H. 1976. Error propagation in determining hydraulic conductivities from successive water content and pressure head profiles. *Soil Sci. Soc. Am. J.* 40:830–6.

Friesen, W. I., and Mikula, R. J. 1987. Fractal dimensions of coal particles. *J. Coll. Interface Sci.* 120:263–71.

Gardner, W. R. 1956. Calculation of capillary conductivity from pressure plate outflow data. *Soil Sci. Soc. Am. Proc.* 20:317–20.

Germann, P. F., and Beven, K. 1985. Kinematic wave approximation to infiltration into soils with sorbing micropores. *Water Resour. Res.* 21: 990–6.

Green, W. H., and Ampt, G. A. 1911. Studies on soil physics: I. Flow of air and water through soils. *J. Agric. Sci.* 4:1–24.

Gregson, K., Hector, D. J., and McGowan, M. 1987. A one-parameter model for the soil water characteristic. *J. Soil Sci.* 38:483–6.

Hills, R. G., Hudson, D. B., and Wierenga, P. J. 1989. Spatial variability at the Las Cruces trench site. In: *Proceedings of the International Workshop on Indirect Methods for Estimating the Hydraulic Properties of Unsaturated Soils,* ed. M. T. van Genuchten, F. J. Leij, and L. J. Lund, pp. 529–38. Riverside: University of California Press.

Hopmans, J. W. 1988. Treatment of spatially variable groundwater level in one-dimensional stochastic unsaturated water flow model. *Agric. Water Mgt.* 15:19–36

Hopmans, J. W. 1989. Stochastic description of field-measured infiltration data. *Trans. ASAE* 32:1987–93.

Hopmans, J. W., Hendricks, J. M. H., and Selker, J. S. In press. Emerging techniques for vadose zone characterizarion. In: *Vadose Zone Hydrology: Cutting across Disciplines*, ed. J. W. Hopmans and M. B. Parlange. Oxford University Press.

IAEA. 1984. *Field Soil-Water Properties Measured through Radiation Techniques.* IAEA-TEC-DOC-312. Vienna: International Atomic Energy Agency.

Jury, W. A., Russo, D., and Sposito, G. 1987. The spatial variability of water and solute transport properties in unsaturated soil. II. Scaling models of water transport. *Hilgardia* 55:33–56.

Klute, A., and Wilkinson, G. E. 1958. Some tests of the similar media concept of capillary flow: I. Reduced capillary conductivity and moisture characteristic data. *Soil Sci. Soc. Am. Proc.* 22:278–81.

Kutílek, M., Zayani, K., Haverkamp, R., Parlange, J.-Y., and Vachaud, G. 1991. Scaling of Richards' equation under invariant flux boundary conditions. *Water Resour. Res.* 27:2181–5.

Lax, P. D. 1972. The formation and decay of shock waves. *Am. Math. Monthly* 79:227–41.

Libardi, P. L., Reichardt, K., Nielsen, D. R., and Biggar, J. W. 1980. Simple field methods for estimating soil hydraulic conductivity. *Soil Sci. Soc. Am. J.* 44:3–7.

Luxmoore, R. J., and Sharma, M. L. 1980. Runoff responses to soil heterogeneity: experimental and simulation comparisons for two contrasting watersheds. *Water Resour. Res.* 16:675–84.
Miller, E. E., and Miller, R. D. 1955a. Theory of capillary flow: I. Experimental information. *Soil Sci. Soc. Am. Proc.* 19:271–5.
Miller, R. D., and Bresler, E. 1977. A quick method of estimating soil water diffusivity functions. *Soil Sci. Soc. Am. J.* 41:1020–2.
Miller, R. D., and Miller, E. E. 1955b. Theory of capillary flow: II. Practical implications. *Soil Sci. Soc. Am. Proc.* 19:267–71.
Nachabe, M. H. 1996. Microscopic capillary length, sorptivity, and shape factor in modeling the infiltration rate. *Soil Sci. Soc. Am. J.* 60:957–62.
Nielsen, D. R., Biggar, J. W., and Erh, K. T. 1973. Spatial variability of field measured soil water properties. *Hilgardia* 42:215–59.
Pachepsky, Y. A., Shcherbakov, R. A., and Korsunskaya, L. P. 1995. Scaling of soil water retention using a fractal model. *Soil. Sci.* 159:99–104.
Philip, J. R. 1955. Numerical solution of equations of the diffusion type with diffusivity concentration-dependent. *Trans. Faraday Soc.* 51:885–92.
Philip, J. R. 1957. Numerical solution of equations of the diffusion type with diffusivity concentration-dependent II. *Australian J. Phys.* 10:29–42.
Rao, P. S. C., Jessup, R. E., Hornsby, A. G., Cassel, D. K., and Pollans, W. A. 1983. Scaling soil microhydrologic properties of Lakeland and Konawa soils using similar media concepts. *Agric. Water Mgt.* 6:681–4.
Reichardt, K., Libardi, P. L., and Nielsen, D. R. 1975. Unsaturated hydraulic conductivity determination by a scaling technique. *Soil Sci.* 120:165–8.
Reichardt, K., Nielsen, D. R., and Biggar, J. W. 1972. Scaling of horizontal infiltration into homogeneous soils. *Soil Sci. Soc. Am. Proc.* 36:241–5.
Richards, L. A., Gardner, W. R., and Ogata, G. 1956. Physical processes determining water loss from soil. *Soil Sci. Soc. Am. Proc.* 20:310–14.
Rockhold, M. L., Rossi, R. E., and Hills, R. G. 1996. Application of similar media scaling and conditional simulation for modeling water flow and tritium transport at the Las Cruces trench site. *Water Resour. Res.* 32:595–609.
Ruark, A. E. 1935. Inspectional analysis: a method which supplements dimensional analysis. *J. Elisha Mitchell Sci. Soc.* 51:127–33.
Russo, D., and Bresler, E. 1980a. Scaling soil hydraulic properties of a heterogeneous field. *Soil Sci. Soc. Am. J.* 44:681–4.
Russo, D., and Bresler, E. 1980b. Field determinations of soil hydraulic properties for statistical analyses. *Soil Sci. Soc. Am. J.* 44:697–702.
Russo, D., and Bresler, E. 1981. Soil hydraulic properties as stochastic processes: I. An analysis of field spatial variability. *Soil Sci. Soc. Am. J.* 45:682–7.
Sharma, M. L., Gander, G. A., and Hunt, C. G. 1980. Spatial variability of infiltration in a watershed. *J. Hydrol.* 45:101–22.
Shepard, J. S. 1993. Using a fractal model to calculate the hydraulic conductivity function. *Soil Sci. Soc. Am. J.* 57:300–7.
Shouse, P. J., Sisson, J. B., de Rooij, G., Jobes, J. A., and van Genuchten, M.T. 1992a. Application of fixed gradient methods for estimating soil hydraulic conductivity. In: *Proceedings of the International Workshop on Indirect Methods for Estimating the Hydraulic Properties of Unsaturated Soils*, ed. M. T. van Genuchten, F. J. Leij, and L. J. Lund, pp. 675–84. Riverside: University of California Press.
Shouse, P. J., Sisson, J. B., Ellsworth, T. R., and Jobes, J. A. 1992b. Estimating in situ unsaturated hydraulic properties of vertically heterogeneous soils. *Soil Sci. Soc. Am. J.* 56:1673–9.

Shouse, P. J., van Genuchten, M. T., and Sisson, J. B. 1991. A gravity-drainage/scaling method for estimating the hydraulic properties of heterogeneous soils. In: *Hydrological Interactions between Atmosphere, Soils and Vegetation*, ed. G. Kienitz, pp. 281–91. Publication no. 204. Wallingford, Oxfordshire, UK: IAHS Press, Institute of Hydrology.

Simmons, C. S., Nielsen, D. R., and Biggar, J. W. 1979. Scaling of field-measured soil-water properties. I. Methodology. II. Hydraulic conductivity and flux. *Hilgardia* 47:74–173.

Sisson, J. B. 1987. Drainage from layered field soils: fixed gradient models. *Water Resour. Res.* 23:2071–5.

Sisson, J. B., Ferguson, A. H., and van Genuchten, M. T. 1980. Simple method for predicting drainage from field plots. *Soil Sci. Soc. Am. J.* 44:1147–52.

Sisson, J. B., and van Genuchten, M. T. 1991. An improved analysis of gravity drainage experiments for estimating the unsaturated soil hydraulic functions. *Water Resour. Res.* 27:569–75.

Snyder, V. A. 1996. Statistical hydraulic conductivity models and scaling of capillary phenomena in porous media. *Soil Sci. Soc. Am. J.* 60:771–4.

Sposito, G. 1990. Lie invariance of the Richards equation. In: *Dynamics of Fluids in Hierarchical Porous Media*, ed. J. Cushman, pp. 327–47. New York: Academic Press.

Sposito, G., and Jury, W. A. 1985. Inspectional analysis in the theory of water flow through unsaturated soil. *Soil Sci. Soc. Am. J.* 49:791–8.

Sposito, G., and Jury, W. A. 1990. Miller similitude and generalized scaling analysis. In: *Scaling in Soil Physics: Principles and Applications*, ed. D. Hillel and D. E. Elrick, pp. 13–22. SSSA special publication no. 25. Madison, WI: Soil Science Society of America.

Swartzendruber, D., and Youngs, E. G. 1974. A comparison of physically-based infiltration equations. *Soil Sci.* 117:165–7.

Tillotson, P. M., and Nielsen, D. R. 1984. Scale factors in soil science. *Soil Sci. Soc. Am. J.* 48:953–9.

Tseng, P.-H., and Jury, W. A. 1993. Simulation of field measurement of hydraulic conductivity in unsaturated heterogeneous soil. *Water Resour. Res.* 29:2087–99.

Tyler, S. W., and Wheatcraft, S. W. 1990. Fractal processes in soil water retention. *Water Resour. Res.* 26:1047–54.

Vogel, T., Cislerova, M., and Hopmans, J. W. 1991. Porous media with linearly variable hydraulic properties. *Water Resour. Res.* 27:2735–41.

Warrick, A. W. 1990. Application of scaling to the characterization of spatial variability in soils. In: *Scaling in Soil Physics: Principles and Applications*, ed. D. Hillel and D. E. Elrick, pp. 39–51. SSSA special publication no. 25. Madison, WI: Soil Science Society of America.

Warrick, A. W., and Hussen, A. A. 1993. Scaling of Richards' equation for infiltration and drainage. *Soil Sci. Soc. Am. J.* 57:15–18.

Warrick, A. W., Mullen, G. J., and Nielsen, D. R. 1977. Scaling field properties using a similar media concept. *Water Resour. Res.* 13:355–62.

Watson, K. K. 1966. An instantaneous profile method for determining the hydraulic conductivity of unsaturated porous materials. *Water Resour. Res.* 2:709–15.

Wendroth, O., Katul, G. G., Parlange, M. B., Puente, C. E., and Nielsen, D. R. 1993. A non-linear filtering approach for determining hydraulic conductivity functions in field soils. *Soil Sci.* 56:293–301.

Youngs, E. G., and Price, R. I. 1981. Scaling of infiltration behavior in dissimilar porous materials. *Water Resour. Res.* 17:1065–70.

# 6

# Scaling Invariance and the Richards Equation

GARRISON SPOSITO

## 6.1 Introduction

A culminating step in the early development of soil-water physics as a predictive science was taken 65 years ago with the appearance of the Richards equation (Richards, 1931). This parabolic partial differential equation describes the temporal and spatial dependences of the volumetric water content during infiltration in a nondeformable, unsaturated porous medium at fixed temperature and applied pressure. The solutions of the Richards equation, whether analytical or numerical, play a key role in the mathematical theory of hydrologic processes near the land surface (Hillel, 1980; Kühnel et al., 1990; Wang and Dooge, 1994; Sposito, 1995).

The physical basis of the Richards equation lies in fundamental postulates about continuum descriptions of multiphase systems and the mechanisms of transport for water in an unsaturated porous medium (Sposito, 1987). The mechanistic postulates invoke empirical "transport coefficients" (e.g., hydraulic conductivity or soil-water diffusivity) whose functional dependence on volumetric water content is presumed known through laboratory or field measurement. The exact nature of the functional dependence differs among porous media of various compositions and textures and therefore among locations in a heterogeneous field soil or sedimentary deposit whose physical properties vary spatially. This medium-specific (or site-specific) character of the transport coefficients adds great complexity to any use of the Richards equation in modeling vadose-zone hydrology (Warrick and Nielsen, 1980).

One option for reducing the complexity is through an application of "similitude analysis" to the Richards equation (Miller, 1980). Similitude analysis is essentially a systematic procedure for casting a differential equation, along with its attendant initial and boundary conditions, into a scaled form, such that it exhibits only scaled dependent and independent variables and therefore has, in principle, scaled solutions. If this scaling can be accomplished, the differential equation is said to be "scale-invariant," which loosely means that the mathematical function that solves the equation does not change in form irrespective of the scale of measurement for the variables on which it depends. For a partial differential equation whose independent variables are time and position coordinates, like the Richards equation, scale invariance implies that the solutions retain the same mathematical dependence on these variables, irrespective of the specific time and space scales over which the solutions are to be applied. Each such scale may represent a certain relative magnification or reduction of time and space, but this kind of size sorting does not affect the mathematical relationship between the solution and the time or space coordinates on which it depends. Porous-media systems that are distinguished solely by differing degrees of space–time magnification or reduction thus are "similar" to one another. The prediction of water movement through such similar porous media, insofar as it can be represented by a single partial differential equation, is not affected by the fact that widely varying space and time scales may exist for water movement in nature.

The formal mathematical apparatus for performing similitude analysis of a partial differential equation (i.e., group theory) is well established (Bluman and Kumei, 1989) and in fact has been applied to the Richards equation (Sposito, 1990; Sposito and Jury, 1990; Yung, Verburg, and Baveye, 1994). Fundamentally, one examines the Richards equation for its response when the volumetric water content, time variable, and position variable are multiplied by positive scale factors. This examination is not trivial, because the transport coefficients in the Richards equation depend strongly on the volumetric water content, and this dependence must surely influence the way in which a transport coefficient will scale when the volumetric water content is scaled. If the transport coefficients do not scale compatibly with the independent variables in the Richards equation, scale invariance of the equation is not possible.

Similitude analysis in soil-water physics began 40 years ago (Miller and Miller, 1956) with a less formal approach than the group-theoretical methods just described in connection with the Richards equation (Sposito, 1990). The arguments given initially were physical in concept, relying on an intuitive picture of porous media at microscopic spatial scales and involving a universal scale factor for soil-water properties based solely on particle size (Miller, 1980). Subsequent researchers (see Chapter 5) recognized the need for a multiplicity of scale factors and for a decoupling of their physical significance from the precise arrangement and granularity of the solid matrix in a conducting porous medium (Warrick and Nielsen, 1980). That revision led to a broader concept of scaling that could be applied readily to water movement in field soils and sediments (Warrick, 1990). A degree of theoretical coherence within this empirical approach to similitude was later brought about by the application of Lie-group techniques to the Richards equation itself (Sposito, 1990). In this chapter, that sequence of developments is discussed in a concise fashion. Emphasis is placed on the inherent symmetry properties of the Richards equation, together with the heuristic search for natural porous media that reflect them.

## 6.2 Experimental Similitude

### *6.2.1 Miller-Miller Similitude*

Miller-Miller similitude is founded on the well-known physical relationship between equivalent pore size (or radius) and matric potential in a porous medium (Miller and Miller, 1956). If $\psi$ [L] is matric potential [taken positive-valued as "soil-water suction" (Hillel, 1980, p. 146)] to which corresponds the pore radius $\ell$ [L], then the equivalence relationship is (Hillel, 1980, p. 47)

$$\psi(\ell) \equiv 2\sigma/\rho g\ell \tag{6.1}$$

where $\sigma$ [$MT^{-2}$] is the surface tension of an air–water interface, $\rho$ [$ML^{-3}$] is the mass density of liquid water, and $g$ [$LT^{-2}$] is gravitational acceleration. Equation (6.1) defines a pore radius that is equivalent to a value of $\psi > 0$. An equivalent distribution of pore sizes is determined by measuring the water-retention curve ["moisture characteristic" (Hillel, 1980, p. 148)], that is, by measuring $\theta$ as a function of $\psi$, where $\theta$ [1] is volumetric water content. A smooth graph of $s \equiv \theta/\theta_s$ versus $\ell$ calculated with equation (6.1) represents the cumulative pore-size distribution function for a porous medium, where $\theta_s$ is the volumetric water content at visible saturation (Danielson and Sutherland, 1986, fig. 18.5).

Two porous media are Miller-similar if a scale factor $\lambda$ [L] can be found for each such that identical values of the dimensionless ratio $\ell/\lambda$ imply geometric similitude in the microscopic arrangements of both pores and solid particles (Miller and Miller, 1956; Miller, 1980). Miller-similar porous media thus have microscopic structures that look identical except for length scale (or magnification), in the same sense that similar triangles in Euclidean geometry look identical except for length scale. This visual self-similarity is exemplified by fractal models of soil structure (Perrier et al., 1995). It is important to note that the scale factor $\lambda$, despite its evident relation to particle size, is applied to the entire porous medium, not just to the solid particles (see Figure 5.1).

Miller-similar porous media are in similar states (Miller and Miller, 1956) whenever they exhibit the same values of $\lambda\psi$, with each $\lambda$ being a characteristic length for a porous medium. Because $\lambda$ applies to both pores and solid particles, and $\theta$ is a ratio of the water-filled-pore volume to the total porous-medium volume, the volumetric water content will have the same value for porous media in similar states (Miller and Miller, 1956). Therefore, the water-retention curves for Miller-similar porous media should have superposable graphs when scaled as $\theta(\lambda\psi)$. Laboratory tests of this scaling relationship indicate that it reasonably describes well-sorted sands, but not natural soils containing a broad range of particle sizes (Miller, 1980; Tillotson and Nielsen, 1984). In field soils, porosity ($\theta_s$) always varies from place to place, and therefore the proposed invariance of $\theta$ while $\lambda$ varies (spatially, as it must) necessarily will fail.

Hydraulic conductivity $K$ [$LT^{-1}$] is brought into Miller-Miller similitude by making the assumption, consistent with equation (6.1), that the mechanisms of water movement in unsaturated porous media involve viscous flow and capillary forces (Miller and Miller, 1956). This physical assumption is made in addition to the postulate of geometric similitude at a microscopic level of resolution. It leads to the scaling relationship

$$K_*(\psi_*) = (1/\lambda^2)K(\lambda\psi) \tag{6.2}$$

where $K_*$ is scaled hydraulic conductivity, and

$$\psi_* = \lambda\psi \tag{6.3}$$

is scaled matric potential. In principle, equation (6.2) can be applied to field soils whose spatial heterogeneity requires a spatially varying $\lambda$. One imagines that such a heterogeneous field soil is a union of approximately homogeneous domains, each of which is represented by a single characteristic length scale $\lambda$. Heterogeneity then is reflected only in the spatial variability of the length scale. If the homogeneous domains are Miller-similar, however, the functional relationships among soil-water properties, as in equation (6.2), will be uniform. Thus, if $K(\psi)$ represents the dependence of the hydraulic conductivity on the matric potential in the neighborhood of some point in a soil, and $K'(\psi')$ represents the same at another spatial point, then a scaling relationship based on equation (6.2),

$$K'(\lambda'\psi')/\lambda'^2 = K(\lambda\psi)/\lambda^2 \tag{6.4}$$

can be used to relate the two hydraulic conductivities, assuming that their functional dependences on $\psi$ have the same mathematical form. If this kind of scaling relationship holds, the study of water movement in field soils is greatly simplified, because the complicated, explicit dependence of soil-water properties on spatial position has been removed. Table 6.1 summarizes the scaling relationships in Miller-Miller similitude. Given that the liquid phase is always water, dependences on liquid-phase properties have been suppressed, as in equation (6.2). Note that the factor relating $K_*$ to $K$ is postulated to be the inverse square of that relating $\psi_*$ to $\psi$. This relation arises from the assumption that the Navier-Stokes equation of fluid mechanics applies to water movement in a soil pore (Miller, 1980, p. 309).

### 6.2.2 Warrick-Nielsen Similitude

Warrick-Nielsen similitude is a generalization of Miller-Miller similitude that focuses directly on the spatial variability of soil-water properties at field scales (Warrick,

Table 6.1. *Scaling relationships in Miller-Miller similitude*

| Soil-water property | Symbol and dimensions | Scaling relationship[a] |
|---|---|---|
| Volumetric water content | $\theta$ [1] | $\theta_* = \theta$ |
| Matric potential | $\psi$ [L] | $\psi_* = \lambda\psi$ |
| Hydraulic conductivity | $K$ [$LT^{-1}$] | $K_* = K/\lambda^2$ |

[a]An asterisk denotes a scaled property that is the same for all Miller-similar porous media or domains in a field soil; $\lambda$ denotes a scale factor for the pores and solid particles combined.

Mullen, and Nielsen, 1977; Warrick and Nielsen, 1980; Warrick, 1990). These soil-water properties usually can be extracted through an inverse analysis of field-plot infiltration data based on the one-dimensional Richards equation (Richards, 1931),

$$\partial\theta/\partial t = \partial[D(\theta)(\partial\theta/\partial z)]/\partial z - (\partial K/\partial\theta)(\partial\theta/\partial z) \tag{6.5}$$

where

$$D \equiv K(\partial\psi/\partial\theta) \tag{6.6}$$

is the soil-water diffusivity (with dimensions $L^2T^{-1}$), and the position coordinate $z$ [L] is restricted to nonnegative values. The functions $D(\theta)$ and $K(\theta)$ are "transport coefficients" that appear in the Buckingham flux law, which traditionally is combined with a differential law of mass balance to derive equation (6.5) (Sposito, 1987). These transport coefficients depend strongly on $\theta$, leading to the well-known nonlinearity of the Richards equation (Hillel, 1980, Ch. 9).

Field experiments for which a mathematical description by a one-dimensional Richards equation is appropriate have been discussed in detail by Nielsen, Biggar, and Erh (1973), Simmons, Nielsen, and Biggar (1979), Warrick and Nielsen (1980), and Rao et al. (1983). In outline, the experiments involve instrumented plots established at a number of sites over a relatively large land area. The field plots are managed as individual irrigated units through which the infiltration of water is monitored by measurements of the volumetric water content and the matric potential. Water movement in soil profiles beneath the field plots is assumed to be described by an area-averaged Richards equation in which only the vertical position coordinate appears, as in equation (6.5).

The scaling interpretation of the field-plot data is based on the assumption that water movement as a whole can be represented by a suitable average of the behaviors observed in individual plots. The averaging process is designed to produce a fieldwide mean Richards equation of the form (Warrick and Amoozegar-Fard, 1979)

$$\partial\theta/\partial t = \partial[D_m(S)(\partial S/\partial z)]/\partial z - (\partial K_m/\partial S)(\partial S/\partial z) \qquad (t > 0) \tag{6.7}$$

where $K_m(S)$ is a mean hydraulic conductivity,

$$D_m(S) = K_m(S)(\partial\psi_m/\partial S) \tag{6.8}$$

is a mean soil-water diffusivity, $\psi_m(S)$ is a mean matric potential, and

$$S \equiv (\theta - \theta_r)/(\theta_s - \theta_r) \tag{6.9}$$

is the relative saturation, with $\theta_r$ being the "residual water content," a measure of the volume of water, immobilized in aggregates or narrow pores, that does not participate in flow on the time scale over which equation (6.5) is applied (Hillel, 1980, p. 212). The fieldwide mean transport coefficients and matric potential are connected to their plot-specific counterparts by the Warrick-Nielsen scaling relationships (Warrick and Nielsen, 1980),

$$\alpha_P\psi_P(S) = \psi_m(S) \tag{6.10a}$$

$$K_P(S)/\omega_P^2 = K_m(S) \tag{6.10b}$$

$$\alpha_P D_P(S)/\omega_P^2 = D_m(S) \tag{6.10c}$$

where $\alpha_P$ and $\omega_P$ are positive scale factors. [The scaling relationship in equation (6.10c) follows from those in equations (6.10a) and (6.10b) and from equation (6.8).] Equations (6.10) are generalizations of equation (6.4), in that $\alpha_P$ and $\omega_P$ are considered to be dimensionless ratios of $\lambda$-type scale factors, but $\alpha_P$ and $\omega_P$ are not required to be equal in value.

The estimation of $\psi_m$, $K_m$, and the set of $\alpha_P$ and $\omega_P$ in terms of measured values of $\psi_P(S)$ and $K_P(S)$ has been discussed by Simmons et al. (1979), Jury, Russo, and Sposito (1987), Hopmans (1987), and Rockhold, Rossi, and Hills (1996). Given sets of $\psi_P$ and $K_P$ data, one can invoke model expressions to describe their $S$ dependence and then fit these expressions, over a range of $S$ values, to a fieldwide set of data subject to equations (6.10a) and (6.10b), along with a normalization constraint on $\alpha_P$ and $\omega_P^2$ (Simmons et al., 1979). Alternatively, one can use measured values of $\psi_P$ and $K_P$, at fixed values of $S$, to estimate parameters in an assumed fieldwide cumulative distribution function for these two soil-water properties, and then use the empirical distributions to calculate $\psi_m$ and $K_m$ as averages subject to equations (6.10a) and (6.10b), along with chosen normalizations of $\alpha_P$ and $\omega_P^2$ (Jury et al., 1987). This latter method requires a larger data set than the former, but it trades the need to postulate explicit model forms of $\psi_P(S)$ and $K_P(S)$ for the less stringent assumption that parametric cumulative distribution functions exist for spatially varying soil-water properties (Warrick and Nielsen, 1980).

Tests of Warrick-Nielsen similitude in field soils [often with $\theta_r$ neglected in equation (6.9)] have been reported by Warrick et al. (1977), Simmons et al. (1979), Russo

Table 6.2. *Scaling relationships in Warrick-Nielsen similitude*

| Soil-water property | Symbol and dimensions | Scaling relationship[a] |
|---|---|---|
| Relative saturation | $S$ [1] | $S_m = S_P{}^b$ |
| Matric potential | $\psi_P$ [L] | $\psi_m = \alpha_P \psi_P$ |
| Hydraulic conductivity | $K_P$ [LT$^{-1}$] | $K_m = K_P/\omega_P^2$ |

[a]The symbol $m$ denotes a fieldwide average, whereas $P$ denotes a field-plot-specific value.
[b]The value of $S$ is held fixed when the Warrick-Nielsen scaling relationships are applied to connect plot-specific ($P$) and fieldwide ($m$) properties (Warrick, 1990).

and Bresler (1980), Sharma, Gander, and Hunt (1970), Warrick and Nielsen (1980), Rao et al. (1983), Jury et al. (1987), Hopmans (1987), Warrick (1990), and Rockhold et al. (1996). Those evaluations of the macroscopic similitude implied by equation (6.10) involved measurement of the scale factors $\alpha_P$, $\omega_P$, and $\theta_s$ ($\theta_r = 0$), along with an examination of their effectiveness in reducing field-measured matric potentials and hydraulic conductivities through equations (6.10a) and (6.10b). Warrick (1990) has stressed the innovation of replacing $\theta$ by $S$ – to avoid the problem of a spatially varying $\theta_s$ – and the need to test equation (6.10) (at fixed values of $S_P = S_m$) without assuming that $\alpha_P$ and $\omega_P$ are equal (Table 6.2) (see also Chapter 5).

Simmons et al. (1979, fig. 5) found a linear-regression relation, with unit slope, between the measured values of $\omega_P$ and $\alpha_P$ in their infiltration field plots at depths of 0.6 and 1.2 m. Very similar results were found by Russo and Bresler (1980) in their analysis of $K(S)$ and $\psi(S)$ for a field soil in Israel and by Hopmans (1987) for watershed soils in The Netherlands. Warrick et al. (1977), however, observed only that $\omega_P$ and $\alpha_P$ were highly correlated statistically for the data sets they analyzed, but were not related linearly. Comparable findings were reported by Sharma et al. (1980) and Rockhold et al. (1996), but Rao et al. (1983) could not even find a statistical correlation between $\omega_P$ and $\alpha_P$. Jury et al. (1987) reanalyzed the data sets developed by Nielsen et al. (1973) and Russo and Bresler (1981), using methods that first removed any nonstochastic spatial variability ("drift") from $\alpha_P$ and $\omega_P$ before tests of statistical correlation were performed. They then found only a weak statistical correlation between the "random parts" of the two scale factors.

## 6.3 Scaling Invariance

### *6.3.1 Pore-Size Distribution*

We have seen experimental similitude evolve from the postulate that the volumetric water content is scale-invariant, while the matric potential and hydraulic conductivity are scaled with a single scale (or magnification) factor for length (Table 6.1), to the postulate that the relative saturation is scale-invariant, while the matric potential and hydraulic conductivity are scaled with different dimensionless scale factors (Table 6.2). To be sure, Miller-Miller similitude was conceived originally to apply only to uniformly packed columns of porous media whose water-flow properties are investigated in the laboratory (Miller, 1980; Youngs, 1990), whereas the broader Warrick-Nielsen similitude was designed specifically to reduce the complexity of soil-water behavior in spatially heterogeneous field soils (Warrick, 1990; Sposito and Jury, 1990). Field-soil spatial variability necessarily requires greater flexibility in the scaling concepts that would attempt to capture and simplify it.

Miller-Miller similitude requires scale invariance of the volumetric water content simply because the scale factor $\lambda$ applies to both the pore space and the solid matrix of a porous medium (Miller, 1980, fig. 12.1). If the water content (or, equivalently, the porosity) is an invariant of scaling, $\lambda$ necessarily must reflect the geometric arrangements of both the pores and the solid particles, because $\theta$ is the ratio of the pore-space volume to the porous medium volume. On the other hand, if only the pore-size distribution is scale-invariant, then $\lambda$ will be associated only with the pore space, as implied already in equation (6.1), and the possibility then will exist for the solid-particle arrangement to have a different scale factor. Indeed, the concept of similar states, based on equation (6.1), actually does not require invariance of the volumetric water content under the scaling of $\psi$, and a valid alternative is to postulate invariance of the pore-size distribution instead (Sposito and Jury, 1990). The conceptual difference involved has simply to do with the physical interpretation of the scale factor $\lambda$. Invariance of the pore-size distribution under scaling implies that similar media will show identical mathematical forms for $S(\lambda\psi)$, but not for $\theta(\lambda\psi)$. This weaker concept of similitude has the advantage of being compatible with the broad range of particle sizes and the nonuniform porosity that are typical of field soils.

Taken as a pore-size distribution, the relative saturation $S$ can be pictured mathematically as a positive-valued function of the (continuous) equivalent pore radius $\ell$ and of the two parameters, $\ell_{\min}$ and $\ell_{\max}$ corresponding to the smallest and largest values of $\ell$ in a given porous medium. Thus,

$$S = S(\ell; \ell_{\min}, \ell_{\max}) \tag{6.11}$$

with the constraints

$$S(\ell_{\min}; \ell_{\min}, \ell_{\max}) = 0 \quad \text{and} \quad S(\ell_{\max}; \ell_{\min}, \ell_{\max}) = 1 \tag{6.12}$$

providing operational definitions of $\ell_{\min}$ and $\ell_{\max}$, as used in a pore-size distribution measurement (Danielson and Sutherland, 1986). Equation (6.11) is subject to the stability condition

$$(\partial S/\partial \ell)_{\ell_{\min}, \ell_{\max}} > 0 \qquad (\ell_{\min} < \ell < \ell_{\max}) \tag{6.13}$$

These physical constraints apply to any explicit mathematical expression introduced on the right side of equation (6.11).

If $S(\ell; \ell_{\min}, \ell_{\max})$ is scale-invariant, then for any $\alpha > 0$,

$$S(\alpha\ell; \alpha\ell_{\min}, \alpha\ell_{\max}) = S(\ell; \ell_{\min}, \ell_{\max}) \tag{6.14}$$

This property is sufficient to permit equation (6.11) to be written as (Hankey and Stanley, 1972, theorem 3)

$$S = F(\ell/\ell_{\max}; \ell_{\min}/\ell_{\max}) \tag{6.15}$$

where the right side is a function defined by replacing the three arguments on the right side of equation (6.11) with the ratios $\ell/\ell_{\max}$, $\ell_{\min}/\ell_{\max}$, and 1, respectively (Hankey and Stanley, 1972). The corresponding changes in equations (6.12) and (6.13) are

$$F(\ell_{\min}/\ell_{\max}; \ell_{\min}/\ell_{\max}) = 0 \tag{6.16}$$

$$F(1; \ell_{\min}/\ell_{\max}) = 1 \tag{6.17}$$

$$[\partial F(u; \ell_{\min}/\ell_{\max})/\partial u]_{\ell_{\min}/\ell_{\max}} > 0 \qquad (\ell_{\min}/\ell_{\max} < u < 1) \tag{6.18}$$

Equation (6.16) suggests that $F$ can be expressed mathematically as the positive difference of a power of $(\ell/\ell_{\max})$ from the same power of $(\ell_{\min}/\ell_{\max})$:

$$F(\ell/\ell_{\max}; \ell_{\min}/\ell) = N\left[(\ell/\ell_{\max})^{\lambda_B} - (\ell_{\min}/\ell_{\max})^{\lambda_B}\right] \tag{6.19}$$

where the exponent $\lambda_B > 0$ because of equation (6.18), and $N > 0$ is a normalization constant to be determined by the adherence of equation (6.19) to equation (6.17). In fact, application of equation (6.17) yields

$$N = 1 - (\ell_{\min}/\ell_{\max})^{\lambda_B} \tag{6.20}$$

and therefore

$$S = \left[(\ell/\ell_{\max})^{\lambda_B} - (\ell_{\min}/\ell_{\max})^{\lambda_B}\right]/\left[1 - (\ell_{\min}/\ell_{\max})^{\lambda_B}\right] \tag{6.21}$$

is a scale-invariant representation of the relative saturation in terms of operationally defined equivalent pore radii (Danielson and Sutherland, 1986).

Reference to equation (6.1) shows that equation (6.21) is equivalent to the water-retention curve,

$$S(\psi) = \left[(\psi_{\min}/\psi)^{\lambda_B} - (\ell_{\min}/\ell_{\max})^{\lambda_B}\right]/\left[1 - (\ell_{\min}/\ell_{\max})^{\lambda_B}\right] \tag{6.22}$$

Table 6.3. *Representative Brooks-Corey parameters for soils of varying textures*[a]

| Texture | $\psi_{\min}^{BC}$ (m) | $\lambda^{BC}$ | Texture | $\psi_{\min}^{BC}$ (m) | $\lambda^{BC}$ |
|---|---|---|---|---|---|
| Sand | 0.0726 | 0.694 | Sandy clay loam | 0.2808 | 0.319 |
| Loamy sand | 0.0869 | 0.553 | Clay loam | 0.2589 | 0.242 |
| Sandy loam | 0.1466 | 0.378 | Silty clay loam | 0.3256 | 0.177 |
| Loam | 0.1115 | 0.252 | Sandy clay | 0.2917 | 0.233 |
| Silt loam | 0.2076 | 0.234 | Silty clay | 0.3419 | 0.160 |
| | | | Clay | 0.3730 | 0.165 |

[a]Data compiled by Rawls and Brakensiek (1995).

where $\psi_{\min} \equiv \psi(\ell_{\max})$. Equation (6.22) is thus a water-retention curve for a porous medium with a scale-invariant pore-size distribution. It reduces to the well-known Brooks-Corey model (Brooks and Corey, 1964) whenever the approximation

$$(\ell_{\min}/\ell_{\max})^{\lambda_B} \ll 1 \tag{6.23}$$

applies:

$$S^{BC}(\psi) = \left(\psi_{\min}^{BC}/\psi\right)^{\lambda^{BC}} \qquad \left(\psi > \psi_{\min}^{BC}\right) \tag{6.24}$$

Milly (1987), Russo (1988), and Rawls, Gish, and Brakensiek (1991) have discussed parameter estimation and other aspects of the application of equation (6.24) to field soils. Table 6.3 lists some representative mean values for $\psi_{\min}^{BC}$ and $\lambda^{BC}$ for hundreds of soils grouped by texture (Rawls and Brakensiek, 1995). Note that $\psi_{\min}^{BC}$ generally increases, whereas $\lambda^{BC}$ generally decreases, as soil texture becomes more clayey. Referring to equation (6.21) [subject to equation (6.23)], we see that this trend implies that as $\ell$ increases toward $\ell_{\max}$, clayey soils should exhibit a sharper increase toward 1.0 in the cumulative pore-size distribution than do sandy soils (Danielson and Sutherland, 1986, fig. 18-5).

Perrier et al. (1996) have recently derived a generic water-retention curve for any porous medium whose pore-size distribution is fractal. Their result can be obtained by specializing equation (6.22) in making the mathematical correspondences

$$\lambda_B \equiv 3 - D, \qquad A/(\theta_s - \theta_r) \equiv 1/\left[1 - (\ell_{\min}/\ell_{\max})^{3-D}\right] \tag{6.25}$$

where $D < 3$ is a fractal dimension, and $A$ is the hypothetical limiting porosity attributable to the fractal pore space in the medium as the radius of the smallest fractal pore goes to zero [Perrier et al., 1996, eq. (5)]. Two important special cases of equation (6.25), discussed fully by Perrier et al. (1996), are the Tyler-Wheatcraft model (revised to three-dimensional space), corresponding to the choice $A = (\theta_s - \theta_r)$, and the

Rieu-Sposito model, corresponding to the choice $A = 1$. The Tyler-Wheatcraft model (Tyler and Wheatcraft, 1990) also requires the inequality in equation (6.22) in order to be self-consistent. Therefore, it is a fractal version of the Brooks-Corey model. With this interpretation, the values of $\lambda^{\mathrm{BC}}$ in Table 6.3 can be associated with a fractal dimension $D \equiv 3 - \lambda^{\mathrm{BC}}$ that ranges from 2.306 (sand) to 2.835 (clay). The Rieu-Sposito model (Rieu and Sposito, 1991), on the other hand, implies that equation (6.22) should be rewritten as

$$\theta(\psi) = (\psi_{\min}/\psi)^{3-D} + \theta_s - 1 \tag{6.26}$$

because a common divisor, $(\theta_s - \theta_r)$, now can be canceled from both sides of the equation. It follows that $\theta(\psi)$, not $S(\psi)$, is the scale-invariant water-content function in the model of Rieu and Sposito (1991). This characteristic of the model, which of course it shares with Miller-Miller similitude, comes about simply because the pore space and the solid matrix in the medium are assumed to scale congruently, with the same associated fractal dimension [Rieu and Sposito, 1991, eq. (2)]. These two examples demonstrate that scale invariance is a more general concept than is fractal self-similarity and that fractal models of the same porous medium are not equivalent, even in respect to the physical significance of fractal scaling (Perrier et al., 1996).

### *6.3.2 The Richards Equation*

A partial differential equation is said to be invariant under mathematical transformations of its dependent and independent variables if it retains the same mathematical form after transformation and if the transformed dependent variable is a solution of the transformed partial differential equation under transformed initial and boundary conditions (Bluman and Kumei, 1989). For the Richards equation [equation (6.5)], the independent variables are spatial position and time. The dependent variable is volumetric water content, in the "mass picture," to be contrasted with the "energy picture," in which the matric potential is the dependent variable (Sposito, 1987).

The invariance of the Richards equation under length- and time-scaling transformations is an issue that can be investigated rigorously with methods in the theory of Lie groups (Bluman and Kumei, 1989). The invariance of equation (6.5) under scaling transformations has been so studied by Sposito (1990). This approach, whose results will be summarized here, provides a systematic framework within which to classify differing perspectives on the scaling of soil-water properties (Sposito, 1995). Yung et al. (1994), following on the results of Sposito (1990), have explored methods of deriving scale-invariant solutions of equation (6.5) using Lie-group analysis. They have also described the use of *Mathematica*® as an alternative to classic analytical methods to establish the full Lie group of the Richards equation [Sposito, 1990, eq. (A4)].

Scaling transformations of the dependent and independent variables in equation (6.5) are defined analogously to the length-scale transformation used in equation (6.14):

$$\theta' = \mu\theta, \qquad t' = \delta t, \qquad z' = \alpha z \tag{6.27}$$

where $\mu$, $\delta$, and $\alpha$ are positive, interrelated scale factors. Application of equations (6.27) to both sides of equation (6.5) produces the scaling relationships (Sposito and Jury, 1985)

$$\partial\theta'/\partial t' = (\mu/\delta)(\partial\theta/\partial t) \tag{6.28a}$$

$$\partial[D'(\theta')(\partial\theta'/\partial z')]/\partial z' = (\mu/\alpha^2)\{\partial[D'(\theta')(\partial\theta/\partial z)]/\partial z\} \tag{6.28b}$$

$$(\partial K'/\partial\theta')(\partial\theta'/\partial z') = (1/\alpha)(\partial K'/\partial\theta)(\partial\theta/\partial z) \tag{6.28c}$$

If the scale factors on the right sides of equations (6.28a–c) are all to be equal to one another, the scaling relationships

$$D'(\theta') = (\alpha^2/\delta)D(\theta), \qquad K'(\theta') = (\mu\alpha/\delta)K(\theta) \tag{6.28d}$$

also must be imposed. Then equations (6.28a–d) can be combined,

$$\begin{aligned}&\partial\theta'/\partial t' - \partial[D'(\theta')(\partial\theta'/\partial z')]/\partial z' - (\partial K'/\partial\theta')(\partial\theta'/\partial z')\\ &\quad = (\mu/\delta)\{\partial\theta/\partial t - \partial[D(\theta)(\partial\theta/\partial z)]/\partial z - (\partial K/\partial\theta)(\partial\theta/\partial z)\}\end{aligned} \tag{6.28e}$$

and the scaled Richards equation on the left side of equation (6.28e) will have the same mathematical form as the unscaled version whenever equation (6.5) holds.

Equations (6.28d) cannot be true for an arbitrary dependence of the transport coefficients on $\theta$, because $\theta$ is not invariant under scaling unless $\mu = 1$ in equation (6.27). It turns out (Sposito, 1990) that equations (6.28d) are valid scaling relationships only if $D(\theta)$ and $K(\theta)$ satisfy the differential equations (Sposito, 1995)

$$d[D(\theta)/(dD/d\theta)]/d\theta = 1/m \qquad (m \neq 0) \tag{6.29a}$$

$$d[(dK/d\theta)/(d^2K/d\theta^2)]/d\theta = 1/m\beta \qquad (\beta \neq 1/2 \text{ or } 1) \tag{6.29b}$$

where $m$ and $\beta$ are constant parameters, and $D(\theta)$ is assumed not to be constant (i.e., $m \neq 0$). The parameter $\beta$ cannot be equal to either $1/2$ or 1 if both time and length are scaled together (Sposito, 1990, table 1). The solutions of equations (6.29) are, for finite $m$,

$$D(\theta) = a(\theta + b)^m \tag{6.30}$$

$$K(\theta) = [ca^\beta/(1 + m\beta)](\theta + b)^{1+m\beta} \tag{6.31}$$

where $a$, $b$, and $c$ are arbitrary constants. Solutions that are exponentials in $\theta$ are also found, if $m \uparrow \infty$ is permitted. The prefactor on the right side of equation (6.31) becomes $[d_1+d_2\theta]$ ($d_1$ and $d_2$ being arbitrary constants) if $m\beta = -2$ in equation (6.29b) (Sposito, 1990). The corresponding mathematical form of the matric potential can be derived by combining equations (6.6), (6.30), and (6.31):

$$\psi(\theta) = [(1 + m\beta)/cm(1 - \beta)](\theta + b)^{m(1-\beta)} \tag{6.32}$$

if $m \neq 0$ is finite, $\beta \neq 1/2$ or 1, and $m\beta \neq -2$. Equations (6.30)–(6.32) specify a $\theta$ dependence of soil-water properties that is consistent with a scale-invariant Richards equation.

Now it is possible to learn how the scale factors in equations (6.27) are related. Given equations (6.28d), (6.30), and (6.31),

$$D'(\theta') \equiv a(\mu\theta + \mu b)^m = \mu^m D(\theta) = (\alpha^2/\delta) D(\theta) \tag{6.33a}$$

$$\begin{aligned} K'(\theta') &\equiv [ca^\beta/(1 + m\beta)](\mu\theta + \mu b)^{1+m\beta} \\ &= \mu^{1+m\beta} K(\theta) = (\mu\alpha/\delta) K(\theta) \end{aligned} \tag{6.33b}$$

It follows that

$$(\alpha^2/\delta)^{1/m} = (\alpha/\delta)^{1/m\beta} \tag{6.33c}$$

and therefore

$$\delta = \alpha^{(2\beta-1)/(\beta-1)} \tag{6.34}$$

Either equation (6.33a) or (6.33b) then yields

$$\mu = \alpha^{1/m(1-\beta)} \tag{6.35}$$

for the scale factor of $\theta$. Equation (6.35) can be introduced into equation (6.32) to demonstrate that

$$\begin{aligned} \psi'(\theta') &= [(1 + m\beta)/cm(1 - \beta)](\mu\theta + \mu b)^{m(1-\beta)} \\ &= \mu^{m(1-\beta)} \psi(\theta) = \alpha\psi(\theta) \end{aligned} \tag{6.36}$$

The scale factor $\alpha$ transforms both the position coordinate and the matric potential. [In equations (6.30)–(6.32), the arbitrary parameter $b$ is assumed to be determined by the datum for $\theta$ and therefore to scale as $\theta$ does.]

The physical implications of these results are seen more easily after applying the constraint

$$\psi(\theta_s) \equiv \psi_{\min} \tag{6.37}$$

to equation (6.32), from which it follows that equation (6.32) becomes

$$\psi(\theta) = \psi_{\min}[(\theta + b)/(\theta_s + b)]^{m(1-\beta)} \tag{6.38a}$$

or

$$\theta = (\theta_s + b)(\psi_{\min}/\psi)^{1/m(\beta-1)} - b \tag{6.38b}$$

Equation (6.38b) becomes the same as equation (6.22) after subtracting $\theta_r$ from both sides, then dividing both sides by $(\theta_s - \theta_r)$, and using the definitions

$$\lambda_B \equiv 1/m(\beta - 1) \qquad (\beta \neq 1/2 \text{ or } 1, m\beta \neq -2) \tag{6.39a}$$

$$b \equiv -\theta_r + \{(\ell_{\min}/\ell_{\max})^{\lambda_B}(\theta_s - \theta_r)/[1 - (\ell_{\min}/\ell_{\max})^{\lambda_B}]\} \tag{6.39b}$$

after noting equation (6.9). Therefore, equation (6.32) is consistent with a scale-invariant relative-saturation variable $S$. In other words, requiring the Richards equation to be scale-invariant is sufficient to make the relative saturation $S$ also scale-invariant.

If, instead of equations (6.39), $\lambda_B \equiv \lambda^{\mathrm{BC}}$, $b \equiv -\theta_r$, equation (6.38b) becomes the Brooks-Corey water-retention curve. If, instead of equations (6.39),

$$3 - D \equiv 1/m(\beta - 1), \qquad b \equiv A - \theta_s \tag{6.40}$$

equation (6.38b) becomes the fractal water-retention curve of Perrier et al. (1996). Thus both of these models are consistent with a scale-invariant Richards equation.

Equation (6.31) can be transformed similarly to equation (6.32) after introducing the constraint

$$K(\theta_s) \equiv K_s \tag{6.41}$$

to derive the result

$$K(\theta) = K_s[(\theta + b)/(\theta_s + b)]^{1+m\beta} \tag{6.42}$$

where $b$ is given by equation (6.39b). Equations (6.38a) and (6.42) then can be combined to derive the result (Sposito and Jury, 1990)

$$K(\theta) = K_s[\psi_{\min}/\psi(\theta)]^{(1+m\beta)/m(\beta-1)} \tag{6.43}$$

Equation (6.43) has the same mathematical form as an empirically based relationship employed by Bresler, Russo, and Miller (1978), Russo and Bresler (1980, 1981), and Jury et al. (1987) in examining the scaling behavior of soil-water properties in many soils. Jury et al. (1987) suggested that the exponent in equation (6.43),

$$\eta \equiv (1 + m\beta)/m(\beta - 1) \tag{6.44}$$

is necessary, in addition to the scale factor $\alpha$ [equation (6.36)], in order to apply scaling analysis comprehensively to soil-water properties in field soils. Equation (6.43) is, in the present case, an outcome from requiring a scale-invariant Richards equation. This equation, in fact, still holds if $m \uparrow \infty$ in equations (6.29), and $K(\theta)$ and $\psi(\theta)$ thereby become exponential functions of $\theta$ (Sposito, 1990).

Equation (6.36) resembles equation (6.10a), the scaling relationship for the matric potential in Warrick-Nielsen similitude. Equation (6.33b) can be made to resemble equation (6.10b), the scaling relationship for hydraulic conductivity, by equating $1/\omega_P^2$ with $\mu^{1+m\beta}$, or, given equation (6.35),

$$\omega_P^{2/\eta} = \mu^{-m(1-\beta)} = \alpha \equiv \alpha_P \tag{6.45}$$

in terms of the scale factor $\alpha_P$ and the Russo-Jury exponent $\eta$ in equation (6.44). Equation (6.45), first derived empirically by Jury et al. (1987), connects the two scale factors in Warrick-Nielsen similitude through the Russo-Jury exponent. As explained by Jury et al. (1987), $\eta$ can vary spatially in a field soil, leading to little or no correlation between the scale factors $\omega_P$ and $\alpha_P$, as has been noted often in field studies (Jury et al., 1987). Bresler et al. (1978), on the other hand, found $\eta \approx 2.6$ for a large number of soils of varying textures, to which the equation

$$K(\psi) = K(\psi_{\min}/\psi)^{\eta} \tag{6.46}$$

was fitted after measurements of soil-water properties. Laboratory measurements of $\eta$ reported by Russo and Bresler (1980) were in the range (2.1, 3.1), whereas field measurements (Russo and Bresler, 1980; Jury et al., 1987) were in the range (2.5, 4.2) for fieldwide averages, with coefficients of variation ranging from 0.2 to 0.5.

Model equations that relate hydraulic conductivity to matric potential are abundant in the literature of soil physics (Hillel, 1980, pp. 209–12). Snyder (1996) recently has shown that most of them are special cases of the integral equation

$$K(\theta_e) = k\theta_e^u \left\{ \int_0^{\theta_e} (\theta_e - \varphi)^v / [(\psi(\varphi)]^w d\varphi \right\}^x \tag{6.47}$$

where $\theta_e$ is an "effective volumetric water content" (e.g., $\theta_e = \theta - \theta_r$), and $k, u, v, w$, and $x$ are constant parameters. [These five parameters are given different symbols by

Snyder (1996).] The values of the exponents in equation (6.47) depend on the specific details of the physical model to which it is applied (Snyder, 1996, table 1), but they can be constrained by imposing the scaling relationships in equations (6.33)–(6.36). Following the same approach as in equations (6.28), we derive from equation (6.47) the result

$$\alpha_P^{-\eta} = \mu^{u+x(v+1)}\alpha_P^{-wx} = \alpha_P^{-[u+x(v+1)]\lambda_B - wx} \tag{6.48}$$

on noting equations (6.39a), (6.44), and (6.45), and the fact that both $\theta_e$ and the variable of integration $\varphi$ are scaled with the factor $\mu$. Therefore, the Russo-Jury exponent is

$$\eta = [u + x(v + 1)]\lambda_B + wx \tag{6.49}$$

{An expression equivalent to equation (6.48) was derived also by Snyder [1996, eq. (19)] for the common special case $wx = 2$.} Russo and Bresler (1980), Jury et al. (1987), and Morel-Seytoux et al. (1996) have recommended the use of equation (6.49) in the simplified "Brutsaert (1967) form," which is obtained by defining

$$u + x(v + 1) \equiv n + 2, \qquad wx \equiv 2 \tag{6.50}$$

where $n$ is an adjustable parameter (not necessarily an integer) that is supposed to reflect correlation between pore size and pore tortuosity (Jury et al., 1987). Values of $n$ in the range $(-1, 2)$ can be inferred from the special cases of equation (6.49) that are summarized by Snyder (1996). Russo and Bresler (1980) found $n$ to take on values in the range $(-1.5, 1.5)$ when adjusted in order to fit equations (6.46) and (6.50) to a variety of soil-water data. Thus, in the context of these computational efforts, and subject to the condition in equation (6.49), equation (6.47) can be regarded as an implicit model form of equation (6.46).

## 6.4 Broken Symmetry

The invariance of the Richards equation under scaling of its independent time and length variables has been shown to yield Warrick-Nielsen similitude (Table 6.2), predicated on transport coefficients $D(\theta)$ and $K(\theta)$ that satisfy equations (6.30) and (6.31). The physical significance of this result is that water movement through field soils whose soil-water properties conform to these latter two equations can, in principle, be described by a scaled Richards equation [equation (6.7)] with fieldwide mean transport coefficients that are related to local transport coefficients by means of the scale factors $\alpha_P$ and $\omega_P$ [equations (6.10)]. Warrick (1990) has stressed the importance of a scaled Richards equation to the solution of flow problems arising in theoretical vadose-zone hydrology. Indeed, once a fieldwide mean Richards equation has been solved under given initial and boundary conditions, the solution can be applied to a specific field site by a judicious choice of scale factors. It is precisely in this sense that scaling invariance reduces the complexity of field-soil spatial heterogeneity.

In equation (6.29b) it is stipulated that the parameter $\beta$ not be equal to 1/2 or 1. If $\beta = 1/2$, it can be demonstrated (Sposito, 1990) that the scale factor $\delta = 1$ in equation (6.27), whereas if $\beta = 1$, $\alpha = 1$ in equation (6.27). Thus, the choice $\beta = 1/2$ or 1 will result in loss of complete scaling invariance for the Richards equation, because either the time variable ($\beta = 1/2$) or the position variable ($\beta = 1$) is left unscaled when the set of transformations in equation (6.27) is applied to equation (6.5).

If the Richards equation does exhibit full scaling invariance, however, there can be no intrinsic time or length scale that characterizes soil-water movement. Indeed, an absence of an intrinsic length scale is at the heart of Miller-Miller similitude, because similar porous media are supposed to be only magnified or reduced versions of one another, with all having similar microscopic appearances, in the same way that similar triangles have similar appearances, except for scale. The scale factor $\lambda$ in Miller-Miller similitude serves only to adjust the degree of magnification or reduction to a "standard image" of the set of Miller-similar porous media (denoted by asterisks in Table 6.1). Likewise, but in a somewhat more abstract sense, field soils that admit Warrick-Nielsen similitude exhibit no intrinsic length scales. Different local subsurface regions in these soils are related, with respect to their soil-water properties, solely by scale factors. In each local region, the scaled water-retention curve shows exactly the same functionality in the relative saturation, and the scaled hydraulic conductivity shows exactly the same dependence on the relative saturation, with both dependences being either power-law or exponential in nature.

If $\beta = 1$, however, $\eta$ in equation (6.44) goes to infinity, and $\alpha_P$ in equation (6.45) must be equal to 1.0 regardless of the value of $\omega_P > 0$. This means that the matric potential [equation (6.36)] and the spatial position [equation (6.27)] are not scaled when equation (6.5) is scaled. The case $\beta = 1$ also requires special treatment with respect to the solution of equation (6.29b) to find $K(\theta)$, and of equation (6.6) to find $\psi(\theta)$ after equation (6.30) and $K(\theta)$ have been introduced (Sposito, 1990, 1995). Sander et al. (1988a,b, 1993) and Kühnel et al. (1990) have discussed and applied the solutions of these equations for the case $\beta = 1$ and $m = -2$, the latter exponent value defining what is termed a Fujita model of the soil-water diffusivity. [For a brief review of the applications of the Fujita model, see Sposito (1995).] For the Fujita model, equation (6.30) becomes

$$D(S) = D_s[(1 - \upsilon)/(1 - \upsilon S)]^2 \qquad (0 < S < 1) \tag{6.51}$$

where $D_s \equiv D_0/(1-v)^2$ and the assignments

$$a \equiv D_0(\theta_s - \theta_r)^2/v^2, \qquad m \equiv -2 \tag{6.52a}$$

$$b \equiv -\theta_r - (\theta_s - \theta_r)/v \qquad (v > 0) \tag{6.52b}$$

have been made. The corresponding solution of equation (6.29b) is (Sposito, 1990, 1995)

$$K(S) = K_s(1-v)S/(1-vS) \tag{6.53}$$

and this result, together with equations (6.6) and (6.51), leads to the matric potential (now renormalized to have negative values),

$$\psi(S) = \psi_0 + (D_s/K_s)\ln[(1-v)S/(1-vS)] \qquad (\psi < 0) \tag{6.54}$$

where $\psi \leq \psi_0 \leq 0$ in this case. Note that equation (6.53) is not a special case of equation (6.31), and equation (6.54) is not a special case of equation (6.32).

Sander et al. (1988b) demonstrated that the Richards equation, with equations (6.51) and (6.53) introduced, and under a constant-flux boundary condition, can be solved analytically. [See also Rogers, Stallybrass, and Clement (1983) and Sposito (1995).] The solution led to physically reasonable predicted moisture profiles. Sander et al. (1988a, 1993) have utilized equation (6.53) also as a mathematical representation of the function $f(\theta)$ $(0 \leq f \leq 1)$ that enters into a partial differential equation describing the effect on water absorption from the movement of a contiguous fluid phase (e.g., air):

$$\partial S/\partial t = \partial[D(S)(\partial S/\partial x)]/\partial x - V(t)(\partial f/\partial x) \tag{6.55}$$

where $V(t)$ is the boundary total fluid flux. Given $f(S) \equiv K(S)/K_s$, based on equation (6.53), Sander et al. (1988a, 1993) presented exact Fujita-model solutions of equation (6.55) for concentration boundary conditions. Those exact solutions permitted an evaluation of approximate equations for water absorption as influenced by concurrent immiscible-fluid movement. Kühnel et al. (1990) developed Fujita models of infiltration, under either constant rainfall flux ("atmosphere control") or surface saturation ("soil control"). Exact solutions of the Richards equation incorporating equations (6.51) and (6.53) were judged to be more realistic than those developed with simpler mathematical forms of $D(S)$ and $K(S)$. Barry and Sander (1991) extended the Fujita model of infiltration based on equations (6.51) and (6.53) to permit a time-varying surface-water flux. These successful applications indicate the practical hydrologic value of equations (6.51), (6.53), and (6.54).

From the point of view of scaling invariance, the practical success of equation (6.53) indicates that broken symmetry, represented by $\alpha_P \equiv 1, \omega_P > 0$ [equations (6.10)], can be found in water movement through unsaturated soils in nature. With equation (6.53) as the model for $K(S)$, the Richards equation no longer will exhibit length-scaling invariance, and the relative saturation $S$ will no longer be invariant under a change in length scale. [Compare equation (6.38a), which applies strictly to a scale-invariant $S$, to equation (6.54).] There must be implicit in equation (6.54) [for $\psi(S)$] an intrinsic length scale that reflects this broken length-scaling symmetry. The intrinsic length scale, in fact, appears in the coefficient of the logarithmic term in equation (6.54). It is equal to the ratio of two soil-specific parameters, $D_s/K_s$, by which the values of $\psi(S)$ are scaled as the relative saturation changes. Its role as a scaling parameter becomes especially transparent after equations (6.53) and (6.54) are combined to derive an equation for $K(\psi)$:

$$K(\psi) = K_s \exp[(K_s/D_s)(\psi - \psi_0)] \tag{6.56}$$

This equation, which differs markedly from equation (6.46), is the well-known defining characteristic of a "Gardner soil" (Gardner, 1958). The inverse of the intrinsic length scale, $D_s/K_s$, thus gives a quantitative measure of the rate of exponential decay of $K(\psi)$ as $\psi$ decreases below $\psi_0$ with decreasing $S$. It reflects directly the lack of length-scaling invariance in Gardner soils. Stochastic models of these soils, it may be noted in passing, arrive at a fieldwide mean Richards equation that differs considerably in formal appearance from equation (6.7) [after equation (6.8) is introduced; see Chapter 8]. Whether this difference reflects only the broken symmetry or is also a consequence of the assumption that soil-water properties can be represented by random fields parameterized by spatial coordinates is not yet apparent.

## 6.5 Concluding Remarks

Let us now summarize the principal physical implications of a scale-invariant Richards equation [equation (6.5)]:

1. Its two transport coefficients, $D(\theta)$ and $K(\theta)$, are power-law (or exponential) functions of the volumetric water content [equations (6.30) and (6.31)].
2. The relative saturation is a scale-invariant quantity [equation (6.21)], leading to a power-law relationship between relative saturation and matric potential [equation (6.22)]. This relationship becomes the Brooks-Corey model of the water-retention curve whenever the ratio of the smallest to the largest conducting pore size in a soil is negligibly small when compared to 1.0 [equation (6.24)]. Physical models of the water-retention curve based on a fractal pore-size distribution are consistent with this generic power-law relationship

between a scale-invariant relative saturation and the soil-water matric potential [equation (6.25)].

3. The scaled Richards equation can be applied to any homogeneous domain in a spatially heterogeneous field soil [equations (6.28)] or to the entire soil in an average sense [equation (6.7)]. The soil-water properties in the domains are related to one another through the Warrick-Nielsen scale factors $\alpha_P$ and $\omega_P$ [equations (6.10)]. These two scale factors, in turn, are connected through a power-law relation involving the Russo-Jury exponent $\eta$ [equation (6.45)]. This exponent also determines the power-law relation between hydraulic conductivity and matric potential in a heterogeneous field soil [equation (6.46)].
4. The symmetry of the Richards equation under scaling can be broken partially by restricting the scaling to the water content and the time variable only [i.e., $\alpha_P = 1$, and equation (6.45) becomes evanescent]. This loss of symmetry will be manifest in the water-retention curve, because the matric potential is no longer a scaled soil-water property. In particular, an intrinsic (i.e., soil-specific) length scale will appear in the water-retention curve and therefore in the relationship between hydraulic conductivity and matric potential. An example is the "Gardner soil," defined by an exponential dependence of hydraulic conductivity on matric potential [equation (6.56)], with the inverse of the exponential-decay parameter defining the intrinsic length scale.

In vadose-zone hydrology, the observation of a power-law dependence or exponential dependence (Simmons et al., 1979) on the volumetric water content among soil-water properties implies that infiltration processes, as described by the Richards equation, will exhibit scaling invariance, a fact that can simplify the understanding of subsurface hydrologic phenomena once fieldwide empirical distribution functions for the scale factors $\alpha_P$ and $\omega_P$ are determined (Warrick and Nielsen, 1980; Hopmans, 1987; Rockhold et al., 1996). For theoretical soil-water physics, a scale-invariant Richards equation offers an opportunity to apply its analytical solutions to a broad variety of spatially heterogeneous field soils, a point often emphasized by Warrick (1990).

The Brooks-Corey model is a simple example of these two strategies. Analytical solutions of the Richards equation for this model [equations (6.24) and (6.46)] are available for physically realizable boundary conditions (Sposito, 1995). The Brooks-Corey parameters $\lambda^{BC}$ and $\psi^{BC}_{\min}$ are known for a broad variety of soil textural classes (Table 6.3). The Russo-Jury exponent for the model can be calculated, for example, with equation (6.50) and $n = 2$, as recommended by Morel-Seytoux et al. (1996) (following a suggestion by Corey), or perhaps with the empirical relationship

$$\eta^{BC} \approx 7.2\lambda^{BC} \tag{6.57}$$

found by Bresler et al. (1978) from applying equations (6.24) and (6.46) to about 20 soils of widely varying textures. Once $\eta^{BC}$ is known, the exponents $m$ and $\beta$ can be readily computed after combining equation (6.39a) and (6.44) to derive the expressions

$$m = (\eta - \lambda_B - 1)/\lambda_B, \qquad \beta = (\eta - \lambda_B)/(\eta - \lambda_B - 1) \tag{6.58}$$

For the estimate of $\eta^{BC}$ favored by Morel-Seytoux et al. (1996), equations (6.58) become

$$m^{BC} \approx 2 + (1/\lambda^{BC}), \qquad \beta^{BC} \approx 2(\lambda^{BC} + 1)/(2\lambda^{BC} + 1) \tag{6.59}$$

whereas for equation (6.57) they are

$$m^{BC} \approx 6.2 - (1/\lambda^{BC}), \qquad \beta^{BC} \approx 6.2\lambda^{BC}/(6.2\lambda^{BC} - 1) \tag{6.60}$$

implying $m^{BC}\beta^{BC} \approx 6.2$ and $6.2\lambda^{BC} \approx \beta^{BC}/(\beta^{BC} - 1)$. Either of these expressions permits the computation of $m$ and $\beta$ with the values of $\lambda^{BC}$ in Table 6.3. Future research must determine what is the optimal relationship between $\eta$ and $\lambda_B$.

## References


Barry, D. A., and Sander, G. C. 1991. Exact solutions for water infiltration with arbitrary surface flux or nonlinear solute adsorption. *Water Resour. Res.* 27:2667–80.

Bluman, G. W., and Kumei, S. 1989. *Symmetries and Differential Equations*. Berlin: Springer-Verlag.

Bresler, E., Russo, D., and Miller, R. D. 1978. Rapid estimate of unsaturated hydraulic conductivity function. *Soil Sci. Soc. Am. J.* 42:170–2.

Brooks, R. H., and Corey, A. T. 1964. *Hydraulic Properties of Porous Media*. Hydrology paper no. 3. Fort Collins: Colorado State University.

Brutsaert, W. 1967. Some methods of calculating unsaturated permeability. *Trans. ASAE* 10:400–4.

Danielson, R. E., and Sutherland, P. L. 1986. Porosity. *Agronomy* 9:443–61.

Gardner, W. R. 1958. Some steady-state solutions of the unsaturated moisture flow equation with application to evaporation from a water table. *Soil Sci.* 85:228–32.

Hankey, A., and Stanley, H. E. 1972. Systematic application of generalized homogeneous functions to static scaling, dynamic scaling, and universality. *Phys. Rev.* 6B:3515–42.

Hillel, D. 1980. *Fundamentals of Soil Physics*. New York: Academic Press.

Hopmans, J. W. 1987. A comparison of various methods to scale soil hydraulic properties. *J. Hydrol.* 93:241–56.

Jury, W. A., Russo, D., and Sposito, G. 1987. Scaling models of water transport. *Hilgardia* 55:33–56.

Kühnel, V., Dooge, J. C. I., Sander, G. C., and O'Kane, J. P. J. 1990. Duration of atmosphere-controlled and of soil-controlled phases of infiltration for constant rainfall at a soil surface. *Ann. Geophys.* 8:11–20.

Miller, E. E. 1980. Similitude and scaling of soil-water phenomena. In: *Applications of Soil Physics*, ed. D. Hillel, pp. 300–18. New York: Academic Press.
Miller, E. E., and Miller, R. D. 1956. Physical theory for capillary flow phenomena. *J. Appl. Phys.* 27:324–32.
Milly, P. C. D. 1987. Estimation of Brooks-Corey parameters from water retention data. *Water Resour. Res.* 23:1085–9.
Morel-Seytoux, H. J., Meyer, P. D., Nachabe, M., Touma, J., van Genuchten, M. T., and Lenhard, R. J. 1996. Parameter equivalence for the Brooks-Corey and van Genuchten soil characteristics: preserving the effective capillary drive. *Water Resour. Res.* 32:1251–8.
Nielsen, D. R., Biggar, J. W., and Erh, K. T. 1973. Spatial variability of field-measured soil-water properties. *Hilgardia* 42:215–60.
Perrier, E., Mullon, C., Rieu, M., and de Marsily, G. 1995. Computer construction of fractal soil structures: simulation of their hydraulic and shrinkage properties. *Water Resour. Res.* 31:2927–43.
Perrier, E., Rieu, M., Sposito, G., and de Marsily, G. 1996. Models of the water retention curve for soils with a fractal pore-size distribution. *Water Resour. Res.* 32:3025–31.
Rao, P. S. C., Jessup, R. E., Hornsby, A. C., Cassel, D. K., and Pollans, W. A. 1983. Scaling soil microhydrologic properties of Lakeland and Konowa soils using similar media concepts. *Agric. Water Mgt.* 6:277–90.
Rawls, W. J., and Brakensiek, D. L. 1995. Utilizing fractal principles for predicting soil hydraulic properties. *J. Soil Water Cons.* 50:463–5.
Rawls, W. J., Gish, T. J., and Brakensiek, D. L. 1991. Estimating soil water retention from soil physical properties and characteristics. *Adv. Soil Sci.* 16:213–34.
Richards, L. A. 1931. Capillary conduction of liquids through porous mediums. *Physics* 1:318–31.
Rieu, M., and Sposito, G. 1991. The water characteristic curve of fragmented porous media and the fractal nature of soil structure. *C. R. Acad. Sci. Paris* 312:1483–9.
Rockhold, M. L., Rossi, R. E., and Hills, R. G. 1996. Application of similar media scaling and conditional simulation for modeling water flow and tritium transport at the Las Cruces trench site. *Water Resour. Res.* 32:595–609.
Rogers, C., Stallybrass, M. P., and Clement, D. A. 1983. On two phase infiltration under gravity and with boundary infiltration: application of a Bäcklund transformation. *Nonlinear Anal. Theory Methods Appl.* 7:785–99.
Russo, D. 1988. Determining soil hydraulic properties by parameter estimation: on the solution of a model for the hydraulic properties. *Water Resour. Res.* 24:453–9.
Russo, D., and Bresler, E. 1980. Field determinations of soil hydraulic properties for statistical analyses. *Soil Sci. Soc. Am. J.* 44:697–702.
Russo, D., and Bresler, E. 1981. Soil hydraulic properties as stochastic processes: I. An analysis of field spatial variability. *Soil Sci. Soc. Am. J.* 45:682–7.
Sander, G. C., Norbury, J., and Weeks, S. W. 1993. An exact solution to the nonlinear diffusion–convection equation for two-phase flow. *Q. J. Mech. Appl. Math.* 46:709–27.
Sander, G. C., Parlange, J.-Y., and Hogarth, W. L. 1988a. Extension of the Fujita solution to air and water movement in soils. *Water Resour. Res.* 24:1187–91.
Sander, G. C., Parlange, J.-Y., Kühnel, V., Hogarth, W. L., Lockington, D. L., and O'Kane, J. P. J. 1988b. Exact nonlinear solution for constant flux infiltration. *J. Hydrol.* 97:341–6.
Sharma, M. L., Gander, G. A., and Hunt, C. G. 1980. Spatial variability of infiltration in a watershed. *J. Hydrol.* 45:101–22.
Simmons, C. S., Nielsen, D. R., and Biggar, J. W. 1979. Scaling of field-measured soil-water properties. *Hilgardia* 47:77–174.

Snyder, V. A. 1996. Statistical hydraulic conductivity models and scaling of capillary phenomena in porous media. *Soil Sci. Soc. Am. J.* 60:771–4.
Sposito, G. 1987. The "physics" of soil water physics. In: *The History of Hydrology*, vol. 3, ed. E. R. Landa and S. Ince, pp. 93–8. Washington, DC: American Geophysical Union.
Sposito, G. 1990. Lie group invariance of the Richards equation. In: *Dynamics of Fluids in Hierarchical Porous Media*, ed. J. H. Cushman, pp. 327–47. London: Academic Press.
Sposito, G. 1995. Recent advances associated with soil water in the unsaturated zone. *Rev. Geophys. (Suppl).* (U.S. National Report to the International Union of Geodesy and Geophysics, 1991–1994). July:1059–65.
Sposito, G., and Jury, W. A. 1985. Inspectional analysis in the theory of water flow through unsaturated soil. *Soil Sci. Soc. Am. J.* 49:791–8.
Sposito, G., and Jury, W. A. 1990. Miller similitude and generalized scaling analysis. In: *Scaling in Soil Physics: Principles and Applications*, ed. D. Hillel and D. E. Elrick, pp. 13–22. Madison, WI: Soil Science Society of America.
Tillotson, P. M., and Nielsen, D. R. 1984. Scale factors in soil science. *Soil Sci. Soc. Am. J.* 48:953–9.
Tyler, S. W., and Wheatcraft, S. W. 1990. Fractal processes in soil water retention. *Water Resour. Res.* 26:1047–54.
Wang, Q. J., and Dooge, J. C. I. 1994. Limit fluxes at the land surface. *J. Hydrol.* 155:429–40.
Warrick, A. W. 1990. Application of scaling to the characterization of spatial variability in soils. In: *Scaling in Soil Physics: Principles and Applications*, ed. D. Hillel and D. E. Elrick, pp. 39–51, Madison, WI: Soil Science Society of America.
Warrick, A. W., and Amoozegar-Fard, A. 1979. Infiltration and drainage calculations using spatially scaled hydraulic properties. *Water Resour. Res.* 15:1116–20.
Warrick, A. W., Mullen, G. W., and Nielsen, D. R. 1977. Scaling field-measured soil hydraulic properties using a similar media concept. *Water Resour. Res.* 13:355–62.
Warrick, A. W., and Nielsen, D. R. 1980. Spatial variability of soil physical properties in the field. In: *Applications of Soil Physics*, ed. D. Hillel, pp. 319–44. New York: Academic Press.
Youngs, E. G. 1990. Application of scaling to soil-water movement considering hysteresis. In: *Scaling in Soil Physics: Principles and Applications*, ed. D. Hillel and D. E. Elrick, pp. 23–37. Madison, WI: Soil Science Society of America.
Youngs, E. G., and Price, R. I. 1981. Scaling of infiltration behavior in dissimilar porous materials. *Water Resour. Res.* 17:1065–70.
Yung, C. M., Verburg, K., and Baveye, P. 1994. Group classification and symmetry reductions of the non-linear diffusion-convection equation $u_t = (D(u)u_x)_x - K'(u)u_x$. *Int. J. Non-linear Mech.* 29:273–8.

# 7

# Scaling of the Richards Equation and Its Application to Watershed Modeling

R. HAVERKAMP, J.-Y. PARLANGE, R. CUENCA, P. J. ROSS, and T. S. STEENHUIS

## 7.1 The Problem of Scaling of Unsaturated Water Flow

Model studies of water transfer in saturated and unsaturated zones within soils have shown that there is an urgent need to assess the adequacy and usefulness of scale matching for the different flow phenomena involved in the interactions among the atmosphere, the land surface, vadose zones, and aquifers. Correct specifications of soil hydrologic processes for both small and large scales are directly dependent on the possibility of parameterizing soil-water fluxes for scales compatible with the grid size of the field scales involved. The difficulty of parameterization for soil-water dynamics arises not only from the nonlinearity of the equations for saturated and unsaturated flows but also from the mismatch between the scale of field measurements and the scale of model predictions. All of the standard measurement methods, such as the entire range of techniques to monitor in situ soil-moisture content and soil-water pressure, provide only point information and highlight the underlying variability in the hydrologic characteristics of soils. The efficiency of parameterization of soil characteristics at different scales depends on clear definitions of the functional relationships and parameters to be measured.

One of the first studies to apply the concept of scaling to the theory of water flow in unsaturated soils was that by Miller and Miller (1955a,b, 1956). They derived scale factors for soil-water properties such as soil-water pressure, hydraulic conductivity, diffusivity, and soil-water transport coefficients by the use of the similar-media concept. Their approach has been the subject of much interest and discussion in the literature (e.g., Simmons, Nielsen, and Biggar, 1979a,b; Miller, 1980; Warrick and Nielsen, 1980; Nielsen, Tillotson, and Vieira, 1983; Tillotson and Nielsen, 1984; Sposito and Jury, 1985; Sposito, 1990), especially with respect to the proper definitions of scale factors for flow equations and soil-water properties, as well as the usefulness of these definitions for estimating soil-water properties at field scale. Subsequent scaling efforts have moved progressively away from the Miller and Miller (1955a,b) original microscopic similar-media concept and have turned toward scaling analyses that are more macroscopic and empirical in nature, based on the idea of functional normalization rather than dimensional techniques (Reichardt, Nielsen, and Biggar, 1972; Warrick, Mullen, and Nielsen, 1977; Simmons et al., 1979a,b; Sharma, Gander, and Hunt, 1980; Russo and Bresler, 1980). The differences between the various scaling techniques have been clearly described in the reviews by Tillotson and Nielsen (1984) and Sposito and Jury (1985) (see also Chapter 5).

Basically, two categories of scaling techniques can be distinguished: physically based techniques, such as dimensional and inspectional analyses, and empirically based techniques, such as the functional-normalization technique. Complete discussions of the different methods have been published by Langhaar (1951) and Kline (1965). As Tillotson and Nielsen (1984) stated, the methods in the former category allow us "to convert a set of physically interrelated dimensional quantities into a set of nondimensional quantities which conserve the original interrelationship for a system that manifests geometric, kinematic or dynamic similarity." The scale factors obtained through such dimensional analysis have definite physical meanings in terms of the system being studied. On the other hand, the methods in the second category (i.e., functional normalization) derive scale factors through least-squares regression analyses relating properties of the two systems in some empirical way. As a generalization, it is evident that the methods in the first category of dimensional and/or inspectional analyses are preferable and are more appropriate to tackle the problem of scaling.

Similitude analysis of partial differential equations using the theory of Lie groups can be applied to the Richards equation. A summary of that approach has been published by Sposito (1995) (see also Chapter 6). The result of such analysis is that soil-water properties must have well-defined analytical forms if the Richards equation is to remain mathematically invariant. It is implicit in this chapter that the soil-water properties discussed will have the proper analytical form, obtained by similitude analysis using Lie groups, even when the scaling of integral properties does not seem to require those analytical forms explicitly. In particular, we consider two limiting cases in which the soil-water diffusivity approaches a delta function, which is also an explicit limit for the analytical forms obtained through similitude analysis.

The aim in Section 7.2 is to address some of the theoretical aspects of the physically significant scale factors when describing the equations for saturated and unsaturated flows by the use of macroscopic *inspectional analysis*; both the concentration-type surface boundary condition (Section 7.2.1) and the flux boundary condition (Section 7.2.2) are considered. Section 7.3 deals with the identification procedures for the different scale parameters from a simple field experiment. This chapter concludes with our perspective regarding the application of this newly emerging similarity approach to the hydrology of large-scale watersheds.

## 7.2 Theoretical Aspects

Inspectional analysis is a dimensional technique that features applications of physical laws along with corresponding initial and/or boundary conditions (Birkhoff, 1960; Sposito and Jury, 1985). As pointed out by Birkhoff (1960), inspectional analysis rests on the fundamental postulate that the invariance of a physical law (e.g., the vadose-zone transfer equation) under a series of scale transformations implies the invariance of all consequences of the law under the same transformations. This principle therefore provides the basis for dynamic similarity in the behaviors of two systems governed by a given physical law (Sposito and Jury, 1985). As knowledge of the flow behavior in an unsaturated soil is of paramount importance for vadose-zone hydrology, rather than knowledge of the soil's hydraulic characteristics (which are nothing else than intermediate mathematical functional relationships used to calculate the flow behavior), development of any dimensional analysis should start from the flow equation itself. Functional-normalization analyses that attempt to coalesce sets of curves for soil-water characteristics into one single reference curve on the basis of simple geometric similarity fail a priori to scale the vadose-zone flow behavior in a general way.

The one-dimensional flow of isothermal water in an unsaturated homogeneous soil (such as infiltration, evaporation, and drainage) can be expressed through the partial differential equation (Richards, 1931)

$$C(\theta)\frac{\partial h}{\partial t} = \frac{\partial}{\partial z}K(\theta)\left(\frac{\partial h}{\partial z} - 1\right) \tag{7.1}$$

where $\theta$ is volumetric water content [$L^3/L^3$], $h$ is soil-water pressure [L] relative to atmospheric pressure ($h \leq 0$), $K(\theta)$ is hydraulic conductivity [L/T] as a function of $\theta$, $z$ is depth [L] taken positive downward, and $t$ is time [T]. Defining the water-retention curve $h(\theta)$ as the nonlinear relationship between volumetric water content $\theta$ and soil-water pressure $h$, $C(\theta)$ is the specific capacity [1/L], given by $C(\theta) = d\theta/dh$. As the focus of this section is not the influence of the water-flow scale factors on the choices of the functional relationships of the soil characteristics, no particular forms of $h(\theta)$ and $K(\theta)$ are considered a priori.

The solution of equation (7.1), subject to given initial and boundary conditions, describes the evolution of the water-content profiles $\theta(z, t)$ as functions of depth and time. The initial condition imposed on $\theta(z, t)$ is

$$\theta(z, 0) = \theta_0(z) \tag{7.2}$$

where $\theta_0$ is the initial water-content value, and the upper boundary condition on $\theta(z, t)$ is either the Dirichlet concentration condition

$$h(0, t) = h_1(t) \tag{7.3}$$

or the Neumann flux condition

$$\left(K - K\frac{\partial h}{\partial z}\right)_{z=0} = q_1(t) \tag{7.4}$$

where $h_1$ is the surface soil-water pressure, and $q_1$ is the flux at the soil surface [L/T]. Values for $h_1$ and $q_1$ can be positive, zero, or negative.

As is the case for most of the macroscopic similitude analyses of the water-flow equation presented in the literature (e.g., Reichardt et al., 1972; Warrick et al., 1977; Simmons et al., 1979a,b; Sharma et al., 1980), a set of scale transformations can be chosen:

$$\begin{aligned} z^* &= \alpha_z z, & t^* &= \alpha_t t, & \theta^* &= \alpha_\theta \theta \\ h^* &= \alpha_h h, & K^* &= \alpha_K K, & C^* &= \alpha_C C \end{aligned} \tag{7.5}$$

where $\alpha_z$ [1/L], $\alpha_t$ [1/T], $\alpha_h$ [1/L], $\alpha_\theta$ [$L^3/L^3$], $\alpha_K$ [T/L], and $\alpha_C = \alpha_h/\alpha_\theta$ [L] are constant parameters that define, respectively, the nondimensional variables $z^*$, $t^*$, $h^*$, $\theta^*$, $K^*$, and $C^*$. Substitution of these scale factors into equations (7.1)–(7.4) allows us to express the flow equation as a nondimensional boundary-value problem. The particular values of the scale factors $\alpha$ for the solution of equation (7.1), along with the chosen initial and boundary conditions and the governing space and time variables, should reflect the effects of the initial and boundary values on the dependent variable. The use of supplementary subjective hypotheses should be avoided, as much as possible, in determining these $\alpha$ values [e.g., by specifying particular preimposed mathematical forms for the soil characteristics: "the Nielsen and/or Warrick similitude," as put by Sposito and Jury (1985)].

In Sections 7.2.1 and 7.2.2 the two different surface boundary conditions are analyzed separately to determine the $\alpha$ values.

### *7.2.1 Concentration-type Surface Boundary Condition*

Although the solution of the Richards equation under general initial and boundary conditions can be obtained only through numerical simulation, exact analytical solutions under idealized conditions are now available. A particular solution of the Richards equation can be found for the case of one-dimensional constant-head infiltration in a semiinfinite soil column with constant initial water content subject to the initial and boundary conditions

$$\begin{aligned} \theta(z,0) &= \theta_0 \quad \text{for } z \geq 0 \\ h(0,t) &= h_1 \quad \text{for } t \geq 0 \end{aligned} \tag{7.6}$$

where $h_1$ is a constant pressure that can take a negative ($h_1 \leq 0$) or positive value equal to the ponded-water depth ($h_{surf}$) imposed at the soil surface ($h_1 = h_{surf} \geq 0$). The

Dirichlet surface boundary condition imposes a constant volumetric water content at the soil surface ($\theta_1$) that is less than or equal to the volumetric water content at natural saturation ($\theta_s$), depending on the value of $h_1$. The cumulative infiltration $I$, expressed in volume per unit surface [$L^3/L^2$], is defined by the integral of the water-content profile $\theta(z, t)$:

$$I(t) = K_0 t + \int_0^{z_f} \theta(z, t)\, dz \tag{7.7}$$

where $z_f(t)$ is the depth of the infiltration wetting front, and $K_0$ is the hydraulic conductivity at the initial water content ($\theta_0$).

As the cumulative-infiltration problem [equation (7.7)] is nothing else than a consequence of the solution of the flow equation (7.1) along with a particular set of initial and boundary conditions, the Birkhoff (1960) postulate guarantees that the scale parameters obtained for the cumulative-infiltration law will apply equally well for the Richards equation. The first step of inspectional analysis is to select a quasi-exact solution for the infiltration problem in order to derive scaling factors from the initial and boundary conditions imposed on the governing equation; see the reviews by Tillotson and Nielsen (1984) and Sposito and Jury (1985) for a step-by-step summary of the standard procedures for the inspectional analysis.

Among the many different analytical solutions for one-dimensional infiltration presented in the literature (e.g., Green and Ampt, 1911; Philip, 1957a,b; Parlange et al., 1982, 1985; Haverkamp et al., 1990) the Green and Ampt (1911) solution is perhaps the least realistic infiltration equation, but certainly the simplest. Because of its simplicity, it represents the ideal case to illustrate the fundamental principles of the scaling theory. For that reason, we begin with the Green and Ampt equation. The second analytical infiltration equation that will be considered is the frequently cited time-series solution of Philip (1957a,b). Even though this series solution is divergent for long times and cumbersome to put into operation, it is definitely useful for identifying some of the essential integral infiltration parameters, such as sorptivity. Full identification of the scale factors $\alpha$ in the equation for unsaturated water flow will be illustrated through the generalized equation for constant-head infiltration developed by Parlange et al. (1982) and Haverkamp et al. (1990).

*The Green and Ampt Solution.* The hypotheses underlying the Green and Ampt approach for positive-head infiltration assume a wetting front in the form of a step function with a one-point hydraulic-conductivity function [$K(\theta) = K_s$] behind the wetting front and a constant driving pressure ($h_{\text{surf}} - h_f$) pulling the wetting front downward through the unsaturated zone. $K_s$ is the hydraulic conductivity at natural saturation, and $h_f$ is a constant equivalent negative water pressure at the wetting front. The corresponding water-retention curve takes the form of a step function, with $\theta = \theta_0$ for $h \leq h_f$, and $\theta = \theta_s$ for $h_f \leq h \leq 0$. The cumulative-infiltration equation corresponding to this configuration is given by

$$I(t) = K_s t + [\theta_s - \theta_0][h_{\text{surf}} - h_f] \ln\left\{1 + \frac{I}{[\theta_s - \theta_0][h_{\text{surf}} - h_f]}\right\} \tag{7.8}$$

and the corresponding infiltration flux $q_1$ at the soil surface can be solved by

$$t = \frac{[\theta_s - \theta_0][h_{\text{surf}} - h_f]}{K_s}\left\{\ln\left[\frac{q_1 - K_s}{q_1}\right] + \frac{K_s}{q_1 - K_s}\right\} \tag{7.9}$$

To represent equations (7.8) and (7.9) in nondimensional form, we should first introduce the concept of "sorptivity," $S_1$ [L/T$^{1/2}$], which characterizes the ability of a soil to absorb water in the absence of gravity and was initially presented by Green and Ampt (1911), the word later being coined by Philip (1957b). For practical convenience we use here the simple but accurate (Elrick and Robin, 1981) estimation of sorptivity given by the expression of Parlange (1975):

$$S_1^2(\theta_s, \theta_0) = \int_{\theta_0}^{\theta_s} [\theta_s + \theta - 2\theta_0] D(\theta)\, d\theta \tag{7.10}$$

where $D(\theta)$ is diffusivity [L$^2$/T], defined by $D(\theta) = K(\theta) dh/d\theta$. For the case of positive-head infiltration, the sorptivity definition has to be reformulated as

$$S_1^2(h_{\text{surf}}, h_0) = S_1^2(\theta_s, \theta_0) + 2K_s h_{\text{surf}}[\theta_s - \theta_0] \tag{7.11}$$

where $h_0$ stands for the initial soil-water pressure $h(\theta_0)$. Hereafter the positive-head sorptivity $S_1(h_{\text{surf}}, h_0)$ will be referred to as $S_+$. For the particular Green and Ampt configuration in which the water-retention curve is represented in the form of a step function, equation (7.11) takes the form

$$S_+^2 = 2K_s[h_{\text{surf}} - h_f][\theta_s - \theta_0] \tag{7.12}$$

The second step of the inspectional analysis (Sposito and Jury, 1985) consists of trying to define the scaling factors $\alpha$ of equation (7.5). To do so, the infiltration solution represented by equations (7.8) and (7.9) is expressed in dimensionless form through the use of the following dimensionless variables:

$$I^* = I\frac{2K_s}{S_+^2} \tag{7.13}$$

$$t^* = t\frac{K_s^2}{S_+^2} \tag{7.14}$$

From equations (7.13) and (7.14) the dimensionless infiltration flux at the soil surface is defined as

$$q_1^* = \frac{dI^*}{dt^*} = \frac{q_1}{K_s} \tag{7.15}$$

Substitution of equations (7.13)–(7.15) into equations (7.8) and (7.9) allows expression the Green and Ampt infiltration solution in the following dimensionless form:

$$I^* = t^* + \ln[1 + I^*] \tag{7.16}$$

and

$$t^* = \frac{1}{q_1^* - 1} - \ln\left[1 + \frac{1}{q_1^* - 1}\right] \tag{7.17}$$

Simultaneous use of equations (7.16) and (7.17) reduces the infiltration problem to a dimensionless boundary-value problem valid for all soil types (obviously, still in the context of the Green and Ampt concept), with uniform initial conditions and constant boundary conditions. This dimensionless solution allows straightforward determination of the specific scale factors for cumulative infiltration $\alpha_I$ and infiltration rate $\alpha_q$. Choosing the scale factors $\alpha_I$ and $\alpha_q$ in a manner similar to the general scale transformations defined previously for equation (7.5),

$$I^* = \alpha_I I \quad \text{and} \quad q_1^* = \alpha_q q_1 \tag{7.18}$$

the cumulative-infiltration scale factor $\alpha_I$ can be directly calculated from equation (7.13) as

$$\alpha_I = \frac{2K_s}{S_+^2} \tag{7.19}$$

and the time scale factor $\alpha_t$ from equation (7.14) as

$$\alpha_t = \frac{2K_s^2}{S_+^2} \tag{7.20}$$

The infiltration-rate scale factor $\alpha_q$ can then be determined from the definition $q_1^* = dI^*/dt^*$:

$$\alpha_q = \frac{\alpha_I}{\alpha_t} = \frac{1}{K_s} \tag{7.21}$$

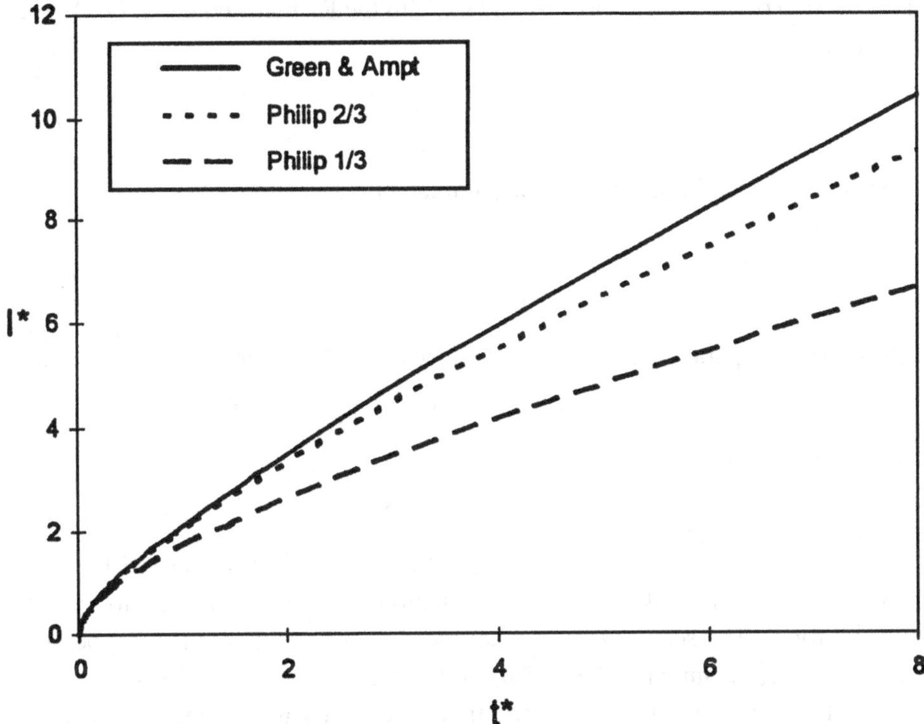


Figure 7.1. Invariant nondimensional infiltration curves ($I^*$) as functions of time ($t^*$) plotted for the solution of Green and Ampt (solid) as well as for the upper (dots) and lower (dash) limits of the two-term expansion for the Philip infiltration equation.

The key point illustrated by the foregoing analysis is that for any Green and Ampt soil the infiltration behavior (cumulative and/or infiltration flux) is fully determined by only two scale factors: $\alpha_I$ (or $\alpha_q$) and $\alpha_t$. Both parameters embody the effects of the soil type and of the initial and boundary conditions. The dimensionless invariant cumulative-infiltration curve $I^*(t^*)$ shown in Figure 7.1 fixes the only "dynamic-similarity class" for which the physical system is said to be macroscopically similar.

Thus far, the scaling factors $\alpha_I$ and $\alpha_q$ still exhibit the specific character related to the type of flow problem analyzed (cumulative constant-head infiltration). To bypass this specificity it is necessary to go down one step in the hierarchy of the water-movement problematic by establishing the set of basic scale factors that can reduce the Richards equation $\theta(z, t)$, rather than those that can reduce the cumulative-infiltration equation $I(t)$. This can be pursued by deconvoluting the specific infiltration scale factors $\alpha_I$ and $\alpha_q$ into the set of scale factors $\alpha_\theta$, $\alpha_h$, $\alpha_K$, and $\alpha_z$.

The conductivity scale factor $\alpha_K$ can be calculated in a straightforward manner by substitution of equation (7.18) into equation (7.15):

$$\alpha_K = \alpha_q = \frac{1}{K_s} \tag{7.22}$$

Introduction of the scale factors $\alpha_t$ and $\alpha_K$ into the Richards equation (7.1) gives the identities

$$\alpha_h = \alpha_z \quad \text{and} \quad \alpha_I = \alpha_\theta \alpha_h \tag{7.23}$$

By choosing the definition for the scale parameter $\alpha_\theta$ as

$$\alpha_\theta = \frac{1}{[\theta_s - \theta_0]} \tag{7.24}$$

the pressure scale parameter $\alpha_h$ and thus the space scale factor $\alpha_z$ are fully defined by

$$\alpha_h = \alpha_z = \frac{2K_s[\theta_s - \theta_0]}{S_+^2} = \frac{1}{[h_{\text{surf}} - h_f]} \tag{7.25}$$

With the four scale factors $\alpha_\theta, \alpha_h$ (equal to $\alpha_z$), $\alpha_K$, and $\alpha_t$ defined by equations (7.22)–(7.25), the soil-water transfer equation can be transformed into a nondimensional form. The scaling factors take into account the effect of soil type, and, at the same time, the initial and boundary conditions endow the dimensionless form of the Richards equation (7.1) with additional flexibility to encompass the water-flow behaviors of diverse physical systems through adjustments to experimentally controllable parameters.

To make this inspectional analysis of the water-transfer and/or infiltration equations useful for practical application, the soil scale factors should be expressed in terms of the classically used soil-characteristic parameters contained in the expressions of $h(\theta)$ and $K(\theta)$. To do so, the scale factors $\alpha_\theta$, $\alpha_h$, and $\alpha_K$ have to be stripped of the effects of their initial and boundary conditions. Letting $\lambda$ stand for the fraction of $\alpha$ that is related solely to a given soil characteristic, the different soil scale factors can be redefined as

$$\alpha_\theta = \lambda_\theta \frac{\theta_s}{[\theta_s - \theta_0]} \quad \text{with } \lambda_\theta = \frac{1}{\theta_s} \tag{7.26}$$

$$\alpha_K = \lambda_K \frac{K_s}{[K_s - K_0]} \quad \text{with } \lambda_K = \frac{1}{K_s} \tag{7.27}$$

and

$$\alpha_h = \lambda_h \frac{[\theta_s - \theta_0]}{\theta_s} \frac{S_1^2(\theta_s, \theta_r)}{S_1^2(\theta_s, \theta_0)} \frac{S_1^2(\theta_s, \theta_0)}{S_+^2} \quad \text{with } \lambda_h = \frac{2K_s\theta_s}{S_1^2(\theta_s, \theta_r)} \tag{7.28}$$

where $S_1^2(\theta_s, \theta_r)$ represents the sorptivity calculated over the complete range of possible water-content values $(\theta_s - \theta_r)$. Although the parameter $\theta_r$ is often associated with the residual water content, theoretically adsorbed on the soil grains, it should rather be considered as a pure mathematical artifact that can easily be set equal to zero. The second and third terms on the right-hand side of equation (7.28) represent the correction for the initial-condition effects, and the last term represents the correction for the surface boundary-condition effects.

For the particular case of Green and Ampt soils, the definitions of the soil scale factors $\lambda_K$ and $\lambda_h$ can be considerably simplified. The limiting hydraulic-conductivity hypothesis with a zero initial conductivity value ($K_0 = 0$) simplifies the conductivity scale factor $\alpha_K$ [equation (7.27)] to $\alpha_K = \lambda_K$, and the hypothesis of the step-function behavior of the water-retention curve reduces the soil-pressure scale factor $\lambda_h$ [equation (7.28)] to the simple expression

$$\lambda_h = -\frac{1}{h_f} \quad \text{with } h_f \leq 0 \tag{7.29}$$

For general field soils where no relaxing hypotheses hold, the definition of equation (7.28) clearly suggests the existence of a close relationship between the water-pressure scale factor $\lambda_h$ and the conductivity scale factor $\lambda_K$ through the sorptivity term. The consequence is rather far-reaching, but is rarely considered in the literature, as it implies that optimization of the parameters for soil-water characteristics over measured field data should be carried out in simultaneous combination with an objective function.

The main point to be recognized from this example is that there is perfect dynamic similarity in the behaviors of soil-water movements in unsaturated Green and Ampt soils governed by the Richards equation. The soil-characteristic scale parameters $\lambda_\theta$, $\lambda_h$, and $\lambda_K$ are directly linked to the water-movement scale factors $\alpha_I$ and $\alpha_t$ (or $\alpha_q$), and the definition of the soil's water-pressure scale factor suggests a definite interdependence between the expressions chosen for the soil's water-retention and hydraulic-conductivity curves.

*The Philip Solution.* The infiltration solution of Philip (1957a,b) was originally developed for the case $h_{\text{surf}} = 0$ and $\theta_1 = \theta_s$, and it is written in terms of a series expansion in powers of $t^{1/2}$ as

$$z(\theta, t) = \varphi(\theta)t^{1/2} + \chi(\theta)t + \psi(\theta)t^{3/2} + \cdots \tag{7.30}$$

where the first term, $\varphi(\theta)$, embodies the influence of the capillary forces on the flow process, and the following terms [e.g., $\chi(\theta)$ and $\Psi(\theta)$] reflect the gravity effect on infiltration. Unfortunately, calculation of the different functions $\varphi(\theta)$, $\chi(\theta)$, and $\Psi(\theta)$ is somewhat tedious and is based on an iterative numerical scheme that is not always precise close to $\theta = \theta_0$ (Haverkamp, 1983).

Integration of equation (7.30) gives the cumulative-infiltration equation $I(t)$:

$$I(t) = S_1(\theta_1, \theta_0)t^{1/2} + [K_0 + S_2(\theta_1, \theta_0)]t + O(t^{3/2})\ldots \quad (7.31)$$

where

$$S_1(\theta_1, \theta_0) = \int_{\theta_0}^{\theta_1} \varphi(\theta)\, d\theta \quad (7.32)$$

and

$$S_2(\theta_1, \theta_0) = \int_{\theta_0}^{\theta_1} \chi(\theta)\, d\theta \quad (7.33)$$

The integral expression for $S_1(\theta_1, \theta_0)$ represents the original sorptivity definition given by Philip (1957b). Equations (7.32) and (7.33) clearly show that the values of $S_1(\theta_1, \theta_0)$ and $S_2(\theta_1, \theta_0)$ are specific for the initial and boundary conditions encountered during each particular infiltration event. The third and higher terms of equation (7.31) can easily be calculated by transforming the time-series expansion into a recurrence series as functions of the first and second terms (Haverkamp et al., 1990).

In theory, the Philip infiltration equation is physically based, which makes it particularly suitable for the possible generation of a similarity solution a priori. However, in practice there is some ambiguity concerning the correct interpretation of equation (7.31) (Haverkamp, 1983):

1. Firstly, it is somewhat awkward to calculate the integrals $S_1(\theta_1, \theta_0)$ and $S_2(\theta_1, \theta_0)$ numerically. Though an accurate and simple estimation of sorptivity has been developed by Parlange (1975) in the form of equation (7.10), no simple approximation exists for the second integral $S_2(\theta_1, \theta_0)$. Generally, the value of $S_2(\theta_1, \theta_0)$ is associated to a constant $A_2$, calculated by curve fitting, with

$$I(t) = A_1 t^{1/2} + [K_0 + A_2]t \quad (7.34)$$

over experimental cumulative-infiltration data, where $A_2$ is confined over the interval $\{[K_1 - K_0]/3\} \leq A_2 \leq \{2[(K_1 - K_0]/3\}$ (Youngs, 1968; Philip, 1969; Talsma and Parlange, 1972). As equation (7.34) deliberately neglects all higher-order terms $[O(t^{3/2})]$ of the original time-series expansion [equation (7.31)], it is evident that the optimized values of $A_1$ and $A_2$ reflect, to some extent, this truncation effect. Consequently, $A_1$ and $A_2$ are just specific ad hoc constants, and it becomes rather delicate to compare them with the physically defined integral parameters $S_1(\theta_1, \theta_0)$ and $S_2(\theta_1, \theta_0)$ (Haverkamp et al., 1988).

2. Secondly, the definitions of $S_1(\theta_1, \theta_0)$ and $S_2(\theta_1, \theta_0)$ were originally developed for the case of infiltration with the surface boundary condition of negative head or zero head ($h_1 \leq 0$ and $\theta_1 \leq \theta_s$). For the case of positive-head infiltration ($h_1 \geq 0$ and $\theta_1 = \theta_s$), the integrals $S_1$ and $S_2$ have to be reformulated. This can easily be done for the sorptivity definition of $S_1$ [equation (7.11)], but for the second integral term $S_2$ it is far more difficult to adjust equation (7.33) (Haverkamp, 1983).
3. Finally, the main problem for application of the Philip (1957a) solution lies in the fact that the time series of equation (7.31) becomes divergent for large times, no matter how many terms are developed. Thus the solution is valid for only a limited time range. The time limit at which the first term in the series solution ceases to dominate is usually set at $t_{\text{grav}}$:

$$t_{\text{grav}} = \left[\frac{S_1(\theta_s, \theta_0)}{K_s - K_0}\right]^2 \tag{7.35}$$

a time for which the gravity forces are supposed to become predominant over the capillary forces during the infiltration process (Philip, 1969).

This last constraint, which is intrinsic to the time-series solution itself, has certainly contributed to the fact that equation (7.31) has not frequently been used for practical purposes. The importance of the solution most probably lies in the definitions of the integral terms $S_1(\theta_1, \theta_0)$ and $S_2(\theta_1, \theta_0)$, which represent two supplementary expressions that sometimes are useful for identification of other less rigorously defined integral soil parameters. For that reason, we choose to present here the exact analytical solution for the second term derived by Fuentes, Haverkamp, and Parlange (1992).

The parameter $S_2(\theta_1, \theta_0)$ is redefined in the form

$$S_2(\theta_1, \theta_0) = K_0 + \frac{1}{3}[K_1 - K_0][1 + \mu(\theta_1, \theta_0)] \tag{7.36}$$

where $\mu$ is expressed as a function of two length scales $\sigma_K$ and $\sigma_D$:

$$\mu(\theta_1, \theta_0) = \frac{\sigma_K(\theta_1, \theta_0)}{\sigma_D(\theta_1, \theta_0)} - 1 \tag{7.37}$$

The length scales $\sigma_K$ and $\sigma_D$ are written as functions of the classic soil characteristics:

$$\sigma_K(\theta_1, \theta_0) = \frac{2}{K_1 - K_0} \int_{\theta_0}^{\theta_1} \frac{K(\theta) - K_0}{K_1 - K_0} \frac{\theta - \theta_0}{\theta_1 - \theta_0} \frac{D(\theta)}{f^2(\theta)} d\theta \tag{7.38}$$

and

$$\sigma_D(\theta_1, \theta_0) = \frac{1}{K_1 - K_0} \int_{\theta_0}^{\theta_1} \frac{1}{3}\left[1 + 2\frac{g(\theta)}{f(\theta)}\right] \frac{\theta - \theta_0}{\theta_1 - \theta_0} \frac{D(\theta)}{f(\theta)} d\theta \tag{7.39}$$

where $f(\theta)$ is the diffusivity-driven "flux-concentration relation" that controls the sorptivity value, expressed by Parlange (1975) as

$$f(\theta) = \frac{2[\theta - \theta_0]}{\theta_1 + \theta - 2\theta_0} \tag{7.40}$$

and $g(\theta)$ is the gravity-driven flux-concentration relation that controls the gravity correction (Philip, 1973), given by

$$g(\theta) = \frac{\theta - \theta_0}{\theta_1 - \theta_0} \tag{7.41}$$

The parameter $\mu$ varies over the interval $0 \leq \mu \leq 1$, depending on the soil type and the governing initial and boundary conditions. The two extreme values $\mu = 0$ and $\mu = 1$ correspond, respectively, to the limiting soil described by Talsma and Parlange (1972), for which $D(\theta)$ and $dK/d\theta$ behave like a Dirac delta function (Parlange, 1977), and the Green and Ampt (1911) soil, for which only $D(\theta)$ behaves like a delta function. Substitution of the extreme $\mu$ values into equation (7.36) gives the experimentally verified interval of variation for $S_2$ mentioned before: $\{[K_1 - K_0]/3\} \leq S_2(\theta_1, \theta_0) \leq \{2[K_1 - K_0]/3\}$.

For the inspectional analysis, we consider the two-term Philip solution [equation (7.31)], with $S_2(\theta_1, \theta_0)$ expressed by equation (7.36):

$$I(t) = S_1(\theta_1, \theta_0)t^{1/2} + \left\{K_0 + \frac{1}{3}[K_1 - K_0][1 + \mu(\theta_1, \theta_0)]\right\}t \tag{7.42}$$

which is supposed to be valid over the limited time range $t < t_{grav}$. Equation (7.42) is transformed to its dimensionless form

$$\Delta I^*(t^*) = (2t^*)^{1/2} + \frac{1}{3}[1 + \mu(\theta_1, \theta_0)]t^* \tag{7.43}$$

by the use of the following dimensionless variables:

$$\Delta I^* = [I - K_0 t]\frac{2[K_1 - K_0]}{S_1^2(\theta_1, \theta_0)} \tag{7.44}$$

and

$$t^* = t\frac{2[K_1 - K_0]^2}{S_1^2(\theta_1, \theta_0)} \tag{7.45}$$

$\Delta I^*$ represents the dimensionless net cumulative infiltration, which is expressed as the difference between the cumulative infiltration $I^*$ and the initial infiltration $I_0^* = K_0^* t^*$ that is necessary to maintain the initial water content.

At this stage it is interesting to verify the validity of the time limit $t_{grav}$. In fact, the nondimensionalization of the infiltration equation allows us to get a handle on the time-threshold value for which the two-term time series of Philip is still reasonably precise. Expressing $t_{\text{grav}}$ [equation (7.35)] in its nondimensional form by the use of equation (7.45) yields $t^*_{\text{grav}} = 2$. Considering the extreme scenario, in the form of a Green and Ampt soil, the upper limit of the two-term time-series expansion of Philip [equation (7.43)], with $K_0 = 0$ and $\mu = 1$, can then be compared with the nondimensional Green and Ampt infiltration equation (7.16) (Figure 7.1). The results clearly show that at $t^* = t^*_{\text{grav}}$ there is a relative error on $I^*$ of 5%. Thus the use of the two-term time series of Philip should be limited to

$$t \leq \frac{1}{2} t_{\text{grav}} \tag{7.46}$$

With the nondimensional form of the Philip infiltration equation being established (valid for different soil types and different initial conditions, but for only the limited time range $t^* \leq t^*_{\text{grav}}$), the specific cumulative-infiltration scale factors [equation (7.18)] can now be derived from the definitions [equations (7.44) and (7.45)]

$$\alpha_I = \frac{2[K_1 - K_0]}{S_1^2(\theta_1, \theta_0)} \tag{7.47}$$

and

$$\alpha_t = \frac{2[K_1 - K_0]^2}{S_1^2(\theta_1, \theta_0)} \tag{7.48}$$

which obviously are almost identical with those established for the Green and Ampt (1911) infiltration solution, the only difference being that for the Philip solution (1957a) the value of $K_0$ is not set equal to zero a priori.

Equations (7.43)–(7.48) show that for short times ($t^* \leq t^*_{\text{grav}}$), the behavior of zero-head and/or negative-head infiltration flow is fully determined by three scale factors: $\alpha_I$, $\alpha_t$, and $\mu$. All three parameters depend upon the soil type and the initial and boundary conditions. This result is illustrated by the fan of invariant dimensionless cumulative-infiltration curves [$\Delta I^*(t^*, \mu)$] shown in Figure 7.2. Each value of $\mu$ fixes an invariant dimensionless net-infiltration curve or "dynamic-similarity class" for which the physical system is said to be macroscopically similar; that is, only those combinations of soil characteristics and initial and boundary conditions that lead to the same value of $\mu$ belong to the same dynamic-similarity class. Therefore, infiltration cases for identical soils, but for different boundary conditions, do not necessarily fall into the same similarity class.

The last and most important step in the dynamic-similarity analysis consists in determining the set of basic scale factors $\alpha_\theta, \alpha_h, \alpha_K$, and $\alpha_z$ [equation (7.5)] that will reduce the Richards equation, rather than the specific Philip infiltration solution.

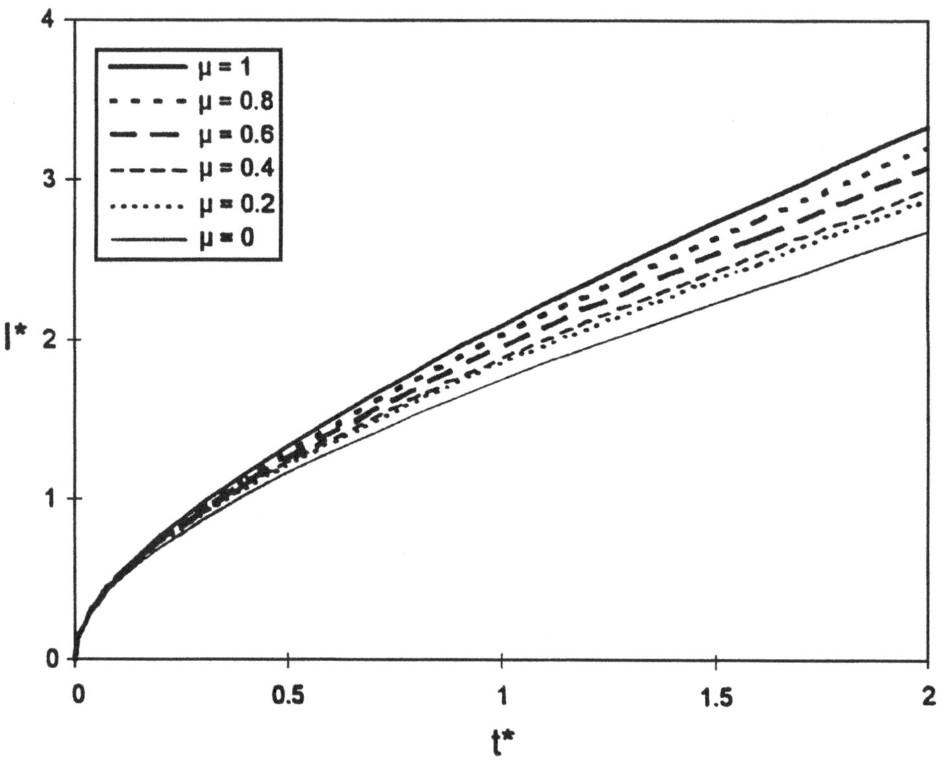


Figure 7.2. Fan of nondimensional zero-head infiltration curves ($\Delta I^*$) calculated by the limited time-series solution [equation (7.43)] as functions of time ($t^*$) and soil parameter $\mu$; time range of validity $t^* \leq t^*_{\text{grav}}$.

Following the same identification procedure explained earlier in the context of the Green and Ampt equation, we find the set of scale factors

$$\alpha_\theta = \lambda_\theta \frac{\theta_s}{\theta_1} \frac{\theta_1}{[\theta_1 - \theta_0]} \quad \text{with } \lambda_\theta = \frac{1}{\theta_s} \tag{7.49}$$

$$\alpha_K = \lambda_K \frac{K_s}{K_1} \frac{K_1}{[K_1 - K_0]} \quad \text{with } \lambda_K = \frac{1}{K_s} \tag{7.50}$$

$$\alpha_h = \lambda_h \frac{\theta_1 K_1 S_1^2(\theta_s, 0)}{\theta_s K_s S_1^2(\theta_1, 0)} \frac{[\theta_1 - \theta_0][K_1 - K_0] S_1^2(\theta_1, 0)}{\theta_1 K_1 S_1^2(\theta_1, \theta_0)} \tag{7.51}$$

$$\text{with } \lambda_h = \frac{2\theta_s K_S}{S_1^2(S_1^2(\theta_s, 0)}$$

and

$$\alpha_z = \alpha_h \tag{7.52}$$

where $\lambda$, once again, identifies the scale factors that are purely soil-related. The final terms on the right-hand side of equations (7.49)–(7.51) represent, respectively, the corrections for the surface boundary-condition effect and the initial-condition effect. It can easily be seen that equations (7.49)–(7.52) are identical with those established for the Green and Ampt infiltration solution [equations (7.26)–(7.28)] once $K_0$ is set equal to zero.

The key point to be recognized from this analysis of the Philip infiltration solution is that, unfortunately, in the behaviors of soil-water movements in unsaturated general field soils governed by the Richards equation, there is no unique dynamic similarity for a given soil type. There exists a multitude of dynamic-similarity classes, depending on the combination of soil type and initial and boundary conditions. The infiltration behavior is defined by only three infiltration scaling factors embodying the effects of soil type and the initial and boundary conditions. However, the basic scale factors of the equation for unsaturated flow and the scale factors that are purely soil-related are identical with those obtained through the dynamic analysis for the Green and Ampt equation.

*The Generalized Solution.* In order to overcome the time-limit problem ($t < t_{\text{grav}}$) that constrains the Philip time-series solution, Parlange et al. (1982) developed a quasi-exact infiltration solution of the Richards equation (7.1) valid over the entire time range $t \in [0, \infty]$. That work was concerned with both zero-head and negative-head surface conditions ($h_1 \leq 0$) and made use of a special double integration procedure for the water-transfer equation. In later studies by Parlange, Haverkamp, and Touma (1985) and Haverkamp et al. (1990), that infiltration solution was extended to a positive-head boundary condition ($h_1 \geq 0$). Without going into many details, we shall give a concise description of that approach.

Considering the case of infiltration into a soil at a uniform initial soil-water pressure $h_0$ (or volumetric water content $\theta_0$) from a surface boundary source at either a constant negative head ($h_1 \leq 0$) or a constant positive head ($h_1 \geq 0$), double integration of the Richards equation allows expression of the water-content profiles $z(\theta, t)$ in the rigorous form

$$z(\theta, t) = z_s(t) + \int_{\theta}^{\theta_s} \frac{D(x)}{-[K(x) - K_0] + F(x, t)[q_1(t) - K_0]} dx \tag{7.53}$$

where $x$ is the integration variable, $z_s$ is the depth to which the soil is considered to be saturated when $h_1 \geq 0$, and $F(\theta, t)$ is the "flux-concentration relation" (Philip, 1973), given by $F(\theta, t) = [q(z, t) - q_0]/[q_1(t) - q_0]$. Obviously, $z_s = 0$ and $\theta_s = \theta_1$ when $h_1 \leq 0$. Integrating $z$ by parts from $\theta_0$ to $\theta_s$ then yields

$$\begin{aligned}[I(t) - K_0 t] = {} & h_{\text{surf}} \frac{[\theta_s - \theta_0]K_s}{q_1(t) - K_s} \\ & + \int_{\theta_0}^{\theta_s} \frac{[\theta - \theta_0]D(\theta)}{-[K(\theta) - K_0] + F(\theta, t)[q_1(t) - K_0]} d\theta \end{aligned} \tag{7.54}$$

where $h_{\text{surf}}$ is the ponded-water depth imposed at the soil surface ($h_1 = h_{\text{surf}} \geq 0$). For the integration of equation (7.54), Parlange et al. (1982) introduced an integral soil parameter $\beta$ expressed as a function of conductivity and diffusivity. The relation was later slightly generalized by Haverkamp et al. (1990) to

$$\frac{[K(\theta) - K_0]}{[K_s - K_0]} = f(\theta)\left[1 - \frac{2\beta(\theta_s, \theta_0)}{S_1^2(\theta_s, \theta_0)} \int_\theta^{\theta_s} \frac{[x - \theta_0]}{f(x)} D(x)\, dx\right] \tag{7.55}$$

where $f(\theta)$ stands for the diffusivity-driven flux-concentration relation $F(\theta, 0)$, given by equation (7.40). Obviously $\beta$ is defined over the interval $0 \leq \beta \leq 1$ and is a function of the soil type and the governing initial and boundary conditions. The lower limit $\beta = 0$ corresponds to the Green and Ampt type of soil, in which $dK/d\theta$ increases much less rapidly with $\theta$ than does the diffusivity, whereas the upper limit $\beta = 1$ corresponds to soils in which $dK/d\theta$ and the diffusivity behave similarly. Hereafter, this latter soil type will be referred to as the "Gardner (1958) soil," after the author who suggested this conductivity behavior. Following equation (7.55), the parameter $\beta$ is defined as a function of volumetric water content ($\theta_0 \leq \theta \leq \theta_s$); as a result, $\beta$ is influenced by the surface boundary condition only when $\theta_1 \leq \theta_s$ or $h_1 \leq 0$. Ross, Haverkamp, and Parlange (1996) showed that $\beta$ is slightly affected by changes in the surface boundary condition $\theta_1$, especially when $\theta_1$ stays close to $\theta_s$ ($\theta_1 \geq 0.75\theta_s$).

The use of equation (7.55) allows solution of the integral term of equation (7.54), resulting in the following cumulative-infiltration equation (Haverkamp et al., 1990):

$$I(q_1, t) = K_0 t + h_{\text{surf}}[\theta_s - \theta_0]\frac{K_s}{q_1 - K_s} + \frac{S_1^2(\theta_s, \theta_0)}{2[K_s - K_0]} \ln\left[1 + \frac{K_s - K_0}{q_1 - K_s}\right] \tag{7.56}$$

However, the cumulative infiltration $I$ can be calculated from equation (7.56) only when the corresponding surface infiltration flux $q_1$ is known. The derivation in time of equation (7.56) yields the necessary flux/time equation, $q_1(t)$:

$$\begin{aligned}2[K_s - K_0]t = {} & \frac{2h_{\text{surf}} K_s[\theta_s - \theta_0]}{q_1 - K_s} + \frac{S_1^2(\theta_s, \theta_0)}{\beta[1 - \beta][K_s - K_0]} \ln\left[1 + \beta\frac{K_s - K_0}{q_1 - K_s}\right] \\ & - \frac{S_1^2(\theta_s, \theta_0) + 2[1 - \beta]h_{\text{surf}} K_s[\theta_s - \theta_0]}{[1 - \beta][K_s - K_0]} \ln\left[1 + \frac{K_s - K_0}{q_1 - K_s}\right]\end{aligned} \tag{7.57}$$

The infiltration solution given by the combined use of equations (7.56) and (7.57) is valid over the entire time range $t \in [0, \infty]$ and is applicable for both positive- and negative-head surface boundary conditions. Although this solution has proved to be extremely precise, its application is cumbersome. For that reason, Barry et al. (1995) presented an extension of the foregoing infiltration solution that improved the applicability of the approach by transforming the implicit combination of equations (7.56) and (7.57) into an explicit infiltration equation, without affecting the precision. However, for the purposes of our inspectional analysis, the computational difficulty is not essential. More important is that the solution is based on the use of parameters with well-defined physical meanings, and it is adjustable for varying initial and boundary conditions.

As in the analysis of the foregoing infiltration equations (Green and Ampt and Philip), the following step in the inspectional analysis is to express equations (7.56) and (7.57) in dimensionless form,

$$\Delta I^*(\Delta q_1^*) = \frac{\gamma}{\Delta q_1^* - 1} + \frac{[1-\gamma]}{\beta} \ln\left[1 + \frac{\beta}{\Delta q_1^* - 1}\right] \tag{7.58}$$

and

$$t^*(\Delta q_1^*) = \frac{\gamma}{\Delta q_1^* - 1} + \frac{[1-\gamma]}{\beta[1-\beta]} \ln\left[1 + \frac{\beta}{\Delta q_1^* - 1}\right] - \frac{[1-\beta\gamma]}{[1-\beta]} \ln\left[1 + \frac{1}{\Delta q_1^* - 1}\right] \tag{7.59}$$

by the use of the following dimensionless variables:

$$\Delta I^* = [I - K_0\, t]\frac{2[K_s - K_0]}{S_+^2} \tag{7.60}$$

$$t^* = t\frac{2[K_s - K_0]^2}{S_+^2} \tag{7.61}$$

and

$$\Delta q_1^* = \frac{d}{dt^*}(\Delta I^*) = \frac{q_1 - K_0}{K_s - K_0} \tag{7.62}$$

with

$$\gamma = \frac{2h_{\text{surf}} K_s[\theta_s - \theta_0]}{S_+^2} \tag{7.63}$$

As in the case of the Philip infiltration equation, $\Delta I^*$ stands for the dimensionless net cumulative infiltration (expressed by the difference between cumulative infiltration $I^*$ and initial cumulative infiltration $I_0^*$), $\Delta q_1^*$ represents the dimensionless net infiltration rate at the soil surface, and the parameter $\gamma$ is a dimensionless surface boundary-condition parameter defined over the interval $0 \leq \gamma \leq 1$. At this stage it is worthwhile to go into more detail about the significance of the parameter $\gamma$. As indicated by equation (7.11), the sorptivity value $S_+$ is composed of two terms: one term calculated over the unsaturated water-content domain ($\theta_0 < \theta < \theta_s$), and one term determined over the saturated domain, where the soil-water pressure is the descriptive parameter ($0 \leq h \leq h_{\text{surf}}$). Following the definition given by equation (7.63), $\gamma$ corresponds to the relative sorptivity fraction calculated over the saturated domain. So for general field soils, $\gamma$ takes into account only the effect of the positive-head boundary condition on the infiltration process (e.g., $\gamma = 0$ when $h_{\text{surf}} = 0$). On the other hand, for Green and Ampt soils, the value of $\gamma$ is systematically equal to unity, as the unsaturated sorptivity fraction is nonexistent by definition.

The combination of equations (7.58) and (7.59) allows us to write the general constant-head infiltration equation in the form

$$\Delta I^* = [1-\beta]t^* + \ln\left[1 + \frac{1}{\Delta q_1^* - 1}\right] + \beta\gamma\left\{\frac{1}{\Delta q_1^* - 1} - \ln\left[1 + \frac{1}{\Delta q_1^* - 1}\right]\right\} \tag{7.64}$$

Equation (7.64) clearly reveals the transition between the different effects of the two types of surface boundary conditions: $h_1 \leq 0$ and $h_1 \geq 0$.

For the case in which $h_1 \leq 0$ (with $\theta_1 \leq \theta_s$), the value of $\gamma$ obviously becomes equal to zero, and equation (7.64) reduces to

$$\Delta I^* = [1-\beta]t^* + \ln\left[1 + \frac{1}{\Delta q_1^* - 1}\right] \tag{7.65}$$

which can easily be transformed to

$$\Delta I^* = [1-\beta]t^* \ln\left[\frac{\exp(\beta \Delta I^*) - 1 + \beta}{\beta}\right] \tag{7.66}$$

Equation (7.66) is the same as the negative-head infiltration equation given by Parlange et al. (1982).

When a soil is of the Green and Ampt type (i.e., $\beta = 0$ and $\gamma = 1$), equation (7.65) reduces to the simple Green and Ampt equation given before as equation (7.16). The other extreme case of the infiltration envelope is given by the so-called Talsma solution (Talsma and Parlange, 1972), which can be obtained by substituting $\beta = 1$ and $\gamma = 0$ into equation (7.65), followed by integration:

$$\Delta I^* = t^* + \left[1 - \frac{1}{\exp(\Delta I^*)}\right] \tag{7.67}$$

Up to now, the validity of the infiltration solution given by equation (7.64) for arbitrary $\beta$ and $\gamma$ has been verified for the specific cases of the infiltration envelope; it is of special interest to analyze the behavior of the solution for limiting time values. For short times, the dimensional form of equation (7.64) can be expanded in a time series of the form (Haverkamp et al., 1990)

$$I = B_1 t^{1/2} + B_2 t + B_3 t^{3/2} + \cdots \tag{7.68}$$

with

$$B_1 = S_+ \tag{7.69}$$

$$B_2 = K_0 + \frac{1}{3}[K_s - K_0][2 - \beta(1-\gamma)] \tag{7.70}$$

$$B_3 = \frac{[K_s - K_0]^2}{9S_+}\{1 + 2[1 - \beta(1-\gamma)] - 2[1 - \beta(1-\gamma)]^2\} \tag{7.71}$$

Equations (7.68)–(7.71) show clearly that the cumulative-infiltration solution expressed by equation (7.64) has the correct time behavior: for $t \to 0$, equation (7.64) behaves like $I = B_1 t^{1/2}$. With increasing times, the influence of gravity becomes more and more important relative to the capillary forces, and that is reflected by the increasing weights of the successive terms $B_2$ and $B_3$ in the infiltration solution. For long times ($t \to \infty$), the expansion of equation. (7.64) takes the form

$$\Delta I^* = t^* + \gamma \ln t^* - \frac{1-\gamma}{1-\beta} \ln \beta - \gamma \ln \gamma \tag{7.72}$$

Equation (7.72) shows that ($\Delta I^* - t^*$) tends toward "ln($t$)" behavior (Green and Ampt, 1911), which becomes more pronounced as the ponded-water depth at the soil surface increases.

The short-time expansion given by equations (7.68)–(7.71) indicates the interaction between $\beta$ and $\gamma$. These two parameters can easily be combined into the single parameter $\beta^*$:

$$\beta^* = \beta[1 - \gamma] \tag{7.73}$$

$\beta^*$ obviously reduces to $\beta$ when $h_1 \leq 0, \theta_1 \leq \theta_s$, and $\gamma = 0$. Using equations (7.11) and (7.63), the term $(1-\gamma)$ can be written as

$$(1-\gamma) = \frac{S_1(\theta_s, \theta_0)^2}{S_+^2} \tag{7.74}$$

This equation indicates that $(1-\gamma)$ simply corrects the soil parameter $\beta$ for the effect of the positive-head boundary condition. By analogy with the two-term Philip solution [equation (7.42)], the value of $\beta^*$ (and hence the values of $\beta$ and $\gamma$) can then be expressed as a function of the classic soil characteristics by use of equation (7.38) (Haverkamp, Arrue, and Soet, 1995):

$$\beta(1-\gamma) = \beta^* = 2 - \frac{2[\theta_s - \theta_0][K_s - K_0]}{S_+^2}\sigma_K(h_{\text{surf}}, h_0) \tag{7.75}$$

This equation is of great help in identifying the different scale and/or shape parameters entering into the specific expressions chosen for the soil characteristics $h(\theta)$ and $K(\theta)$, such as the van Genuchten (1980) and Brooks and Corey (1964) equations.

Use of the term $\beta(1-\gamma)$ also allows for development of a precise approximation of equation (7.64) in the form

$$\Delta I^* = [1 - \beta(1-\gamma)]t^* + \ln\left\{\frac{\exp[\beta(1-\gamma)\Delta I^*] + \beta(1-\gamma) - 1}{\beta(1-\gamma)}\right\} \tag{7.76}$$

Although this equation is not as precise as equation (7.64) (i.e., a maximum error of 2% on $\Delta I^*$), it has the advantage of using only two dependent variables, $\Delta I^*$ and $t^*$, instead of the three ($\Delta I^*$, $t^*$, and $\Delta q_1^*$) in equation (7.64).

On the basis of the nondimensional form of the generalized-head infiltration equation, the specific cumulative-infiltration scale factors can be derived from equations (7.60) and (7.61):

$$\alpha_I = \frac{2[K_1 - K_0]}{S_+^2} \tag{7.77}$$

and

$$\alpha_t = \frac{2[K_1 - K_0]^2}{S_+^2} \tag{7.78}$$

As expected, these scale factors are almost identical with those established for the Philip infiltration solution, the only difference being that for the generalized solution the sorptivity value is not restricted to the non-ponding case.

The foregoing analysis shows that the infiltration-flow behavior, under either constant positive or negative head, for all different types of soil, is fully determined by four scale factors: $\alpha_I$, $\alpha_t$, $\beta$, and $\gamma$. All four scale factors depend upon the soil type and the initial and boundary conditions, where $\beta$ is specific for unsaturated flow, and $\gamma$ applies only for a positive-head boundary condition. This result is illustrated by the fan of invariant dimensionless cumulative-infiltration curves [$\Delta I^*(t^*, \beta, \gamma)$] shown in Figure 7.3. Each combination of $\beta$ and $\gamma$ values fixes a dynamic-similarity class

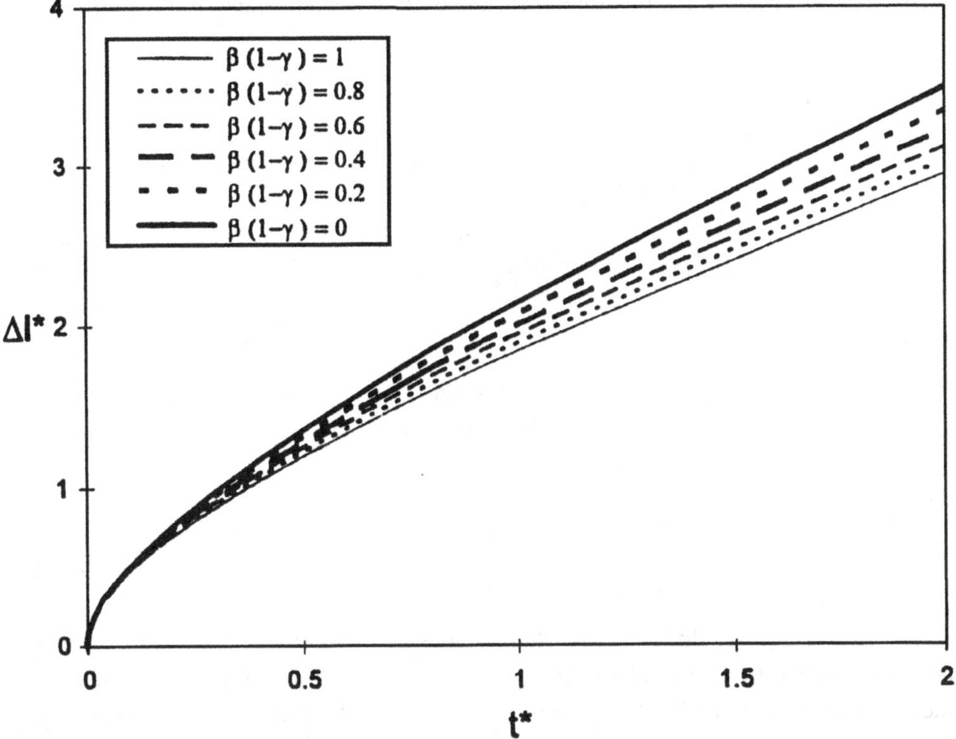


Figure 7.3. Fan of nondimensional infiltration curves ($\Delta I^*$) calculated by the generalized-head infiltration equation [equation (7.64)] as functions of time ($t^*$) and the parameters $\beta$ and $\gamma$.

for which the physical system is said to be macroscopically similar. Even though a fixed relation between $\beta$ and $\gamma$ can be suggested through the form of equation (7.73), which will bring the number of descriptive scale parameters back from four to three [as shown by equation (7.76)], there still will be a multitude of similarity classes for the infiltration behavior, depending on the combination of soil type and initial and boundary conditions. Nevertheless, the envelope of possible similarity classes is bounded by the two solutions of Green and Ampt [equation (7.16)] and Talsma equation (7.67)]. Each of these solutions is obtained for a specific soil type (i.e., the Green and Ampt soil and the Gardner soil). For each of these extreme cases there exists a unique dynamic similarity in the behavior of infiltration flow for which the physical system is macroscopically similar. This is of great importance when dealing with large-scale hydrologic modeling of water catchments, because it fixes the two extreme scenarios on which decision making can be based.

The final step of the dynamic analysis consists in the determination of the set of basic scale factors $\alpha_\theta$, $\alpha_h$, $\alpha_K$, and $\alpha_z$ [equation (7.5)] to reduce the Richards equation. Following the same procedure as given for the Green and Ampt and Philip solutions, the set of scale factors can be defined for the case of positive-head infiltration as

$$\alpha_\theta = \lambda_\theta \frac{\theta_s}{[\theta_s - \theta_0]} \quad \text{with } \lambda_\theta = \frac{1}{\theta_s} \tag{7.79}$$

$$\alpha_K = \lambda_K \frac{K_s}{[K_s - K_0]} \quad \text{with } \lambda_K = \frac{1}{K_s} \tag{7.80}$$

$$\alpha_h = \lambda_h \frac{[\theta_s - \theta_0][K_s - K_0] S_1^2(\theta_s, 0)}{\theta_s K_s S_1^2(\theta_s, \theta_0)} \frac{S_1^2(\theta_s, \theta_0)}{S_+^2} \tag{7.81}$$

with

$$\lambda_h = \frac{2\theta_s K_s}{S_1^2(\theta_s, 0)}$$

and

$$\alpha_z = \alpha_h \tag{7.82}$$

where $\lambda$, once again, identifies the scale factors that are solely soil-related. For the case of negative-head infiltration (i.e., $\theta_1 \leq \theta_s$, $h_1 \leq 0$, and $K_1 \leq K_s$), the scale factors become identical with those given earlier for the Philip infiltration solution [equations (7.49)–(7.52)].

The essence of the analysis of the generalized constant-head infiltration equation is that there exists no unique dynamic similarity in the behavior of soil-water movement in unsaturated general field soils governed by the Richards equation. Instead, there is a multitude of dynamic-similarity classes, depending on the combination of soil type and initial and boundary conditions. In its most general form, the positive- or negative-head infiltration behavior is defined by three infiltration scaling factors embodying the effects of soil type and initial and boundary conditions. For the two particular cases that correspond to the Green and Ampt soil and the Gardner soil, each has a unique similarity solution for which the physical system is macroscopically similar. Because these two solutions are the bounds of the envelope of all possible similarity classes, they become of great use for water-catchment modeling at large scales. The scale factors that are solely soil-related are identical with those obtained through the dynamic analysis for the Green and Ampt and Philip equations.

### *7.2.2 Flux-type Surface Boundary Condition*

Boundary conditions at the surface ($z = 0$) can assume many complex forms. Thus far we concentrated on the constant-head boundary condition at the surface. The purpose of this section is to show that the scaling parameters introduced for constant-head infiltration apply naturally to the flux boundary condition as well.

For instance, at ponding time ($t_p$), when water begins accumulating at the soil surface, simple relations between the flux at ponding ($q_p$) and the cumulative infiltration ($I_p$) are observed (Smith and Parlange, 1977). That is, for a Green and Ampt soil,

$$[q_p - K_s]I_p \cong \frac{S_1^2(\theta_s, \theta_0)}{2} \tag{7.83}$$

or, for a Gardner soil, which earlier led to the Talsma infiltration solution (Talsma and Parlange, 1972),

$$\frac{2I_p K_s}{S_1^2(\theta_s, \theta_0)} \cong \ln \frac{q_p}{[q_p - K_s]} \tag{7.84}$$

For those two expressions, as well as for more complex expressions given by Smith and Parlange (1977), the flux at the soil surface ($q_1$) is clearly scaled by $\alpha_K = 1/K_s$, and the cumulative infiltration by $\alpha_I = 2K_s/[S_1(\theta_s, \theta_0)]^2$; hence time is scaled by $\alpha_t = 2[K_s]^2/[S_1(\theta_s, \theta_0)]^2$. We easily observe that these scale factors, $\alpha_K$, $\alpha_1$, and $\alpha_t$, are identical with those determined earlier for the case of constant-head infiltration (Section 7.2.1).

Another important example is provided by an exact solution of the Richards equation that applies equally well for infiltration and drainage and for arbitrary soil properties (Ross and Parlange, 1994). In that case we can look in detail at the shape of the water-content profile $\theta(z, t)$. For instance, the water-content profile during infiltration in a soil behaving as a Green and Ampt soil is given by

$$At - z = \frac{S_1^2(\theta_s, 0)K(\theta)}{2\theta_s K_s\{A[\theta - \theta_0] - K(\theta)\}} \tag{7.85}$$

where $A$ is a constant such that at the surface where $\theta(z = 0, t) = \theta_1$, the flux is given by

$$q_1 = A[\theta_1 - \theta_0] \tag{7.86}$$

The evolution of the volumetric water content at the soil surface, $\theta_1(t)$, and of the surface flux, $q_1(t)$, is given by

$$At = \frac{S_1^2(\theta_s, 0)K_1}{2\theta_s K_s\{A[\theta_1 - \theta_0] - K_1\}} \tag{7.87}$$

so that time, $t$, is scaled once again by $\alpha_t = 2[K_s]^2/[S_1(\theta_s, \theta_0)]^2$, and depth, $z$ by $\alpha_z = 2K_s[\theta_s - \theta_0]/[S_1(\theta_s, \theta_0)]^2$.

If we analyze the evolution of the water-content profiles $\theta(z, t)$ in Gardner-behaving soils, the results become (Selker, Parlange, and Steenhuis, 1992a)

$$At - z = \frac{S_1^2(\theta_s, 0)}{2\theta_s K_s} \ln \frac{A[\theta - \theta_0]}{\{A[\theta - \theta_0] - K(\theta)\}} \tag{7.88}$$

with, again, $\theta_1(t)$ being given for $z = 0$. The scaling for this equation is identical with the scaling for the Green and Ampt result.

The foregoing two cases were considered for the infiltration problem. A simple sign change permits the use of the same fundamental solution for drainage (Ross and Parlange, 1994) without affecting the scaling.

Selker et al. (1992a) showed that the same exact solution can be used to describe the structures of preferential paths or fingers that result from unstable wetting fronts, thus leading to the same scaling. Unstable wetting fronts were first studied in the laboratory in the form of a less permeable horizon overlying a highly permeable soil (Hill and Parlange, 1972), as first observed in the field by Starr et al. (1978). Later, Glass, Steenhuis, and Parlange (1989) pointed out that hysteresis effects were crucial in understanding both the formation and persistence of fingers. Ritsema et al. (1996) showed that the fingered flow patterns could also occur in water-repellent soils, and Selker et al. (1992a,b) studied their structure and found that the presence of a less permeable layer at the soil surface was not always necessary.

In the simplest case, for fairly dry soils, the fingers enter the soil with a water content close to saturation and a pressure near the water-entry value (Hillel and Baker, 1988). Because of hysteresis effects (Liu, Steenhuis, and Parlange, 1994), the fingers widen rapidly as the initial water content of the soil increases. In the approach of Parlange and Hill (1976), finger widths are estimated as

$$d_F = \frac{\kappa S_1^2(\theta_F, \theta_0)}{K_F[\theta_F - \theta_0][1 - (q_1/K_F)]} \tag{7.89}$$

where $\kappa$ depends on the dimensionality of the finger and is equal to $\pi$ in two dimensions (Parlange and Hill, 1976) and is equal to 4.8 in three dimensions (Glass, Parlange, and Steenhuis, 1991). The subscript $F$ refers to values at the fingertip, and $q_1$ is the flux per unit area applied at the soil surface. Hence for this phenomenon of fingering, the width scales as $K_s[\theta_F - \theta_0]/[S_1(\theta_F, \theta_0)]^2$, and the flux as $1/K_F$.

Once again we recognize that for a variety of problems involving flux conditions, the same scale factors that were observed for the head conditions enter for the extreme soil behaviors of a Green and Ampt soil and a Gardner soil.

## 7.3 Identification Procedure

The cumulative-infiltration data $[I_{\exp}(t)]$, together with the surface boundary condition ($h_{\text{surf}}$), are the only data supposed to be known from field measurements.

The identification procedure aims to determine the scaling factors $\alpha_I$, $\alpha_t$, $\beta$, and $\gamma$ [equation (7.64)] from these input data. Speaking generally, the optimization of four parameters on a monotone-increasing function (such as the cumulative-infiltration curve) does not guarantee a priori the uniqueness of the solution (Haverkamp, 1983). When starting up the optimization procedure, the initial guess at the parameter values is crucial to ensure the physical correctness of the final results. For the particular case of equation (7.64), the inverse problem appears to be highly ill-posed, with extremely small values for $\beta$ and $\gamma$. It shows that the forms of the different dimensionless cumulative-infiltration curves given by Figure 7.3 are very similar, indicating, at the same time, that the differences between the various similarity classes are extremely small. The consequence is that, from a practical viewpoint, the hypothesis of a unique dynamic similarity in the behaviors of soil-water movements is close to reality.

In order to overcome the problem imposed by the presence of many parameters, the identification procedure is built in three successive steps: (1) a first step that aims to estimate (guess) accurate values for $\alpha_I$ and $\alpha_t$; (2) a second step that determines first estimates for the parameters $\beta$ and $\gamma$ by use of the dimensional form of the close approximation given by equation (7.76) together with the values of $\alpha_I$ and $\alpha_t$ calculated in the first step; (3) the full optimization of the four parameters $\alpha_I$, $\alpha_t$, $\beta$, and $\gamma$ by use of the dimensional form of the infiltration solution given by equations (7.56) and (7.57). For conciseness, we choose to illustrate the procedure using both the Green and Ampt solution [equation (7.16)] and the Talsma solution [equation (7.67)] to estimate $\alpha_I$ and $\alpha_t$. These results are of particular interest as they determine the invariant upper and lower bounds of the envelope of all possible similarity classes and hence the extreme parameter values that are of great use for water-catchment studies where stochastic modeling is needed.

Both the Green and Ampt equation [equation (7.16)] and the Talsma equation [equation (7.67)] are defined by only two unknown parameters: $\alpha_I$ and $\alpha_t$. When optimizing two unknown parameters, it can easily be shown that there exists one unique solution for a monotonous-increasing curve such as cumulative infiltration as a function of time. For the Green and Ampt equation [equation (7.16)], where $K_0$ is set equal to zero a priori, we choose to identify the parameters $\alpha_I$ and $K_s$. Because sorptivity $S_+^2$ is proportional to $K_s$ [equation (7.12)], the ratio $[2S_1^2/K_s]$ (coefficient $\alpha_I$) is equal to $[h_{\text{surf}} - h_f][\theta_s - \theta_0]$ and hence is independent of $K_s$. The scale factor $\alpha_t$ is then determined by $\alpha_t = K_s\alpha_I$. For the Talsma equation [equation (7.67)], the same parameters $\alpha_I$ and $[K_s - K_0]$ are optimized.

At this stage it is important to note that there is a significant difference between the identification procedure used for positive-head infiltration data and that used for negative-head infiltration data. For the case of zero- or negative-head infiltration ($h_1 \le 0$ and $\theta_1 \le \theta_s$), the combined term $[K_s - K_0]$ cannot be deconvoluted into separate values of $K_s$ and $K_0$ a priori, as conductivity enters systematically through the factor $[K_s - K_0]$. However, for positive-head infiltration ($h_1 > 0$ and $\theta_1 > \theta_s$), the conductivity term enters into the definition of sorptivity through two different

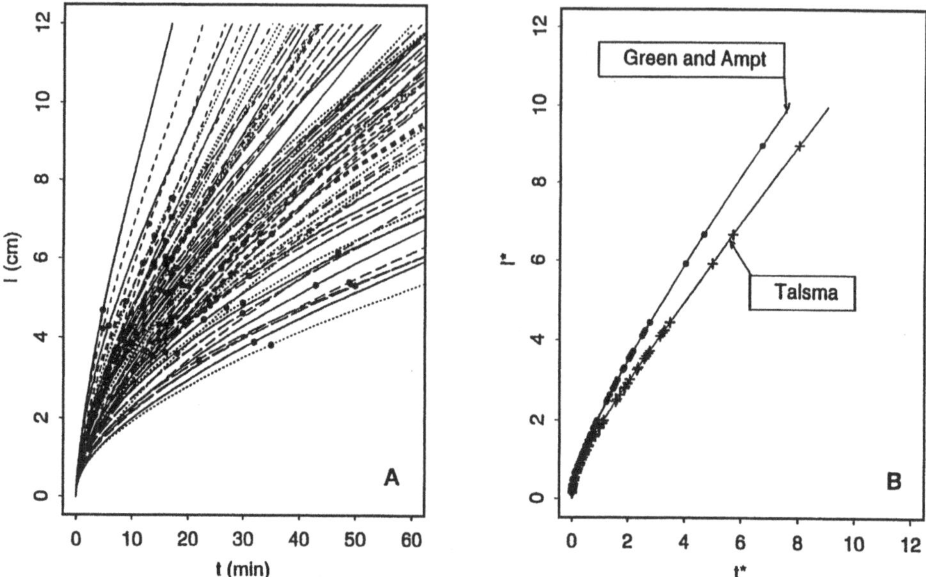


Figure 7.4. Fan of 100 different cumulative-infiltration curves measured over an area of 100 km$^2$ before (A) scaling and after (B) scaling.

forms: the unsaturated-flow contribution, which again is proportional to $[K_s - K_0]$, and the positive-head contribution, which is proportional to $K_s$. Consequently, only for positive-head infiltration experiments can the values of $K_s$ and $K_0$ be determined separately.

Some of the results from this first identification procedure are illustrated in Figures 7.4 and 7.5. In the context of the European International Project on Climate and Hydrological Interactions between the Vegetation, the Atmosphere and the Land Surface (ECHIVAL), which was developed by members of the European research community jointly with the Commission of the European Communities, an infiltration field experiment was carried out in the desertification-threatened area of La Mancha in Spain. Over an area of 100 km$^2$, with a grid mesh of $1 \times 1$ km, 100 cumulative-infiltration curves were measured (Haverkamp et al., in press). The results are shown in Figure 7.4a. For each of the 100 curves, the parameters $\alpha_I$ and $\alpha_t$ were determined by use of the Green and Ampt [equation (7.16)] and Talsma [equation (7.67)] solutions. The scaled results, $I^*(t^*)$, in Figure 7.4b show the unique similarity for the two infiltration solutions. All of the 100 points fit perfectly well each of the two infiltration curves. The local values of $\alpha_I$ and $\alpha_t$ were then used to calculate the local values of $K_s$, shown in Figure 7.5. For simplicity, the initial conductivity value, $K_0$, was taken equal to zero (a realistic assumption for the extremely dry conditions of Spain during the summer months). Although the local $K_s$ values calculated by the Green and Ampt solution and the Talsma solution represent, respectively, the upper and lower limits of the expected conductivity values, the differences are small

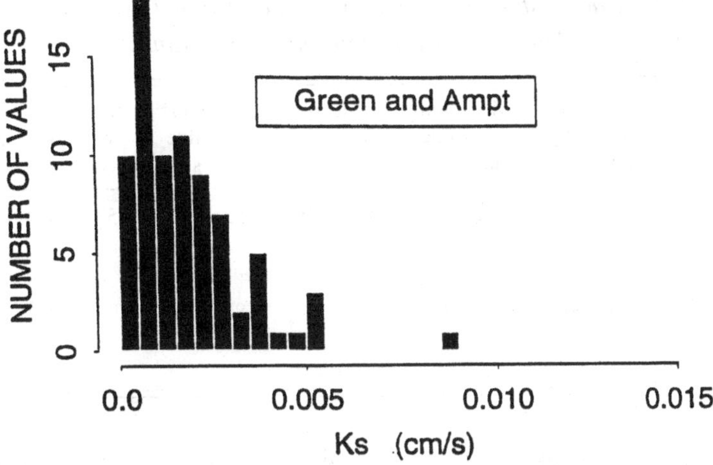


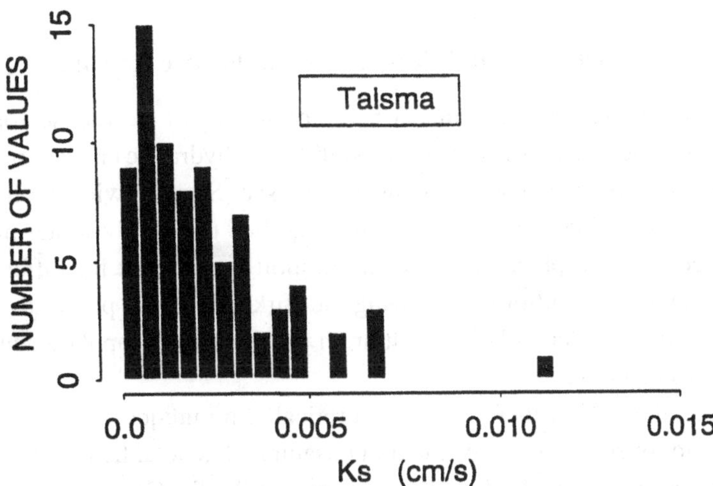


Figure 7.5. Histogram calculated for both distributions of $K_s$ values using either the Green and Ampt or the Talsma infiltration equation.

(Table 7.1 and Figure 7.5). The spatial distributions for the Green and Ampt and Talsma approaches are almost identical (Figure 7.5); for both cases the conductivity values are log-distributed, as would be expected from the work of Nielsen, Biggar, and Erh (1973) (Figure 7.6). The median conductivity values, which are of great importance for large-scale aggregation models, are of the same order of magnitude.

The results discussed here clearly demonstrate the efficiency of the scaling theory. With a minimum of measurement points for $I_{\exp}(t)$, precise information can be obtained for the local hydraulic characteristics (such as $K_s$) and their possible variations (upper and lower limits). Large areas can be surveyed in short times (e.g., the entire cumulative-infiltration measurement campaign over the area of 100 km$^2$

Table 7.1. *Characteristic conductivity values calculated over the spatial distribution of locally determined $K_s$ values*

| Envelope | Minimum $K_s$ value ($cm \cdot s^{-1}$) | Maximum $K_s$ value ($cm \cdot s^{-1}$) | Mean $K_s$ value ($cm \cdot s^{-1}$) | Median $K_s$ value ($cm \cdot s^{-1}$) | Standard deviation ($cm \cdot s^{-1}$) | Variation coefficient (%) |
|---|---|---|---|---|---|---|
| Green and Ampt [equation (7.16)] | $2.05 \cdot 10^{-3}$ | $8.70 \cdot 10^{-3}$ | $1.88 \cdot 10^{-3}$ | $1.53 \cdot 10^{-3}$ | $1.52 \cdot 10^{-3}$ | 99.3 |
| Talsma solution [equation (7.67)] | $1.94 \cdot 10^{-4}$ | $1.10 \cdot 10^{-2}$ | $2.30 \cdot 10^{-3}$ | $1.86 \cdot 10^{-3}$ | $1.93 \cdot 10^{-3}$ | 104.8 |

discussed earlier did not take more than 3 weeks). The theory permits us to obtain clear information on the spatial distribution of the $K_s$ values and hence on the spatial distribution of characteristics such as soil porosity and soil structure.

## 7.4 Conclusions and Perspectives on Future Research

This chapter has presented a new theory based on scaling of the vadose-zone water-flow equation, rather than on scaling of the soil's static hydraulic characteristics. The approach is based on a dynamic-similarity analysis: Starting with the hypothesis that the Richards (1931) flow equation can be applied to describe water movement in unsaturated soil, the phenomenon of infiltration is analyzed for different types of surface boundary conditions. Following the Birkhoff (1960) postulate, the scale factors obtained for the cumulative-infiltration problem can be applied equally well to the Richards equation.

It has been shown that there exists, theoretically, no unique dynamic similarity for the behavior of soil-water movement in unsaturated general field soils. A unique dynamic similarity is found for only two particular soils: the Green and Ampt (1911) soil and the Gardner (1958) soil. These two soils define the limits of the envelope of possible similarity classes that exist for general field soils. However, for practical purposes, the hypothesis of unique dynamic similarity is an accurate approximation.

Generally speaking, the phenomenon of infiltration can be described by three scale factors ($\alpha_I$, $\alpha_t$, and $\beta^*$) that take into account the effects of initial conditions, boundary conditions, and soil type. Only for the Green and Ampt soil and the Gardner soil is the infiltration determined by only two scale factors. Going down one step in the hierarchy of the water-movement problematic, a set of three basic scale factors ($\alpha_\theta$, $\alpha_h$, and $\alpha_K$) is defined for the Richards equation. The space ($\alpha_z$) and time ($\alpha_t$) scale factors are given as functions of the basic scale factors ($\alpha_\theta$, $\alpha_h$, and $\alpha_K$). Once again, the basic scale factors are expressed as functions of the initial and boundary conditions and of the soil type. Stripping the $\alpha$ scale factors of their

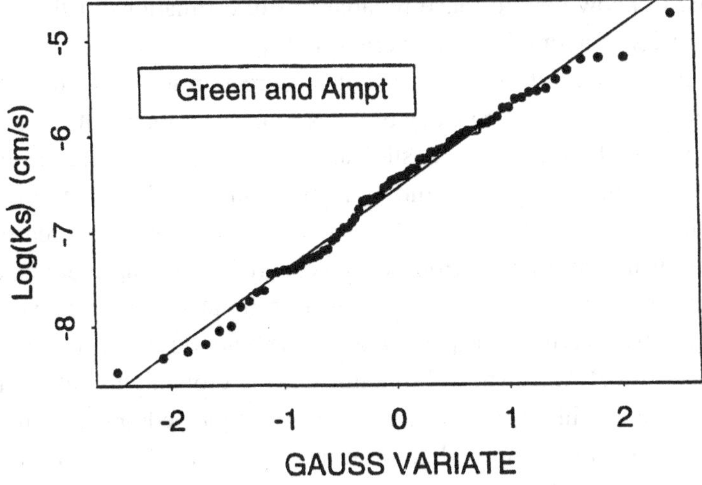


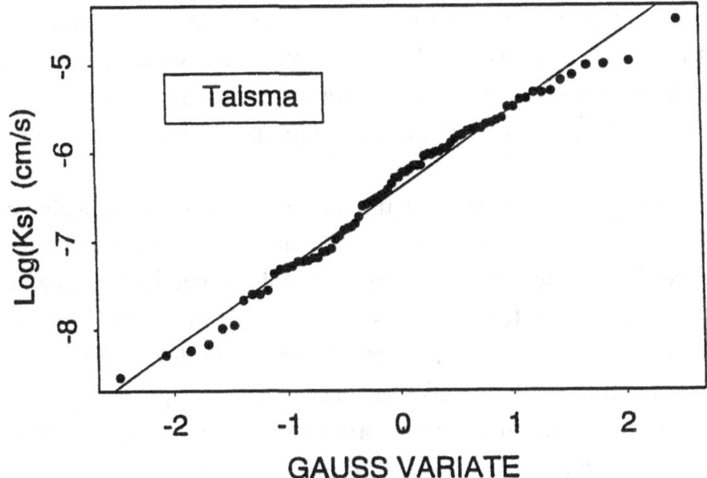


Figure 7.6. Distribution curves of $K_s$ values using either the Green and Ampt or the Talsma infiltration equation.

initial and boundary conditions yields the scale factors that are related solely to the soil ($\lambda_\theta$, $\lambda_h$, and $\lambda_K$). The definitions of these soil-related scale factors show that there exists a definite dependence between the expressions chosen for the water-retention curves and the hydraulic-conductivity curves, implying that optimization of the parameters for soil-water characteristics over measured field data should be carried out simultaneously with a combined objective function.

The implications of this newly developed theory are very promising, as it opens up two distinct research directions that will be crucial for successful modeling of water hydrology at large scales:

1. We need to know how to make reliable in situ estimations of the parameters to describe soil hydraulic characteristics at a great number of grid locations over a catchment, with realistic use of resources in terms of time and costs. It is virtually impossible to carry out classic measurements of the parameters for a soil's hydraulic characteristics at a sufficiently high frequency over a complete catchment, as the number of grid points involved is too large. It is increasingly obvious that although some soil parameters might reasonably be estimated using information from soil surveys (e.g., the shape parameters in the function for a soil's hydraulic characteristics estimated from soil texture), other parameters (such as the scale parameters in the functions for soil characteristics, which are strongly linked to soil structure and therefore highly subject to spatial variability) are unlikely to be estimated from simple pedotransfer functions.

   As illustrated by the Spanish infiltration experiment, the approach presented here seems to be a highly promising tool for pragmatic determination of the parameters for local soil characteristics. Of course, the deconvolution procedure for the specific infiltration scale factors (such as $\alpha_I$) into the basic soil scale factors ($\lambda_\theta$, $\lambda_h$, and $\lambda_K$) must be addressed in more detail, but the first results obtained for 50 soils allows us to put into evidence the functional relationship that exists between the water-retention and hydraulic-conductivity shape parameters, as well as the methodology to obtain the water-retention scale factor.
2. The mismatch that exists between the scale (local $\approx 1\ m^2$) on which our current knowledge of vadose-zone water-transfer processes is based and the scale for which the hydrologic-catchment and regional-climate models (i.e., the typical micro-meteorological fetch is of the order of $10^3\ m^2$) are trying to make predictions is well known. To implement parameters for a soil's water-transfer processes in large-scale models in a successful way, upscaling procedures will be required for either the parameterization of the soil characteristics or the governing equation for soil-water transfer.

   The invariance and near-unique dynamic similarity of the scaled infiltration equations make these scaling factors particularly appropriate to advance the research on upscaling methodologies for soil hydraulic properties and flow phenomena while taking into account their spatial variability.

## References

Barry, D. A., Parlange, J.-Y., Haverkamp, R., and Ross, P. J. 1995. Infiltration under ponded conditions: 4. An explicit predictive infiltration formula. *Soil Sci.* 160:8–17.

Birkhoff, G. 1960. *Hydrodynamics*, 2nd ed. Princeton University Press.

Brooks, R. H., and Corey, C. T. 1964. *Hydraulic Properties of Porous Media*. Hydrology paper no. 3. Fort Collins: Colorado State University.

Elrick, D. E., and Robin, M. J. 1981. Estimating sorptivity of soils. *Soil Sci.* 132:127–33.

Fuentes, C., Haverkamp, R., and Parlange, J.-Y. 1992. Parameter constraints on closed form soil water relationships. *J. Hydrol.* 134:117–42.

Gardner, W. R. 1958. Some steady state solutions of the unsaturated moisture flow equation with application to evaporation from a water table. *Soil Sci.* 85:228–32.

Glass, R. J., Parlange, J.-Y., and Steenhuis, T. S. 1991. Immiscible displacement in porous media: stability analysis of three-dimensional, axisymmetric disturbances with application to gravity-driven wetting front instability. *Water Resour. Res.* 27:1947–56.

Glass, R. J., Steenhuis, T. S., and Parlange, J.-Y. 1989. Mechanism for finger persistence in homogeneous, unsaturated, porous media: theory and verification. *Soil Sci.* 148:60–70.

Green, W. H., and Ampt, G. A. 1911. Studies in soil physics: I. The flow of air and water through soils. *J. Agric. Sci.* 4:1–24.

Haverkamp, R. 1983. Resolution de l'equation de I'infiltration de l'eau dans le sol. Approaches analytiques et numeriques. These de Docteur d'Etat. Universite Scientifique et Medicale de Grenoble.

Haverkamp, R., Arrue, J. L., and Soet, M. 1995. Soil physical properties within the root zone of the vine area of Tomelloso. Local and spatial standpoint. In: *Final Report of Project II, EFEDA II*, ed. F. Martin de Santa Olalla, Ch. 3. Albacete, Spain: University of Castilla–La Mancha.

Haverkamp, R., Kutílek, M., Parlange, J.-Y., Rendon, L., and Krejca, M. 1988. Infiltration under ponded conditions: 2. Infiltration equations tested for parameter time dependence and predictive use. *Soil Sci.* 145:317–29.

Haverkamp, R., Parlange, J.-Y., Starr, J. L., Schmitz, G., and Fuentes, C. 1990. Infiltration under ponded conditions: 3. A predictive equation based on physical parameters. *Soil Sci.* 149:292–300.

Haverkamp, R., Ross, P. J., Fuentes, C., Barry, D. A., Parlange, J.-Y., and Baron, T. T. In press. Infiltration under ponded conditions: 5. Prediction of infiltration parameters with changing initial condition. *Soil Sci.* (submitted).

Hill, D. E., and Parlange, J.-Y. 1972. Wetting front instability in homogeneous soils. *Soil Sci. Soc. Am. Proc.* 36:697–702.

Hillel, D., and Baker, R. S. 1988. A descriptive theory of fingering during infiltration into layered soils. *Soil Sci.* 146:51–6.

Kline, S. J. 1965. *Similitude and Approximation Theory*. New York: McGraw-Hill.

Langhaar, H. L. 1951. *Dimensional Analysis and Theory of Models*. New York: Wiley.

Liu, Y., Steenhuis, T. S., and Parlange, J.-Y. 1994. Closed form solution for finger width in sandy soils at different water contents. *Water Resour. Res.* 30:949–52.

Miller, E. E. 1980. Similitude and scaling of soil-water phenomena. In: *Applications of Soil Physics*, ed. D. Hillel, pp. 300–18. New York: Academic Press.

Miller, E. E., and Miller, R. D. 1955a. Theory of capillary flow: I. Practical implications. *Soil Sci. Soc. Am. J.* 19:267–71.

Miller, E. E., and Miller, R. D. 1955b. Theory of capillary flow: II. Experimental information. *Soil Sci. Soc. Am. J.* 19:271–5.

Miller, E. E., and Miller, R. D. 1956. Physical theory for capillary flow phenomena. *J. Appl. Phys.* 27:324–32.

Nielsen, D. R., Biggar, J. W., and Erh, K. T. 1973. Spatial variability of field measured soil water properties. *Hilgardia* 35:491–506.

Nielsen, D. R., Tillotson, P. M., and Vieira, S. R. 1983. Analyzing field-measured soil-water properties. *Agric. Water Mgt.* 6:93–109.

Parlange, J.-Y. 1975. On solving the flow equation in unsaturated soils by optimisation: horizontal infiltration. *Soil Sci. Soc. Am. Proc.* 39:415–18.

Parlange, J.-Y. 1977. A note on the use of infiltration equations. *Soil Sci. Soc. Am. J.* 41:654–5.

Parlange, J.-Y., Haverkamp, R, and Touma, J. 1985. Infiltration under ponded conditions: 1. Optimal analytical solution and comparison with experimental observations. *Soil Sci.* 139:305–11.
Parlange, J.-Y., and Hill, D. E. 1976. Theoretical analysis of wetting front instability in soils. *Soil Sci.* 122:236–9.
Parlange, J.-Y., Lisle, I., Braddock, R. D., and Smith, R. E. 1982. The three parameter infiltration equation. *Soil Sci.* 133:337–41.
Philip, J. R. 1957a. The theory of infiltration: 1. The infiltration equation and its solution. *Soil Sci.* 83:345–57.
Philip, J. R. 1957b. The theory of infiltration: 4. Sorptivity and algebraic infiltration equations. *Soil Sci.* 84:257–64.
Philip, J. R. 1969. The theory of infiltration. *Adv. Hydrosci.* 5:215–305.
Philip, J. R. 1973. On solving the unsaturated flow equation: 1. The flux–concentration relation. *Soil Sci.* 116:328–35.
Reichardt, K., Nielsen, D. R., and Biggar, J. W. 1972. Scaling of horizontal infiltration into homogeneous soils. *Soil Sci. Soc. Am. Proc.* 36:241–5.
Richards, L. A. 1931. Capillary conduction of liquids through porous media. *Physics* 1: 318–33.
Ritsema, D. J., Steenhuis, T. S., Parlange, J.-Y., and Dekker, L. W. 1996. Predicted and observed finger diameters in field soils. *Geoderma* 70:185–96.
Ross, P. J., Haverkamp, R, and Parlange, J.-Y. 1996. Calculating parameters of infiltration equations from soil hydraulic functions. *Trans. Porous Media* 24:315–39.
Ross, P. J., and Parlange, J.-Y. 1994. Comparing exact and numerical solutions of Richards equation for one-dimensional infiltration and drainage. *Soil Sci.* 157:341–4.
Russo, D., and Bresler, E. 1980. Scaling soil hydraulic properties of a heterogeneous field. *Soil Sci. Soc. Am. J.* 44:681–4.
Selker, J., Parlange, J.-Y., and Steenhuis, T. S. 1992a. Fingered flow in two dimensions: 2. Predicting finger moisture profile. *Water Resour. Res.* 28:2523–8.
Selker, J., Steenhuis, T. S., and Parlange, J.-Y. 1992b. Wetting front instability in homogeneous sandy soils under continuous infiltration. *Soil Sci. Soc. Am. J.* 56:1346–50.
Sharma, M. L., Gander, G. A., and Hunt, C. G. 1980. Spatial variability of infiltration in a watershed. *J. Hydrol.* 45:101–22.
Simmons, C. R., Nielsen, D. R., and Biggar, J. W. 1979a. Scaling of field-measured soil-water properties: I. Methodology. *Hilgardia* 47:77–102.
Simmons, C. R., Nielsen, D. R. and Biggar, J. W. 1979b. Scaling of field-measured soil-water properties: II. Hydraulic conductivity and flux. *Hilgardia* 47:103–73.
Smith, R. E., and Parlange, J.-Y. 1977. Optimal prediction of ponding. *Trans. ASAE* 20:493–6.
Sposito. G. 1990. Lie invariance of the Richards equation. In: *Dynamics of Fluids in Hierarchical Porous Media*, ed. J. Cushman, pp. 327–47. New York: Academic Press.
Sposito, G. 1995. Recent advances associated with soil water in the unsaturated zone. *Rev. Geophys.* (*Suppl.*) July:1059–65.
Sposito, G., and Jury, W. A. 1985. Inspectional analysis in the theory of water flow through unsaturated soil. *Soil Sci. Soc. Am. J.* 49:791–8.
Sposito, G., and Jury, W. A. 1990. Miller similitude and generalized scaling analysis. *Scaling in Soil Physics*, ed. D. Hillel and D. E. Elrick, pp. 13–22. Madison, WI: Soil Science Society of America.
Starr, J. L., de Roo, H. C., Frink, C. R., and Parlange, J.-Y. 1978. Leaching characteristics of a layered field soil. *Soil Sci. Soc. Am. J.* 42:386–91.
Talsma, T., and Parlange, J.-Y. 1972. One-dimensional vertical infiltration. *Austr. J. Soil Res.* 10:143–50.

Tillotson, P. M., and Nielsen, D. R. 1984. Scale factors in soil science. *Soil Sci. Soc. Am. J.* 48:953–9.

van Genuchten, M. T. 1980. A closed form equation for predicting the hydraulic conductivity of unsaturated soils. *Soil Sci. Soc. Am. J.* 44:892–8.

Warrick, A. W., Mullen, G. J., and Nielsen, D. R. 1977. Scaling field-measured soil hydraulic properties using a similar media concept. *Water Resour. Res.* 13:355–62.

Warrick, A. W., and Nielsen, D. R. 1980. Spatial variability of soil physical properties in the field. In: *Applications of Soil Physics*, ed. D. Hillel, pp. 319–44. New York: Academic Press.

Youngs, E. G. 1968. An estimation of sorptivity for infiltration studies from moisture moment considerations. *Soil Sci.* 106:157–63.

# 8

# Scale Issues of Heterogeneity in Vadose-Zone Hydrology

T.-C. J. YEH

## 8.1 Introduction

The hydrologic properties of the vadose zone often exhibit high degrees of spatial variability over a range of scales because of the heterogeneous nature of geologic formations. For laboratory-scale problems (i.e., small cores, soil columns, and sandboxes), variations in pore size, pore geometry, and tortuosity of pore channels are the major sources of heterogeneity. They are called laboratory-scale heterogeneities. Microstratification, foliation, cracks, and roots are also some possible heterogeneities at this scale. As our problem scale increases to that of a field, stratification or layering in a geologic formation becomes the dominant heterogeneity, often classified as field-scale heterogeneity. At an even larger problem scale, regional-scale heterogeneity consists in variations in geologic formations or facies. Variations among sedimentary basins are then categorized as global-scale heterogeneities.

The fundamental theories for flow and solute transport through porous media have essentially been derived for laboratory-scale heterogeneities. When we attempt to apply these theories to the vadose zone, comprising heterogeneities on many different scales, we encounter a scale issue. That is, these theories, suitable for the laboratory-scale problem, may not be applicable to problems at other scales. To deal with this issue, two approaches have evolved: the systems approach and the physical approach. The systems approach treats the vadose zone as a low-pass filter, and its governing principle is determined by the relationship between its input and output histories (e.g., Jury, Sposito, and White, 1986). The physical approach relies on the upscaling of laboratory-scale theories to the vadose zone. Although the systems approach has been widely used by soil scientists, it is often criticized for its empiricism and lack of physical principles. Also, it is known to be limited to non-point-source problems or those related to the integrated behaviors of a system (e.g., the average concentration of nitrate in the return flow at an irrigation drain, or the breakthrough into the water table beneath an irrigated field). Because the systems approach requires knowledge of input and output histories and model calibrations, flow and tracer experiments must be carried out at a given site prior to making predictions. Further, a calibrated-systems model for the vadose zone at a given depth under given conditions often will be found unsuitable for other depths and different conditions (e.g., Butters, Jury, and Ernst, 1989; Butters and Jury, 1989; Rŏth et al.,1991).

Although such systems approaches are practical tools for predicting water flow and pollutant transport through thin vadose zones to the water table, or to irrigation drains from agricultural fields, their utility for general hydrogeologic problems is limited. Hydrogeologic problems involve vadose zones that are tens or hundreds of meters in thickness. The inputs to such vadose zones are small compared with the scales of such hydrogeologic settings. Yet groundwater hydrologists have to focus on the spatial and temporal evolution of flow and the spread of solutes over the vadose zone and regional aquifers (Stephens, 1996). For all of the aforementioned reasons, the following discussion will concentrate on the physical approach, which has been widely used by groundwater hydrologists. Moreover, this chapter will present only the author's point of view regarding the scale issue and the various approaches to the problems of heterogeneity in the vadose zone.

## 8.2 Flow and Transport at the Laboratory Scale

Before we examine the physical approach to vadose-zone hydrology, some fundamental concepts associated with the laboratory-scale problem must be revisited, for they have not been well explained in the past. Additionally, these concepts are essential to our discussion because of the parallelism involved in the upscaling concept for the vadose zone.

### *8.2.1 Control Volume, Darcian Continua, and Representative Elementary Volume*

To study water flow and solute transport in porous media at the laboratory scale, hydrologists have relied on a continuum concept. Flow in porous media takes place through a complex network of interconnected pores or openings. To describe such an intricate network in any exact mathematical manner is practically impossible. Thus hydrologists have had to abandon the basic equation governing fluid flow at the pore-scale level (the Navier-Stokes equation). As a result, hydrologists consider only the average flow behavior over a certain volume of a porous medium, which must be of greater volume than several pores. This volume over which the flow is averaged is then defined as a control volume (CV). Using this CV approach, we essentially bypass both the molecular level (at which one concentrates on what happens to each fluid molecule) and the pore level (at which one studies the flow patterns within a pore and between pores). Therefore our observations of flow in porous media move to the macroscopic level, at which only averages of phenomena over the control volume are considered. A property defined at a point in our mathematical models thus represents the average, varying smoothly in space, such that the differential calculus applies. The medium and flow are then considered as the Darcian continuum. This continuum concept is parallel to the continuum hypothesis in fluid mechanics and other branches of sciences.

If we apply the CV approach to a porous medium at the laboratory scale (e.g., a sandbox packed with uniform sand grains), we can classify the medium as heterogeneous or homogeneous, depending on the size of the CV. If the size of the CV is small, its properties are likely to vary spatially; the medium will be called heterogeneous, although the spatial variations of its properties will be smooth. If progressively larger CVs are used, we may find that the value of a property will tend toward a constant value everywhere in the medium (i.e., translational invariance). These volumes are then defined as representative elementary volumes (REVs), and the medium can be classified as a homogeneous medium. Specifically, if an REV can be defined, then a property assessed over that REV of the medium must represent that property over the entire medium. To do that, the size of the REV must be large enough to include all representative heterogeneities in the medium. This REV concept is analogous to the minimum sample volume used in statistical analysis, which must be sufficiently large such that the statistics for samples in the volume will represent a given population. In addition, the REV must be small enough that spatially continuous functions can be used to describe the properties of the medium. As an example, the size of an REV for a sandbox must be several times the average size of the sand grains, but smaller than the size of the sandbox itself.

The size of an REV is scale-dependent. That is, one can define an REV for a laboratory-scale problem, but that REV may not be applicable to field-scale problems. Furthermore, for some properties of a medium, the existence of an REV will be time-dependent. For medium properties such as porosity and bulk density that are not associated with flow, the existence of the REV is independent of time. On the other hand, the REV for properties related to flow processes (such as hydraulic conductivity) can be defined only when the REV requirement with respect to flow is met. That is, the flow process must take sufficient time to experience many heterogeneities in the medium. For example, the wetting front from a point source in a sandbox must evolve to a certain size before the REV concept can be applied. This time-dependent phenomenon is nonetheless obscured by the fact that in a homogeneous soil column, the size of the heterogeneity is on the order of grain diameters, and the time required for the flow to reach the REV is insignificant compared with the time scale of our experiments. Thus, the time-dependent phenomenon is often neglected.

### 8.2.2 Governing Equations for Flow in Variably Saturated Porous Media

Strictly speaking, water movement in variably saturated porous media is related to the movement of air. A rigorous analysis of water flow and solute transport in such media should consider the flows of both water and air. However, the movement of air can be ignored in many cases (especially those related to point-source problems) to simplify the analysis. Based on the continuum concept and the CV approach, the governing equation for water flow in variably saturated media at the laboratory scale can therefore be expressed as (Bear, 1972)

$$\nabla[K(\psi)\cdot\nabla(\psi+z)] = [C(\psi)+\epsilon S_s]\frac{\partial\psi}{\partial t} \tag{8.1}$$

where $z$ corresponds to the vertical direction, and $K(\psi)$ is the hydraulic conductivity, which depends on the soil-water pressure head $\psi$. If the porous medium is fully saturated, the pressure head is positive, and it is negative when the medium is unsaturated. The moisture-capacity term $C(\psi)$ represents the change in moisture content per unit change in negative pressure when the geologic medium is partially saturated. It corresponds to the slope of the water-release curve or the moisture–pressure-head relationship $\theta(\psi)$ of a given soil at given values for the soil-water pressure head. When a soil is fully saturated, the change in water storage associated with a given change in pressure is denoted by the specific-storage term $S_s$, which is related to the compressibility of the porous medium and the water. In equation (8.1), $\epsilon$ is a saturation index for the porous medium, for the sake of convenience in mathematics. It is zero when the medium is unsaturated, and unity if the medium is fully saturated. This governing flow equation is commonly called the Richards equation if $S_s$ is omitted.

The Richards equation is based on the continuum assumption. It ignores discontinuous water films in individual pores and assumes that they are continuous, on the average, over the entire medium. If a point in the medium is said to be unsaturated, that means that the pores inside the CV at that point are unsaturated, on the average; some pores may be fully saturated, whereas others may be unsaturated. Similarly, hydraulic conductivity, soil-water pressure head, moisture capacity, and water content defined at a "point" in equation (8.1) also represent averages over a CV, encompassing many pores in the medium.

The hydraulic properties $K(\psi)$ and $C(\psi)$ for an unsaturated medium in equation (8.1) vary with the degree of saturation or moisture content and pressure head. Mathematical formulas (e.g., Brooks and Corey, 1966) are often employed to describe their dependence on soil-water pressure and moisture content, although a tabulation of relationships between the properties and pressure is sometimes used also. One popular formula is the exponential model (Gardner, 1958)

$$K(\psi) = K_s \exp(\beta\psi) \tag{8.2}$$

$$\theta(\psi) = (\theta_s - \theta_r)\exp(\beta\psi) + \theta_r$$

where $K_s$ is the hydraulic conductivity at saturation, $\beta$ is the pore-size-distribution parameter (representing the rate of reduction in conductivity as the soil desaturates), $\theta_s$ is the moisture content at saturation, and $\theta_r$ is the residual moisture content. In spite of its popularity, this model fits the observed $K(\psi)$ and $\theta(\psi)$ data over only a limited range of pressure-head values. Other widely used models for $K(\psi)$ and $\theta(\psi)$ are those by Mualem (1976) and van Genuchten (1980):

$$K(\psi) = K_s \frac{\{1 - (\alpha\psi)^{n-1}[1 + (\alpha\psi)^n]^{-m}\}^2}{[1 + (\alpha\psi)^n]^{m/2}} \tag{8.3}$$
$$\theta(\psi) = (\theta_s - \theta_r)[1 + (\alpha\psi)^n]^{-m} + \theta_r$$

in which $\alpha$, $n$, and $m$ are soil parameters, with $m = 1 - 1/n$. These models are valid over ranges of pressure values broader than that for the exponential model (van Genuchten and Nielsen, 1985). Use of these mathematical models allows us to categorize porous media on the basis of parameters such as $\alpha$, $\beta$, $n$, $\theta_s$, $\theta_r$, and $K_s$. For example, coarse-textured soils have large values for $\alpha$, $n$, $\beta$, and $K_s$, and fine-textured soils have small values (e.g., Stephens, Lambert, and Watson, 1987). The values for these parameters are not necessarily unique for a given geologic medium, because of hysteretic behavior in the $K(\psi)$ and $\theta(\psi)$ relationships. They vary according to the wetting and drying histories of the medium. Measurement of a moisture-release curve is generally easier than measurement of unsaturated hydraulic conductivity. As a result, the $\alpha$, $\beta$, and $n$ values derived from the moisture-release curve are often assumed to be the same as those for unsaturated hydraulic conductivity, although they may in fact be different (Yeh and Harvey, 1990). Furthermore, equations (8.2) and (8.3) are frequently used to extrapolate from measured conductivity values to provide estimates for very dry conditions in which direct measurement of hydraulic conductivity is beyond our ability (Khaleel, Relyea, and Conca, 1995). Notice that the values for the parameters $\alpha$, $\beta$, $n$, $\theta_s$, $\theta_r$, and $K_s$ are constant in space if the medium is homogeneous, and vary otherwise.

### *8.2.3 Dispersion Concept in Saturated Porous Media*

The flow behavior described by the Richards equation represents an average over a CV. Variations in flow behavior caused by heterogeneities at scales smaller than the CV are omitted. Generally, neglecting the effects of small-scale variations does not affect the estimate of the quantity of water flow in a laboratory sand column. However, these small-scale variations can have a profound impact on solute transport in the column. They represent the effects of slow-flow and fast-flow channels, where solutes are likely to travel, causing the spread of solutes. Therefore, the concept of hydrodynamic dispersion is used to include their effects in fully saturated porous media. That is, concentration fluxes from the fast- and slow-flow channels are embraced in a transport equation as a dispersive flux. Fick's law is then employed to relate this flux to the concentration gradient $\nabla C$ and dispersion coefficient $D$. The solute-transport equation thus includes a convective term, $q \cdot \nabla C$, and a dispersive term, $\nabla(D \cdot \nabla C)$. Whereas the former term represents the mass flux resulting from the average flow velocity $q$, the latter depicts that from variations in velocity about the average.

This dispersion concept is similar to that of the molecular diffusion in chemistry, and hydrodynamic dispersion in surface-water hydrology. However, chemical diffusion is caused by the random motions of molecules, and hydrodynamic dispersion in rivers results from the velocity variations caused by channel roughness and the shear effect (Fischer et al., 1979). On the other hand, dispersion in porous media is attributable to the velocity variations in porous media caused by laboratory-scale heterogeneities. In spite of these differences, both diffusion and dispersion concepts are subject to the same REV requirement for flow. That is, molecules of a tracer must collide with a sufficient number of other molecules so that an REV for their random motions can be defined and thus Fick's law for molecular diffusion will apply. Similarly, a tracer plume in a sandbox must travel a large distance to encounter a sufficient number of heterogeneities so that an REV for the random displacements of tracer particles can be established and Fick's law for dispersion will apply. The distance required for definition of such an REV in a sandbox is several times the diameter of the sand grains. Because that distance is much smaller than the dimensions of the experimental setup, the time required to establish an REV is short and often is ignored.

One should note that whereas the REV requirement for dispersion ensures the existence of a constant dispersion coefficient under uniform flow conditions, the REV requirement for flow ensures constant hydraulic conductivity over the medium. The concept of the two different REVs is important: The former REV allows us to adopt Fick's law, and the latter permits the use of the convective term $q \cdot \nabla C$ with a constant $q$ value over the entire medium.

In addition, many laboratory experiments have shown that the dispersion coefficient embedded in the aforementioned dispersion concept varies linearly with the average water velocity in fully saturated porous media. The constant of proportionality is defined as dispersivity, which can be related to the characteristic length scale of heterogeneities (the average size of the significant heterogeneity within the REV). For example, dispersivity can be related to the average grain diameter in a uniformly packed soil column through a power law: $\alpha = cd^n$, where $\alpha$ is the dispersivity, $c$ is a constant, $d$ is the mean grain diameter, and $n$ is a constant that is approximately equal to unity for most porous media (Bear, 1972). Dispersivity is generally considered a physical property of porous media.

Note that the dispersion concept is designed to capture the overall effect of heterogeneities overlooked by the average flow, but it is a continuum approach. In other words, the dispersion concept does not inform us about the locations of fast- and slow-flow channels in a sand column, but reflects their overall effect on the concentration averaged over an REV.

### 8.2.4 Governing Equations for Solute Transport in Variably Saturated Media

The classic convection–dispersion concept and equation for saturated flow have been adopted to account for mixing during solute transport in variably saturated porous media (Bear, 1972). It is expressed as

$$\nabla(D_{ij} \cdot \nabla C) - q \cdot \nabla C = \theta \frac{\partial C}{\partial t} \tag{8.4}$$

where $C$ is the concentration of the solute, $D_{ij}$ is the dispersion-coefficient tensor, and $q_i$ are the specific discharge components. The dispersion coefficient is generally defined as

$$D_{ij} = (\alpha_L - \alpha_T)\frac{q_i q_j}{q} + \alpha_T q \delta_{ij} + D_m \tag{8.5}$$

in which $\alpha_L$ and $\alpha_T$ are the longitudinal and transverse dispersivities, respectively, $q = (q_i \cdot q_i)^{1/2}$ and is the magnitude of specific discharge, $\delta_{ij}$ is the Kronecker delta ($\delta_{ij} = 1$ if $i = j$, and $\delta_{ij} = 0$ otherwise), and $D_m$ is molecular diffusion, which is generally small and can be omitted. This linear relationship between dispersivity and specific discharge is an extension of the relationship for solute transport in saturated porous media. Very few physical experiments (Wilson and Gelhar, 1974) or theoretical investigations have attempted to verify relationship (8.5) for unsaturated porous media, because of difficulties in design and implementation of the experiment.

Despite many simplifying assumptions, the modeling of flow and solute transport based on equations (8.1)–(8.4) has been satisfactory for many laboratory soil-column and sandbox experiments using packed uniform sand or glass beads (e.g., Nielsen and Biggar, 1961; Krupp and Elrick, 1968; Vauclin, Khanji, and Vachaud, 1979; Bond, 1986). These equations have been regarded as embodying the fundamental principle for predicting water flow and solute transport in the vadose zone. Thus, mathematical predictions for water flow and solute transport in a laboratory experiment require solution of equation (8.1), with specified initial and boundary conditions, to obtain the distributions for soil-water pressure head and moisture content. The specific discharge is then determined from Darcy's law, along with the distribution of soil-water pressure head. With the specific-discharge information, equation (8.4) is solved for spatial-concentration distributions at different times.

### 8.2.5 Scale of Measurement

The pressure head in equation (8.1) and the concentration in equation (8.4) represent the average pressure head and concentration of a chemical species over the REV, which are not necessarily the same as those measured in a single pore. To be consistent with equations (8.1) and (8.4), our measurement scale for head and concentration must be at least as large as the size of the REV. Generally, the difference between the pressure head within a pore and the average pressure head is small in fully saturated porous media, and thus this requirement is not critical. On the other hand, the difference can be significant for unsaturated media, depending on how close they are to saturation. Nevertheless, the ceramic cup of a tensiometer usually is long enough to contact many pores and thus register the average pressure. For tracer experiments in soil columns, this requirement is also satisfied: First, the size of the tracer source and the column length and diameter in the experiment often are much larger than the dimension of the heterogeneity (i.e., variations in pore size and geometry). Consequently, after the tracer is displaced over a distance encompassing several pores, its behavior is well described by equation (8.4), reflecting the fact that the REV for dispersion has been reached. More importantly, concentration breakthroughs are always collected at the end of the column, representing averages over the cross-sectional area, which is equal to or greater than the REV. Thus, our observations are consistent with the REV assumption behind equation (8.4), and success in predicting tracer movements in laboratory-scale problems using equation (8.4) has been widely reported. However, exceptions do occur, such as extremely long tailing in the breakthrough, not described by equation (8.4). They may reflect the fact that an REV for the tracer has not yet been reached, because of nonuniform packing. Such nonuniform packing introduces significant heterogeneities (Herr, Shäfer, and Spitz,1989) whose effects are too strong to be averaged out over the length of the column. Similarly, limited flow paths, due to partial saturation of the soil column, may increase the time to reach the REV. Thus a dead-end pore model (Coats and Smith, 1964) has been proposed to reproduce the tailing. Another possible explanation is the inconsistency between the measurement scale and the REV. For instance, a flush-mounted lysimeter in a soil column may not sample a volume that is equivalent to the REV embedded in equation (8.4).

## 8.3 Flow and Solute Transport in the Vadose Zone

After recognizing the continuum assumption behind the principles of flow and solute transport in porous media for laboratory-scale problems, we proceed to discussion of flow and solute transport in field-scale problems (i.e., the vadose zone). For analyzing flow and solute transport in the vadose zone, two methodologies, based on the theories for laboratory-scale problems, have been employed, namely, classic analysis and stochastic analysis.

### *8.3.1 Classic Analysis*

The classic ("deterministic") analysis, commonly used by practitioners in hydrology, assumes that equations (8.1) and (8.4), developed for laboratory-scale problems, are valid for the vadose zone. On the basis of that assumption, equivalent homogeneous and heterogeneous approaches have emerged. The first approach assumes that a heterogeneous vadose zone can be treated as an equivalent homogeneous zone with fictitious hydraulic properties that are constant in space. This approach is analogous to the REV approach for a laboratory-scale problem. Here, the REV is scaled up to include heterogeneities at many different scales in the vadose zone. It can be equal to or much larger than the REV defined for a laboratory-scale problem, depending on the scale of heterogeneity in the vadose zone. We shall refer to it as the field-scale REV (FSREV). Traditionally, the fictitious hydraulic properties are obtained by conducting large-scale hydraulic tests and then by applying inverse approaches to identify their values (such as the instantaneous-profile method)(Watson, 1966). Alternatively, values from many small-scale tests can be averaged to represent the fictitious properties (i.e., the arithmetic, geometric, or harmonic mean of conductivity values obtained from core samples). These fictitious properties are then used as input parameters for equations (8.1) and (8.4) to predict water flow or transport of solutes in the vadose zone. They are usually referred to as effective hydraulic properties.

The heterogeneous approach, on the other hand, visualizes the vadose zone as a collection of many elements with different hydraulic properties. The hydraulic properties of each element are derived from the available hydrogeologic information. For instance, groundwater hydrologists may use well logs and geologic information to delineate large-scale geologic features such as layering, stratifications, formations, and fault zones. Hydraulic properties, either measured at some locations or reported in the literature for similar geologic materials, are then assigned to each element by interpolation or extrapolation. This approach aims at describing the behavior of water flow or transport of contaminants at higher resolutions than those of the homogeneous approach. It employs the CV concept instead of the FSREV assumption for the entire vadose zone.

Whether treating the vadose zone as a homogeneous or heterogeneous medium, the classic analysis faces many difficulties. First, the validity of equations (8.1) and (8.4) for the equivalent homogeneous vadose zone has not yet been demonstrated. Because input sources are small compared with the size of the vadose zone, it can be argued, on the basis of our discussion of the laboratory-scale problem, that water and contaminant plumes must travel a great distance to encounter all heterogeneities at different scales before an FSREV can be defined. This distance may be greater than the thickness of the vadose zone. As a result, an FSREV may never be reached. Second, even if a large FSREV for a given vadose zone can be defined, no certain means is available to determine the effective properties of the equivalent homogeneous vadose zone using data from large-scale hydraulic tests. Relying on small-scale hydraulic tests, one would obtain different values for the hydraulic parameters at various parts of the vadose zone. Then, a theoretically rigorous means to average these different values would be needed. Most important of all, often we do not know what we are predicting using the effective property. Are the predicted behaviors unbiased estimates of the processes in the vadose zone, or estimates that do not bear any statistical significance?

Additionally, the sample sizes encompassed by our monitoring devices [e.g., sample volumes for a tensiometer, lysimeter, time-domain reflectrometer probe (TDR), or neutron probe] generally are much smaller than the FSREV. Predictions based upon an assumption of homogeneity can be expected to deviate from our point observations unless our observations have focused on integrated or averaged behaviors. Then, a measure of the difference between a prediction and our point observation is imperative for making decisions (e.g., selecting the location for a pump-and-treat well system, or determining the extent of contaminations).

The aforementioned issues related to the FSREV are critical, and the classic homogeneous approach has no means to address those issues. Using only a limited amount of data and a general geohydrologic description of the vadose zone, the heterogeneous approach confronts the same issues, though it does not invoke the FSREV assumption for the entire vadose zone.

### *8.3.2 Stochastic Analysis*

In attempts to resolve the issues faced by the classic analysis, numerous stochastic theories have been developed, relying on stochastic representations of the heterogeneities in the vadose zone. Generally, characterizing the spatial distributions of hydraulic properties at the laboratory scale in a vadose zone requires numerous measurements, takes considerable time, and involves great expense. Such an intensive sampling effort is practically impossible. A statistical description (mean, standard deviation, and probability distribution) of the variabilities in the hydraulic properties at a given site becomes the alternative (e.g., Russo and Bouton, 1992). Because the hydraulic properties at every location in the vadose zone will vary, they are best treated as stochastic processes in space or spatial random functions (e.g., Yeh, 1992).

To illustrate stochastic representation of the spatial variabilities of hydrologic parameters, we shall use the example of saturated hydraulic-conductivity data along a vertical borehole. The value for the hydraulic conductivity at a point $x_0$, along the borehole can be envisioned as the value for one of the many possible geologic materials that may have been deposited at that point. Thus, the hydraulic conductivity at that point is a random variable, $K(x_0, \omega)$. The symbol $\omega$ indicates that there can be many possible values for $K$ at $x_0$. Similarly, the hydraulic-conductivity values at other locations along the borehole are random variables. As a result, the hydraulic-conductivity values for the entire depth of the borehole can be considered as a collection of many random variables in space. Thus, if the conductivity is measured at locations $x_1, x_2, x_3, \ldots, x_n$, then $K(x_1, \omega)$ is a random variable, as is $K(x_2, \omega)$ and so on out to $K(x_n, \omega)$. Each has a probability distribution, and furthermore the probability distributions may be interrelated. The chance of finding a particular sequence of hydraulic-conductivity values along the borehole, $K(x, \omega_1)$, depends not only on the probability distribution of the hydraulic conductivity at one location but also on those at other locations. This implies that actual hydraulic-conductivity values along the borehole are one possible sequence of $K(x, \omega_1)$ out of all the possible sequences $K(x, \omega)$. In the vocabulary of stochastic processes, the probability of finding the sequence is defined as the joint probability distribution. These possible sequences are called an ensemble, and the realization refers to one possible sequence.

Determining the probability of occurrence of a particular sequence of random variables requires knowledge of the joint probability distribution (JPD) for these random variables. This JPD is completely defined only if the probabilities for all possible sequences of $K(x, \omega)$ values along a borehole are known. Obviously, the JPD is not available in real-life situations, because hydraulic-conductivity values sampled along a borehole represent only one realization out of the ensemble. Therefore, one must resort to simplifying assumptions, namely, stationarity and ergodicity.

Stationarity (or strict stationarity) implies that the JPD for a stochastic process is translationally invariant. The ergodicity assumption means that through observation of the spatial variation of a single realization of a stochastic process, the JPD for the process can be determined. To apply the stochastic representation to a field, the assumption of ergodicity must be adopted as a working hypothesis.

A JPD is characterized by its statistical moments. Stationarity thus implies that all the moments are constant in space, and it is a very stringent assumption. Because most stochastic analyses are limited to the first and second moments (mean and covariance function, respectively) of a stochastic process, an assumption of weak or second-order stationarity is often invoked. Second-order stationarity implies that the mean is a constant and that the covariance function depends only on the separation distance, which is the distance that separates any two samples in the calculation of the covariance function. This assumption allows us to characterize a stochastic process by using only its mean and covariance function.

The covariance function of a stochastic process is the product of its variance and its autocorrelation function. Whereas the variance characterizes the variability of the spatial stochastic process around its mean value, the autocorrelation function describes the persistence of the process in space, which can be further defined by the correlation scale. Intuitively, the correlation scale of a hydraulic property can be interpreted as the average dimension of the heterogeneity (e.g., dimension of the layering, stratification, inclusions) in a geologic medium. This can be attributed to the fact that the values for the hydraulic property measured at intervals less than the dimension of the heterogeneity will be similar, and the correlation between these values will be high. On the other hand, the correlation will be low if measurements were taken at intervals greater than this dimension, since values of these measurements very likely represent the values of the property of different geologic materials. Thus, the covariance function is a statistical measure of the degree of heterogeneity and the spatial structure of hydraulic properties.

The means and covariance functions for hydrogeologic properties in the vadose zone consequently become the tools for characterizing spatial variability. One must, however, recognize that this representation does not specify the values for the properties at any particular locations in the vadose zone, but provides a way to quantify their spatial variability. That is, we know only where the mean value of the property lies, how widely the property values spread around the mean value, and how these values are correlated in space statistically.

Once we accept the concept of stochastic representation of hydraulic properties, we can discuss stochastic analysis of flow and solute transport in the vadose zone. In the following sections, stochastic theories of flow and solute transport in the vadose zone will be categorized as unconditional and conditional effective-parameter theories.

*Unconditional Effective-Parameter Theory* The unconditional effective-parameter theory can be viewed as equivalent to the homogeneous approach of the classic analysis set in a stochastic context. It attempts to address issues raised in the foregoing discussion about the equivalent homogeneous approach: (1) the governing equations for flow and solute transport for the equivalent homogeneous vadose zone, (2) the relationship between the hydraulic properties at the laboratory scale and the effective hydraulic properties for the equivalent homogeneous medium, (3) uncertainties associated with the prediction based upon the effective parameter.

Consider a steady, vertical infiltration through the vadose zone that consists of an infinite number of soil elements. Each element has an isotropic hydraulic-conductivity function $K(\psi)$, and the steady flow is described by Darcy's law: $\mathbf{q} = -K(\psi)\nabla(\psi + z)$. Suppose the hydraulic properties $K(\psi)$ can be considered to constitute a second-order stationary stochastic process. Then, according to the stochastic concept, there will be an infinite number of realizations of such hydraulic-conductivity fields and corresponding steady-pressure-head fields. As a result, Darcy's law becomes a stochastic partial differential equation. To take an average over the ensemble, the variables in Darcy's equation are decomposed into means and perturbations: $K(\psi) = \langle K(\psi)\rangle + k(\psi)$, and $\psi = \langle\psi\rangle + h$, where $\langle\,\rangle$ denotes the expected value. Substituting these into the equation and taking the expected value of the equation, the mean for Darcy's law over the ensemble can be expressed as

$$\langle\mathbf{q}\rangle = -\langle K(\psi)\rangle \cdot \nabla(\langle\psi\rangle + z) - \langle k(\psi) \cdot \nabla h\rangle \tag{8.6}$$

Using a first-order analysis, Yeh, Gelhar, and Gutjahr (1985a,b) showed that the second term in equation (8.6), $\langle k(\psi) \cdot \nabla h\rangle$, is proportional to the mean gradient vector, $\nabla\langle J\rangle = \nabla(\langle\psi\rangle + z)$. Therefore, equation (8.6) can be rewritten in the following form:

$$\langle\mathbf{q}\rangle = -K^e(\langle\psi\rangle) \cdot \nabla\langle J\rangle \tag{8.7}$$

This equation is similar to Darcy's law for a laboratory-scale problem, but $\langle \mathbf{q} \rangle$, $\langle \psi \rangle$, and $\nabla \langle J \rangle$ in equation (8.7) represent quantities averaged over the ensemble. The constant of proportionality, $K^e$, between $\langle \mathbf{q} \rangle$ and $\nabla \langle J \rangle$ is referred to as the unconditional effective unsaturated hydraulic conductivity (UEUHC). It is the sum of the ensemble mean hydraulic conductivity $\langle K(\psi) \rangle$ and a product term, $\langle \nabla [k(\psi) \cdot \nabla h] \rangle \nabla \langle J \rangle^{-1}$. Thus, the UEUHC differs from the mean hydraulic conductivity: It depends on flow and can create conductivity anisotropy. On the basis of this definition, the UEUHC is viewed as the ability of a fictitious medium to transmit an amount of water $\langle \mathbf{q} \rangle$ under the ensemble mean hydraulic gradient. If these ensemble averages $\langle \mathbf{q} \rangle$ and $\nabla \langle J \rangle$ are equal to the averages over an FSREV (spatial average), then the UEUHC is equivalent to the effective hydraulic conductivity defined in the FSREV for an equivalent homogeneous medium. That is, ergodicity in terms of flow is valid for the vadose zone as a whole. The governing steady-flow equation for the equivalent homogeneous vadose zone then takes the following form:

$$\nabla [K^e(\langle \psi \rangle) \cdot \nabla (\langle \psi \rangle + z)] = 0 \tag{8.8}$$

with boundary conditions specified in terms of mean pressure heads or fluxes. Again, equation (8.8) has a form identical with that of the flow equation (8.1) for a laboratory-scale problem, implying equation (8.1) can be scaled up to field-scale problems. Nonetheless, a theoretically consistent approach must be employed in determining the effective conductivity for a field site (one realization of a stationary process).

One way to derive a theoretically consistent UEUHC is to use an inverse approach, as demonstrated by Yeh (1989), or the unit-mean-gradient method employed by Yeh and Harvey (1990). This approach determines the parameter value such that it will minimize the mean-square error between the pressure head at our point observation and the average head simulated for a given specific discharge. As a result, the fictitious conductivity value satisfies equation (8.8) for a given specific discharge and produces an unbiased mean-head distribution.

Another approach is to find a theoretical relationship between the conductivity at the laboratory scale and the UEUHC. Yeh et al. (1985a,b) analyzed steady-state vertical infiltration in unbounded random porous media under unit-mean-gradient conditions. They assumed that the unsaturated hydraulic conductivity could be depicted by the Gardner model, equation (8.2), at the laboratory scale, with the parameters $\ln K_s$ (natural logarithm of $K_s$) and $\beta$ in equation (8.2) as three-dimensional stochastic processes. These stochastic processes can be expressed as means and perturbations: $\ln K_s = F + f$, and $\beta = A + a$. The correlation scales for the covariance functions of the perturbations in the horizontal directions are assumed to be equal ($\lambda_x = \lambda_y$), and the vertical scale ($\lambda_z$) is much smaller than the horizontal scales. Based on these assumptions, an approximate expression for the anisotropy ratio for the UEUHC can be written as

$$\frac{K^{eh}(\langle\psi\rangle)}{K^{ev}(\langle\psi\rangle)} = \exp\left[\sigma^2_{\ln K}\right] = \exp\left[\frac{\sigma_f^2}{1+A\lambda_{fz}} + \frac{\sigma_a^2\langle\psi\rangle^2}{1+A\lambda_{az}}\right] \tag{8.9}$$

where $\sigma^2_{\ln K}$ is the variance of the log unsaturated hydraulic conductivity (which depends on the mean pressure head $\langle\psi\rangle$), $a$ and $f$ are assumed independent of each other, and $\lambda_{fz}$ and $\lambda_{az}$ denote the correlation scales for $f$ and $a$ in the vertical direction, respectively. Notice that the mean gradient is assumed unity (i.e., $J_z = 1$, and $J_x = J_y = 0$) in the analysis. The assumption of unbounded porous media implicitly assumes that the mean flux is known. Similar expressions have been derived by Green and Freyberg (1995).

As indicated by equation (8.9), the hydraulic anisotropy depends not only on the assumption of porous media but also on the degree of mean saturation (or $\langle\psi\rangle$). Such an anisotropy implies that the horizontal effective conductivity becomes greater than the vertical when the mean soil-water pressure head $\langle\psi\rangle$ becomes more negative (i.e., soil becomes drier). Thus, significant lateral migration of water and contaminants can occur in the vadose zone (Yeh et al., 1985c; McCord, Stephens, and Wilson, 1991; Stephens, 1996).

Following the described procedure, the ensemble mean-flow equation under transient and unsaturated conditions can be written as

$$\begin{aligned}\nabla[\langle K(\psi)\rangle \cdot \nabla(\langle\psi\rangle + z)] + \langle\nabla[k(\psi)\,\cdot\,\nabla h]\rangle \\ = \langle C(\psi)\rangle\frac{\partial\langle\psi\rangle}{\partial t} + \left\langle c(\psi)\frac{\partial h}{\partial t}\right\rangle\end{aligned} \tag{8.10}$$

Again, it is assumed that the product term can be combined with the ensemble mean conductivity to form an unconditional effective conductivity as in equation (8.8). If the second term on the right-hand side of equation (8.10) is proportional to $\partial\langle\psi\rangle/\partial t$, it can then be lumped with the term for mean moisture capacity to form an effective moisture capacity. The governing equation for flow in an equivalent homogeneous medium is then expressed as

$$\nabla[K^e(\langle\psi\rangle) \cdot \nabla(\langle\psi\rangle + z)] = \langle C^e(\psi)\rangle\frac{\partial\langle\psi\rangle}{\partial t} \tag{8.11}$$

The form of this equation is again identical with that of equation (8.1) if the compressibility of water and porous media is omitted. But the equation uses unconditional effective hydraulic conductivity and moisture capacity and predicts the mean pressure head and mean flux. Mantoglou and Gelhar (1987b) employed this approach to examine the effective conductivity for transient unsaturated flow. Desbarats (1995) derived an expression for the pressure–saturation relationship for equivalent homogeneous media.

Because the unconditional effective hydraulic property predicts the mean- pressure-head distribution $\langle\psi\rangle$, the deviation of the true head from the mean can be quantified by the head variance. A closed form for head variance for steady infiltration has been derived by Yeh et al. (1985a,b), and it can be generalized as

$$\sigma_h^2(\langle\psi\rangle) = J_z^2 \sigma_f^2 \lambda_{fz}^2 \rho_f^2 G(\rho_f, A\lambda_{fz}, J_z) + J_z^2 \sigma_a^2 \lambda_{az}^2 \langle\psi\rangle^2 \rho_a^2 G(\rho_a, A\lambda_{az}, J_z) \tag{8.12}$$

where $\rho_f$ and $\rho_a$ are the aspect ratios (the ratio of the horizonal to the vertical correlation scale) for $\ln K_s$ and $\beta$, respectively. The function $G$ is the integral described by Yeh et al. [1985b, eg. (5)]. Green and Freyberg (1995) derived a slightly different form for the head variance.

Equation (8.12) predicts that the head variance will change with the mean head value $\langle\psi\rangle$. The variability of the soil-water pressure head in the field increases as the soil becomes dry, and decreases as the soil approaches saturation. Recent numerical experiments and laboratory and field observations (e.g., Yeh, Gelhar, and Wierenga, 1986; Hopmans, Schukking, and Torfs, 1988; Greenholtz et al., 1988; Yeh, 1989; Ünlü, Nielsen, and Biggar, 1990; Yeh and Harvey, 1990; Herkelrath, Hamburg, and Murphy, 1991; Stephens, 1996) qualitatively support this finding. A detailed physical explanation for the mean-dependent head variance has been given by Yeh (1989). Equation (8.12) also shows that as the soil becomes drier (large negative values for $\langle\psi\rangle$), the influence of the variability of $\beta$ increases. In turn, the correlation structure for the soil-water pressure varies with $\langle\psi\rangle$. At small values of $\langle\psi\rangle$ (wet conditions) the correlation scale for the pressure is dominated by the correlation scale of $\ln K_s$, and at large values of $\langle\psi\rangle$ (dry conditions) the correlation scale is controlled by that of $\beta$. In addition, the head variation predicted by a three-dimensional model is much smaller than that predicted by a one-dimensional model. This implies that a three-dimensional model is more appropriate for analyzing flow and solute transport in the field than is a one- or two-dimensional model.

Head variance also increases with the aspect ratio (Figure 8.1). For flow through the vadose zone toward aquifers, the aspect ratio is likely to be much greater than unity; the variance in soil-water pressure head during downward percolation is likely to be large. Such high variability in soil-water pressure can induce great lateral and preferential movements of water or contaminants, resulting in complex multidimensional flow phenomena in the vadose zone (Harter and Yeh, 1996a,b; Kung, 1993).

The temporal evolution of head variance was investigated by Mantoglou and Gelhar (1987a), and Ünlü et al. (1990) used a Monte Carlo simulation technique to study the head variance during infiltration. Using a numerical first-order-approximation

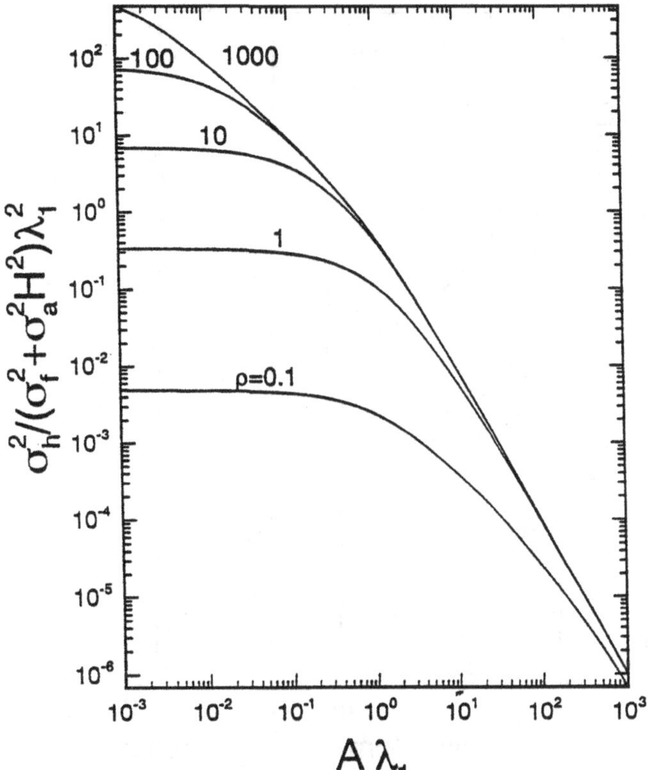


Figure 8.1. Normalized head variance for three-dimensional flow as a function of $A\lambda_z$ with different aspect ratios, assuming that ln $K_s$ and $\beta$ have the same correlation structure.

method, Ferrante and Yeh (1995) investigated the propagation of head and flux variances for a moisture pulse in stratified, one-dimensional, random porous media. They found that during the propagation of the moisture pulse, large head and flux variances were always associated with the mean pressure gradients at the mean wetting and drying fronts (Figure 8.2). In addition, the head variances were greater than that of the initial steady-head distribution, and flux variance was amplified by the soil heterogeneity as the pulse moved toward the water table. Although the head and flux variances in three-dimensional random porous media are expected to be smaller than those in one-dimensional media, the general behavior will remain the same.

In terms of practicality, those studies provided a framework for upscaling laboratory-scale measurements of unsaturated hydraulic properties to the effective properties for the large-scale vadose zone. For instance, at a given field site where few core samples are available, one can use those samples to determine variances and means for ln $K_S$ and $\beta$ and their correlation scales. Then the effective unsaturated hydraulic conductivity for the field site can be estimated, and thus the mean head can be determined. Once the mean head is determined, the head variance can be evaluated. This head variance is a measure of the discrepancy between the true head field and

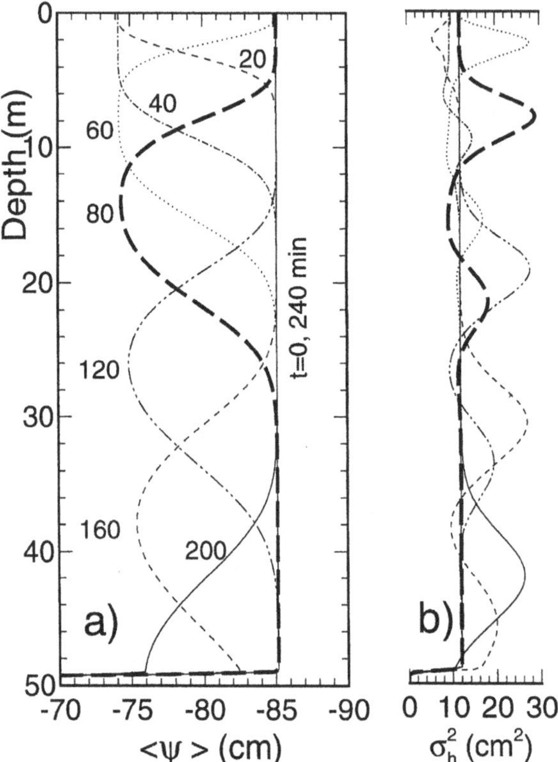


Figure 8.2. Mean pressure-head profiles at 0, 20, 40, 60, 80, 120, 160, and 240 minutes after infiltration of a pulse of water toward the water table (left column). Head variances associated with the mean profiles are shown in the right column.

the predicted mean, representing the effect of unmodeled vadose-zone heterogeneity. Polmann et al. (1991) demonstrated this application in predicting wetting-front movements at a field site. The estimated effective hydraulic conductivity can also be used to predict mean migration paths for contaminant plumes resulting from the average flow. Such an approach was used by McCord et al. (1991) to simulate field tracer experiments.

The analyses that have been described have been valuable for many practical problems, but one must recognize the limitations of such methods. First, they assumed that the source of flow was much larger than the scale of the heterogeneity and that the flow was either steady or quasi-steady, uniform, downward flow. Based on our discussion of the REV concept for laboratory-scale heterogeneity, it would seem likely that the unconditional effective hydraulic property would equal the effective hydraulic property of an equivalent homogeneous medium under these conditions. That is, the ergodicity assumption for flow is valid. For general problems that involve transient flow from a small source (much smaller than the horizontal scale of heterogeneity), one can speculate that this assumption may not be valid until the flow encounters a sufficient number of significant heterogeneities. Therefore, effective properties will be time-dependent until an FSREV for the flow has been reached. The time required to attain an FSREV may be very long. Dagan (1982) reached a similar conclusion on the basis of a theoretical analysis of saturated flow through random aquifers. Under field conditions, boundary conditions for the vadose zone can vary significantly, involving upward and downward movements, lateral redistribution of water, and hysteresis effects. Because a rigorous mathematical analysis of field conditions is intractable, we can only hypothesize that the ergodicity condition for flow likely will not be met. The formulas for the unconditional effective properties nevertheless serve as our best approximation.

One must also recognize that even if the ergodicity assumption can be met, the discrepancy between our observations and the predicted mean head can be large. Because of the large head variance, practitioners should be aware of the fact that a few point observations of the head in a field may not give sufficient information about the general movement of water in the vadose zone. This point is crucial, because recharge and hydraulic-property estimates often are derived from a limited number of field observations. These estimates are consequently inconsistent with the theory and thus biased.

In addition to the effective parameters for the mean flow, the stochastic analysis derives effective parameters for solute transport. In similarity to the laboratory-scale problem, a hydrodynamic-dispersion concept is adopted to include the effect of velocity variations around the mean. Consider solute movements in a heterogeneous vadose zone, being visualized as a collection of many porous blocks with different hydraulic properties. Suppose that these hydraulic properties can be treated as stochastic processes, and movement of the solute in each block can be described by the convective flow alone (omitting dispersion within each block). Then $q$, $C$, and $\theta$ in equation (8.4) are spatial stochastic processes. The ensemble mean transport equation becomes

$$\langle q \rangle \cdot \nabla \langle C \rangle + \langle q' \cdot \nabla C' \rangle = -\langle \theta \rangle \frac{\partial \langle C \rangle}{\partial t} - \left\langle \theta' \frac{\partial C'}{\partial t} \right\rangle \tag{8.13}$$

where a prime indicates the perturbation. If we can express $\langle q' \cdot \nabla C' \rangle$ as $\nabla(D^e \cdot \nabla \langle C \rangle)$, as in G. I. Talyor's shear-flow analysis (Fischer et al., 1979), and assume that the last term in equation (8.13) is proportional to $\partial \langle C \rangle / \partial t$, then we have an ensemble mean convection–dispersion equation for the vadose zone:

$$\nabla \left( D_{ij}^e \cdot \nabla \langle C \rangle \right) - \langle q \rangle \cdot \nabla \langle C \rangle = \theta^e \frac{\partial \langle C \rangle}{\partial t} \tag{8.14}$$

Again, this ensemble mean equation has the form of equation (8.4) for the laboratory-scale problem. The effective-dispersion-coefficient tensor is denoted by $D_{ij}^e$, and $\theta^e$

is the effective moisture content. Because the FSREV involved in equation (8.14) is greater than that in the laboratory-scale problem, the dispersion caused by variations in hydraulic conductivity at scales smaller than this FSREV is then called macrodispersion. The effective-dispersion tensor is called the macrodispersion tensor, and dispersivity the macrodispersivity. Based on G. I. Taylor's analysis of shear-flow dispersion, it is easy to see that equation (8.14) is valid for describing the distribution of a tracer plume only if the plume has been displaced for a large distance and has experienced enough velocity variation. With this restriction in mind, application of equation (8.14) to the vadose zone requires knowledge of $\langle q \rangle$ and macrodispersivities. The mean discharge can be derived from equation (8.7), but a relationship between macrodispersivities and the spatial variability of hydraulic conductivity at the laboratory scale is needed. This relationship for transport in saturated aquifers was derived by Gelhar and Axness (1983) and Dagan (1984). Mantoglou and Gelhar (1985) derived approximate expressions for macrodispersivities in unsaturated porous media. They showed that the longitudinal macrodispersivity, under unit-mean-gradient conditions, can be written as

$$A_{11}(\langle \psi \rangle) = \sigma^2_{\ln K} \lambda / \gamma^2 \tag{8.15}$$

where $A_{11}$ is the longitudinal macrodispersivity at large time, $\lambda$ is the correlation scale, and $\gamma$ is a flow correlation factor that depends on the direction of the mean flow and the orientation of the heterogeneity. Detailed discussions of the components of the macrodispersivity tensor have been given by Mantoglou and Gelhar (1985) and Russo (1993). The macrodispersivity, equation (8.15), represents an effective parameter of the FSREV for a plume that is much larger than the size of the heterogeneity under steady-state uniform flow. Again, it theoretically exists only when the tracer plume has been displaced for a large distance in geologic formations. Before establishment of the macrodispersion regime, use of equation (8.14) for the mean concentration is generally unwarranted.

Another effective approach for predicting solute transport in porous media examines the evolution of the second spatial moment of a contaminant plume. Extending Taylor's theorem of diffusion (Taylor, 1922), Dagan (1987) used first-order perturbation analysis of the stochastic equation for steady-state flow and derived the groundwater-velocity covariance function for spatially correlated random hydraulic-conductivity fields: Assuming that tracer particles moved along with the velocity fields, mathematical expressions for the second spatial moment of the tracer particles at any given time were derived. The second moment represents the spatial-displacement variance of particle position around the mean and, in turn, the "size" of a tracer plume. Note that the second moment of a plume does not predict the shape of the concentration plume and does not explicitly assume the validity of Fick's law. That is, this approach eliminates the FSREV requirement for dispersion. For cases where the initial size of the plume is much smaller than the scale of the heterogeneity, the location of the centroid of the plume will not be predicted correctly by the mean velocity. This is attributed to the fact that the FSREV scale for movement of the centroid of the plume has not been reached. To apply the moment approach to this type of problem, the second moment must be corrected by subtracting variation in the centroid position of the plume. This concept has been explained in detail by Csanady (1973) and Fischer et al. (1979). Dagan (1990) and Rajaram and Gelhar (1993a,b) developed formulas for this purpose for uniform, steady-state, saturated flow.

The moment analysis for saturated aquifers can be adopted for unsaturated porous media if flow is steady and is under gravity drainage (unit-mean-gradient) conditions. In these cases, the variance of unsaturated hydraulic conductivity will be the controlling factor for the macrodispersivity. Therefore, formulas for solute transport in the unsaturated zone will be similar to those for saturated flow. The variance of $\ln K_s$ is replaced with the variance of unsaturated conductivity $\sigma^2_{\ln K}$, and the mean flux with the effective unsaturated hydraulic conductivity in the mean flow direction ($J_z = 1$). Detailed derivations of the formulas have been given by Russo (1995).

The concentration variance is a measure of the deviation of the real concentration distribution $C(x, t)$ from the mean concentration $\langle C(x, t)\rangle$. For solute transport in aquifers, expressions for the concentration variance have been developed by Vomvoris and Gelhar (1990), Kapoor and Gelhar (1994a,b), and others. The concentration variance is found to be directly proportional to the mean concentration gradient and variance and the correlation scales of log-hydraulic conductivity, and inversely proportional to local dispersivity values. The concentration variance can be large, depending on the magnitude of the mean concentration gradient. After a plume has been displaced for a large distance, and when the mean concentration gradient is small, the variance will be small, and thus the macrodispersion approach will produce satisfactory results. Kapoor and Gelhar (1994a,b) and Kapoor and Kitanidis (1996) have also shown that the coefficient of variation for the concentration increases at early times and decreases over a characteristic variance residence time, because of local dispersion. This local dispersion represents the effects of velocity variations at scales smaller than that of the REV defined for a laboratory-scale problem. Although no similar formula has been developed for solute transport in the vadose zone, the variance of concentration in the vadose zone is expected to behave in a manner similar to that in aquifers, if we consider a gravity-drainage scenario. The major difference will lie in the fact that the variance will depend on the mean soil-water pressure. Such a result implies that for dry soils, the mean concentration distribution may be significantly different from that observed in a field site.

Equation (8.15) allows practitioners to estimate macrodispersivity from knowledge of the variation of unsaturated hydraulic-conductivity values at the laboratory scale. For this reason, the macrodispersivity approach is a practical tool for predicting the migration of pollutants in the vadose zone without extensive site characterization. Whereas the macrodispersivity approach may be restricted to thick vadose zones because of the FSREV assumption, the moment approach predicts the spatial moment of the mean concentration distribution without invoking the FSREV assumption for dispersion. It thus is more appealing than the macrodispersivity approach, because the thickness of the vadose zone usually is on the order of a few meters – too thin to allow the development of the FSREV for macrodispersivities.

In spite of their practicality, both the macrodispersivity and moment approaches have been criticized for their first-order approximation (Cushman, 1983; Loaiciga and Mariño, 1990; Harter, 1994). For cases where a large variance in unsaturated hydraulic conductivity is expected, the accuracy of the first-order approximation is questionable. Although this is a legitimate concern, these approaches have proved useful for improving our understanding of the general behavior of solute movements in the vadose zone. From a practitioner's point of view, the critical issue is that the ensemble average concentration distribution or moments may deviate significantly from those of a real plume, even if the approach is mathematically flawless.

The assumption of stationarity is another common criticism of the effective-parameter approach. The hydraulic-conductivity field in the vadose zone likely is nonstationary because of the presence of heterogeneities of different scales. This criticism again is valid from a theoretician's point of view, but it is not critical in terms of the applicability of the approach. For example, the REV approach for the laboratory scale has already included heterogeneities at many different scales (e.g., variations in molecules, surface roughness of solids, and pore geometry), and it has been successfully applied to numerous laboratory experiments. Similarly, the stochastic analysis is also suitable for investigating the effects of multiscale heterogeneities. Here the correlation scale represents the size of the most significant heterogeneity, and the dispersion represents the effects of heterogeneities at scales smaller than the correlation scale. Nevertheless, the key question is whether or not the vadose zone of our interest is large enough to include a sufficient number of the most significant heterogeneities so that an FSREV can be defined. Even if the vadose zone is thick enough, the scale of our interest (objective) or the scale of our measurement often prevents us from using the effective-parameter approach for the entire vadose zone. Specifically, our interests often are limited to scales much smaller than the scale of the most significant heterogeneity. As in the laboratory-scale problem, the assumption of homogeneity will produce large uncertainties (especially in a vadose zone with a wide range of scales of heterogeneity). This uncertainty can be reduced if and only if the integrated behavior is our primary interest or the travel time of a plume is much longer than the characteristic variance residence time (Kapoor and Kitanidis, 1996). For a vadose zone with multiscale heterogeneity, the residence time may be too long at our scale of interest. A similar conclusion was reached by Rajaram and Gelhar (1995) for solute transport in aquifers with multiscale heterogeneity.

Very few well-controlled, large-scale, field tracer experiments have been conducted in the vadose zone. To illustrate the effect of multiscale heterogeneity on solute transport at a scale smaller than the FSREV, we shall examine the findings from a tracer experiment in a coastal sandy aquifer in Georgetown, South Carolina (Mas-Pla, 1993; Yeh et al., 1995). That experiment monitored the evolution of a chloride-tracer plume in an aquifer under a steady-flow field, created by an injection and a pumping well. Snapshots of a chloride plume (at 10, 30, and 50 hours after the injection) along a vertical section between the two wells are shown in Figure 8.3. The tracer was injected uniformly through the injection well, but the plume split into two parts as time progressed. A small portion of the plume moved slowly at the upper portion of the aquifer, and the major portion moved rapidly along the bottom of the aquifer. The splitting of the plume was caused mainly by the stratification and low-permeability inclusions in the middle part of the aquifer. The variability within each layer only slightly affected the overall behavior of the plume. Without including these large-scale features, the effective-property approach cannot mimic this behavior unless a strong density effect is considered, which is not the case in real aquifers. Similarly, the moment analysis cannot depict the splitting of the plume. The second moment of a plume with a single concentration peak is similar to that of a plume with multiple peaks. In spite of the complex flow regime, Mas-Pla et al. (1992) found that the integrated behavior, such as the concentration breakthrough measured at the withdrawal well, could be easily reproduced using even an unrealistic one-dimensional model.

On the basis of the Georgetown findings for saturated aquifers, it can be seen that when the macrodispersion concept is applied to the vadose zone, it suffers from the same difficulties or even more severe difficulties than those encountered with aquifers. First, when a soil becomes dry, it can become highly heterogeneous. Subsequently, many preferential channels can develop, and a contaminant plume in the vadose zone can split into many smaller plumes (Harter and Yeh, 1996b). Thus the predictions for the moment analysis will no longer be meaningful. Secondarily, the thickness of the vadose zone is generally much smaller than the distance required for the development of macrodispersivity. The mean plume based on the macrodispersivity concept is likely very different from the real one. Temporal variations in boundary conditions, as well as water-table position, redistribution, and hysteresis effects, may ease the problem. For instance, the redistribution process may smooth out the variabilities in pressure head and concentration. A decrease in the mean pressure near the water table can cause reductions in their variabilities. The scale of our interest, however, remains the factor that ultimately limits the use of the effective-parameter concept. In other words, the unconditional effective-parameter approach does not provide predictions at resolutions of our interest. Nevertheless, these theoretical developments over the past decade have significantly advanced our understanding of the effects of heterogeneities on flow and solute transport in the vadose zone.

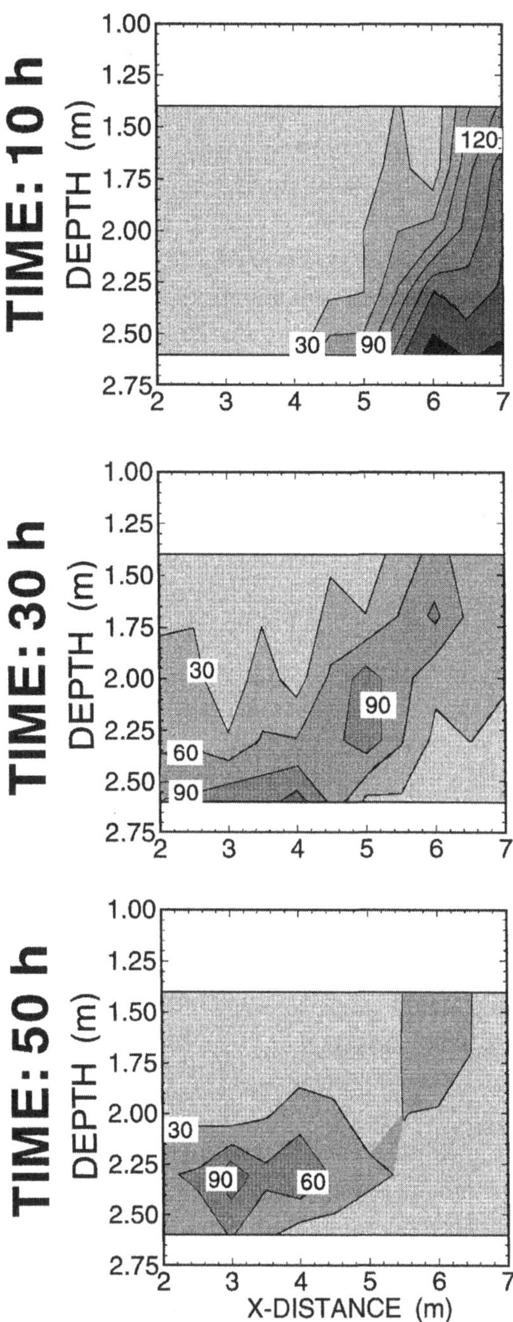


Figure 8.3. Observed chloride-plume distributions along a vertical cross section, cutting through the injection and the pumping well.

*Conditional Effective-Parameter Theory* Because the unconditional effective-parameter theory does not seem to achieve the resolution demanded by our interest, the conditional theory becomes the alternative. Whereas the goal of this theory parallels that of the heterogeneous approach in the classic analysis, it aims at predicting conditional means and variances of flow and transport processes. Unlike the unconditional theory, it avoids the FSREV assumption but regards flow and porous media as continua. "Conditioning" means that our estimations of parameters and predictions of processes incorporate our knowledge of the properties of porous media or the behaviors of flow and transport processes at sample locations.

Consider the conditional ensemble mean equation for transient flow:

$$\begin{aligned}&\nabla[\langle K_c(\psi)\rangle \cdot \nabla(\langle \psi_c\rangle + z)] + \langle \nabla[k_c(\psi)\cdot \nabla h_c]\rangle \\ &\quad = \langle C_c(\psi)\rangle \frac{\partial \langle \psi_c\rangle}{\partial t} + \left\langle c_c(\psi)\frac{\partial h_c}{\partial t}\right\rangle\end{aligned} \tag{8.16}$$

in which $\langle K_c(\psi)\rangle$ and $\langle \psi_c\rangle$ are the conditional means for unsaturated hydraulic conductivity and pressure head, respectively, with $k_c(\psi)$ and $h_c$ corresponding to their conditional perturbations. Again, the mean equation contains products of perturbations, $\langle \nabla[k_c(\psi)\cdot \nabla h_c]\rangle$ and $\langle c_c(\psi)(\partial h_c/\partial t)\rangle$. These two terms will be zero if $K(\psi)$ and $C(\psi)$ or $\psi$ are known everywhere (i.e., their perturbations are zero). Under these conditions, the conditional mean equation is equivalent to the Richards equation for heterogeneous media in the classic analysis. For other situations, their contributions can be grouped to form the conditional effective hydraulic conductivity and moisture capacity if the perturbation terms are proportional to $\langle \psi_c\rangle$. The conditional mean flow equation can thus be written as

$$\nabla\left[K_c^e(\langle \psi_c\rangle)\cdot \nabla(\langle \psi_c\rangle + z)\right] = C_c^e(\langle \psi_c\rangle)\frac{\partial \langle \psi_c\rangle}{\partial t} \tag{8.17}$$

in which $K_c^e$ and $C_c^e$ denote the conditional effective conductivity and capacity fields. That is, these fields, in conjunction with $\langle \psi_c\rangle$, satisfy the continuity equation (8.17) and honor our measurements at the sample location. In following discussions, a conditioning mean approach that uses primary information [e.g., $K(\psi), \theta(\psi)$, and $C(\psi)$] measured at sample locations will be discussed first. Methods for conditioning using both primary and secondary information (e.g., $\psi$ and $\theta$) will be examined afterward.

Suppose that one attempts to investigate water and solute movements in the vadose zone using a two-dimensional, vertical-profile, finite-element, flow-and-solute-transport model. For accuracy or other reasons, a 2,000-node finite-element mesh is generated for the entire profile. But only 50 measured $K(\psi)$ and $\theta(\psi)$ values scattered around the entire profile are available. To extrapolate the measurements to the remaining elements, many mathematical tools can be used. Because exact determination of the hydraulic properties for the remaining elements is impossible, it is plausible to select a tool that can provide the conditional mean as our best estimate. Such a tool will ensure that our estimate will be unbiased and will have the minimum variance. It also will guarantee that measurement values at sample locations are honored and that the underlying general structure of the property field is preserved. Kriging in geostatistics is one mathematical tool to accomplish this goal.

The theory of kriging has been extensively documented in many textbooks (e.g., Journel and Huijbregts, 1978; de Marsily, 1986). In principle, kriging is a best linear unbiased estimate (BLUE) of a stochastic process. It uses a linear combination of the products of weighting factors and measurements of the process at sample locations to estimate the process at the other location. The weights are derived by minimizing the mean square error of the estimate and can be determined from knowledge of only the autocovariance of the stochastic process. If the stochastic process is Gaussian, then the kriging estimate is equivalent to the conditional expectation of the process. If it is not, we can only say that it is the best linear approximation of the conditional expectation (Priestley, 1981). Besides, kriging can derive the conditional variance, a measure of the deviation of our mean estimate from the true value.

Applications of the kriging technique to vadose-zone hydrology problems have been demonstrated by Rockhold, Rossi, and Hills (1996). Generally, estimates for the conditional mean and variance of a hydraulic property are determined first by kriging. Numerical models for flow and solute transport are then used with the conditional mean property to derive the corresponding hydraulic head, velocity, and concentration distributions.

Another approach to derive conditional mean fields is the conditional Monte Carlo simulation. It differs from kriging in the sense that it generates a large number of conditional realizations of hydraulic-property fields, instead of the conditional mean field. Each realization of the property field honors the measurement value at the sample location and preserves the spatial statistics of the vadose zone. The average of many conditional realizations at a given point $x$ is equal to the kriging estimate, and the variance is the kriging variance. The complete theory of a conditional simulation procedure based on kriging and a superposition technique can be found in the work of Matheron (1973) and Journel and Huijbregts (1978).

Applications of this simulation technique to vadose-zone problems would involve routing conditioned hydraulic-property fields through numerical flow and transport models to obtain corresponding realizations of soil-water pressure, velocity, and concentration fields. By analysis of the statistics of these fields, ensemble means and variances for soil-water pressure, velocity, and concentration thus can be derived. Compared with the unconditional Monte Carlo simulation, the conditional simulation incorporates measurements at sample locations. Consequently, reduction in the variance is anticipated if many measurements are used. This approach is generally thought to be more realistic than the unconditional Monte Carlo simulation. Although it still requires a substantial computational resource, the number of realizations needed to stabilize the ensemble mean and variance is generally less than that for the unconditional Monte Carlo simulation, because of the effect of conditioning (Harter and Yeh, 1996b).

Because of difficulties in solving the Richards equation, applications of the simulation to vadose-zone problems are limited. A hybrid analytical and numerical method recently developed by Harter and Yeh (1993) alleviates some difficulties for steady-infiltration problems. With this method, they applied this simulation technique to investigate the effects of conductivity measurements on the predictions of flow and solute transport in the unsaturated zone. They found significant reductions in the uncertainty only if many measurements of unsaturated hydraulic properties were included.

It is a well-known fact that the more primary information used, the more accurate the conditional simulation. However, measurements of hydraulic properties for unsaturated porous media are time-consuming and costly. Detailed characterization of the vadose zone by direct measurements of hydraulic conductivities is a formidable task, but information about the soil-water pressure head and water content can be collected with relative ease in most shallow and unconsolidated vadose zones with inexpensive tools (e.g., tensiometers, neutron probes, TDRs, and electric-resistivity tomography). The poorly sorted alluvial deposits, conglomerates, and solid rock masses composing the vadose zone in the western regions of United States often prohibit the use of pressure-measurement devices. In that case, water content may be the only information that can be collected in large quantities. For these reasons, it becomes necessary to take advantage of the abundance of information on soil-water pressures and degrees of saturation to improve our estimates for unsaturated hydraulic properties in the field. This parameter-estimation task is the so-called stochastic inverse problem (Yeh et al., 1996).

Inverse problems have been a major focus of groundwater hydrology during the past decade. Numerous mathematical models have been developed to estimate transmissivity, $T$, for aquifers with given scattered hydraulic head, $\phi$, and transmissivity measurements (Yeh, 1986). One popular method is the minimum-output-error-based (MOEB) approach. Application of this approach to variably saturated flow is limited because of the complex nonlinear nature of the Richards equation. Kool and Parker (1988), Mishra and Parker (1989), and Russo et al. (1991) applied it to problems of one-dimensional unsaturated flow and solute transport, with the goal of estimating parameter values for unsaturated porous media in laboratory columns. Because the solution of the inverse problem is inherently nonunique and the approach is a regression, the identity of the estimated parameter values thus may be undefined, and uncertainty in the estimate cannot be addressed. To resolve these problems, a coconditional mean estimate is most appropriate (i.e., estimation of the mean value for a parameter that is conditioned on the observed primary and secondary information). A possible method to derive the coconditional mean is the geostatistical inverse approach (cokriging).

Cokriging has been applied to estimate water content in the vadose zone, using data sets on water content, soil-water pressure head, soil-surface temperature, and soil texture (e.g., Vauclin et al., 1983; Yates and Warrick, 1987; Mulla, 1988). Little attention has been directed toward application of this method to the inverse problem in the vadose zone (i.e., estimating unsaturated hydraulic-conductivity parameters using data sets on soil-water pressure head and water content). Harter and Yeh (1996b) applied the cokriging technique to investigate the effects of measurements of soil-water pressure head and unsaturated hydraulic conductivity on solute transport under unit-mean-gradient conditions. Yeh and Zhang (1996) developed a flexible cokriging technique to estimate parameters for unsaturated hydraulic conductivity. Their technique can be applied to nonstationary properties and flow fields, and it can utilize both soil-water pressure and water saturation as secondary information to improve the estimates for unsaturated hydraulic-conductivity parameters. A brief discussion of their method follows.

Suppose that the natural log of saturated hydraulic conductivity, $\ln K_s$, and the pore-size-distribution coefficient, $\ln \beta$, in equation (8.2) are stochastic processes with means $E[\ln K_s] = F(x)$ and $E[\ln \beta] = A(x)$ and perturbations $f(x)$ and $a(x)$, respectively. Similarly, the soil-water pressure $\psi$ and saturation $\Theta$ are also considered as stochastic processes that can be expressed as $\psi(x) = H(x) + h(x)$ and $\Theta(x) = S(x) + s(x)$, where $H(x) = E[\psi(x)]$ and $S(x) = E[\Theta(x)]$ are the means and $h(x)$ and $s(x)$ are the perturbations, respectively. Assume that we have the following: $n_f$ observed saturated hydraulic conductivities $f(x_i)$, where $i = 1, 2, \ldots, n_f$; $n_a$ observed pore-size-distribution coefficients $a(x_j)$, where $j = 1, 2, \ldots, n_a$; $n_h$ soil-water-pressure measurements $h(x_k)$, where $k = 1, 2, \ldots, n_h$; $n_s$ sampled saturations $s(x_l)$, where $l = 1, 2, \ldots, n_s$. We want to estimate $f$ and $a$ at locations where no samples are available. Our desired estimates are the means for $f$ and $a$ fields conditioned on the measurements of $f$ and $a$.

Assuming that $f$, $a$, $h$, and $s$ are jointly normal, the conditional mean estimates for $f(x)$ and $a(x)$ at locations $\mathbf{x}_0$ can be expressed by the linear combination of the weighted observed values of $f$, $a$, $h$, and $s$:

$$f_{\text{co}}(\mathbf{x}_0) = \sum_{i=1}^{n_f} P_{fi} f(\mathbf{x}_i) + \sum_{k=1}^{n_h} Q_{fk} h(\mathbf{x}_k) + \sum_{l=1}^{n_s} R_{fl}\, s(\mathbf{x}_l) \tag{8.18}$$

$$a_{\text{co}}(\mathbf{x}_0) = \sum_{j=1}^{n_a} P_{aj} a(\mathbf{x}_j) + \sum_{k=1}^{n_h} Q_{ak} h(\mathbf{x}_k) + \sum_{l=1}^{n_s} R_{al} s(\mathbf{x}_l)$$

Here, $f$ and $a$ are uncorrelated. In equation (8.18), $f_{\text{co}}$ and $a_{\text{co}}$ are the cokriged values for $f$ and $a$ at the unsampled location $\mathbf{x}_0$. $P_{fi}$, $Q_{fk}$, and $R_{fl}$ are cokriging weights for $f$ estimates with respect to the measurements of $f$, $h$, and $s$; $P_{aj}$, $Q_{ak}$, and $R_{al}$ are cokriging weights for $a$ estimates with respect to the samples of $a$, $h$, and $s$. The $i$, $j$, $k$, and $l$ are the indices for observed $f$, $a$, $h$, and $s$, respectively. These weights are selected so that the estimations expressed by equation (8.18) will have the minimal variances. Details have been given by Yeh and Zhang (1996).

Based on the above approach, Yeh and Zhang (1996) showed that unsaturated hydraulic parameters in a heterogeneous vadose zone could be reasonably identified if large amounts of information on the soil-water pressure and water saturation were used. The first column in Figure 8.4 shows the true $\ln K_s$ distribution in a hypothetical vadose zone with an area of 7 m $\times$ 7m ( 35 $\times$ 35 elements) and the values obtained by the geostatistical method using different approaches under the condition of nonuniform steady-state flow. The second column shows the corresponding $\ln \beta$ fields. Whereas Approach I used the $\ln K_s$ and $\ln \beta$ data sets (primary information) at sampling locations (represented by the rectangles) and kriging, Approach II used both primary information and pressure-head data at sampling locations (circles in the figure) to estimate the true fields. In Approach III, primary information and degree-of-saturation data at sample locations (squares in the figure) were used. The last row (Approach IV) represents the case where primary information, pressure head, and degree of saturation were used together in the geostatistical inverse model to estimate the fields. As illustrated in this figure, sufficient information about pressure head and degree of saturation (200 out of 1,225 elements) can improve the estimates for $\ln K_s$ and $\ln \beta$ fields significantly.

It was also found that improvements in estimates varied with the cross-correlation between primary and secondary information. For unsaturated conditions, the value of the cross-correlation between $f$ and $h$ at any given separation distance decreased as $\langle \psi \rangle$ became more negative, implying that the drier the soil, the lower the correlation between $f$ and $h$. Consequently, measurements of $h$ did not improve the estimate for $\ln K_s$ under dry conditions. On the contrary, the value of the cross-correlation between $a$ and $h$ increased as the soil became less saturated, and thus measurements of $h$ improved the estimate for $\ln \beta$. As $\langle \psi \rangle$ became less negative (approaching zero) or the soil neared saturation, the cross-correlation between $f$ and $h$ increased, while that between $a$ and $h$ dropped. Measurements of $h$ thus improved the estimate for $\ln K_s$, but not that for $\ln \beta$. The behavior of the cross-correlation between $f$ and $h$ at full saturation was the same as the result reported by Mizell, Gutjahr, and Gelhar (1982) for saturated flow. It is independent of the mean pressure head, and the cross-correlation between $a$ and $h$ becomes zero.

Although these findings are interesting, the geostatistical inverse approach has its theoretical limitations. In general, the relationships among $f$, $a$, $h$, and $s$ are nonlinear. As the soil becomes less saturated, the variance of the unsaturated hydraulic conductivity grows. The nonlinearity becomes stronger (Yeh et al., 1985a,b; Harter and Yeh, 1996a,b). Such a strong nonlinearity in the flow equation implies that in general, $h$ will not be normal, and $f$, $a$, and $h$ will not be jointly normal, even if

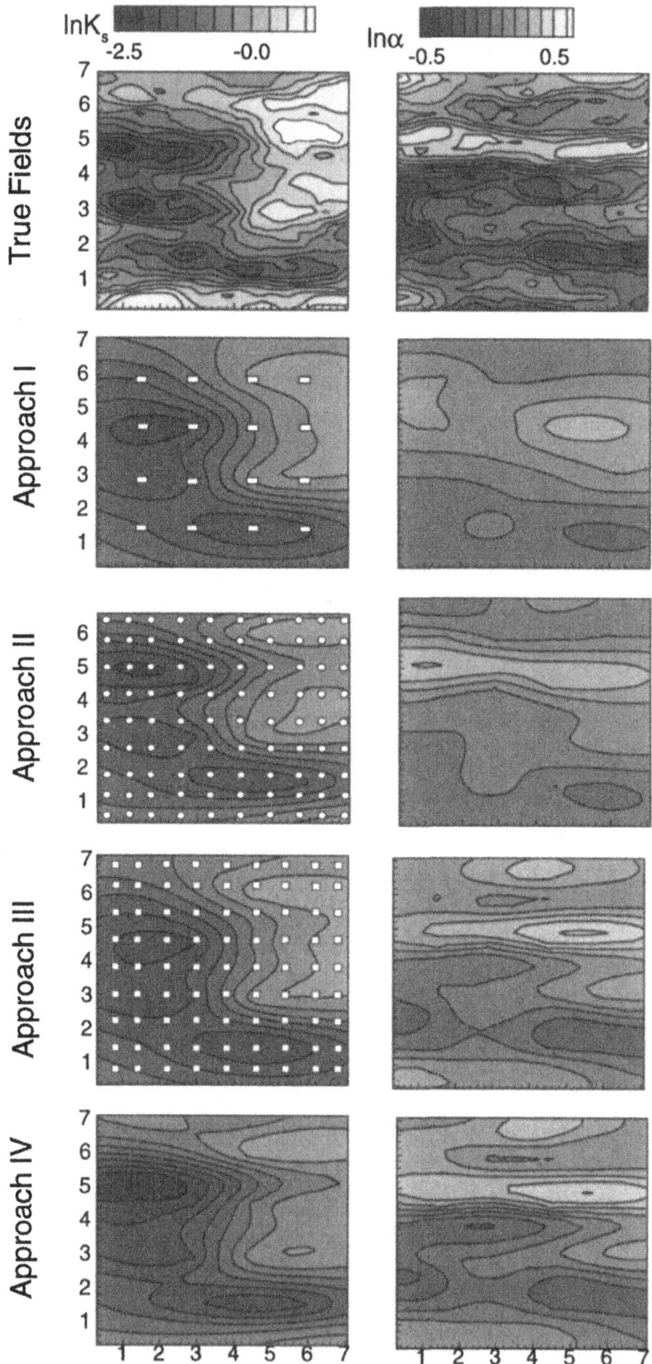


Figure 8.4. Comparisons of true ln $K_s$ and ln $\beta$ fields with those estimated by various approaches.

$f$ and $a$ are normal. In addition, the cross-covariance and covariance required in cokriging generally are derived from a first-order linearized version of the governing flow equation (8.1). The use of linear geostatistical inverse techniques consequently may not produce optimal results even if a large amount of secondary information is incorporated. This problem is bound to be exacerbated if the nonlinearity of the flow equation is strong, as with nonuniform flow, or if the variance of the log of the unsaturated hydraulic conductivity is large. Yeh, Jin, and Hanna (1996) have recently demonstrated this problem for saturated-flow problems.

To alleviate this problem, an iterative geostatistical approach based on successive linear estimators has been developed by Zhang and Yeh (1997), which can be expressed as

$$\begin{aligned} Y_c^{(r+1)}(\mathbf{x}_0) &= Y_c^{(r)}(\mathbf{x}_0) + \sum_{k=1}^{n_h} \lambda_k(\mathbf{x}_k)\left[\psi_k^*(\mathbf{x}_k) - \psi_k^{(r)}(\mathbf{x}_k)\right] \\ &\quad + \sum_{l=1}^{n_s} \mu_l(\mathbf{x}_l)\left[\Theta_l^*(\mathbf{x}_l) - \Theta_l^{(r)}(\mathbf{x}_l)\right] \\ Z_c^{(r+1)}(\mathbf{x}_0) &= Z_c^{(r)}(\mathbf{x}_0) + \sum_{k=1}^{n_h} \zeta_k(\mathbf{x}_k)\left[\psi_k^*(\mathbf{x}_k) - \psi_k^{(r)}(\mathbf{x}_k)\right] \\ &\quad + \sum_{l=1}^{n_s} \eta_l(\mathbf{x}_l)\left[\Theta_l^*(\mathbf{x}_l) - \Theta_l^{(r)}(\mathbf{x}_l)\right] \end{aligned} \tag{8.19}$$

where $r$ is the iteration index, and $Y_c^{(r)}$ and $Z_c^{(r)}$ are the estimates of the conditional means of $\ln K_s$ and $\ln \beta$ at iteration $r$, respectively. Measurements of pressure head and water saturation at the sample location $k$ are denoted by $\psi_k^*$ and $\Theta_k^*$, respectively. These successive linear estimators are unbiased. If $r = 0$, $Y_c^{(0)}$ and $Z_c^{(0)}$ correspond to the cokriged $\ln K_s$ and $\ln \beta$ fields, respectively. If $r > 0$, new estimates are obtained by adding the weighted sums of $[\psi^* - \psi^{(r)}]$ and $[\Theta^* - \Theta^{(r)}]$ at sample locations to the estimates at the previous iteration. $\psi^{(r)}$ and $\Theta^{(r)}$ represent the simulated soil-water pressure head and saturation values at the sample locations. They are derived from solving the Richards equation with given boundary conditions and $\ln K_s$ and $\ln \beta$ fields estimated from the previous iteration.

The coefficients, $\lambda_k$, $\mu_l$, $\zeta_k$, and $\eta_l$ are weights that vary with iterations. At each iteration, they are determined in a way similar to the cokriging technique that ensures minimal mean square error (MSE) of the estimates. Similarly, the MSE associated with $Z_c^{(r+1)}$ can be derived (Zhang and Yeh, 1997). Once the new cokriging coefficients $\lambda_k$, $\mu_l$, $\zeta_k$, and $\eta_l$ are evaluated, $Y_c^{(r+1)}$ and $Z_c^{(r+1)}$ can be calculated using equation (8.19).

As the $\ln K_s$ and $\ln \beta$ fields are improved at each iteration, the difference in $[\psi^* - \psi^{(r)}]$ and $[\Theta^* - \Theta^{(r)}]$ will become progressively smaller, and thus the values of $Y_c$ and

$Z_c$ will become stable. In addition, $\sigma_y^2$ and $\sigma_z^2$ (variances of the estimated $\ln K_s$ and $\ln \beta$ fields) will gradually approach constants. These constant values should be greater than the variances of the cokriging fields, but less than those of the true fields. To end the iterative process, the absolute values of the differences in $\sigma_y^2$ and $\sigma_z^2$ between two successive iterations are examined. If the differences are less than prescribed tolerances, the iteration stops. Otherwise, a new field for soil-water pressure head, $\psi$, is obtained by solving the Richards equation based on the newly obtained $Y_c$ and $Z_c$; a new saturation field $\Theta$ is determined based on the newly solved $\psi$ field and the newly obtained $Z_c$ field. Residual covariance and cross-covariance (Zhang and Yeh, 1997) are then evaluated, and in turn the new coefficients and new estimates. At the end of iteration, the residual covariance can be used to address the uncertainty associated with the estimates for $\ln K_s$, $\ln \beta$, head, and saturation fields.

The results of this iterative geostatistical inverse approach for two-dimensional problems appear promising. Comparisons of the known hypothetical $\ln K_s$ and $\ln \beta$ fields and those derived from the classic geostatistical approach and the iterative approach are illustrated in Figure 8.5. Based on this figure, the iterative approach evidently yields maps of $\ln K_s$ and $\ln \beta$ fields that are much closer to the known fields than the maps produced by the conventional approach. It clearly depicts the high- and low-permeability zones that can significantly affect predictions for transport of contaminants in the soil.

Studies by Yeh and Zhang (1996) and Zhang and Yeh (1997) have provided insights into the inverse problem. Whereas the correlation structure embedded in the geostatistical inverse model defines the generic heterogeneity pattern for a given field site, a few measurements of primary variables tailor this pattern to produce a site-specific heterogeneity map. This map is then improved by use of the available secondary information. The degree of improvement, however, will depend on the cross-correlation between primary information and secondary information and on the model describing their relationship. The iterative estimator circumvents problems of the linear predictor and thus reveals more detailed heterogeneities.

Moreover, the conditional stochastic approach attempts to derive the conditional mean parameter field, which theoretically is the unique solution to the inverse problem. Its result, however, is limited to approximate conditional fields if the medium is strongly heterogeneous, because an approximate conditional mean flow equation must be used. If the medium is mildly heterogeneous, the approach can yield the conditional mean field (Yeh et al., 1996; Zhang and Yeh, 1997). Similarly, it is conceivable that starting with a cokriged field as the initial guess, the MOEB inverse model may lead to a solution that is very close to the conditional mean field. Therefore, we are a step closer to resolving the nonuniqueness problem associated with classic inverse problems.

Whereas moisture-content and pressure measurements have been found to improve the estimate of unsaturated hydraulic conductivity, studies of inverse problems with aquifers (e.g., Harvey and Gorelick, 1995; Copty and Rubin, 1995) have shown that

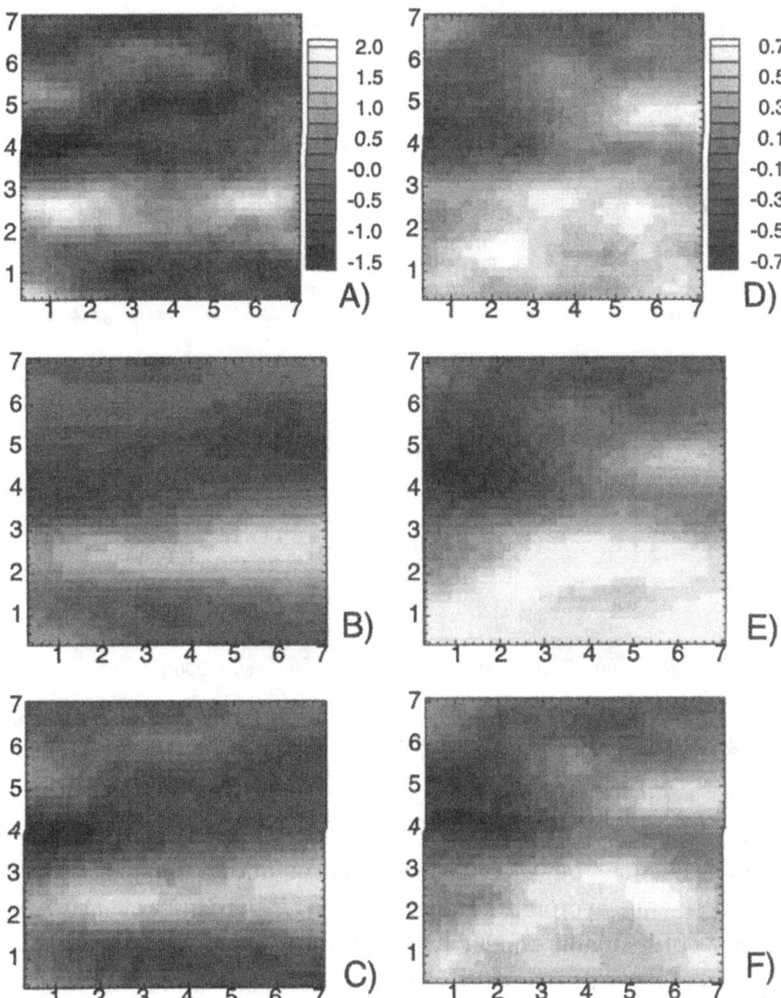


Figure 8.5. Comparisons of true ln $K_s$ field (A) with those estimated by classic cokriging (B) and by the iterative geostatistical approach (C), using the same amount of primary and secondary information. Similarly, the true ln $\beta$ field (D) and the estimates by classic cokriging (E) and by the iterative geostatistical approach (F).

the concentrations, travel times for solutes, and seismic data are useful information for hydrologic inversion. Thus, one can speculate that data on concentrations and travel times will also lead to improvements in the inverse modeling of the vadose zone.

Finally, the coconditional Monte Carlo simulation is another tool for coconditional analysis. It is similar to the conditional Monte Carlo simulation, except that both primary information and secondary information are used to generate coconditional realizations of the fields for unsaturated hydraulic properties. Details of the method have been given by Harter and Yeh (1996b). The concept of coconditional simulation is appealing, but its implementation is difficult because of the nonlinearity

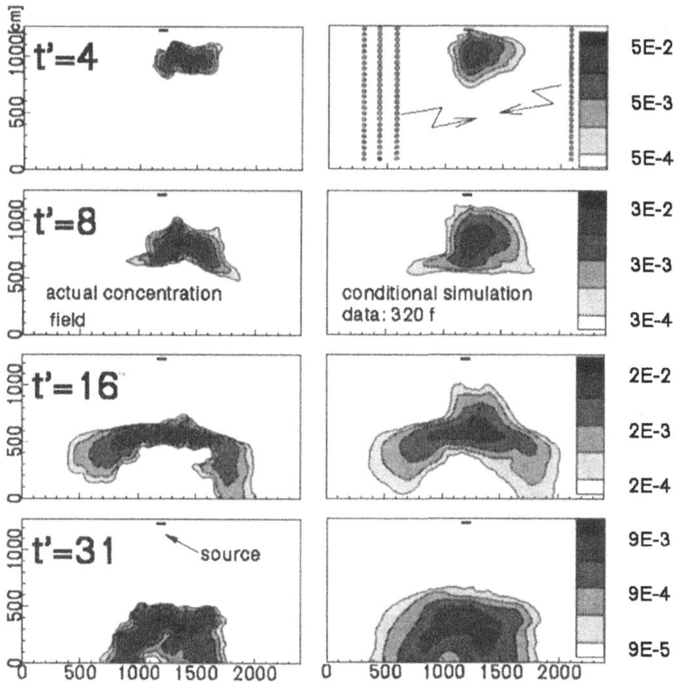


Figure 8.6. Illustrations of actual concentration fields (first column) and those estimated by conditional simulation using 320 $f$ measurements (second column) at dimensionless times (4, 8, 16, 31) after introduction of a slug of tracer in a heterogeneous vadose zone under steady-state flow under a unit-mean-gradient condition.

of the Richards equation. Few applications of coconditional simulation have been reported in vadose-zone hydrology. Harter and Yeh (1996b) investigated the effects of measurements of hydraulic conductivity and soil-water pressure heads on solute transport in the vadose zone. The first column in Figure 8.6 shows the evolution of a tracer plume at different dimensionless times ($t$ = 4, 8, 16, and 31) in a hypothetical vadose zone, with the variance of the unsaturated hydraulic conductivity equal to 3.2. The effect of 320 saturated conductivity measurements on the simulated mean plume using the conditional approach is illustrated in the second column in Figure 8.6. The effect of 320 head measurements is shown in the first column of Figure 8.7, demonstrating that head measurements can improve the prediction of the plume distribution, but the improvement is not as significant as that using conductivity measurements. Combined measurements of saturated conductivity and pressure head always provide the best prediction of the plume distribution, as illustrated in the second column of Figure 8.7. The effect of conditioning is minimal if the variance of the unsaturated hydraulic conductivity is small.

It should be pointed out that the foregoing findings are preliminary because of the assumptions of a unit mean gradient condition, an exponential function for unsaturated hydraulic conductivity, and steady-state flow. Nevertheless, the findings

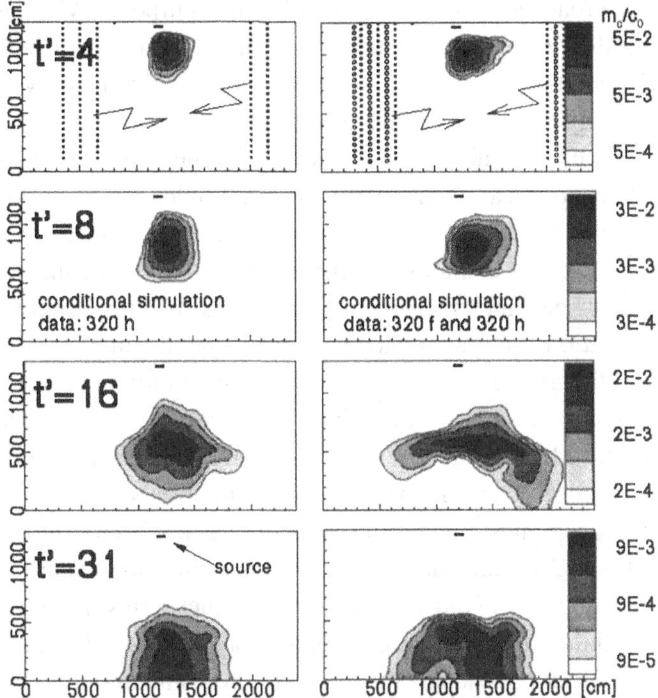


Figure 8.7. Illustrations of the concentration fields by conditional simulation using 320 $h$ measurements (first column) and those using 320 $f$ and $h$ measurements (second column) at dimensionless times (4, 8, 16, 31).

demonstrate the practical usefulness of locally measured data when incorporated into stochastic analysis as conditional information.

These aforementioned conditional approaches evidently are promising tools that can lead to predictions at high resolutions, but they are not free of theoretical difficulties. Suppose that, based on scatter measurements in a given vadose zone, exact conditional means for $K(\psi)$ and $C(\psi)$ can be derived from kriging; the boundary and initial mean conditions of the vadose zone are known exactly. With these conditional mean fields and boundary and initial conditions, equation (8.16) will not yield the conditional mean head, $\langle \psi_c \rangle$, unless the products of the conditional perturbation terms in equation (8.16) are zero or are known. If $K(\psi)$ and $C(\psi)$ in the entire vadose zone are known exactly, the product terms will be zero. Otherwise, these terms must be evaluated, but they are intractable at this moment. Therefore, the head field resulting from a numerical model using kriged $K(\psi)$ and $C(\psi)$ fields represents a conditional effective head field in the sense that it satisfies the flow equation and the boundary and initial conditions. Because the effective-head field is not the conditional mean, its uncertainty cannot be properly addressed.

Using both primary information and secondary information to improve our estimates for the coconditional means of $K(\psi)$, $C(\psi)$, and $\psi$ is rational. Both cokriging and the iterative approach are appropriate tools for this purpose. Whereas the linear assumption embedded in cokriging restricts its application to mildly heterogeneous vadose zones, the iterative geostatistical approach relaxes the linearity assumption. Nevertheless, both approaches face the problem arising from the unknown product terms in the conditional mean flow equation. On the one hand, cokriging may yield approximate conditional conductivity and head fields, but they do not satisfy the mass-balance principle and can create problematic velocity fields. Although this problem can be resolved by using the cokriged conductivity field and a flow model to derive a consistent head field, the simulated head field will not honor the observed heads at sample locations. The simulated head is essentially an effective-head field, instead of the conditional mean. On the other hand, the iterative approach yields head and conductivity fields that satisfy the continuity equation, but they are coconditional effective-head and conductivity fields. Besides, both conditional and coconditional mean approaches produce only the mean flow. An additional conditional effective-macrodispersion concept must be introduced, and it has not been explored before.

Conditional Monte Carlo techniques, using primary information or both primary and secondary information, avoid the unknown product terms and do not require the use of the conditional effective macrodispersion, but they demand significant computational resources. Although the coconditional simulation takes advantage of potentially useful secondary information, it faces the same problem as cokriging because of the linear-predictor and superposition methods. Thus the resultant coconditional realizations of head and conductivity fields will be different from the true coconditional ones. If one generates the coconditional realizations of conductivity fields first and then solves the flow equation with the conductivity fields, simulated head fields will not agree with the head measurement at the sample location, and therefore they are not coconditional head fields. The problem will be exacerbated if the vadose zone is highly heterogeneous. Development of a new generation of iterative coconditional Monte Carlo simulation approaches (e.g., Hanna and Yeh, in press) that can generate realizations of self-consistent coconditional conductivity and head fields seems imminent. The new approaches will demand even greater computation effort, compared with the classic noniterative one. Only rapid advances in computing technology can resolve such a problem.

For practical field problems, these theoretical difficulties with the coconditional approach may not be critical. The important issue is that many costly boreholes are required in order to gather sufficient information on water content and pressure to facilitate this approach. For this reason, use of the approach likely will be confined to research communities, unless cost-effective tools for collecting the information become available.

Recent geophysical studies may have shed light on a solution to this problem. Such studies have shown that electric resistivity tomography (ERT) can be used to detect changes in moisture content over a large volume of geologic media in a cost-effective manner. Briefly, the ERT method is a direct-current resistivity technique in which the electric potentials generated by a current source placed either on the earth's surface or in the subsurface are measured at a similarly positioned receiver. These measured potentials are sensitive to the bulk electrical properties of the subsurface, which in turn are primarily diagnostic of porosity, the amount and connectivity of pore fluids, and the chemistry of pore fluids. Thus ERT is very sensitive to the amount of pore fluid (i.e., the saturation), and it is a natural for remotely detecting changes in moisture content and concentration in the vadose zone.

Because of its sensitivity to fluid saturation and chemistry, ERT has been used for monitoring water leaks from storage ponds (Park and Van, 1993), monitoring fluid flow in an infiltration test (Pfeifer and Anderson, 1996), subsurface imaging of an industrial-waste site (Morgan et al., 1990), and monitoring remediation by air sparging (Schima, LaBrecque, and Lundegard, 1994; LaBrecque et al., 1996). Three-dimensional images of subsurface resistivity can also provide valuable information on geologic structures that control water flows and the movements of contaminants.

To this end, clearly the recent advances in conditional stochastic methods are giving us more detailed information about subsurface heterogeneity than ever before. On the other hand, advances in ERT allow us to cost-effectively amass large quantities of data on water content. Subsequently, new methods that will combine the merits of these tools may finally allow predictions at high resolutions.

## 8.4 Conclusion

The unconditional effective-parameter approach has addressed intractable issues associated with the homogeneous approach in classic analysis. It derives the mean flow and transport equation for the vadose zone and provides estimates of effective hydraulic properties and variances of predictions. Furthermore, it advances our understanding of the scale effects of heterogeneity on flow and transport. If the size of a given domain in the vadose zone is much greater than the correlation scale of the most significant heterogeneity, the FSREV is expected to be definable. Then the unconditional effective-parameter theory should provide an unbiased estimate of the spatial mean behavior of the system, implying that the ergodicity assumption is met. This unbiased estimate will be in good agreement with our observation if the scale of our interest is greater than or equal to the FSREV. That is, we are interested in the integrated behaviors (i.e., moments, or breakthroughs) of moisture or concentration distributions at large times, when they have already encountered a sufficient number of heterogeneities. Then the deviation between our observation and prediction will be small. On the other hand, if the scale of our interest is smaller than the FSREV, the discrepancy will depend upon the magnitude and size of the heterogeneity and the mean flow and transport behaviors (i.e., mean hydraulic or concentration gradients). For cases where the domain of the problem does not encompass many of the most significant heterogeneities, an FSREV will not be reached. The unconditional effective-parameter theory will produce a biased estimate of the mean behavior of the system, which likely will disagree with our observation even if the scale of our interest is larger than the FSREV. Thus, the conditional effective-parameter approach will be the only viable approach.

Among all the various conditional effective-parameter approaches, the coconditional approach using unsaturated hydraulic conductivity, soil-water pressure, moisture content, concentration, and geologic and geophysical information is most appealing and deserves further development. Regardless of how much uncertainty the coconditional approach can reduce, incorporation of all of the available information in the prediction is rational. The uncertainty will be gradually reduced as more data become available. Thus, this approach is a promising mathematical tool that can bring our prediction one step closer to reality. One must recognize that even if a site is completely characterized, our ability to predict the actual plume is limited to its "bulk" behavior, because of measurement, interpolation, modeling, and numerical errors (Yeh et al., 1995). Therefore, the iterative geostatistical method may provide a good approximation of the mean flow path. To address uncertainties, the coconditional Monte Carlo simulation is a logical choice, although its estimates are uncertain themselves.

For many field problems where the vadose zone consists of hard rocks hundreds or thousands of feet in thickness, the usefulness of the stochastic approach is limited. The collection of statistically meaningful estimates of moments for $K(\psi)$ and $\theta(\psi)$ curves will demand a great number of deep wells and samples, which are seldom affordable. A classic analysis may be the only choice. Thus, identifying large-scale geologic structures and units using geologic and geophysical well logs is the essential step. Representative samples can then be taken from each unit for measurements of $K(\psi)$ and $\theta(\psi)$ relationships. Though this information is general and insufficient, its combination with a three-dimensional model will produce a more realistic result than the unconditional effective-parameter approach. In fact, such an approach is essentially a coconditional effective-parameter approach, at least in a qualitative sense.

Although the stochastic approach deserves further development, we must recognize that all stochastic methods are tools based on statistics that can provide only best unbiased estimates. No matter how elegant and flawless a stochastic theory is, it never reduces the uncertainty in our predictions. Only large amounts of data can lessen the uncertainty and make the stochastic result statistically meaningful. Improving our knowledge of the spatial distributions of hydrologic properties in the vadose zone thus becomes the only solution to the spatial-variability problem. For this reason, cost-effective site-characterization techniques that can sense a large volume of the geologic formation and can detect the pattern of the significant heterogeneity in the vadose zone must be developed. Exact knowledge of hydraulic-property values generally is less important than knowledge of the spatial pattern of a property, because our predictions will always be limited to the "bulk" behavior. Finally, efficient numerical algorithms for solving three-dimensional nonlinear equations and faster and less expensive computational tools are needed to advance our ability of predicting flow and solute transport in the vadose zone at high resolutions.

## 8.5 Acknowledgments

This work was supported in part by DOE/Sandia contract AV-0655#1, NSF grant EAR-9317009, USGS grant 1434-92-G-2258, and grant ES04949 from the National Institute of Environmental Health Sciences. Comments on the first draft of the chapter by Alec Desbarats, Thomas Harter, Bailing Li, Daniel Tartakovsky, and Jim Smith are greatly appreciated. Finally, many thanks go to Edward Sudicky for recommending me to be a contributor for this volume.

## References

Bear, J. 1972. *Dynamics of Fluids in Porous Media.* New York: Dover.

Bond, W. J. 1986. Velocity-dependent hydrodynamic dispersion during unsteady, unsaturated soil water flow: experiments. *Water Resour. Res.* 22:1881–9.

Brooks, R. H., and Corey, A. T. 1966. Properties of porous media affecting fluid flow. *J. Irrig. Drain. Div., ASCE* 92:61–88.

Butters, G. L., and Jury, W. A. 1989. Field scale transport of bromide in an unsaturated soil. 2. Dispersion modeling. *Water Resour. Res.* 25:1583–9.

Butters, G. L., Jury, W. A., and Ernst, F. F. 1989. Field scale transport of bromide in an unsaturated soil. 1. Experimental methodology and results. *Water Resour. Res.* 25:1575–81.

Coats, K. H., and Smith, B. D. 1964. Dead-end pore volume and dispersion in porous media. *Soc. Pet. Eng. J.* 4:73–84.

Copty, N., and Rubin, Y. 1995. A stochastic approach to the characterization of lithofacies from surface seismic and well data. *Water Resour. Res.* 31:1673–86.

Csanady, G. T. 1973. *Turbulent Diffusion in the Environment.* Dordrecht: Reidel.

Cushman, J. H. 1983. Comment on "Three-dimensional stochastic analysis of macrodispersion in aquifers" by L. W. Gelhar and C. L. Axness. *Water Resour. Res.* 19:1641–2.

Dagan, G. 1982. Analysis of flow through heterogeneous random aquifers. 2. Unsteady flow in confined formations. *Water Resour. Res.* 18:1571–85.

Dagan, G. 1984. Solute transport in heterogeneous porous formation. *J. Fluid Mech.* 145:151–77.

Dagan, G. 1987. Theory of solute transport by groundwater. *Ann. Rev. Fluid Mech.* 19:183–215.

Dagan, G. 1990. Transport in heterogeneous porous formations: spatial moments, ergodicity and effective dispersion. *Water Resour. Res.* 26:1281–90.

de Marsily, G. 1986. *Quantitative Hydrogeology: Groundwater Hydrogeology for Engineers.* Orlando: Academic Press.

Desbarats, A. J. 1995. Upscaling capillary pressure–saturation curves in heterogeneous porous media. *Water Resour. Res.* 31:281–8.

Ferrante, M., and Yeh, T.-C. J. 1995. *Stochastic Analysis of Water Flow in Heterogeneous Unsaturated Soils under Transient Conditions.* Technical report. HWR95-040. Department of Hydrology and Water Resources, University of Arizona.

Fischer, H. B., List, E. J., Koh, R. C. Y., Imberger, J., and Brooks, N. H. 1979. *Mixing in Island and Coastal Waters.* San Diego: Academic Press.

Gardner, W. R. 1958. Some steady state solutions of unsaturated moisture flow equations with applications to evaporation from a water table. *Soil Sci.* 85:228–32.

Gelhar, L. W., and Axness, C. L. 1983. Three-dimensional stochastic analysis of macrodispersion in aquifers. *Water Resour. Res.* 19:161–80.

Green, T. R., and Freyberg, D. L. 1995. State-dependent anisotropy: comparisons of quasi-analytical solutions with stochastic results for steady gravity drainage. *Water Resour. Res.* 31:2201–12.

Greenholtz, D. E., Yeh, T.-C. J., Nash, M. S. B., and Wierenga, P. J. 1988. Geostatistical analysis of soil hydrologic properties in a field plot. *J. Contam. Hydr.* 3:227–50.

Hanna, S., and Yeh, T.-C. J. In Press. Estimation of coconditional moments of transmissivity, hydraulic head, and velocity fields. *Adv. Water Resour.*

Harter, T. 1994. Unconditional and conditional simulation of flow and transport in heterogeneous, variably saturated porous media, Ph.D dissertation, Department of Hydrology and Water Resources, University of Arizona.

Harter, T. and Yeh, T.-C. J. 1993. An efficient method for simulating steady unsaturated flow in random porous media: using an analytical perturbation solution as initial guess to a numerical model. *Water Resour. Res.* 29:4139–49.

Harter, T. and Yeh, T.-C. J. 1996a. Stochastic analysis of solute transport in heterogeneous, variably saturated soils. *Water Resour. Res.* 32:1585–96.

Harter, T. and Yeh, T.-C. J. 1996b. Conditional stochastic analysis of solute transport in heterogeneous, variably saturated soils. *Water Resour. Res.* 32:1597–610.

Harvey, C. F., and Gorelick, S. 1995. Mapping hydraulic conductivity: sequential conditioning with measurements of solute arrival time, hydraulic head, and local conductivity. *Water Resour. Res.* 31:1615–26.

Herkelrath, W. N., Hamburg, S. P., and Murphy, F. 1991. Automatic, real-time monitoring of soil moisture in a remote field area with time-domain reflectometry. *Water Resour. Res.* 27:857–64.

Herr, M., Shäfer, G., and Spitz, K. 1989. Experimental studies of mass transport in porous media with local heterogeneities. *J. Contam. Hydr.* 4:127–37.

Hopmans, J. W., Schukking, H. and Torfs, P. J. J. F. 1988. Two-dimensional steady state unsaturated water flow in heterogeneous soils with autocorrelated soil hydraulic properties. *Water Resour. Res.* 24:2005–17.

Journel, A. G., and Huijbregts, C. H. J. 1978. *Mining Geostatistics.* New York: Academic Press.

Jury, W. A., Sposito, G., and White, R. E. 1986. A transfer function model of solute movement through soil. 1. Fundamental concepts. *Water Resour. Res.* 22:243–7.

Kahleel, R., Relyea, J. F., and Conca, J. L. 1995. Evaluation of van Genuchten-Mualem relationships to estimate unsaturated hydraulic conductivity at low water contents. *Water Resour. Res.* 31:2659–68.

Kapoor, V., and Gelhar, L. W. 1994a. Transport in three-dimensionally heterogeneous aquifers. 1. Dynamics of concentration fluctuations. *Water Resour. Res.* 30:1775–88.

Kapoor, V., and Gelhar, L. W. 1994b. Transport in three-dimensionally heterogeneous aquifers. 2. Predictions and observations of concentration fluctuations. *Water Resour. Res.* 30:1789–802.

Kapoor, V., and Kitanidis, P. K. 1996. Concentration fluctuations and dilution in two-dimensional periodic heterogeneous porous media. *Trans. Porous Media.* 22:91–119.

Kool, J. B., and Parker, J. C. 1988. Analysis of the inverse problem for transient unsaturated flow. *Water Resour. Res.* 24:817–30.

Krupp, H. K., and Elrick, D. E. 1968. Miscible displacement in an unsaturated glass bead medium. *Water Resour. Res.* 4:809–15.

Kung, K.-J. S. 1993. Laboratory observation of the funnel flow mechanism and its influence on solute transport. *J. Environ. Qual.* 22:91–102.

LaBrecque, D. J., Ramirez, A. L., Daily, W. D., Binley, A. M., and Schima, S. A. 1996. ERT monitoring of environmental remediation processes. *Measurement Sci. Tech.* 7:375–83.

Loaiciga, H. A., and Mariño, M. A. 1990. Error analysis and stochastic differentiability in subsurface flow modeling. *Water Resour. Res.* 26:2897–982.

McCord, J. T., Stephens, D. B., and Wilson, J. L. 1991. The importance of hysteresis and state-dependent anisotropy in modeling variably saturated flow. *Water Resour. Res.* 27:1501–18.

Mantoglou, A., and Gelhar, L. W. 1985. *Large Scale Models of Transient Unsaturated Flow and Transport.* R. M. Parsons Laboratory technical report 299. Dept. of Civil Engineering, Massachusetts Institute of Technology.

Mantoglou, A., and Gelhar, L. W. 1987a. Stochastic modeling of large-scale transient unsaturated flow systems. *Water Resour. Res.* 23:37–46.

Mantoglou, A., and Gelhar, L. W. 1987b. Capillary tension head variance, mean soil moisture content, and effective specific moisture capacity of transient unsaturated flow in stratified soils. *Water Resour. Res.* 23: 47–56.

Mas-Pla, J. 1993. Modeling the transport of natural organic matter in heterogeneous porous media. Analysis of a field-scale experiment at the Georgetown site, S.C. Ph.D. dissertation, Department of Hydrology and Water Resources, University of Arizona.

Mas-Pla, J., Yeh, T.-C. J., McCarthy, J. F., and Williams, T. M. 1992. A forced gradient tracer experiment in a coastal sandy aquifer, Georgetown site, South Carolina. *Ground Water* 30:958–64.

Matheron, G. 1973. The intrinsic random functions and their applications. *Adv. Appl. Probab.* 5:438–68.

Mishra, S., and Parker, J. C. 1989. Parameter estimation for coupled unsaturated flow and transport. *Water Resour. Res.* 25:385–96.

Mizell, S. A., Gutjahr, A. L., and Gelhar, L. W. 1982. Stochastic analysis of spatial variability in two-dimensional steady groundwater flow assuming stationary and non-stationary heads. *Water Resour. Res.* 19:1853–67.

Morgan, F. D., Simms, J., Aspinall,W., and Shepherd, J. 1990. Volume determination of buried sand using d.c. resistivity; an engineering geophysics case history. Presented at the annual meeting of the Society of Exploration Geophysicists.

Mualem, Y. 1976. A new model for predicting the hydraulic conductivity of unsaturated porous media. *Water Resour. Res.* 12:513–22.

Mulla, D. J. 1988. Estimating spatial patterns in water content, matric suction, and hydraulic conductivity. *Soil Sci. Soc. Am. J.* 52:1547–53.

Nielsen, D. R., and Biggar, J. W. 1961. Miscible displacement in soils: 1. Experimental information. *Soil Sci. Soc. Am. Proc.* 25:1–5.

Park, S. K., and Van, G. P. 1993. Inversion of pole-pole data for 3-D resistivity structure beneath arrays of electrodes. *Geophys. J. Int.* 114:12–20.

Pfeifer, M. C., and Anderson, H. T. 1996. DC-resistivity array to monitoring fluid flow at the INEL infiltration test. In: *Proceedings of the 1995 Symposium on the Application of Geophysics to Engineering and Environmental Problems.* pp. 709–18. Englewood, CO: Environmental and Engineering Geophysical Society.

Polmann, D. J., McLaughlin, D., Luis, S., Gelhar, L. W., and Ababou, L. 1991. Stochastic modeling of large-scale flow in heterogeneous unsaturated soils. *Water Resour. Res.* 27:1447–58.

Priestley, M. B. 1981. *Spectral Analysis and Time Series.* New York: Academic Press.

Rajaram, H., and Gelhar, L. W. 1993a. Plume scale-dependent dispersion in heterogeneous aquifers. 1. Lagrangian analysis in a stratified aquifer. *Water Resour. Res.* 29:3249–60.

Rajaram, H., and Gelhar, L. W. 1993b. Plume scale-dependent dispersion in heterogeneous aquifers. 2. Eulerian analysis and three-dimensional aquifers. *Water Resour. Res.* 29:3261–76.

Rajaram, H., and Gelhar, L. W. 1995. Plume scale-dependent dispersion in aquifers with a wide range of scales of heterogeneity. *Water Resour. Res.* 31:2469–82.

Rockhold, M. L., Rossi, R. E., and Hills, R. G. 1996. Application of similar media scaling and conditional simulation for modeling water flow and tritium transport at the Las Cruces trench site. *Water Resour. Res.* 32:595–610.

Roth, K., Jury, W. A., Flühler, H., and Attinger, W. 1991. Transport of chloride through an unsaturated field soil. *Water Resour. Res.* 27:2533–41.

Russo, D. 1993. Stochastic modeling of macrodispersion for solute transport in a heterogeneous unsaturated porous formation. *Water Resour. Res.* 29:383–97.

Russo, D. 1995. Stochastic analysis of the velocity covariance and the displacement covariance tensors in partially saturated heterogeneous anisotropic porous formations. *Water Resour. Res.* 31:1647–85.

Russo, D., and Bouton, M. 1992. Statistical analysis of spatial variability in unsaturated flow parameters. *Water Resour. Res.* 28:1911–25.

Russo, D., Bresler, E., Shani, U., and Parker, J. 1991. Analyses of infiltration events in relation to determining soil hydraulic properties by inverse problem methodology. *Water Resour. Res.* 27:1361–73.

Schima, S. A., LaBrecque, D. J., and Lundegard, P. D. 1994. Using resistivity tomography to track air sparging. In: *Proceedings of the 1994 Symposium on the Application of Geophysics to Engineering and Environmental Problems*, pp. 757–74. Englewood, CO: Environmental and Engineering Geophysical Society.

Stephens, D. B., 1996. *Vadose Zone Hydrology.* Boca Raton, FL: Lewis Publishers.

Stephens, D. B., Lambert, K., and Waston, D. 1987. Regression models for hydraulic conductivity and field test of borehole permeameter. *Water Resour. Res.* 23:2207–14.

Taylor, G. I. 1922. Diffusion by continuous movements. *Proc. London Math. Soc.* A20:196–211.

Ünlü, J., Nielsen, D. R., and Biggar, J. W. 1990. Stochastic analysis of unsaturated flow: one-dimensional Monte Carlo simulations and comparison with spectral perturbation analysis and field observations. *Water Resour. Res.* 26:2207–18.

van Genuchten, M. T. 1980. A closed-form equation for predicting the hydraulic conductivity of unsaturated soils. *Soil Sci. Soc. Am. J.* 44:892–8.

van Genuchten, M., and Nielsen, D. R. 1985. On describing and predicting the hydraulic properties of unsaturated soils. *Ann. Geophys.* 3:615–28.

Vauclin, M., Khanji, D., and Vachaud, G. 1979. Experimental and numerical study of transient, two-dimensional unsaturated–saturated water table recharge problem. *Water Resour. Res.* 15:1089–101.

Vauclin, M., Vieira, D. R., Vachaud, G., and Nielsen, D. R. 1983. The use of cokriging with limited field soil observations. *Soil Sci. Soc. Am. J.* 47:175–84.

Vomvoris, E. G., and Gelhar, L. W. 1990. Stochastic analysis of the concentration variability in a three-dimensional heterogeneous aquifer. *Water Resour. Res.* 26:2591–602.

Watson, K. K. 1966. An instantaneous profile method for determining the hydraulic conductivity of unsaturated porous materials. *Water Resour. Res.* 2:709–15.

Wilson, J. L., and Gelhar, L. W. 1974. *Dispersive Mixing in a Partially Saturated Porous Media.* Ralph M. Parsons Laboratory technical report 191. Massachusetts Institute of Technology.

Yates, S. R., and Warrick, A. W. 1987. Estimating soil water content using cokriging. *Soil Sci. Soc. Am. J.* 51:23–30.

Yeh, T.-C. J. 1989. One-dimensional steady state infiltration in heterogeneous soils. *Water Resour. Res.* 25:2149–58.

Yeh, T.-C. J. 1992. Stochastic modelling of groundwater flow and solute transport in aquifers. *Hydrol. Proc.* 5:369–95.

Yeh, T.-C. J., Gelhar, L. W., and Gutjahr, A. L. 1985a. Stochastic analysis of unsaturated flow in heterogeneous soils. 1: Statistically isotropic media. *Water Resour. Res.* 21:447–56.

Yeh, T.-C. J., Gelhar, L. W., and Gutjahr, A. L. 1985b. Stochastic analysis of unsaturated flow in heterogeneous soils. 2: Statistically anisotropic media with variable alpha. *Water Resour. Res.* 21:457–64.

Yeh, T.-C. J., Gelhar, L. W., and Gutjahr, A. L. 1985c. Stochastic analysis of unsaturated flow in heterogeneous soils, 3: Observations and applications. *Water Resour. Res.* 21:465–72.

Yeh, T.-C. J., Gelhar, L. W., and Wierenga, P. J. 1986. Observations of spatial variability of soil-water pressure in a field soil. *Soil Sci.* 142:7–12.

Yeh, T.-C. J., and Harvey, D. J. 1990. Effective unsaturated hydraulic conductivity of layered sand. *Water Resour. Res.* 26:1271–9.

Yeh, T.-C. J., Jin, M. H., and Hanna, S. 1996. An iterative stochastic inverse method: conditional effective transmissivity and hydraulic head fields. *Water Resour. Res.* 32:85–92.

Yeh, T.-C. J., Mas-Pla, J., Williams, T. M., and McCarthy, J. F. 1995. Observation and three-dimensional simulation of chloride plumes in a sandy aquifer under forced-gradient conditions. *Water Resour. Res.* 31:2141–57.

Yeh, T.-C. J., and Zhang, J. 1996. A geostatistical inverse method for variably saturated flow in the vadose zone. *Water Resour. Res.* 32:2757–66.

Yeh, W. W-G. 1986. Review of parameter identification procedure in groundwater hydrology: the inverse problem. *Water Resour. Res.* 22:95–108.

Zhang, J., and Yeh, T.-C. J. 1997. An iterative geostatistical inverse method for steady flow in the vadose zone. *Water Resour. Res.* 33:63–71.

# 9

# Stochastic Modeling of Scale-dependent Macrodispersion in the Vadose Zone

DAVID RUSSO

## 9.1 Introduction

Quantitative field-scale descriptions of chemical transport in the unsaturated (vadose) zone are essential for improving our basic understanding of the transport process in near-surface geologic environments and for sharpening the predictive tools that in turn will be used to predict the future spread of pollutants in these environments. The traditional approach to modeling transport processes in the vadose zone has been to model water flow and solute transport by using the macroscopic physical and chemical properties of the soil that vary in a deterministic manner, obey physical and chemical laws, and are expressed in the form of partial differential equations.

One of the distinctive features of a natural formation at the field scale, however, is the spatial heterogeneity of its properties (e.g., Nielsen, Biggar, and Erh, 1973; Russo and Bresler, 1981; Ünlü, Kavvas, and Nielsen, 1989; Russo and Bouton, 1992). This spatial heterogeneity is generally irregular, and it occurs on a scale beyond the scope of laboratory samples. These features have distinct effects on the spatial distributions of solute concentrations that result from transport through the heterogeneous porous formations, as has been observed in field experiments (e.g., Schulin et al., 1987; Butters, Jury, and Ernst, 1989; Ellsworth et al., 1991; Roth et al., 1991) and demonstrated by simulation (e.g., Russo, 1991; Russo, Zaidel, and Laufer, 1994; Tseng and Jury, 1994) of solute transport in unsaturated, heterogeneous soils.

Inasmuch as a deterministic description of a heterogeneous soil is, for practical purposes, impossible, a fundamental question is how to develop predictive models that can incorporate the impacts of the field-scale spatial variabilities of soil properties on vadose-zone flow and transport. In the following sections, a few advances in this area will be presented and analyzed. The plan of this chapter is as follows: The basic concepts and definitions are given in Section 9.2. Different approaches to modeling solute transport in heterogeneous porous media are discussed briefly in Section 9.3, and statistical assessment of the velocity of steady-state, unsaturated flow is discussed in Section 9.4, using a stochastic continuum framework. Transport of a passive solute under ergodic and nonergodic conditions is analyzed in Section 9.5, using a general Lagrangian framework. Section 9.6 concludes and summarizes this chapter.

## 9.2 Basic Concepts and Definitions

At the field scale, the soil properties affecting transport are subject to uncertainty owing to their inherently erratic nature and to a paucity of measurements. In the stochastic approach adopted in this chapter, uncertainty is set in a mathematical framework by regarding the relevant formation properties (e.g., hydraulic-conductivity and water-retention functions) as random space functions (RSFs). As a consequence, the flow and transport equations are of a stochastic nature, and the dependent variables (e.g., fluid pressures and saturations, contaminant concentrations) are also RSFs. The aim of the stochastic approach, therefore, is to evaluate the statistical moments of the variables of interest, given the statistical moments of the formation properties. In general, this is a formidable task, and usually its scope is restricted to finding the first two moments: the mean value and the two-point covariance.

In a stochastic framework, a given heterogeneous soil property, $p(\mathbf{x})$, where $\mathbf{x}$ is the spatial-coordinate vector, is treated as if it were a sample (or a realization) drawn at random from an ensemble of physically plausible RSFs, $P(\mathbf{x})$. The ensemble concept is convenient for defining the statistical properties of $P(\mathbf{x})$. Physically, the ensemble mean can be understood as the arithmetic average of all possible property values that can occur at a given spatial point, under the same external conditions. The ensemble mean is the minimum-variance, unbiased estimate of the actual property, provided that this property can be described by a probability distribution with finite moments (Jazwinski, 1978). The ensemble variance is a measure of the uncertainty associated with this estimate. However, because only one realization of $P(\mathbf{x})$ will be available in practice, the ergodic hypothesis must be invoked. According to this hypothesis (Lumley and Panofsky, 1964), inferences about the statistical structure of $P(\mathbf{x})$ can be based on a substitution: replacing the ensemble averages with spatial averages obtained from a single realization of $P(\mathbf{x})$.

To simplify matters, it is assumed in this chapter that $P(\mathbf{x})$ is a second-order stationary RSF defined by a constant mean, independent of the spatial position $\langle P(\mathbf{x})\rangle$, and by a two-point covariance $C_{pp}(\mathbf{x}, \mathbf{x}') = \langle [P(\mathbf{x}) - \langle P(\mathbf{x})\rangle][P(\mathbf{x}') - \langle P(\mathbf{x})\rangle]\rangle$ that in turn depends on the separation vector $\boldsymbol{\xi} = \mathbf{x} - \mathbf{x}'$, and not on $\mathbf{x}$ and $\mathbf{x}'$ individually. The heterogeneous porous formation can be visualized as being composed of a three-dimensional structured arrangement of blocks of different soil materials that may exhibit specific sizes, but are not completely regular. The two-point covariance $C_{pp}(\boldsymbol{\xi})$, therefore, depends on the amount of overlap between these blocks. For example, when the overlapping blocks are allowed to vary randomly in size, the covariance can be described by the exponential model

$$C_{pp}(\boldsymbol{\xi}) = \sigma_p^2 \exp(-\xi') \tag{9.1}$$

where $\boldsymbol{\xi}' = (\mathbf{x} - \mathbf{x}')/\mathbf{I}_p$ is the scaled separation vector, $\xi' = |\boldsymbol{\xi}'|$; $\mathbf{I}_p = (I_{p1}, I_{p2}, I_{p3})$ are the correlation scales of $P(\mathbf{x})$, defined as $I_{pi} = \int_0^\infty C_{pp}(\xi_i)\, d\xi_i / C_{pp}(0)$ $(i = 1, 2, 3)$; $\sigma_p^2$ is the variance of $P(\mathbf{x})$.

Before concluding this section, it is convenient to address various characteristic length scales relevant to flow and transport processes in the vadose zone. The first scale, $L$, characterizes the extent of the transport domain in the direction of the mean flow; this length scale determines the travel distance (e.g., the distance to the water table). The second scale, $\boldsymbol{\ell} = (\ell_1, \ell_2, \ell_3)$, is the one that characterizes the extent of the solute inlet; its lateral components ($\ell_2$ and $\ell_3$) determine, roughly, the lateral extent of the solute body over which entities of interest can be averaged spatially. The third scale is the correlation scale, $\mathbf{I}_p = (I_{p1}, I_{p2}, I_{p3})$, or the scale of the heterogeneity, which determines, roughly, the distances at which property variations cease to be correlated. The fourth scale is the macroscopic capillary length scale, $\lambda$, which determines the relative magnitudes of the capillary forces in unsaturated flow; it can be regarded as a natural length scale of the unsaturated soil (Raats, 1976). As will be shown later, each of these length scales plays an important role regarding flow and transport processes in the vadose zone.

## 9.3 Stochastic Modeling of Transport in the Vadose Zone

Stochastic models of solute transport focus on the ensemble macrodispersive flux due to random variations in the solute concentration $c$ and the specific discharge $\mathbf{q}$ (Gelhar and Axness, 1983). The ensemble macrodispersive flux accounts for differences among the trajectories of random concentration replicates over the ensemble. It dominates the ensemble mean concentration field and should be distinguished from the macroscopic spread of a single plume for particular site-specific applications. As pointed out by Sposito, Jury, and Gupta (1986), a prediction of the spreading of a single plume, based on ensemble statistics, may be inappropriate for particular site-specific applications unless the relevance of ensemble behavior to the single-field behavior can be established explicitly. As will be shown later, ergodicity (i.e., the interchange between ensemble and spatial averages) can be achieved if the lateral extent of the solute input zone is sufficiently large compared with the scale of the heterogeneity in the transverse directions (i.e., $\ell_i \gg I_{pi}, i = 2, 3$).

Stochastic transport models can be classified as either mechanistic or nonmechanistic. "Mechanistic" is taken here to imply that the model incorporates the most fundamental mechanisms of the process, as understood at present [e.g., describing local flow and transport by the Richards equation and by the convection–dispersion equation (CDE), respectively]. "Nonmechanistic" implies that the model disregards the internal physical mechanisms that contribute to the transport process and concentrates on the relationships between an input function and an output response function.

In subsurface hydrology, stochastic mechanistic models of solute transport at the field (formation) scale have been developed (e.g., Dagan, 1982, 1984; Gelhar and

Axness, 1983; Neuman, Winter, and Newman, 1987). These models have been restricted to saturated groundwater flow. Although they depend on similar physical principles, groundwater transport and vadose-zone transport differ with respect to the length scale of the flow domain, the flow regime, and the direction of the principal flow components relative to the porous formation strata. Theoretical developments in groundwater-transport modeling have focused on steady-state flow, in which the only relevant flow parameter is saturated conductivity $K_s$, which in turn is an inherent formation property.

Water flow in the vadose zone (which is transient and can also be involved with wetting and drying cycles) is much more complicated than groundwater saturated flow; the relevant flow parameters – hydraulic conductivity $K$ and water capacity $S$ – depend on a few formation properties (e.g., saturated conductivity and pore-size distribution) and on flow-controlled variables (water saturation $\Theta$ and pressure head $\psi$) in a highly nonlinear fashion. Consequently, under unsaturated flow, the evaluation of the effects on transport of variations in the properties of the porous medium is extremely complex and requires several simplifying assumptions regarding both the structure of the formation heterogeneity and the flow regime.

Most of the existing stochastic models of vadose-zone transport treat the heterogeneous soil as though it were composed of a series of isolated streamtubes (vertically homogeneous, independent soil columns) with different velocities. These models are either mechanistic (e.g., Dagan and Bresler, 1979; Bresler and Dagan, 1981; Destouni and Cvetkovic, 1989, 1991) or nonmechanistic (e.g., Jury, 1982). In this chapter, however, we shall focus our attention on another group of stochastic, mechanistic transport models that are based on the stochastic continuum description of flow developed by Yeh, Gelhar, and Gutjahr (1985a,b) and a general Lagrangian description of transport (e.g., Russo, 1993, 1995a,b). In this approach, quantification of solute transport in heterogeneous porous formations is accomplished in a two-stage approach: The first stage involves relating the statistical moments of the probability density function (PDF) of the velocity to those of the properties of the formation; the second stage involves relating the statistical moments of the PDF for the particle displacement to those of the velocity PDF.

## 9.4 Statistics of the Velocity

Beginning from the stochastic continuum approach of Yeh et al. (1985a,b), we consider an unbounded flow domain of a partially saturated, heterogeneous soil with a three-dimensional, statistically anisotropic structure in a cartesian coordinate system $(x_1, x_2, x_3)$, with $x_1$ directed vertically downward. In order to relate the statistical moments of the PDF of the Eulerian velocity to those of the properties of the heterogeneous porous formation, the following assumptions are employed: (1) The local, steady-state unsaturated flow obeys Darcy's law and continuity. (2) The local relationships between the unsaturated conductivity $K$ and the capillary pressure head

$\psi$ are isotropic, given by the expression (Gardner, 1958)

$$K(\psi, \mathbf{x}) = K_s(\mathbf{x})\exp[-\alpha(\mathbf{x})|\psi|] \tag{9.2}$$

where $\alpha$ is a soil parameter, viewed as the reciprocal of the macroscopic capillary length scale $\lambda$. (3) Both formation properties, $\log K_s$ and $\alpha$, are multivariate normal (MVN) RSFs (Lumley and Panofsky, 1964), ergodic over the region of interest, characterized at second order by the constant means $F = E[\log K_s]$ and $A = E[\alpha]$ and by the statistically anisotropic exponential covariance functions $C_{ff}(\boldsymbol{\xi})$ and $C_{aa}(\boldsymbol{\xi})$, respectively, given by equation (9.1), with $p = f$ or $a$, where $f = \log K_s - F$ and $a = \alpha - A$ are the perturbations of the $\log K_s$ and $\alpha$ fields, respectively. (4) For a given mean pressure head $H = \langle\psi(\mathbf{x})\rangle$, variations in $\Theta$ are small enough to be excluded from the analysis of the flow.

Equation (9.2) implies that for a given mean capillary pressure head $H$, log-unsaturated conductivity $\log K$, being a linear combination of $\log K_s$ and $\alpha$, is also MVN. By employing the aforementioned assumptions and by linearization of Darcy's law for unsaturated flow and for a given mean water saturation $\Theta$, the first-order approximation of the velocity perturbation is given (Russo, 1995b) by

$$u_i(\Theta) = [K_g(\Theta)/n\Theta]\{J_i y(\Theta) + [\partial h(\Theta)/\partial x_i]\} \tag{9.3}$$

where $K_g(\Theta) = \exp[Y(\Theta)]$ is the geometric mean conductivity, $Y(\Theta) = F - AH(\Theta)$ is the mean log-conductivity, $n$ is porosity (considered as a deterministic constant), $y(\Theta) = \log K(\Theta) - Y(\Theta) = f - aH(\Theta) - Ah(\Theta)$ and $h(\Theta) = \psi(\Theta) - H(\Theta)$ are the log-conductivity and the pressure-head perturbations, respectively, $J_i$ is the mean head gradient vector, and $i = 1, 2, 3$.

Inasmuch as for a given $\Theta$, $\log K(\Theta)$ is MVN, and $\psi(\Theta)$, through the linearization of the flow equation, is a linear function of $\log K(\Theta)$, it follows that for a given $\Theta$, $\psi(\Theta)$ is also MVN, and by equation (9.3), for a given $\Theta$, so is the velocity. Using equation (9.3), for a given $\Theta$, zero-order approximation of the mean velocity $U_i = \langle u_i(\mathbf{x})\rangle$ $(i = 1, 2, 3)$ and first-order approximation of the velocity covariance $u_{ij}(\boldsymbol{\xi}) = \langle u'_i(\mathbf{x})u'_j(\mathbf{x}')\rangle$ $(i, j = 1, 2, 3)$, where $u'_i(\mathbf{x}) = u_i(\mathbf{x}) - U_i$, are

$$U_i(\Theta) = K_g(\Theta)J_i/n\Theta \tag{9.4}$$

$$u_{ij}(\boldsymbol{\xi};\Theta) = \left[\frac{K_g(\Theta)}{n\Theta}\right]^2 \left[J_iJ_jC_{yy}(\boldsymbol{\xi};\Theta) - J_i\frac{\partial C_{hy}(-\boldsymbol{\xi};\Theta)}{\partial \xi_j} + J_j\frac{\partial C_{hy}(\boldsymbol{\xi};\Theta)}{\partial \xi_i} + \frac{\partial^2 C_{hh}(\boldsymbol{\xi};\Theta)}{\partial \xi_i \xi_j}\right] \tag{9.5}$$

where $C_{yy}(\boldsymbol{\xi};\Theta)$ and $C_{hh}(\boldsymbol{\xi};\Theta)$ are the covariances of the perturbations of $\log K$

and $\psi$, respectively, and $C_{hy}(\boldsymbol{\xi}; \Theta)$ is the cross-covariance between perturbations of $\psi$ and $\log K$.

Explicit expressions for the dependence of $C_{hh}(\boldsymbol{\xi}; \Theta)$, $C_{yy}(\boldsymbol{\xi}; \Theta)$, and $C_{hy}(\boldsymbol{\xi}; \Theta)$ on the statistics of the formation properties and on the mean water saturation are given elsewhere (Yeh et al., 1985a,b; Russo, 1993; 1995a,b, 1997a). In particular, the log-conductivity variance $\sigma_y^2(\Theta)$ and its correlation scales $I_{yi}(\Theta) = [1/\sigma_y^2(\Theta)] \int_0^\infty C_{yy}(\xi_i; \Theta)\, d\xi_i$ $(i = 1, 2, 3)$ are given (Russo, 1997b) as

$$\sigma_y^2(\Theta) = \sigma^2(\Theta) - A^2\sigma_h^2(\Theta) \tag{9.6a}$$

$$I_{yi}(\Theta) = \frac{\sigma^2(\Theta)I_i(\Theta) - A^2C_{hh}(\xi_i; \Theta)}{\sigma^2(\Theta) - A^2\sigma_h^2(\Theta)} \tag{9.6b}$$

where $\sigma_h^2(\Theta)$ is the variance of $\psi(\Theta)$, and $\sigma^2(\Theta)$ and $I_i(\Theta)$ $(i = 1, 2, 3)$ are given by

$$\sigma^2(\Theta) = \sigma_f^2 + H^2(\Theta)\sigma_a^2 - 2H(\Theta)\sigma_{fa}^2 \tag{9.6c}$$

$$I_i(\Theta) = \frac{\sigma_f^2 I_{fi} + H^2(\Theta)\sigma_a^2 I_{ai} - 2H(\Theta)\sigma_{fa}^2 I_{fai}}{\sigma_f^2 + H^2(\Theta)\sigma_a^2 - 2H(\Theta)\sigma_{fa}^2} \tag{9.6d}$$

where $\sigma_f^2$ and $\sigma_a^2$ and $I_{fi}$ and $I_{ai}$ $(i = 1, 2, 3)$ are the respective variances and correlation scales for $\log K_s$ and $\alpha$, $\sigma_{fa}^2$ is the cross-variance between perturbations of $\log K_s$ and $\alpha$, and $I_{fai} = 2I_{fi}I_{ai}/(I_{fi} + I_{ai})$ $(i = 1, 2, 3)$.

As was discussed by Russo (1997b), because of the dependence of $\mathbf{I}_y(\Theta)$ on $C_{hh}(\boldsymbol{\xi}; \Theta)$, in the directions perpendicular to the mean flow, $I_{yi}(\Theta)$ $(i = 2, 3)$ could be negative and therefore physically meaningless, whereas in the direction of the mean flow, $I_{y1}(\Theta)$ is positive. For a given $\Theta$ and given statistics of the $\log K_s$ and $\alpha$ fields, because of the dependence of $C_{yy}(\boldsymbol{\xi}; \Theta)$ on $C_{hh}(\boldsymbol{\xi}; \Theta)$, $\sigma_y^2(\Theta)$ decreases with increasing $A = \langle\alpha(\mathbf{x})\rangle = \lambda^{-1}$ and increasing $\rho_i = I_i/I_1$ $(i = 1, 2, 3)$, whereas the converse is true for $I_{y1}(\Theta)$. At the small $A$ limit, $A \to 0$, however, $\sigma_y^2(\Theta) \to \sigma^2(\Theta)$ and $\mathbf{I}_y(\Theta) \to \mathbf{I}(\Theta)$, independent of $A$ and $\rho_i$ $(i = 2, 3)$.

Considering heterogeneous porous formations of axisymmetric anisotropy (i.e., $I_{pv} = I_{p1}$, $I_{ph} = I_{p2} = I_{p3}$, where $p = f$ or $a$), with mean flow in the vertical direction, perpendicular to the formation bedding [i.e., $\mathbf{J} = (J_1, 0, 0)$], components of the velocity covariance tensor, equation (9.5), for the general case in which the separation vector does not coincide with the mean gradient vector $\mathbf{J}$ have been calculated (Russo, 1995b) using spectral-representation techniques. In the following, some of the findings from those calculations are presented and discussed for the case in which the correlation scales for $\log K_s$ and $\alpha$ are identical, so that the correlation scales [equation (9.6d)] $I_v = I_{pv}$ and $I_h = I_{ph}$ $(p = f$ or $a)$ are independent of $\Theta$.

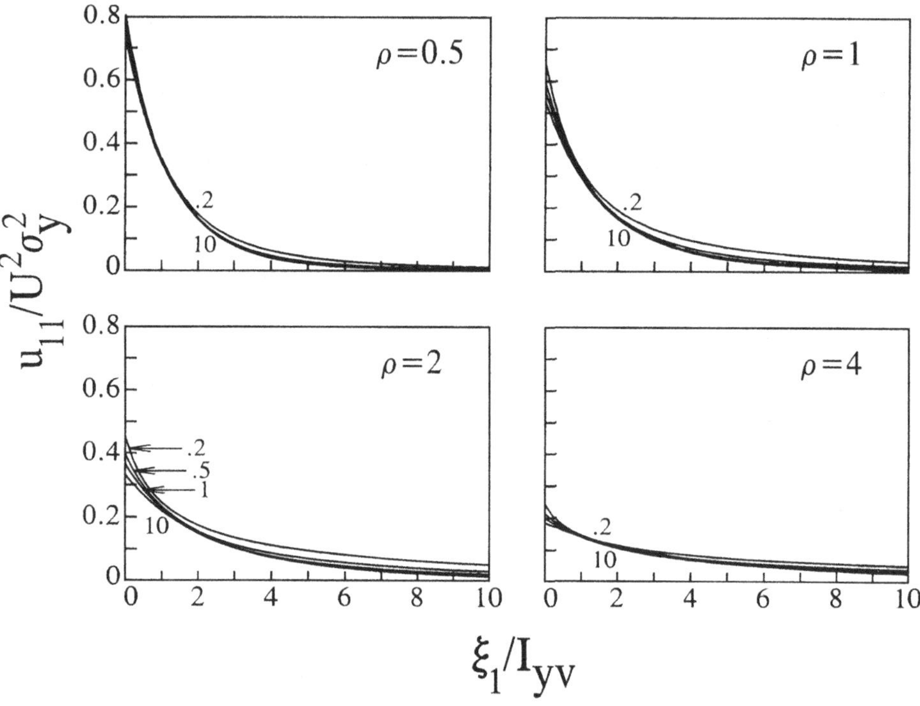


Figure 9.1. Longitudinal component ($u'_{11}$) of the scaled velocity covariance $u'_{ij} = u_{ij}/U\sigma_y^2$ as a function of the scaled separation distance in the direction of the mean flow $\xi'_1 = \xi_1/I_{yv}$ for selected values of $\rho = I_h/I_v$ and $\eta = \lambda/I_v$ (denoted by the numbers labeling the curves).

In Figures 9.1 and 9.2, the principal components of the scaled velocity covariance, independent of $\Theta$ [i.e., $u'_{ii}(\boldsymbol{\xi}') = u_{ii}(\boldsymbol{\xi}'; \Theta)/\sigma_y^2(\Theta)U^2$ $(i = 1, 2, 3)$, where $U = |\mathbf{U}|$], are depicted as functions of the scaled separation distance in the direction parallel to the mean flow, $\xi'_1 = \xi_1/I_{yv}$, where $I_{yv} = I_{y1}$, for selected values of the length-scale ratios $\eta = \lambda/I_v$ and $\rho = I_h/I_v$. It is demonstrated in these figures that the longitudinal component of the scaled velocity variance $u'_{11}(0)$ decreases, while its transverse components $u'_{ii}(0)$ $(i = 2, 3)$ increase with increasing $\eta$. On the other hand, $u'_{11}(0)$ is a monotonically decreasing function of $\rho$, while $u'_{ii}(0)$ $(i = 2, 3)$ are nonmonotonic functions of $\rho$. Furthermore, the persistence of $u'_{11}(\xi'_1)$ increases with increasing $\rho$ and decreasing $\eta$ (Figure 9.1) whereas the persistence of $u'_{ii}(\xi'_1)$ $(i = 2, 3)$ increases with increasing $\rho$ and $\eta$ (Figure 9.2).

Interpretation of the behavior of the principal components of $u'_{ij}(\xi'_1)$ depicted in Figures 9.1 and 9.2 is facilitated by referring to the first-order approximation of the velocity fluctuation $u_i$ $(i = 1, 2, 3)$ [equation (9.3)]. For mean flow parallel to the $x_1$ axis, because the conductivity, equal at first order to $[K_g(\Theta)/n\Theta][1 + y(\Theta)]$, varies along the streamtubes, the first term of $u_1$, namely $[K_g(\Theta)/n\Theta][J_i y(\Theta)]$, is responsible for the variations in the velocity in the direction of the mean flow, while the first terms of $u_2$ and $u_3$ are zero. On the other hand, the second term of $u_i$ $(i = 1, 2, 3)$, namely, $-[K_g(\Theta)/n\Theta][\partial h(\Theta)/\partial x_i]$ $(i = 1, 2, 3)$, is responsi-

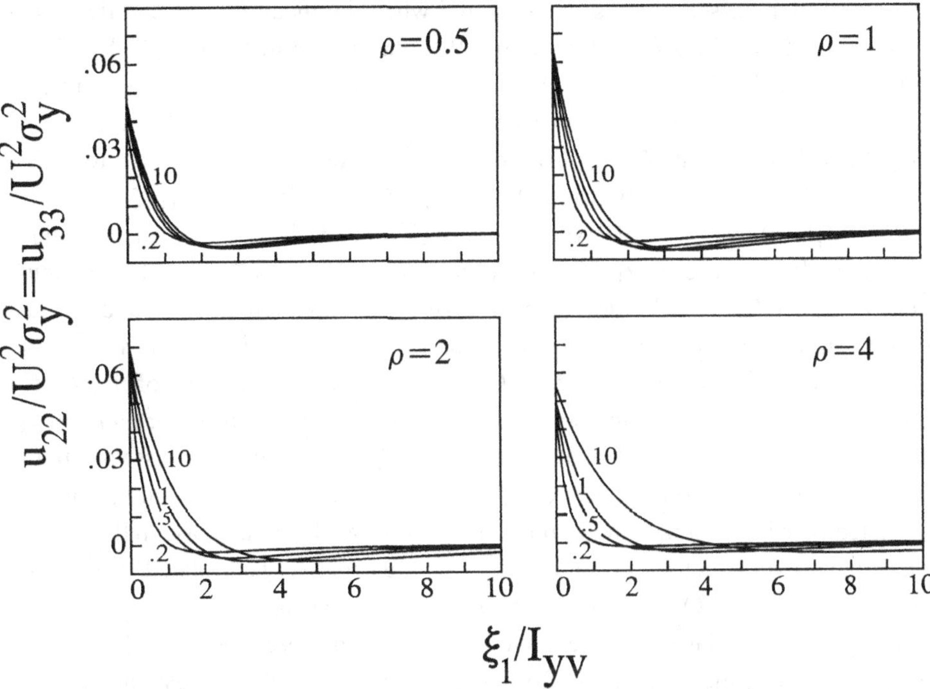


Figure 9.2. Transverse components ($u'_{33} = u'_{22}$) of the scaled velocity covariance $u'_{ij} = u_{ij}/U\sigma_y^2$ as functions of the scaled separation distance in the direction of the mean flow $\xi'_1 = \xi_1/I_{yv}$ for selected values of $\rho = I_h/I_v$ and $\eta = \lambda/I_v$ (denoted by the numbers labeling the curves).

ble for the winding of the streamtubes and for the changes in their cross sections (Dagan, 1989). When the mean flow is vertical and $I_v$ is kept constant, an increasing $\rho$ expresses an increase in the size of the typical flow barriers in a direction normal to the mean flow. In other words, the possibility that a streamline might circumvent subdomains of low conductivity will decrease as $\rho$ increases and concurrently the distance over which the velocity is correlated in the direction of flow will increase as $\rho$ increases. Consequently, for a given $\eta$, the first term of $u_1$ decreases and the second term of $u_i$ ($i = 1, 2, 3$) increases with increasing $\rho$. On the other hand, for fixed $I_v$ the increase in $\eta$ expresses an increase in the macroscopic capillary length scale $\lambda$ (i.e., a transition from a coarse-textured soil material, associated with negligible capillary forces, to a fine-textured soil material, associated with significant capillary forces). Hence, for a given $\rho$, both the first term of $u_1$ and the second term of $u_i$ ($i = 1, 2, 3$) increase with increasing $\eta$. Consequently, the $u_{ii}(0)$ ($i = 1, 2, 3$) increase with increasing $\eta$. The transverse components of the scaled velocity variance $u'_{ii}(0) = u_{ii}(0; \Theta)/\sigma_y^2(\Theta)U^2$ ($i = 2, 3$) also increase with increasing $\eta$ (Figure 9.2). On the other hand, because the second term of $u_1$ increases with increasing $\eta$ more slowly than its first term, the longitudinal component of the scaled velocity variance $u'_{11}(0) = u_{11}(0; \Theta)/\sigma_y^2(\Theta)U^2$ decreases with increasing $\eta$ (Figure 9.1).

For given length-scale ratios $\rho$ and $\eta$, and when the mean flow is parallel to the $x_1$ axis [i.e., $\mathbf{J} = (J_1, 0, 0)$], in equation (9.5) the first term on its right-hand side (RHS) vanishes for $i, j = 2, 3$, and both $\partial C_{hh}(\boldsymbol{\xi}; \Theta)/\partial \xi_1$ and $C_{hy}(\boldsymbol{\xi}; \Theta)$ vanish at $\xi_1 = 0$ and $\xi_1 \to \infty$. Furthermore, the last is antisymmetric in $\xi_1$. Consequently, both $\partial C_{hh}(\boldsymbol{\xi}; \Theta)/\partial \xi_1$ and $C_{hy}(\boldsymbol{\xi}; \Theta)$ drop out of equation (9.5) when the integration of equation (9.5) over $\xi_1$ is performed up to $\xi_1 \to \infty$. Hence $\int_0^\infty u_{11}(\boldsymbol{\xi}; \Theta)\, d\xi_1 = U^2\sigma_y^2(\Theta) I_{yv}(\Theta)$ and $\int_0^\infty u_{ii}(\boldsymbol{\xi}; \Theta)\, d\xi_1 = 0$ $(i = 2, 3)$ irrespective of the values of $\rho$ or $\eta$. In the case of $u_{11}(\boldsymbol{\xi}; \Theta)$, because the integral $\int_0^\infty u_{11}(\xi; \Theta)\, d\xi_1$ must equal $U^2\sigma_y^2(\Theta) I_{yv}(\Theta)$, the decrease in $u_{11}(0; \Theta)$ with increasing $\rho$ and $\eta$ must be compensated by an increase in the separation distance over which $u_1(x_1)$ and $u_1(x_1 + \xi_1)$ are positively correlated. Consequently, the persistence of $u_{11}(\xi_1; \Theta)$ increases with increasing $\rho$ and decreasing $\eta$ (Figure 9.1). On the other hand, in the case of $u_{ii}(\boldsymbol{\xi}; \Theta)$ $(i = 2, 3)$, because the integral $\int_0^\infty u_{ii}(\boldsymbol{\xi}; \Theta)\, d\xi_1$ must vanish, the increase in the separation distance over which $u_i(x_1)$ and $u_i(x_1 + \xi_1)$ are positively correlated with increasing $\eta$ and $\rho$ must be compensated by an increase in the separation distance over which $u_i(x_1)$ and $u_i(x_1 + \xi_1)$ are negatively correlated. Hence, the persistence of $u_{ii}(\xi_1; \Theta)$ $(i = 2, 3)$ increases with increasing $\eta$ and $\rho$ (Figure 9.2).

As will be shown later, the role played by the length-scale ratios $\rho = I_h/I_v$, $\eta = \lambda/I_v$, and $L/I_v$ in the unsaturated flow (Figures 9.1 and 9.2) manifests itself in the transport. This is so because when pore-scale dispersivity is neglected, the movement and spreading of passive solutes at the field scale are determined by the velocity field.

## 9.5 Transport of Passive Solutes

### *9.5.1 Spatial Moments of the Solute Body*

Consider the advection-dominated transport of a passive, conservative solute by steady flow in a heterogeneous, partially saturated porous formation with mean water saturation $\Theta$. It is assumed that the solute does not exchange with the solid phase of the soil and does not undergo degradation (chemical or biological) or volatilization. The finite body of the passive solute is visualized as being made up of many indivisible particles, initially located at $\mathbf{x} = \mathbf{a}$, within the volume $V_0$. The resident solute concentration $c$ (defined as the mass of the solute per unit volume of aqueous solution) is related to the particle displacement $\mathbf{X} = \mathbf{X}(t, \mathbf{a}; \Theta)$ through the relationships (Dagan, 1989)

$$c(\mathbf{x}, t; \Theta) = (m/n\Theta)\delta[\mathbf{x} - \mathbf{X}(t, \mathrm{a}; \Theta)] \tag{9.7}$$

where $m$ is the solute mass, and $\delta$ is the Dirac delta function.

Because of the complex heterogeneity of the formation's hydraulic properties, however, the point values for $c$ are subject to considerable uncertainty. Consequently, we focus our attention on larger-scale, integrated measures of the solute transport, such as the spatial moments of the distribution of the point values for $c$, because these moments are subject to a much lesser degree of uncertainty than are the point values.

For a finite body of a passive solute, with initial condition $c(\mathbf{x}, t) = c_0$ in an initial volume $V_0$ and water saturation $\Theta$, the spatial moments of the distribution of $c$ are given (Dagan, 1989, 1991) as

$$M = \int n\Theta c \, d\mathbf{x} = n\Theta c_0 V_0, \tag{9.8}$$

$$\mathbf{R} = \frac{1}{M} \int n\Theta c\mathbf{x} \, d\mathbf{x} = \frac{1}{V_0} \int_{V_0} \mathbf{X}(t, \mathbf{a}; \Theta) \, d\mathbf{a}$$

$$\begin{aligned} S_{ij} &= \frac{1}{M} \int n\Theta(x_i - R_i)(x_j - R_j)c \, d\mathbf{x} \\ &= \frac{1}{V_0} \int_{V_0} [X_i(t, \mathbf{a}; \Theta) - R_j][X_j(t, \mathbf{a}; \Theta) - R_j] \, d\mathbf{a} \end{aligned} \tag{9.9}$$

where $M$ is the total mass of the solute, $\mathbf{R}(t; \Theta) = (R_1, R_2, R_3)$ is the coordinate of the centroid of the solute plume, and $S_{ij}(t; \Theta)$ $(i, j = 1, 2, 3)$ are second spatial moments, proportional to the moments of inertia of the plume.

Under ergodic conditions for the spatial moments, assumed to prevail if $\ell_i \gg I_{pi}$ $(i = 2, 3)$, $\mathbf{R} \approx \langle \mathbf{R} \rangle = \mathbf{a} + \mathbf{U}t$ (where $\mathbf{x} = \mathbf{a}$ is the centroid of $V_0$), and $S_{ij}(t; \Theta) \approx \langle S_{ij}(t; \Theta) \rangle = S_{ij}(0; \Theta) + X_{ij}(t, 0; \Theta)$. Hence, under these conditions, the one-particle-trajectory statistical moments to be discussed in the next section characterize the spatial moments of the solute body.

### 9.5.2 One-Particle-Trajectory Statistics in Ergodic Transport

Adopting a general Lagrangian framework (e.g., Dagan, 1984), the transport is described in terms of the motions of indivisible solute particles that are convected by the fluid. Neglecting pore-scale dispersion, the trajectory of a solute particle is related to the velocity field by the fundamental kinematic relationship

$$\frac{d\mathbf{X}}{dt} = \mathbf{u}(\mathbf{X}) \quad \text{for } t > 0, \qquad \mathbf{X} = \mathbf{a} \quad \text{for } t = 0 \tag{9.10}$$

where $t$ is time, $\mathbf{X} = \mathbf{X}(t, \mathbf{a})$ is the trajectory of a particle that at $t = 0$ is at $\mathbf{X} = \mathbf{a}$, and $\mathbf{u}$ is the Eulerian velocity vector.

In order to relate the statistical moments of the PDF of the particle displacement to those of the velocity PDF, the following assumptions are employed: (1) Lagrangian and Eulerian stationarity and homogeneity of the velocity field, (2) given statistics of the velocity field, (3) small fluctuations in particle displacements about the mean trajectory, and (4) large Peclet numbers (i.e., local dispersion is omitted).

Under the aforementioned assumptions, for ergodic conditions and for a particle of a solute injected into the flow field at time $t = 0$ and location $\mathbf{x} = \mathbf{a}$, the solution of equation (9.10) is

$$\mathbf{X}(t, \mathbf{a}; \Theta) = \mathbf{a} + \mathbf{U}t + \int_0^t \mathbf{u}(\mathbf{a} + \mathbf{U}t'; \Theta)\, dt' \tag{9.11}$$

For fixed $\mathbf{a}$, the first two moments of $\mathbf{X}$ [equation (9.11)], the particle-displacement mean $\langle \mathbf{X}(t; \Theta) \rangle$ and, by a first-order approximation in the velocity variance, the particle-displacement covariance $X_{ij}(t; \Theta) = \langle X_i'(t; \Theta) X_j'(t; \Theta) \rangle$ $(i, j = 1, 2, 3)$ at time $t$ and water saturation $\Theta$, where $\mathbf{X}' = \mathbf{X} - \langle \mathbf{X} \rangle$ is the fluctuation, are given (Dagan, 1989) by

$$\langle \mathbf{X}(t; \Theta) \rangle = \mathbf{U}(\Theta)t \tag{9.12a}$$

$$X_{ij}(t; \Theta) = 2 \int_0^t (t - \tau) u_{ij}(\mathbf{U}\tau; \Theta)\, d\tau \tag{9.12b}$$

where $\mathbf{U} = (U_1, U_2, U_3)$ and $u_{ij}$ $(i, j = 1, 2, 3)$ are given by equations (9.4) and (9.5), respectively, and for simplicity the particle location at $t = 0$ is taken here as $\mathbf{a} = 0$.

Equations (9.12a) and (9.12b) are of a general nature, independent of flow conditions (i.e., whether the flow is saturated or unsaturated), consistent with the linearization of the flow equation, and they have been used in the past to model transport under saturated groundwater flow (e.g., Dagan, 1984, 1988) and vadose-zone flow (e.g., Russo, 1993; 1995b). It should be emphasized, however, that equation (9.12b) is restricted to flow regimes associated with relatively large Peclet numbers $UI_1/D_L$, $UI_2/I_1D_T$, and $UI_3/I_1D_T$ (Dagan, 1988), where $D_L$ and $D_T$ are the longitudinal and transverse components of the pore-scale-dispersion tensor.

For an MVN Eulerian velocity field, that restriction also applies to $\mathbf{X}(t; \Theta)$, the PDF of which is completely defined by the mean $\langle \mathbf{X}(t; \Theta) \rangle$ and the covariance $X_{ij}(t; \Theta)$ $(i, j = 1, 2, 3)$. From a physical point of view, the ratios $X_{ij}/2t$ can be regarded as the apparent dispersion coefficients that will lead to the same $X_{ij}$ as the actual time-dependent coefficients in the solution of the convection–dispersion equation with constant coefficients. For $t \to \infty$, these ratios tend to the macrodispersion coefficients obtained from

$$D_{ij}(t; \Theta) = \frac{1}{2}\left[\frac{dX_{ij}(t; \Theta)}{dt}\right] \qquad (i, j = 1, 2, 3) \tag{9.13}$$

Components of the displacement covariance tensor $X_{ij}$ [equation (9.12b)] and the macrodispersion tensor $D_{ij}$ $(i, j = 1, 2, 3)$ [equation (9.13)] for mean flow in an arbitrary direction with respect to the principal axes of the formation heterogeneity were calculated (Russo, 1995b) by integrating the respective components of the

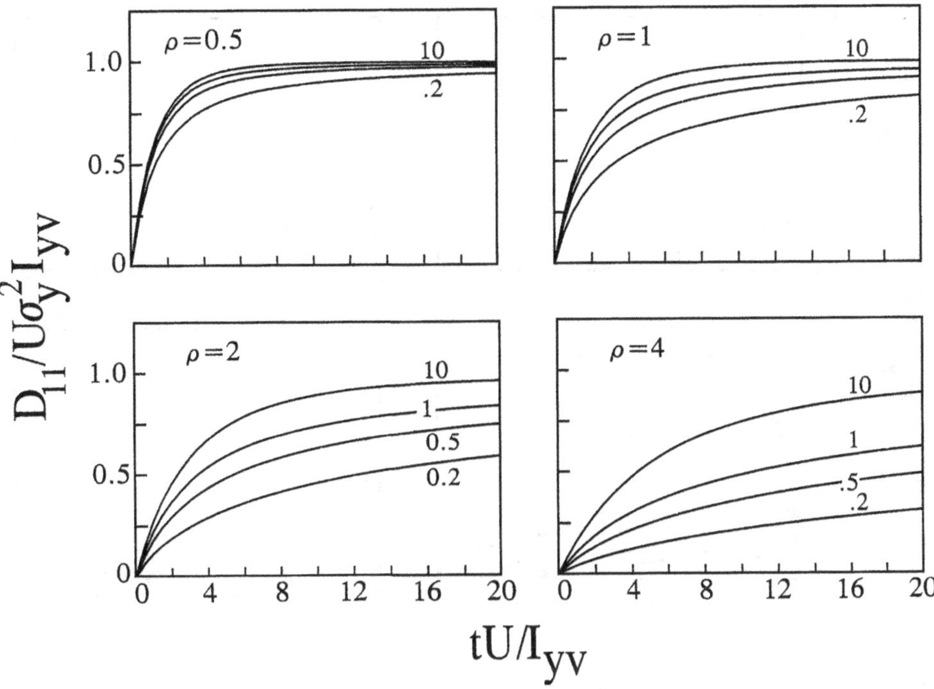


Figure 9.3. Longitudinal component ($D'_{11}$) of the scaled macrodispersion coefficient tensor $D'_{ij} = D_{ij}/U\sigma_y^2 I_{yv}$ as a function of the scaled travel time $t' = tU/I_{yv}$ for selected values of $\rho = I_h/I_v$ and $\eta = \lambda/I_v$ (denoted by the numbers labeling the curves).

velocity covariance tensor [equation (9.5)] along a line parallel to the mean gradient. In the following, some of the results of those calculations are presented and discussed.

Scaled forms of the principal components of equation (9.13), independent of water saturation $D'_{ii} = D_{ii}/\sigma_y^2(\Theta)I_{yv}(\Theta)U$ $(i = 1, 2, 3)$, are illustrated graphically in Figures 9.3 and 9.4 as functions of the scaled travel time $t' = tU/I_{yv}$, for the case in which $\mathbf{J}$ coincides with the $x_1$ axis, for selected values of $\rho = I_h/I_v$ and $\eta = \lambda/I_v$. For given $\rho$ and $\eta$, $D'_{ii}$ $(i = 1, 2, 3)$ describe a continuous transition from a convection-dominated transport process to a convection-dispersion transport process; that is, $D'_{11}$ is a monotonically increasing function of $t'$, approaching a constant value at the large-$t'$ limit, $t' \to \infty$, whereas $D'_{ii}$ $(i = 2, 3)$ are nonmonotonic functions of $t'$, vanishing at both $t' \to 0$ and $t' \to \infty$. In other words, for a short time, when the solute first invades the flow system, the principal components of the particle-displacement covariance tensor $X_{ii}$ $(i = 1, 2, 3)$ are proportional to the square of the travel time $t'$, and those of $D'_{ii}$ $(i = 1, 2, 3)$ are proportional to $t'$. After a relatively long time, when the solute body has traveled a few tens of log $K$ correlation scales, however, $X_{11}$ increases linearly with $t'$, whereas the transverse components $X_{ii}$ $(i = 2, 3)$ approach constant asymptotic values. Hence, $D'_{ii}$ $(i = 1, 2, 3)$ tend to their asymptotic values, with $D_{11}$ being the only nonzero component. For given

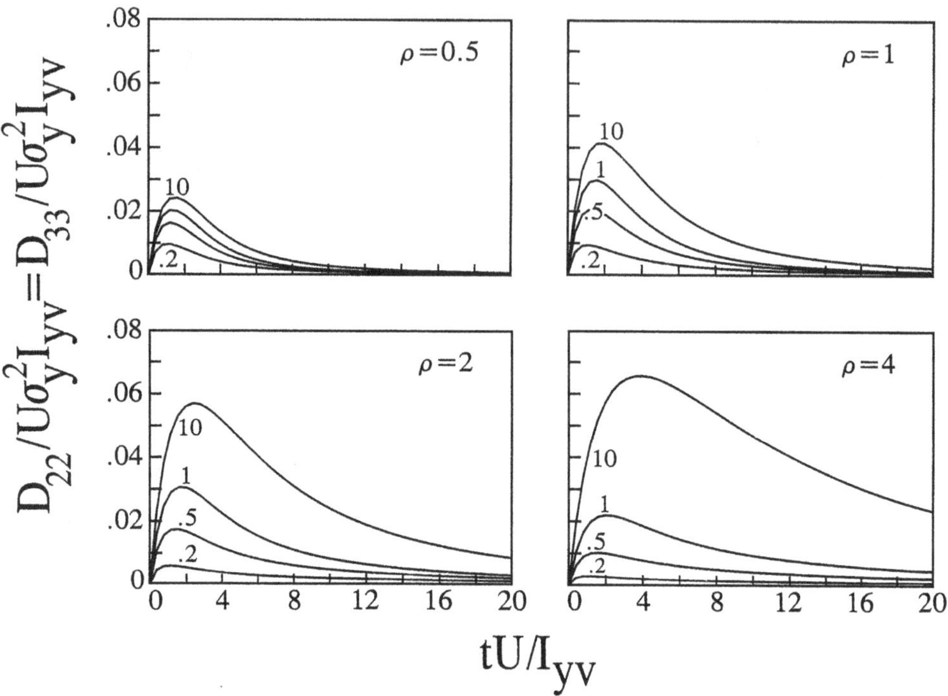


Figure 9.4. Transverse components ($D'_{33} = D'_{22}$) of the scaled macrodispersion coefficient tensor $D'_{ij} = D_{ij}/U\sigma_y^2 I_{yv}$ as functions of the scaled travel time $t' = tU/I_{yv}$ for selected values of $\rho = I_h/I_v$ and $\eta = \lambda/I_v$ (denoted by the numbers labeling the curves).

$\rho$ and $\eta$, the behavior of $D'_{ii}(t')$ $(i = 1, 2, 3)$ (Figures 9.3 and 9.4) stems from the shapes of the respective components of the velocity covariance in the direction of the mean flow (Figures 9.1 and 9.2). In other words, along the $x_1$ axis, because of the continuity considerations discussed in Section 9.4, $u'_{11}(\xi_1)$ is positive and has a finite correlation scale, whereas $u'_{ii}(\xi_1)$ $(i = 2, 3)$ must exhibit negative values as well and have zero correlation scales.

The evolution of $X'_{11}$ at a rate faster than linearly in $t'$ persists longer as $\rho$ increases and as $\eta$ decreases. Consequently, when the flow is parallel to $x_1$, $D'_{11}$ approaches its asymptotic value (unity) faster as $\rho$ decreases and as $\eta$ increases. On the other hand, the rate at which $X'_{ii}$ $(i = 2, 3)$ approach their asymptotic values increases as both $\rho$ and $\eta$ decrease. Hence, $D'_{ii}$ $(i = 2, 3)$ approach their asymptotic values (zero) faster as both $\rho$ and $\eta$ decrease. The behavior of the principal components of equation (9.13) (Figures 9.3 and 9.4) demonstrates the mutual influences of the formation heterogeneity and the capillary forces on solute spreading under unsaturated flow. For fixed $I_{yv}$, solute spreading in the longitudinal direction is diminished in formations in which the lateral extent of flow barriers, normal to the mean flow, is large (large $\rho$) and the capillary forces are small (small $\eta$). On the other hand, the enhanced transverse solute spreading due to increasing $\rho$ is balanced by decreasing $\eta$.

### 9.5.3 Nonergodic Transport

In the case in which the length-scale ratio $\ell_i/I_{pi}$ $(i = 2, 3)$ is not large enough to ensure ergodicity, the spatial moments of the solute plume [equations (9.8) and (9.9)] in each realization may differ from their ensemble mean. To characterize transport under such circumstances, $\mathbf{R}$ [equation (9.8)] and $S_{ij}$ $(i, j = 1, 2, 3)$ [equation (9.9)] are regarded as random variables, represented by their statistical moments (Dagan, 1989; 1990). Under the conditions specified earlier, and for a stationary velocity field, the first two moments of the centroid trajectory $\mathbf{R}$, the mean $\langle \mathbf{R}(t; \Theta) \rangle$, and the covariance tensor $R_{ij}(t, \ell; \Theta) = \langle R'_i(t; \Theta) R'_j(t; \Theta) \rangle$ $(i, j = 1, 2, 3)$, where $\mathbf{R}' = \mathbf{R} - \langle \mathbf{R} \rangle$ is the fluctuation, are given (Dagan, 1991) as

$$\langle \mathbf{R}(t; \Theta) \rangle = \mathbf{a} + \mathbf{U}(\Theta) t \tag{9.14a}$$

$$R_{ij}(t, \ell; \Theta) = \frac{1}{V_0^2} \int_{V_0} \int_{V_0} X_{ij}(t, \mathbf{b}; \Theta)\, d\mathbf{a}'\, d\mathbf{a}'' \tag{9.14b}$$

where $X_{ij}(t, \mathbf{b}; \Theta)$ $(i, j = 1, 2, 3)$ is the covariance tensor of the trajectories of two particles originating from points $\mathbf{a}'$ and $\mathbf{a}''$ in $V_0$, and $\mathbf{b} = \mathbf{a}' - \mathbf{a}''$.

In a similar manner, the expected value of $S_{ij}$ $(i, j = 1, 2, 3)$ from equation (9.9) is given (Kitanidis, 1988; Dagan, 1990, 1991) by

$$\langle S_{ij}(t; \Theta) \rangle = S_{ij}(0; \Theta) + X_{ij}(t, 0; \Theta) - R_{ij}(t, \ell; \Theta) \tag{9.15}$$

where $S_{ij}(0; \Theta)$ is the initial second spatial moment of the plume.

The fundamental relationship (9.15) shows that the trajectory covariance with respect to the mean centroid $X_{ij}(t, 0; \Theta)$ is equal to the sum of the covariance with respect to the realization centroid $\langle S_{ij}(t, \Theta) \rangle$ and the covariance of the centroid trajectory $R_{ij}(t, \ell; \Theta)$. Using equation (9.15), the expected value of the effective macrodispersion tensor $D_{ij}$ $(i, j = 1, 2, 3)$, defined by $D_{ij} = \frac{1}{2}(d\langle S_{ij} \rangle / dt)$, is given (Dagan, 1991) by

$$\begin{aligned} D_{ij}(t, \ell; \Theta) &= \frac{1}{2} \frac{dX_{ij}(t, 0; \Theta)}{dt} - \frac{1}{2} \frac{dR_{ij}(t, \ell; \Theta)}{dt} \\ &= D_{ij}(t; \Theta) - \frac{1}{2} \frac{dR_{ij}(t, \ell; \Theta)}{dt} \end{aligned} \tag{9.16}$$

where $D_{ij}(t; \Theta)$ is given by equation (9.13).

Under ergodic conditions, $D_{ij}(t, \ell; \Theta) \to D_{ij}(t; \Theta)$. The tendency to ergodicity can be assessed according to the ratio $\langle S'_{ij}(t; \Theta)^2 \rangle / \langle S_{ij}(t; \Theta) \rangle^2$, where $S'_{ij}(t; \Theta) = S_{ij}(t; \Theta) - \langle S_{ij}(t; \Theta) \rangle$ $(i, j = 1, 2, 3)$. Its computation in terms of $X_{ij}$, however, is very complicated, even if $\mathbf{X}$ is assumed to be Gaussian (Dagan, 1990). Following Dagan (1991), therefore, the tendency of $D_{ij}(t, \ell; \Theta)$ [equation (9.16)] to

$D_{ij}(t;\Theta)$ $(i, j = 1, 2, 3)$ [equation (9.13)] can be used as a measure of the approach to ergodic conditions.

Inasmuch as $D_{ij}(t;\Theta)$ was evaluated (Figures 9.3 and 9.4), evaluation of $D_{ij}(t, \boldsymbol{\ell};\Theta)$ [equation (9.16)] requires only the evaluation of $R_{ij}(t, \boldsymbol{\ell};\Theta)$, which in turn is related to the covariance of the two particles' trajectories, $X_{ij}(t, \mathbf{b};\Theta)$. From the first-order approximation of $\mathbf{X}'$ [equation (9.11)], $X_{ij}(t, \mathbf{b};\Theta)$ $(i, j = 1, 2, 3)$ is given (Dagan, 1991) as

$$X_{ij}(t, \mathbf{b};\Theta) = \int_0^t \int_0^t u_{ij}[U(t' - t'') + b_1, b_2, b_3;\Theta]\, dt'\, dt'' \tag{9.17a}$$

That is,

$$\frac{dX_{ij}(t, \mathbf{b};\Theta)}{dt} = \int_0^t \{u_{ij}[U(t - t') + b_1, b_2, b_3;\Theta] + u_{ij}[U(t' - t) + b_1, b_2, b_3;\Theta]\}\, dt' \tag{9.17b}$$

The effective macrodispersion tensor $D_{ij}(t, \boldsymbol{\ell};\Theta)$ $(i, j = 1, 2, 3)$ is obtained by substituting equation (9.17a) into equation (9.14b), computing the first derivative of the resultant with respect to time, and substituting into equation (9.16). In the following, we consider the case in which the extent of the solute source in the longitudinal direction is negligibly small compared with its extent in the transverse directions. For a square solute-input zone defined by $0 < x_2 < \ell_2$ and $0 < x_3 < \ell_3$, $D_{ij}(t, \boldsymbol{\ell};\Theta)$ $(i, j = 1, 2, 3)$ is given (Russo, 1997c) by

$$D_{ij}(t, \ell_2, \ell_3;\Theta) = \int_0^t u_{ij}(Ut', 0, 0;\Theta)\, dt' - \frac{4}{\ell_2^2 \ell_3^2} \int_0^{\ell_2} \int_0^{\ell_3} \int_0^t (\ell_2 - b_2) \times (\ell_3 - b_3) u_{ij}(Ut', b_2, b_3;\Theta)\, dt'\, db_2\, db_3 \tag{9.18}$$

Note that as in the case of ergodic transport, when the mean flow is parallel to the $x_1$ axis [i.e., $\mathbf{J} = (J_1, 0, 0)$], the off-diagonal components of equation (9.18) vanish. Using equation (9.18), the limit values for the principal components of $D_{ij}(t, \ell;\Theta)$ can be easily estimated. With $\ell = \ell_2 = \ell_3$, at the small-$\ell$ limit, $\ell \to 0$, the second term on the RHS of equation (9.18) equals the first one, and $D_{ii}(t, \ell;\Theta)$ $(i = 1, 2, 3)$ vanish. At the large-$\ell$ limit, $\ell \to \infty$, the second term on the RHS of equation (9.18) vanishes, and $D_{ii}(t, \ell;\Theta)$ approaches $D_{ii}(t;\Theta)$. In other words, if the transverse extent of the solute body is sufficiently large, the solute body tends to disperse according to the ergodic limit. As was discussed by Dagan (1991), this stems from the fact that particles making up the solute body that move along remote streamlines no longer have correlated trajectories.

The effective-macrodispersion for mean flow parallel to the $x_1$ axis was calculated (Russo, 1997c) by integrating the RHS of equation (9.18) by means of the velocity covariance tensor [equation (9.5)]. In the following, some of the results of those calculations are presented and discussed. The principal components of the nondimensional effective-macrodispersion tensor [i.e., $D'_{ii}(t', \ell) = D_{ii}(t, \ell; \Theta)/\sigma_y^2(\Theta) I_{yv}(\Theta) U$ $(i = 1, 2, 3)$], independent of $\Theta$, are depicted in Figures 9.5 and 9.6 as functions of the dimensionless travel time $t' = tU/I_{yv}$ for selected values of the length-scale ratios $\delta = \ell/I_h = \ell/(\rho I_v)$, $\rho = I_h/I_v$, and $\eta = \lambda/I_v$. It is clearly demonstrated in these figures that for given $\rho$, $\eta$, and $t'$, $D'_{ii}(t', \ell)$ $(i = 1, 2, 3)$ are controlled by the length scale of the solute body in the transverse directions, relative to the length scale of the formation heterogeneity in these directions (i.e., by the ratio $\delta = \ell/I_h$). In other words, the principal components of equation (9.18) can reach their ergodic limits [i.e., $D_{ii}(t', \ell; \Theta)/D_{ii}(t'; \Theta) \to 1$ $(i = 1, 2, 3)$] if $\delta$ is sufficiently large ($\delta \gg 10$). For smaller values of $\delta$, $D_{ii}(t', \ell; \Theta)$ $(i = 1, 2, 3)$ decrease with decreasing $\delta$. At the small-$\delta$ limit, $\delta \to 0$, the solute body degenerates into an indivisible particle that does not disperse at all [i.e., $D_{ii}(t', \ell; \Theta) = 0$ $(i = 1, 2, 3)$]. Note that because $d[\langle S_{ii}(t, \ell; \Theta)\rangle]/dt + d[R_{ii}(t, \ell; \Theta)]/dt$ is constant [equation (9.16)], the decay of $D_{ii}(t', \ell; \Theta)$ with decreasing $\delta$ means that the variability in velocity manifested in the centroid trajectory compensates for the reduction in $D_{ii}(t, \ell; \Theta)$ $(i = 1, 2, 3)$.

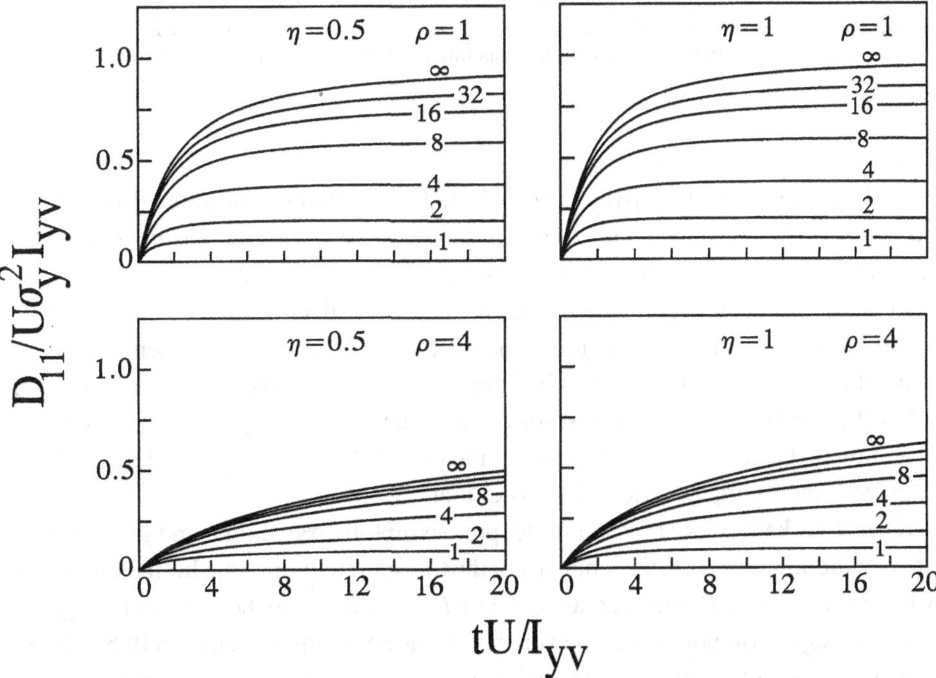


Figure 9.5. Longitudinal component ($D'_{11}$) of the scaled macrodispersion tensor $D'_{ij} = D_{ij}/U\sigma_y^2 I_{yv}$ as a function of the scaled travel time $t' = tU/I_{yv}$ for various values of the scaled solute-body initial size $\delta = \ell/I_h$ (denoted by the numbers labeling the curves) for selected values of $\rho = I_h/I_v$ and $\eta = \lambda/I_v$.

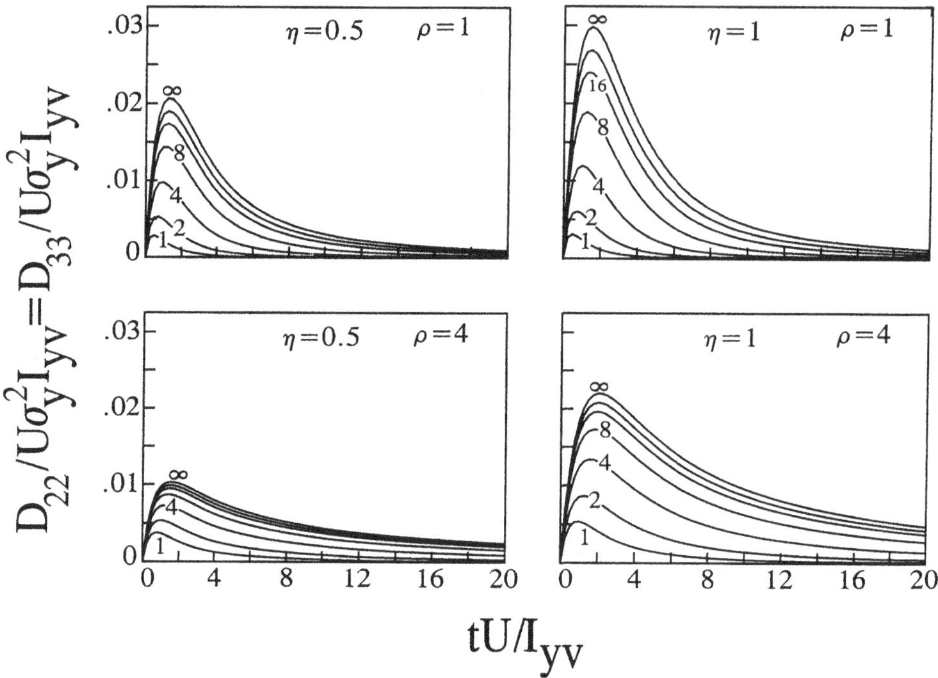


Figure 9.6. Transverse components ($D'_{33} = D'_{22}$) of the scaled macrodispersion tensor $D'_{ij} = D_{ij}/U\sigma_y^2 I_{yv}$ as functions of the scaled travel time $t' = tU/I_{yv}$ for various values of the scaled solute-body initial size $\delta = \ell/I_h$ (denoted by the numbers labeling the curves) for selected values of $\rho = I_h/I_v$ and $\eta = \lambda/I_v$.

As the counterparts of $D'_{ii}(t', \ell)$ $(i = 1, 2, 3)$ for ergodic conditions (Figures 9.3 and 9.4), for given finite $\ell$ and $t'$, $D'_{11}(t', \ell)$ decreases with increasing $\rho$ and decreasing $\eta$ (Figure 9.5), and the increase in $D'_{ii}(t', \ell)$ $(i = 2, 3)$ with increasing $\rho$ is compensated by decreasing $\eta$ (Figure 9.6). For given $\rho$ and $\eta$ and finite pre-asymptotic $t'$, the longitudinal component of equation (9.18) tends to reach its ergodic limit with increasing $\ell$ (Figure 9.5) faster than the transverse components of equation (9.18) (Figure 9.6). This stems from the fact that for given $\rho$ and $\eta$ and a given separation distance in the direction of the mean flow $\xi_1 = x_1 - x'_1$, $u'_{ii}(\boldsymbol{\xi})$ $(i = 2, 3)$ decay with increasing $\xi_2$ (and/or $\xi_3$) faster than $u'_{11}(\boldsymbol{\xi})$.

When $I_v$ is kept constant, for a finite, pre-asymptotic travel time and given lateral extent of the solute body $\ell$, the tendency of the transport to its ergodic limit is enhanced by increasing $I_h = \rho I_v$ and decreasing $\lambda = \eta I_v$. This is due to the fact that for a given $\xi_1$, the principal components of the velocity covariance tensor [equation (9.5)] decay with increasing $\xi_2$ (and/or $\xi_3$) at a rate that increases with increasing $\rho$ and decreasing $\eta$. The effects of both $\rho$ and $\eta$ on the tendency of the transport to its ergodic limit, however, decrease with increasing travel time. As will be shown later, they vanish at the large time limit, at which the transport approaches Fickian behavior.

### *9.5.4 Asymptotic Macrodispersion Coefficient*

When the travel distance $L$ is large enough, compared with $I_{yv}$, a plume can reach its asymptotic Fickian behavior. The asymptotic constant-macrodispersion tensor $D_{ij}(\ell;\Theta)$ $(i, j = 1, 2, 3)$ associated with this case is calculated from equation (9.18), taking the limit as $t \to \infty$. In line with the discussion in Section 9.4, when the mean flow is parallel to the $x_1$ axis [i.e., $\mathbf{J} = (J_1, 0, 0)$] and the integration of equation (9.18) is performed up to $t \to \infty$, the only nonzero component of the resultant asymptotic macrodispersion tensor is $D_{11}(\ell;\Theta)$. Thus, integration of equation (9.18) over $t$, for $t \to \infty$ and $i = j = 1$, yields

$$D_{11}(\ell_2, \ell_3; \Theta) = U \int_0^{\infty} C_{yy}(\xi_1, 0, 0; \Theta)\, d\xi_1 - \frac{4U}{\ell_2^2, \ell_3^2} \int_0^{\ell_2} \int_0^{\ell_3} \int_0^{\infty} (\ell_2 - b_2) \times (\ell_3 - b_3) C_{yy}(\xi_1, b_2, b_3; \Theta)\, d\xi_1\, db_2\, db_3 \tag{9.19}$$

Note that upon integration over $t$, the first term on the RHS of equation (9.19) reduces to $U\sigma_y^2(\Theta)I_{yv}(\Theta)$, which in turn is identical with the asymptotic $D_{11}(\Theta)$ under ergodic transport conditions. Using equation (9.19), the limit values of $D_{11}(\ell;\Theta)$ can be easily estimated. With $\ell = \ell_2 = \ell_3$, at the small-$\ell$ limit, $\ell \to 0$, the second term on the RHS of equation (9.19) equals the first one, and $D_{11}(\ell;\Theta)$ vanishes. At the large-$\ell$ limit, $\ell \to \infty$, the second term on the RHS of equation (9.19) vanishes, and $D_{11}(\ell;\Theta)$ approaches $D_{11}(\Theta)$. In other words, also at the large time limit, for which the transport approaches Fickian behavior, the solute body tends to disperse according to the ergodic limit if the extent of the solute body in the transverse directions is sufficiently large.

However, unlike the pre-asymptotic, time-dependent $D_{11}(t, \ell; \Theta)$, which tends to $D_{11}(t;\Theta)$ with increasing $\ell$ in a manner that depends on both $\rho$ and $\eta$ (Figures 9.5 and 9.6), the constant, asymptotic $D_{11}(\ell;\Theta)$ tends to $D_{11}(\Theta)$ with increasing $\ell$, independently of both $\rho$ and $\eta$. This is illustrated in Figure 9.7, in which the scaled asymptotic macrodispersion coefficient $D'_{11}(\ell) = D_{11}(\ell;\Theta)/U\sigma_y^2(\Theta)I_{yv}$ is plotted against the ratio $I_h/\ell$. Clearly the asymptotic $D'_{11}(\ell)$ can reach its ergodic limit (i.e., $D'_{11} \to 1$) if $\delta = \ell/I_h$ is sufficiently large, independently of both $\rho$ and $\eta$.

### *9.5.5 Effect of Mean Water Saturation on Macrodispersion*

In earlier sections of this chapter we addressed the principal components of the scaled macrodispersion tensor $D'_{ij}(t') = D_{ij}(t';\Theta)/U\sigma_y^2(\Theta)I_{yv}(\Theta)$ $(i, j = 1, 2, 3)$, which are independent of water saturation $\Theta$. Before concluding this chapter, we should address the effect of $\Theta$ on $D_{ij}(t, \Theta)$, which is due to its effects on the log-unsaturated conductivity variance $\sigma_y^2(\Theta)$ and the correlation scale $I_{yv}(\Theta)$ [equation (9.6)]. Under ergodic conditions, and for given $\eta = \lambda/I_v$ and $\rho = I_h/I_v$, denoting $D^*_{ij}(t^*;\Theta^*) =$

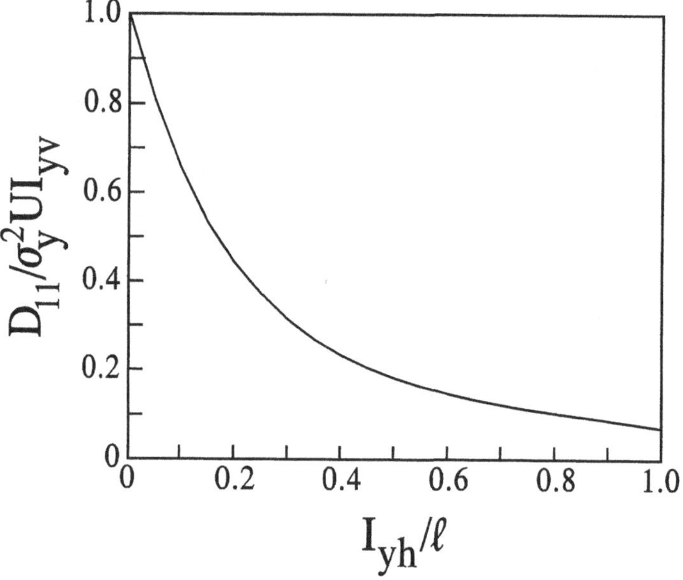


Figure 9.7. Longitudinal component ($D'_{11}$) of the scaled asymptotic macrodispersion tensor $D'_{ij} = D_{ij}/U\sigma_y^2 I_{yv}$ as a function of $\delta^{-1} = I_h/\ell$.

$UD'_{ij}(t')\sigma_y^2(\Theta^*)I_{yv}(\Theta^*)$ as the macrodispersion tensor at a reference water saturation $\Theta^*$, where $\xi^* = \xi' I_{yv}(\Theta^*)$ and $t^* = t' I_{yv}(\Theta^*)/U$, then the macrodispersion tensor at an arbitrary water saturation $\Theta$ can be expressed (Russo, 1995a,b) as

$$D_{ij}(t;\Theta) = R_1(\Theta)D^*_{ij}\{[R_2(\Theta)t^*];\Theta^*\} \tag{9.20a}$$

where

$$R_1(\Theta) = \sigma_y^2(\Theta)I_{yv}(\Theta)/\sigma_y^2(\Theta^*)I_{yv}(\Theta^*) \tag{9.20b}$$

$$R_2(\Theta) = I_{yv}(\Theta)/I_{yv}(\Theta^*) \tag{9.20c}$$

When $\eta \to \infty$, $D'_{ij}$ is essentially independent of $\eta$, and the reference water saturation can be taken as $\Theta^* = 1$, with $\sigma_y^2(\Theta^*) = \sigma_f^2$ and $I_{yv}(\Theta^*) = I_{fv}$. The coefficient $R_1(\Theta)$ determines the magnitudes of $D_{ij}(t;\Theta)$, and the coefficient $R_2(\Theta)$ determines the rate of growth of $D_{ij}(t;\Theta)$ with travel time, or how far the center of mass has to travel for $D_{ij}(t;\Theta)$ to approach their asymptotic values $D^\infty_{ij}(\Theta)$, as compared with the flow conditions at the reference water saturation $\Theta^*$.

A detailed analysis of the dependence of the coefficients $R_1$ and $R_2$ [equation (9.20)] on mean water saturation and on the statistics of the log $K_s$ and $\alpha$ fields in terms of the cross-correlation coefficient between log $K_s$ and $\alpha$, $\rho_{fa}$, and the ratios between their correlation scales $R_s = I_{av}/I_{fv}$ and variabilities $R_v = A^2\sigma_a^2/\sigma_f^2$, was presented by Russo (1995a) for the case in which $\eta \to \infty$. For given statistics of the log $K_s$ and

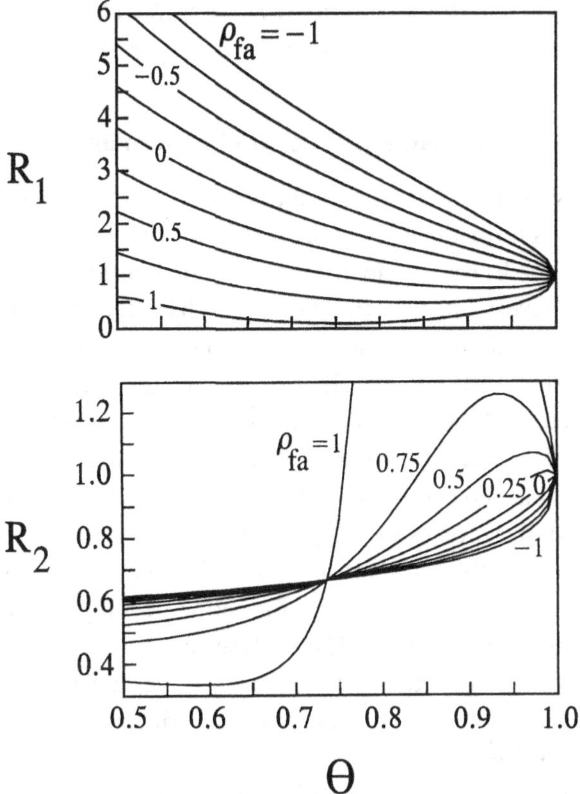


Figure 9.8. Dimensionless coefficients $R_1$ [equation (9.20b)] and $R_2$ [equation (9.20c)] as functions of water saturation $\Theta$ for various cross-correlation coefficients $\rho_{fa}$ between $\log K_s$ and $\alpha$ (denoted by the numbers labeling the curves). Calculations were performed for $\Theta^* = 1$, $\sigma_f^2 = 1$, $I_{fv} = 0.25$ m, $I_{av}/I_{fv} = 0.5$, $A^2\sigma_a^2/\sigma_f^2 = 0.5$, and $A = 0.1\,\text{m}^{-1}$.

$\alpha$ fields, the results of this analysis (Figure 9.8) reveal that for $\rho_{fa} \leq 0$, the coefficient $R_1$ is a monotonically decreasing function, and $R_2$ a monotonically increasing function, of water saturation $\Theta$. For $\rho_{fa} > 0$, both coefficients are nonmonotonic functions of $\Theta$.

This means that for a heterogeneous porous formation of given statistics, for $\rho_{fa} \leq 0$ the magnitudes of the components of $D_{ij}$ in unsaturated flow increase with decreasing $\Theta$. For $\rho_{fa} > 0$, the components of $D_{ij}$ increase with decreasing $\Theta$, when $\Theta$ is smaller than $\Theta_c(H_c)$, where $H_c$ is given (Russo, 1995a) by $H_c = 2\rho_{fa}/[\sqrt{R_v}(1 + R_s)]$. Furthermore, for $\eta \to \infty$, the components of $D_{ij}$ are larger than their counterparts in saturated flow. For $I_{av} = I_{fv}$, $R_2$ is independent of $\Theta$, and the rate at which $D_{ij}$ approach $D_{ij}^{\infty}$ is independent of $\rho_{fa}$; for $\eta \to \infty$, $D_{ij} \to D_{ij}^{\infty}$ at the same rate as their counterparts in saturated flow. For $I_{av} \neq I_{fv}$ and $\rho_{fa} \leq 0$, the rate at which $D_{ij} \to D_{ij}^{\infty}$ increases with decreasing $\Theta$ when $I_{av} < I_{fv}$, whereas the converse is true for $I_{av} > I_{fv}$. The same is true for $\rho_{fa} > 0$ when $\Theta$ is smaller than

$\Theta_c(H_c)$, where $H_c$ is given (Russo, 1995a) by $H_c = \sqrt{R_v}\{[(1-R_s^2)^2+4\rho_{fa}R_s^2]^{1/2} - (1-R_s^2)\}/2\rho_{fa}R_s^2$, and the converse is true when $\Theta$ is greater than $\Theta_c$.

## 9.6 Summary and Concluding Remarks

This chapter discusses the problem of solute transport through partially saturated, heterogeneous porous formations. First-order analysis, based on a stochastic continuum presentation of the Eulerian velocity and a general Lagrangian description of the transport, is used to investigate the effects of a few characteristic length scales of the heterogeneous transport domain on the principal components of the macrodispersion tensor in unsaturated flow. The first scale, $L$, characterizes the extent of the transport domain in the direction of the mean flow; the second scale, $\ell$, characterizes the lateral extent of the solute-inlet zone; the third scale, $\mathbf{I}_p$, characterizes the correlation structure of the heterogeneous formation; the fourth scale, $\lambda$, determines the relative magnitude of the capillary forces in unsaturated flow.

The analyses focus on the transport of a passive solute for the case in which the mean flow is vertical, perpendicular to the formation bedding, and the solute-input zone lies in the transverse directions. The analyses can be extended to the transport of reactive (sorptive) solutes (Russo, 1997a) and to the general case in which the mean gradient vector $\mathbf{J}$ does not coincide with the principal axes of the formation heterogeneity (Russo, 1995b).

It is demonstrated in this chapter that as in groundwater transport, (1) the transport can approach its ergodic limit if the characteristic length scale of the lateral extent of the solute-input zone $\ell$ is sufficiently large compared with the length scale of the heterogeneity in the transverse direction (i.e., $\ell \gg I_h$), and (2) the transport can approach Fickian behavior if the extent of the transport domain in the direction of the mean flow is sufficiently large compared with the length scale of the heterogeneity in the longitudinal direction (i.e., if $L \gg I_v$).

Distinctive to vadose-zone transport are the flow regime (unsaturated flow and the role of capillary forces) and the direction of the mean flow (vertical, perpendicular to the formation bedding). Consequently, in vadose-zone transport, both of the length-scale ratios $\rho = I_h/I_v$ and $\eta = \lambda/I_v$ influence the solute spreading. The ratio $\rho$ determines the scale of the heterogeneity in a direction perpendicular to the mean flow, relative to its counterpart in the direction of the mean flow, and the ratio $\eta$ determines the characteristic length scale of the unsaturated flow relative to that of the heterogeneity in the direction of the mean flow. The main results of the analyses presented in this chapter can be summarized as follows:

1. For both ergodic and nonergodic transport conditions and for given mean water saturation $\Theta$ and finite travel time, the magnitude of $D_{11}$, the longitudinal component of the macrodispersion tensor, diminishes when $I_h$ is sufficiently

large and $\lambda$ is sufficiently small compared with $I_v$. On the other hand, the magnitudes of the transverse components $D_{ii}$ $(i = 2, 3)$ diminish when both $I_h$ and $\lambda$ are sufficiently small compared with $I_v$.

2. For both ergodic and nonergodic transport conditions and for given $\Theta$, the tendency of $D_{11}$ to its asymptotic limit with increasing travel distance is delayed when $I_h$ is sufficiently large and $\lambda$ is sufficiently small compared with $I_v$, whereas that of $D_{ii}$ $(i = 2, 3)$ is delayed when both $I_h$ and $\lambda$ are sufficiently small compared with $I_v$.
3. Under nonergodic transport conditions and for a given $\Theta$, the tendency of the principal components of the pre-asymptotic macrodispersion tensor to their ergodic limits with increasing size of the solute-inlet zone $\ell$ is enhanced when $I_h$ is sufficiently large and $\lambda$ is sufficiently small compared with $I_v$. The effects of both $\rho = I_h/I_v$ and $\eta = \lambda/I_v$ on the tendency of the transport to its ergodic limit, however, decrease with increasing travel time and vanish at the large time limit, at which the transport approaches Fickian behavior.
4. For given $\eta$ and $\rho$, the effects of mean water saturation $\Theta$ on the magnitudes of the components of the macrodispersion tensor and on the rate at which they approach their asymptotic limits depend on the statistics of the formation properties, $\log K_s$ and $\alpha$, and on the cross-correlation between the two properties.

One of the major conclusions of this chapter is that in unsaturated flow, solute spreading in the longitudinal direction is expected to diminish in formations of large stratification and coarse-textured soil material. On the other hand, the expected enhanced transverse solute spreading caused by increasing stratification could be balanced by coarse-textured soil material. Furthermore, the travel distance required for the longitudinal component of the macrodispersion tensor to approach its asymptotic value can be exceedingly large, particularly in formations of large stratification and coarse-textured soil material with small capillary forces. In addition, nonzero transverse components of the macrodispersion tensor can persist over exceedingly large travel distances, particularly in formations of large stratification and fine-textured soil material characterized by appreciable capillary forces. Hence, in many practical situations of vadose-zone transport, the typical travel distance $L$ can be small compared with the travel distance required for the plume to reach its asymptotic, Fickian behavior.

The aforementioned results highlight the important role played by the soil parameter $\alpha$. Unfortunately, detailed experimental information on the spatial behavior of $\alpha$ (and $\log K_s$) (e.g., Ünlü et al., 1990; Russo and Bouton, 1992; White and Sully, 1992) is limited. Furthermore, controlled, field-scale transport experiments that could be used to validate existing stochastic transport models are also limited in number. There is an immediate need, therefore, for additional field studies to characterize the spatial variability in soil properties relevant to vadose-zone transport and for controlled, field-scale transport experiments (and simulations) to further validate existing stochastic transport models.

Before concluding, it must be emphasized that there are limitations to the approach described in this chapter. One of the limitations is due to the small-perturbation, first-order approximations of the velocity covariance tensor and the macrodispersion tensor. Consequently, the results are formally limited to formations with log-unsaturated conductivity variances smaller than unity. Another limitation is due to the fact that components of the velocity covariance tensor and the macrodispersion tensor presented here were derived for an unbounded flow domain associated with a constant mean head gradient **J**. For bounded flow situations, such as a flow approaching a water table, in which **J** is not constant, however, the applicability of the results of this approach will be restricted to situations in which **J** varies slowly over the log-conductivity correlation scale (i.e., for $\eta > 10$).

There are other factors that can also limit the applicability of the results of this approach: the simplifying assumptions regarding the spatial covariance of the heterogeneous formation [the exponential model (9.1) with axisymmetric anisotropy], the statistics of the relevant formation properties and the flow-controlled attributes (statistically homogeneous), the local flow (steady-state flow), the local unsaturated conductivity [equation (9.2)], the neglect of the variations in water saturation, and the transport (the neglect of the fluctuations of the particle displacement about the mean trajectory and the neglect of pore-scale dispersion).

The simplifying assumption that for a given $H$ the variations in $\Theta$ will be small enough to be excluded from the analysis of flow and transport in partially saturated formations was tested recently in a theoretical framework (Russo, in press). Results of those analyses suggest that this assumption may be satisfied in formations in which $\rho = I_h/I_v$, $C_{aa}(0)/A^2C_{ff}(0)$, and $J_i/J_1$ $(i = 2, 3)$ are relatively small, $\eta = \lambda/I_v$ and $\Theta$ are relatively large, and $\sigma_{fa}^2 < 0$.

## Acknowledgments

This is contribution number 1919-E, 1996 series, from the Agricultural Research Organization, The Volcani Center, Bet Dagan, Israel. The author is grateful to Mr. Asher Laufer for technical assistance during this study.

## References

Bresler, E., and Dagan, G. 1981. Convective and pore scale dispersive solute transport in unsaturated heterogeneous fields. *Water Resour. Res.* 17:1683–9.

Butters, G. L., Jury, W. A., and Ernst, F. F. 1989. Field scale transport of bromide in an unsaturated soil. 1. Experimental methodology and results. *Water Resour. Res.* 25:1575–81.

Dagan, G. 1982. Stochastic modeling of groundwater flow by unconditional and conditional probabilities. 2. The solute transport. *Water Resour. Res.* 18:835–48.
Dagan, G. 1984. Solute transport in heterogeneous porous formations. *J. Fluid Mech.* 145:151–77.
Dagan, G. 1988. Time-dependent macrodispersion for solute transport in anisotropic heterogeneous aquifers. *Water Resour. Res.* 24:1491–500.
Dagan, G. 1989. *Flow and Transport in Porous Formations*. Berlin: Springer-Verlag.
Dagan, G. 1990. Transport in heterogeneous porous formations: spatial moments, ergodicity and effective dispersion. *Water Resour. Res.* 26:1281–90.
Dagan, G. 1991. Dispersion of a passive solute in non-ergodic transport by steady velocity fields in heterogeneous formations. *J. Fluid Mech.* 233:197–210.
Dagan, G., and Bresler, E. 1979. Solute transport in unsaturated heterogeneous soil at field scale. 1. Theory. *Soil Sci. Soc. Am. J.* 43:461–7.
Destouni, G., and Cvetkovic, V. 1989. The effect of heterogeneity on large scale solute transport in the unsaturated zone. *Nordic Hydrol.* 20:43–52.
Destouni, G., and Cvetkovic, V. 1991. Field scale mass arrival of sorptive solute into the groundwater. *Water Resour. Res.* 27:1315–25.
Ellsworth, T. R., Jury, W. A., Ernst, F. F., and Shouse, P. J. 1991. A three-dimensional field study of solute transport through unsaturated layered porous media. 1. Methodology, mass recovery and mean transport. *Water Resour. Res.* 27:951–65.
Gardner, W. R. 1958. Some steady state solutions of unsaturated moisture flow equations with application to evaporation from a water table. *Soil Sci.* 85:228–32.
Gelhar, L. W., and Axness, C. 1983. Three-dimensional stochastic analysis of macrodispersion in aquifers. *Water Resour. Res.* 19:161–90.
Jazwinski, A. H. 1978. *Stochastic Processes and Filtering Theory.* San Diego: Academic Press.
Jury, W. A. 1982. Simulation of solute transport using a transfer function model. *Water Resour. Res.* 18:363–8.
Kitanidis, P. K. 1988. Prediction by the method of moments of transport in a heterogeneous formation. *J. Hydrol.* 102:453–73.
Lumley, J. L., and Panofsky, A. 1964. *The Structure of Atmospheric Turbulence.* New York: Wiley.
Neuman, S. P., Winter, C. L., and Newman, C. N. 1987. Stochastic theory of field-scale Fickian dispersion in anisotropic porous media. *Water Resour. Res.* 23:453–66.
Nielsen, D. R., Biggar, J. W., and Erh, K. H. 1973. Spatial variability of field-measured soil-water properties. *Hilgardia* 42:215–60.
Raats, P. A. C. 1976. Analytic solutions of a simplified equation. *Trans. ASAE* 19:683–9.
Roth, K., Jury, A. W., Flühler, H., and Attinger, W. 1991. Field-scale transport of chloride through an unsaturated soil. *Water Resour. Res.* 27:2533–41.
Russo, D. 1991. Stochastic analysis of vadose-zone solute transport in a vertical cross section of heterogeneous soil during nonsteady water flow. *Water Resour. Res.* 27:267–83.
Russo, D. 1993. Stochastic modeling of macrodispersion for solute transport in a heterogeneous unsaturated porous formation. *Water Resour Res.* 29:383–97.
Russo, D. 1995a. On the velocity covariance and transport modeling in heterogeneous anisotropic porous formations. II. Unsaturated flow. *Water Resour. Res.* 31:139–45.
Russo, D. 1995b. Stochastic analysis of the velocity covariance and the displacement covariance tensors in partially saturated heterogeneous anisotropic porous formations. *Water Resour. Res.* 31:1647–58.
Russo, D. 1997a. On the transport of reactive solutes in partially saturated heterogeneous anisotropic porous formations. *J. Contam. Hydr.* 24:345–62.

Russo, D. 1997b. On the estimation of parameters of log-unsaturated conductivity covariance from solute transport data. *Adv. Water Resour.* 20:191–205.

Russo, D. 1997c. A note on nonergodic transport of a tracer solute in partially saturated anisotropic heterogeneous porous formations. *Water Resour. Res.* 32:3623–8.

Russo, D., 1998. Stochastic analysis of flow and transport in unsaturated heterogeneous porous formation: effects of variability in water saturation. *Water Resour. Res.* (in press).

Russo, D., and Bouton, M. 1992. Statistical analysis of spatial variability in unsaturated flow parameters. *Water Resour. Res.* 28:1911–25.

Russo, D., and Bresler, E. 1981. Soil hydraulic properties as stochastic processes: I. An analysis of field spatial variability. *Soil Sci. Soc. Am. J.* 45:682–7.

Russo, D., Zaidel, J., and Laufer, A. 1994. Stochastic analysis of solute transport in partially-saturated heterogeneous soil: I. Numerical experiments. *Water Resour. Res.* 30:769–79.

Schulin, R., van Genuchten, M. T., Flühler, H., and Ferlin, P. 1987. An experimental study of solute transport in a stony field soil. *Water Resour. Res.* 23:1785–94.

Sposito, G., Jury, W. A., and Gupta, V. K. 1986. Fundamental problems in the stochastic convection–dispersion model of solute transport in aquifers and field soils. *Water Resour. Res.* 22:77–88.

Tseng, P. H., and Jury, W. A. 1994. Comparison of transfer function and deterministic modeling of area-average solute transport in heterogeneous field. *Water Resour. Res.* 30:2051–64.

Ünlü, K., Kavvas, M. L., and Nielsen, D. R. 1989. Stochastic analysis of field measured unsaturated hydraulic conductivity. *Water Resour. Res.* 25:2511–19.

Ünlü, K., Nielsen, D. R., Biggar, J. W., and Morkoc, F. 1990. Statistical parameters characterizing the spatial variability of selected soil hydraulic properties. *Soil Sci. Soc. Am. J.* 54:1537–47.

White, I., and Sully, M. J. 1992. On the variability and use of the hydraulic conductivity alpha parameter in stochastic treatments of unsaturated flow. *Water Resour. Res.* 28:209–13.

Yeh, T.-C., Gelhar, L. W., and Gutjahr, A. L. 1985a. Stochastic analysis of unsaturated flow in heterogeneous soils. 1. Statistically isotropic media. *Water Resour. Res.* 21:447–56.

Yeh, T.-C., Gelhar, L. W., and Gutjahr, A. L. 1985b. Stochastic analysis of unsaturated flow in heterogeneous soils. 2. Statistically anisotropic media with variable $\alpha$. *Water Resour. Res.* 21:457–64.

# 10

# Dilution of Nonreactive Solutes in Heterogeneous Porous Media

VIVEK KAPOOR and PETER KITANIDIS

## 10.1 Introduction

A primary objective of mathematical modeling of mass transport in groundwater is to predict the rate of dilution of a tracer or soluble contaminant. "Dilution" is the process of redistributing a solute mass over a larger volume of water, which results in a reduction in the maximum concentration. For example, decontamination is often incomplete, and a crucial question is whether or not dilution can reduce the concentration of a pollutant below the mandated level within a certain period or across a given travel path. The problem of dilution is closely related to the issue of mixing in groundwater. For example, the rate of aerobic degradation of hydrocarbons in groundwater is often controlled by the availability of oxygen and thus by the rate of mixing of the contaminated water in the plume with the ambient oxygenated water.

It is difficult to predict the rates of dilution in heterogeneous geologic formations, because those rates depend on the conditions of flow and the heterogeneity of the formation. In this chapter we shall discuss the mechanisms for dilution of nonreactive (conservative) solutes, review recent research, and present findings from mathematical modeling that illustrate how the interaction between heterogeneous advection and local dispersion controls dilution.

## 10.2 Fundamentals

The transport of dilute solutions of nonreactive solutes is governed by the advection–dispersion (A–D) equation:

$$\frac{\partial(nc)}{\partial t}+\frac{\partial(nv_ic)}{\partial x_i}-\frac{\partial}{\partial x_i}\left(nd_{ij}\frac{\partial c}{\partial x_j}\right)=0 \tag{10.1}$$

where $c$ is the concentration (mass per unit volume of fluid), $v_i$ is the seepage velocity in direction $i$, $n$ is the porosity, and $d_{ij}$ is the local dispersion (or microdispersion) coefficient in the direction pair $ij$, where $i,j$ = 1, 2, 3. Summation over an index appearing exactly twice in a term is implied in equation (10.1) and in the following equations. The velocity vector and the local dispersion tensor generally vary in space. The A–D equation embodies the principle of mass conservation and the transport mechanisms of advection and hydrodynamic dispersion.

Description of transport through a differential equation unfortunately beclouds the scale dependence of the concentration. Concentration may appear to be associated with a "point," but it is really associated with a finite volume, being defined as $m_{V_S}/V_S$, where $m_{V_S}$ is solute mass in the "support" volume of solution $V_S$. In run-of-the-mill hydrogeologic applications, the size of $V_S$ is left undefined, which hinders interpretation of results and comparison with data, given that the same equation can be used whether the concentration is defined over a volume of a few cubic centimeters or several cubic meters. For example, the concentration in a model may represent concentrations averaged over the depth of a formation, whereas the concentration measurements may have been obtained from syringe-size samples. The size of $V_S$ has special significance in the context of dilution and mixing. The larger the volume $V_S$, the more uniform the concentration and the more thorough the *apparent* dilution and mixing.

The velocity and dispersion coefficients are also scale-dependent quantities, attendant on the concentration. The velocity $v_i$ is effectively the average velocity of the fluid in the support volume, provided, of course, that certain conditions are met, as discussed by Kitanidis (1992). Much has been made of the scale dependence of dispersion coefficients, which are viewed by many practitioners as the archetypes of fudge factors in groundwater-transport modeling. How can one trust, after all, a parameter that can vary so widely depending on the scale of the problem? But, in principle, dispersion coefficients do not differ from other quantities in hydrogeology; they are all scale-dependent, though some more dramatically than others. The severe scale dependence of dispersion coefficients is nothing to fret about if one recognizes the concomitant scale dependence of the concentration. For example, a local dispersivity of 0.1 cm is not inconsistent with a macrodispersivity of 1 m, because the former may refer to a concentration with a support volume less than a liter, and the latter to a concentration with a support volume of the order of cubic meters. These concentrations may differ markedly because of the slow rate of mixing in groundwater.

Understanding the importance of the support volume is requisite for a fruitful study of dilution. In practice, the support volume should be chosen in accordance with the application that motivated the mass-transport study in the first place. To fix ideas, although it is not necessary for the analysis that follows, consider that the concentration is defined at the laboratory scale, which can be defined as corresponding to a support volume of the order of 1 liter. The reasons for this choice include the following:

1. Constitutive parameters can be measured in the laboratory.
2. In bioremediation, mixing at a small scale is important.

3. Samples taken in the field to demonstrate regulatory compliance have a small volume.
4. Contaminant exposure to a small pumping well is of concern.

With these stipulations, equation (10.1) is a valid representation of the transport processes.

The local dispersion coefficients are well-defined quantities, according to theory (e.g., Mei, 1992) and experiment; some of the experimental evidence has been summarized by Eidsath et al. (1983). For a locally homogeneous and isotropic medium, the classic Scheidegger equations (Bear, 1972) furnish the following form for the dispersion tensor:

$$d_{ij} = \left(\alpha_T (v_i v_i)^{1/2} + d^m\right)\delta_{ij} + \frac{\alpha_L - \alpha_T}{(v_i v_i)^{1/2}} v_i v_j \tag{10.2}$$

where $(v_i v_i)^{1/2}$ is the velocity magnitude, $\alpha_L$ is the longitudinal local dispersivity, $\alpha_T$ is the transverse local dispersivity, $d^m$ is the effective coefficient of molecular diffusion for the solute in the porous medium, and $\delta_{ij}$ is the Kronecker delta. The molecular-diffusion coefficient for the solute in the porous medium tends to make small contributions to local dispersion coefficients when the pore-scale Peclet number is large (i.e., the time scale characteristic of molecular diffusion over the pore size is greater than the time scale characteristic of advection).

The dimensions of the porous specimen in which hydraulic conductivity and dispersivity are inferred in laboratory tests can be employed to define an operational representative elementary volume (REV) (Figure 10.1). The hydraulic conductivity is the average pore-scale flow velocity times the porosity within an REV due to a unit gradient of hydraulic head. The seepage velocity $v_i$ is simply the average pore-scale flow velocity. The local dispersivities $\alpha_L$ and $\alpha_T$ measure mass fluxes generated by a unit gradient of concentration and normalized by the mean flow velocity. Local dispersion coefficients reveal macroscopically the dispersive effects of the heterogeneity of pore-scale flow velocity within an REV.

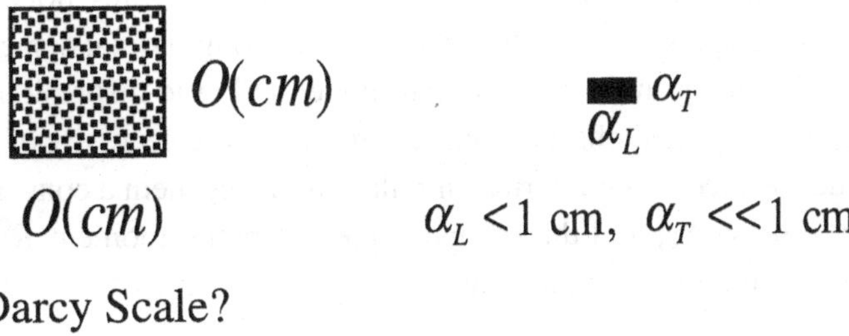


Figure 10.1. The Darcy scale, or the representative elementary volume (REV), and local dispersivities.

Because variability at a scale smaller than the REV has already been accounted for in the definitions of the local hydraulic conductivity and the local dispersion coefficients, the conductivity and the accompanying seepage velocity should not vary at or below the scale of the REV. This consideration precludes the exponential covariance model for log-conductivity, for example. That model has been used widely in stochastic groundwater modeling, but entails variability at any scale, however small. The diffusive effects of its small-scale variability are accounted for in the local dispersion coefficients. Additionally, the exponential model results in an unbounded mean squared flow gradient (Kapoor, 1997), and is, in principle, inconsistent with Darcy's equation, which is obtained through volume averaging of Stokes flow over finite samples of a porous medium, and which becomes meaningless when the averaging volume size becomes smaller than the scale of a representative pore.

Although the local dispersion coefficient can vary in space, we focus on the influence of the spatially variable seepage-velocity field $v_i(\mathbf{x})$, where $\mathbf{x}$ indicates a point in space, and a constant local dispersion coefficient $d_{ij}$ and porosity $n$, for which equation (10.1) becomes

$$\frac{\partial c}{\partial t} + \frac{\partial v_i c}{\partial x_i} - d_{ij}\frac{\partial^2 c}{\partial x_i \partial x_j} = 0 \tag{10.3}$$

## 10.3 Macrodispersion

Much of the recent work on scale problems in solute transport in aquifers has been motivated by the patent disparity between the small local dispersion coefficients measured in the laboratory and the large spreading rates observed in the field (Pickens and Grisak, 1981). Theoretical studies (Schwartz, 1977; Gelhar, Gutjahr, and Naff, 1979; Smith and Schwartz, 1980; Matheron and de Marsily, 1980; Dieulin et al., 1981; Dagan, 1982, 1984, 1987; Gelhar and Axness, 1983; Gelhar, 1986; Sudicky, 1986; Sposito and Barry, 1987; Neuman, Winter, and Newman, 1987; Kitanidis, 1988) have shown that the field-scale spreading can be ascribed to variability in hydraulic conductivity and the attendant variability in seepage velocity. This large-scale spreading, which is also known as macrodispersion, is typically quantified by macrodispersion coefficients or macrodispersivities. Although it is still debated whether or not and when it is appropriate to use large dispersivities, practitioners have long adopted large dispersivities in their models (e.g., longitudinal dispersivity about equal to the grid spacing in their numerical models). Large dispersion coefficients can be justified by empirical observations; additionally, the necessity of reducing the number of nodes needed in a numerical simulation makes them a convenient choice.

The center of mass $X_i$, and the second centered spatial moment $R_{ij}^2$ of a solute concentration distribution are defined as

$$X_i \equiv \frac{\int_V x_i c\,(\mathbf{x}, t) dV}{\int_V c\,(\mathbf{x}, t) dV}, \qquad R_{ij}^2 \equiv \frac{\int_V (x_i - X_i)(x_j - X_j) c\,(\mathbf{x}, t) dV}{\int_V c\,(\mathbf{x}, t) dV} \tag{10.4}$$

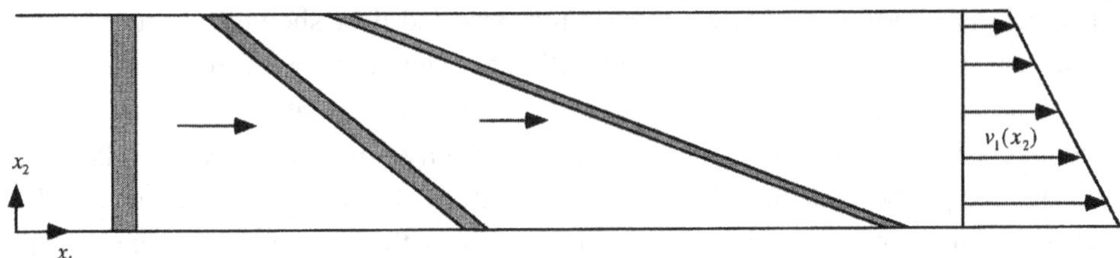


Figure 10.2a. Illustration of the influence of a stratified velocity field (the $x_1$ velocity is only a function of $x_2$) on a marked blob of fluid. A simple example is a linearly varying velocity. The influence of such a velocity in distorting the solute body is illustrated. The growth of the $x_1$ spatial second moment of the solute body, $R_{11}^2$, is not accompanied by any increase in the volume occupied by the solute body.

Figure 10.2b. Illustration of the influence of a heterogeneous divergence-free velocity field on a marked blob of fluid. Because of the volume-preserving nature of the velocity field, the advective distortions of the solute body, which can be accompanied by an increase in its spatial second moments, cannot cause any dilution (Kitanidis, 1994), irrespective of the nature of the heterogeneity and/or uncertainty of the velocity field.

where $V$ denotes the domain. The macrodispersion tensor is defined as half the time rate of change of the tensor of the second centered spatial moment of the plume: $0.5dR_{ij}^2/dt$. Consider some simple cases that will allow us to juxtapose second moments or macrodispersion coefficients and dilution rates. Consider that the $x_1$ velocity varies linearly with $x_2$ (Figure 10.2a), and assume tentatively that the local dispersion can be neglected, because the local dispersivities are small. A band of solute introduced into such a flow would be distorted as shown in Figure 10.2a. That distortion would be accompanied by an increase in the $x_1$ spatial second moment of the solute body $R_{11}^2$, and the macrodispersion coefficient $0.5dR_{11}^2/dt$ would increase linearly with time. However, the concentrations inside the distorted band would remain unchanged, and the volume occupied by the solute would not vary. Therefore, spatial second moments per se do not quantify dilution, although they may be good measures of spreading. Under more realistic conditions of spatially random heterogeneity, the complicated three-dimensional velocity variations could severely distort a blob of marked fluid, as shown in Figure 10.2b, but still the content (area or volume) of the isoconcentration lines or surfaces would remain unchanged. These simple examples demonstrate the following:

1. Spatial second moments and macrodispersion coefficients are not intrinsically related to dilution rates.
2. Pure advection, no matter how heterogeneous and/or uncertain, cannot effect dilution.

If local dispersion is neglected, then in the stratified flow shown in Figure 10.2a the vertical thickness of the solute band will keep decreasing with time under pure advection; likewise, in the case of Figure 10.2b, the isoconcentration lines will become increasingly ragged as a consequence of velocity nonuniformity. But the creation of precipitous concentration gradients confutes the hypothesis of negligible local dispersion, because the local dispersive flux is the product of concentration gradients and local dispersion coefficients. Thus, local dispersion coefficients should not be neglected, even though they may be small.

The macrodispersivity is the sum of an advection-dominated term plus the local dispersivity (e.g., Gelhar and Axness, 1983; Kitanidis, 1988); the latter term is relatively small and often is neglected. Because researchers have been preoccupied with spreading rates in heterogeneous aquifers, an incorrect impression may have been created that local dispersion can always be neglected. But, in principle, there can be no dilution without local dispersion. Therefore, the analysis of dilution should account for aquifer heterogeneity and local dispersion.

## 10.4 How to Quantify Dilution in Heterogeneous Media

To predict the maximum concentration or degree of dilution, it suffices, in principle, to solve the A–D equation at the appropriate scale. However, the use of finite-difference or finite-element methods to get an accurate solution of the A–D equation in three dimensions with longitudinal dispersivity of the order of 1 cm or less, and a much smaller transverse dispersivity, is a computational task beyond the scope of many practical applications. In practice, it is likely that groundwater modelers will continue to use grids of the order of meters and implicitly (through "numerical dispersion") or explicitly employ large dispersivities in order to meet the grid and Peclet-number requirements of numerical techniques (Ames, 1992), justified to some extent in order to account for dispersive transport due to velocity variability at subgrid scales.

The spatial second moment does not quantify the volume occupied by the solute and thus is not necessarily a good predictor of dilution. Is there an alternative means to quantify the volume of the aquifer occupied by the solute? Can we define measures that will reflect the fact that advective transport alone does not result in attenuation of concentrations, which we intuitively refer to as dilution? We shall discuss two related ways of addressing this issue: (1) through global measures of the volume occupied by solute (Kitanidis, 1994) and (2) through measures of concentration variability (Kapoor and Gelhar, 1994a,b). We shall examine the dilution index $E$, the reactor ratio $I$, the concentration variance $\sigma_c^2$, and the coefficient of variation CV $= \sigma_c/\bar{c}$ (which is the standard deviation $\sigma_c$ divided by the mean $\bar{c}$ ).

These measures can be computed for multidimensional heterogeneous porous media, with appropriately defined initial and boundary conditions, and in the case of variance, provided that a "mean" or "scaled-up" concentration is defined. After review of basic definitions, we shall explore an example that will clarify these points.

### *10.4.1 Dilution Index and Reactor Ratio*

Kitanidis (1994) defined the *dilution index* or *volume* $E$ as follows:

$$E(t) = \exp\left[-\int_V p(\mathbf{x}, t) \ln[p(\mathbf{x}, t)]\, dV\right] \tag{10.5}$$

$$p(\mathbf{x}, t) = \frac{c(\mathbf{x}, t)}{\int_V c(\mathbf{x}, t)\, dV}, \qquad \int_V c(\mathbf{x}, t) = \frac{M}{n}$$

where $M$ is the total mass of solute. The definition of $E$ is not contingent upon any specific model of hydraulic-conductivity heterogeneity or conditions of flow and transport. The dimensions of $E$ are $[\mathrm{L}^m]$, where $m$ is the number of spatial dimensions. In a macroscopic sense, $E$ serves as a measure of the volume occupied by the solute and tends to vary inversely to the maximum concentration $c_p$ (i.e., $E \propto 1/c_p$). Thus, unlike the second spatial moments, which measure spreading, $E$ is a measure of dilution.

For divergence-free velocity fields, Kitanidis (1994) demonstrated from equation (10.3) that

$$\frac{d \ln E}{dt} = d_{ij} \int_V p \frac{\partial \ln p}{\partial x_i} \frac{\partial \ln p}{\partial x_j} dV \tag{10.6}$$

That is, the rate of increase of $E$ is controlled by the product of local dispersion and concentration gradients. Therefore, for the illustrations in Figure 10.2, $E$ is invariant if local dispersion is neglected, no matter how heterogeneous and/or uncertain the flow field, in contrast to the spatial second moment of the solute body, which can increase rapidly.

In many applications, what really matters is how the actual $E$ compares with $E_{\max}$, the maximum (or "complete-dilution") $E$ possible under certain conditions (which may be implied by a given macroscopic model). Kitanidis defined the reactor ratio as

$$I = \frac{E}{E_{\max}} \tag{10.7}$$

The reactor ratio is a dimensionless number, a ratio of "reactor" volumes, that takes values between zero and unity and is an indicator of how much the peak of the mean concentration underestimates the actual peak concentration. The reactor ratio can be tailored to any specific transport problem, at any scale. For a localized solute plume with given second-moment tensor in a large domain, $E_{\max}$ is computed from a Gaussian concentration distribution with the same second moments (Kitanidis, 1994). Thierrin and Kitanidis (1994) evaluated the dilution index and the reactor ratio for two well-known field tracer tests and showed that the reactor ratio is a rough estimate of the degree of underestimation of the peak concentration when the equivalent homogeneous-medium model with macrodispersivity coefficients is used. The same analysis concluded that the maximum concentration was underestimated even after hundreds of days since the start of the experiment, indicating that considerable concentration variability persisted.

For transport in a two-dimensional domain with no-flux boundaries at the top and bottom, and for a given $x_1$ spatial second moment $R_{11}^2$ (Figure 10.3), the appropriate $E_{\max}$ corresponds to a solute concentration that is uniform over the vertical dimension and Gaussian in the $x_1$ dimension with spatial second moment $R_{11}^2$. Then $E_{\max}(t) = (2e\pi)^{1/2} H R_{11}(t)$(Kapoor and Kitanidis, 1996).

As we shall see in the specific example, if $I \ll 1$, the second moment describes only the spatial extent of the solute, not the dilution, on account of severe irregularity in the concentration distribution.

### 10.4.2 Concentration Variance and Coefficient of Variation

In practical assessments of mass transport it is presumed that models that employ coarse computational grids and large effective-dispersion coefficients compute some spatial average of the concentration field. The question that naturally arises is how the actual (i.e., defined on the support volume over which the constitutive parameters of

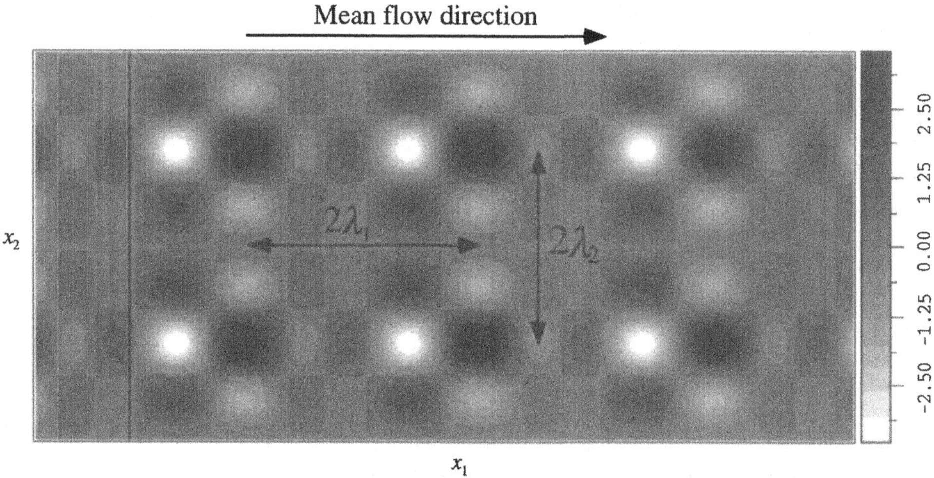


Figure 10.3a. Hydraulic-conductivity field (case II, Table 10.1). The deviations of the logarithms of the hydraulic conductivity around the mean of the logarithm, $\tilde{f}$ (Appendix), are shown. The vertical line denotes the center of mass of the initial solute concentration, which was uniform in $x_2$, with the maximum initial concentration being 1 unit. The domain spans an integral multiple of $2\lambda_2$ in $x_2$ ($H = 2m\lambda_2$; $m = 1, 2, \ldots$) and is of a large longitudinal ($x_1$) extent.

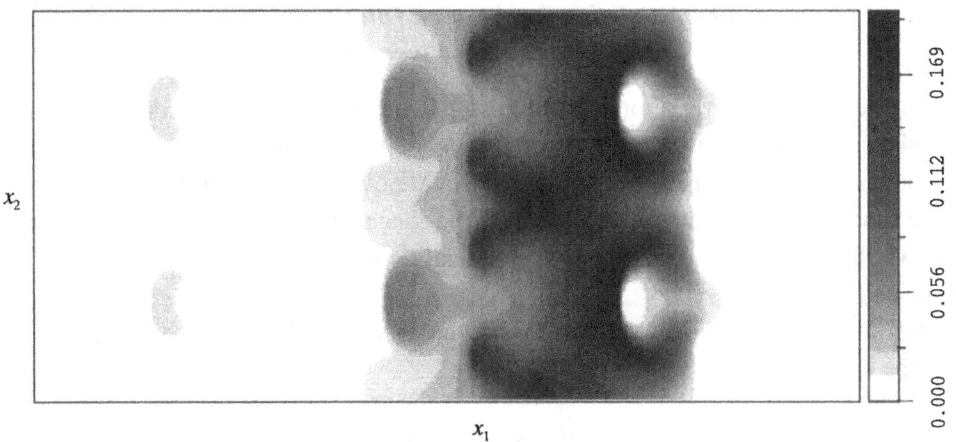


Figure 10.3b. Concentration field (case II, $\alpha/\lambda_1 = 0.000625$, $t = 70.3$ days = 11 VRT). The reactor ratio is 0.82, and the CV at the center of mass is 0.16. Initially the concentration was uniform over $x_2$, and $R_{11}$ was equal to 10 cm.

the porous medium are inferred) concentrations vary about these average or scaled-up values.

In this study, we shall focus on a specific simple example. Consider that the actual flow-and-transport domain is two-dimensional, with impermeable boundaries at the top and bottom; the actual concentration varies in two dimensions and is subject to dispersion rates given by the small local dispersion coefficients. At the same time, a practical model is one-dimensional, employs large dispersion coefficients, and calculates a depth-averaged $\bar{c}(x_1, t)$. A bar over a quantity indicates its average over the depth of the domain. The actual concentration $c(x_1, x_2, t)$ differs from $\bar{c}(x_1, t)$ because of the slow mixing in the vertical direction. A useful complement to the mean concentration would be a measure of the overall difference between the actual and mean concentrations. Such a measure is the concentration variance $\sigma_c^2(x_1, t) \equiv \overline{(c - \bar{c})^2}$. When the coefficient of variation $\mathrm{CV}(x_1, t) = \sigma_c(x_1, t)/\bar{c}(x_1, t)$ is much less than unity, the mean concentration $\bar{c}(x_1, t)$ is a good predictor of the actual concentration $c(x_1, x_2, t)$.

How does the CV change with time? What controls its spatial–temporal evolution? Does the CV increase unboundedly with time because of heterogeneous advection, or does local dispersion halt its increase and eventually cause it to decrease with time? With a few exceptions, such as the work of Kapoor and Gelhar (1994a,b) and Kapoor and Kitanidis (1996), these questions have not received attention in the subsurface-research community, as macrodispersion continues to be the main object of study, and the topic of many increasingly ornate theories. We think that it is not possible to make sound assessments of dilutions and levels of contaminant exposure without knowing the CV, no matter how well we understand the growth of the spatial second moment.

### *10.4.3 Relationship between Dilution Index and Concentration Fluctuations*

A large coefficient of variation indicates that patches of solute at high concentrations persist, and the actual maximum concentration may be significantly higher than the mean predicted by the practical model. Thus, intuitively, we expect that the dilution index and the reactor ratio should depend on the concentration coefficient of variation. As we shall see, the analysis and the numerical experiments bear out the validity of this insight.

For the two-dimensional domain considered in Section 10.4.2, the dilution index can be written as

$$E(t) = \exp\left[ - \left\| \int_0^H \frac{c(\mathbf{x}, t)}{M/n} \ln \left( \frac{c(\mathbf{x}, t)}{M/n} \right) dx_2 \right\| \right] \tag{10.8}$$

The double verticals indicate integration of its argument over $x_1$. Using the Taylor expansion about the mean concentration,

$$\ln \left( \frac{c}{M/n} \right) = \ln \left( \frac{\bar{c}}{M/n} \right) + \sum_{j=1}^{\infty} \frac{(-1)^{j+1}}{j} \left( \frac{c'}{\bar{c}} \right)^j \tag{10.9}$$

where the prime indicates the difference between the actual and the mean: $c' = c - \bar{c}$. The dilution index can be expressed as

$$E(t) = \exp\left[ - \left\| H \frac{\bar{c}}{M/n} \left[ \ln \left( \frac{\bar{c}}{M/n} \right) + \sum_{j=2}^{\infty} \left( \frac{(-1)^{j+1}}{j} + \frac{(-1)^j}{j-1} \right) \frac{\overline{c'^j}}{\bar{c}^j} \right] \right\| \right] \tag{10.10}$$

Equation (10.10) reveals the relationship between the dilution index and concentration fluctuations. If vertical averages for the different powers of the concentration perturbation $\overline{c'^j}$ can be calculated, and assuming that the series converges, the dilution index can be calculated through equation (10.10). A useful approximation, which becomes asymptotically exact as the coefficient of variation tends to zero, is obtained by retaining only the leading-order terms (Kapoor and Kitanidis, 1996):

$$E(t) = \exp\left[ - \left\| H \frac{\bar{c}}{M/n} \left[ \ln \left( \frac{\bar{c}}{M/n} \right) + \frac{(\mathrm{CV})^2}{2} \right] \right\| \right] \tag{10.11}$$

According to this result, given two concentration fields with identical $\bar{c}$ values and different CV values, the one with the larger CV will have a smaller dilution index.

Additionally, if $\bar{c}$ approaches a Gaussian distribution, then

$$I(t) = \exp\left[-\left\| H \frac{\bar{c}}{M/n} \left[\frac{(\text{CV})^2}{2}\right]\right\|\right] \tag{10.12}$$

The crucial question of how CV changes with time has been addressed by Kapoor and Gelhar (1994a,b) and Kapoor and Kitanidis (1996) and will also be discussed here.

## 10.5 Concentration Variance for a Two-Dimensional Periodic Medium

Numerical simulations will illustrate the points made previously and will show how concentration variance and dilution vary over time. Consider a two-dimensional steady-seepage field $v_i = q_i/n = v\delta_{i1} + v_i'(\mathbf{x})$, where $v$ is the $x_2$ average ($0 \leq x_2 \leq H$) velocity, and $v_i'$ is the velocity fluctuation from the mean, which is modeled as periodic, as given in the Appendix to this chapter. The local dispersion tensor is diagonal, with $d_{11} = v\alpha_L$ and $d_{22} = v\alpha_T$, where $\alpha_L$ and $\alpha_T$ are the longitudinal and transverse local dispersivities. Table 10.1 presents the details of the two cases examined here. The cases have the same periodicity ($2\lambda_1 \times 2\lambda_2$) and the same mean squared intensity of fluctuations of $\ln K$, $\sigma_f^2$. The difference is that case II has two modes of variability, and case I has only one mode. As a result, in case II, the flow field has more fine-scale structure than in case I. For simplicity, the local dispersivity is taken to be isotropic ($\alpha_L = \alpha_T = \alpha$) in the simulations.

Table 10.1. *Flow parameters*[a]

| Parameter | Symbol | Value |
|---|---|---|
| Hydraulic conductivity: | | |
| Geometric mean | $K_G$ | 1,000 cm/d |
| Standard deviation of $\ln K$ | $\sigma_f$ | 1 |
| Maximum $x_1$ half wavelength | $\lambda_1$ | 100 cm |
| Maximum $x_2$ half wavelength | $\lambda_2$ | 10 cm |
| Case I | | |
| Number of modes | $k_1^m = k_2^m$ | 1 |
| Amplitude | $a(1,1)$ | 2 |
| Case II | | |
| Number of modes | $k_1^m = k_2^m$ | 2 |
| Amplitude | $a(1,1) = a(1,2) = a(2,1) = a(2,2)$ | 1 |
| Porosity | $n$ | 0.3 |
| Mean hydraulic gradient | $J$ | 0.001 |

[a]For hydraulic conductivity and flow-field description, see the Appendix.

### 10.5.1 Fluctuation Dissipation Function $\varepsilon$

Following Kapoor and Gelhar (1994a,b), from the transport equation (10.3) we can derive one equation for the mean concentration and one for the concentration variance:

$$\frac{\partial \bar{c}}{\partial t} + v\frac{\partial \bar{c}}{\partial x_1} + \frac{\partial \overline{c'v_1'}}{\partial x_1} - v\alpha_L \frac{\partial^2 \bar{c}}{\partial x_1^2} = 0 \tag{10.13}$$

$$\frac{\partial \sigma_c^2}{\partial t} + v\frac{\partial \sigma_c^2}{\partial x_1} + \frac{\partial \overline{c'^2 v_1'}}{\partial x_1} - v\alpha_L \frac{\partial^2 \sigma_c^2}{\partial x_1^2} = -2\overline{c'v_1'}\frac{\partial \bar{c}}{\partial x_1} - 2d_{ij}\overline{\frac{\partial c'}{\partial x_i}\frac{\partial c'}{\partial x_j}} \tag{10.14}$$

Variables with bars are $x_2$ averages [e.g., $\bar{c}(x_1, t) = (1/H)\int_0^H c(x_1, x_2, t)\, dx_2$], and primed variables represent variations around the average values (e.g., $c' = c - \bar{c}$). The mean concentration equation (10.13) has been extensively analyzed, and the dispersive flux $\overline{c'v_1'}$ is the cause of "macrodispersion" of the mean concentration. Surely the concentration variance is significant whenever the dispersive flux is significant, because $\sigma_c^2 \geq (\overline{c'v_1'})^2/\overline{v_1'^2}$ (Schwarz inequality). Kapoor and Gelhar (1994a) showed that the term $\overline{c'^2 v_1'}$ in the variance equation results in macrodispersion of the concentration variance, just as $\overline{c'v_1'}$ causes macrodispersion of the mean concentration. Therefore the terms on the left-hand sides of equations (10.13) and (10.14) quantify similar transport effects for the mean and variance, respectively.

The right-hand side of the variance equation (10.14) quantifies the process of creation and destruction of variance by local dispersion. The fluctuation dissipation function $\varepsilon$ is defined as

$$\varepsilon = 2d_{ij}\overline{\frac{\partial c'}{\partial x_i}\frac{\partial c'}{\partial x_j}} = 2v\alpha_L \overline{\left(\frac{\partial c'}{\partial x_1}\right)^2} + 2v\alpha_T \overline{\left(\frac{\partial c'}{\partial x_2}\right)^2} \tag{10.15}$$

The function $\varepsilon$ represents the rate of destruction of concentration variance by local dispersion, as made evident from an interpretation of equation (10.14) as an A–D reaction equation for $\sigma_c^2$. For the zero-local-dispersion (ZLD) case (or the pure-advection case), $\varepsilon = 0$.

In the particular periodic flow field adopted here, $\partial c/\partial x_2 = 0$ at $x_2 = k\lambda_2$ ($k = 0, 1, 2, 3 \ldots$); therefore the mean squared $x_2$ derivatives of the concentration perturbation ($c' = c - \bar{c}$) are positive-definite in $\sigma_c^2$. In particular, it follows (Kapoor and Kitanidis, 1996) from the calculus of variations that

$$\overline{\left(\frac{\partial c'}{\partial x_2}\right)^2} \geq \frac{\sigma_c^2}{(\lambda_2/\pi)^2} \tag{10.16}$$

This variant of the Poincaré inequality provides a lower bound on the fluctuation dissipation rate:

$$\varepsilon \geq \frac{2v\alpha_T}{(\lambda_2/\pi)^2}\sigma_c^2 \tag{10.17}$$

This inequality underlines the cardinal role of local dispersion in the variance evolution, because it shows that no matter how small the local dispersivity, dropping the local dispersion altogether in assessing the concentration variance (ZLD case) would result in neglecting at least a first-order decay term in the variance budget, equation (10.14), and would result in a qualitatively incorrect description of variance at large times. From this bound we can derive a rigorous proof that asymptotically, $\|\sigma_c^2\| \to 0$, no matter how intense the velocity variability or how small the local dispersion coefficients. In sharp contrast, ZLD theories predict an asymptotically nondecreasing $\|\sigma_c^2\|$.

### 10.5.2 Concentration Microscales

The concentration microscales, $\Delta_i^c$, are defined as

$$\overline{\left(\frac{\partial c'}{\partial x_i}\right)^2} \equiv \frac{\sigma_c^2}{(\Delta_i^c)^2} \tag{10.18}$$

Therefore the fluctuation dissipation function can be expressed as

$$\varepsilon = \chi\sigma_c^2, \qquad \chi = \frac{2v\alpha_L}{(\Delta_1^c)^2} + \frac{2v\alpha_T}{(\Delta_2^c)^2} \tag{10.19}$$

That is, the variance decay is first-order, with coefficient $\chi$. The characteristic time scale for the decay of variance, named the variance residence time (VRT), is VRT $= \chi^{-1}$. The concentration microscales and VRT, as defined, can vary in $x_1$ and $t$. Nevertheless, for simplification, we consider the overall (spatially integrated) microscales $\hat{\Delta}_i^c$:

$$\left\|\overline{\left(\frac{\partial c'}{\partial x_i}\right)^2}\right\| \equiv \frac{\|\sigma_c^2\|}{(\hat{\Delta}_i^c)^2} \tag{10.20}$$

where the double verticals indicate integration of its argument over $x_1$. In numerical simulations (Kapoor and Kitanidis, 1996) it was observed that as the solute sampled the flow heterogeneity, the concentration microscales changed with time and approached a constant large-time asymptotic value. The discussion will focus on these large-time values. Associated with these spatially integrated concentration microscales is an estimate of the overall VRT:

$$\mathrm{VRT} \equiv \left[\frac{2v\alpha_L}{(\hat{\Delta}_1^c)^2} + \frac{2v\alpha_T}{(\hat{\Delta}_2^c)^2}\right]^{-1} \tag{10.21}$$

The significance of the large-time asymptotic concentration microscales is that at large time, the time scale over which local dispersion destroys concentration fluctuations, the VRT, approaches a constant value. It is illustrated in the results presented next that understanding VRT holds the key to understanding and developing practical methods to predict dilution in heterogeneous porous media.

## 10.6 Results

### *10.6.1 Concentration Microscales*

Figure 10.4 shows the longitudinal and transverse large-time values for the concentration microscales as functions of the ratio of the local dispersivity to the half wavelength in the direction of flow (the inverse of that ratio being interpreted as a type of Peclet number, which ranged between 200 and 1,600 in the results presented here). Some of the significant characteristics of the concentration microscales are as follows:

1. The concentration microscales reflect the anisotropy of the porous media. If $\lambda_1 \gg \lambda_2$, then $\hat{\Delta}_1^c \gg \hat{\Delta}_2^c$.

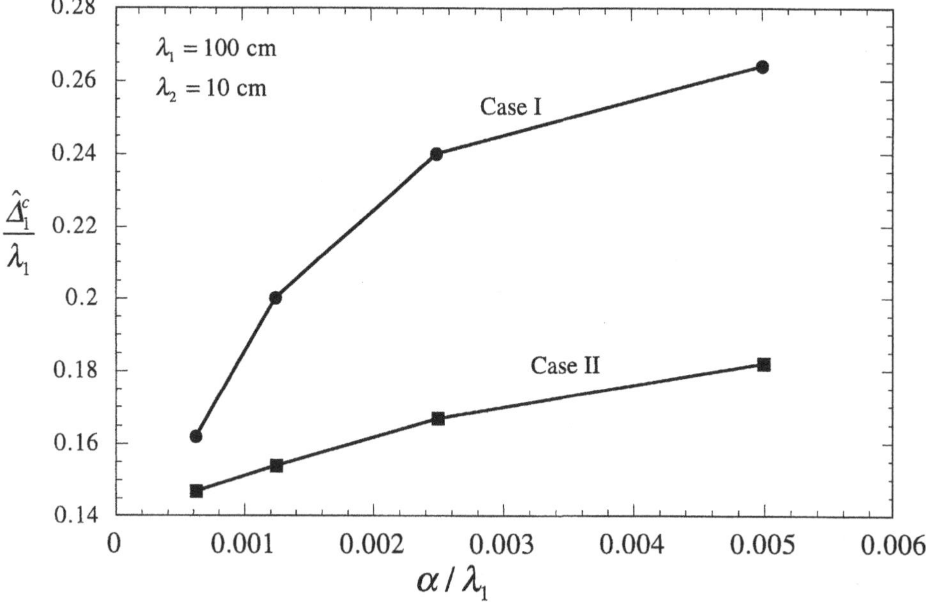


Figure 10.4a. Concentration microscale in the $x_1$ direction. Case I has less fine-scale variability than case II (Table 10.1).

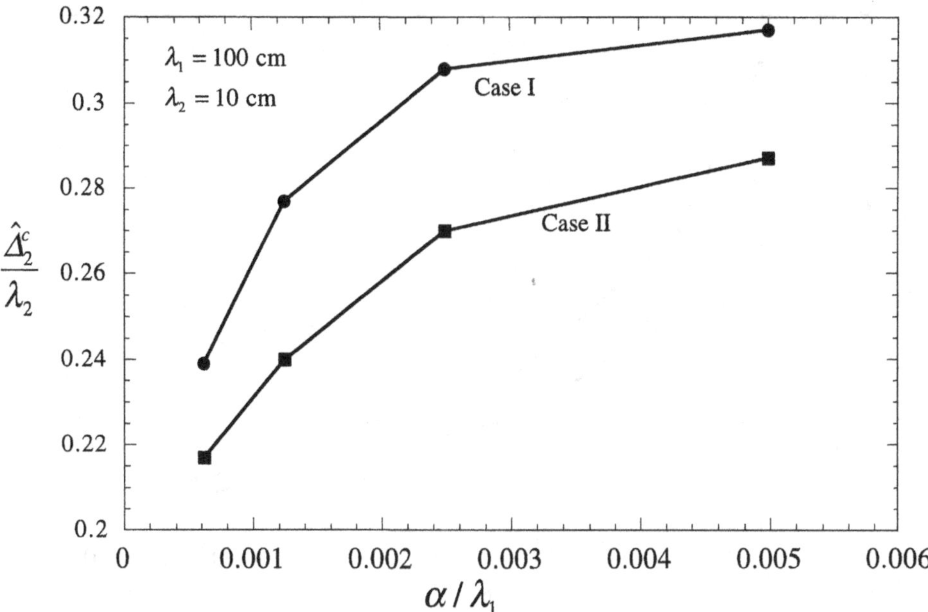


Figure 10.4b. Concentration microscale in the $x_2$ direction. Case I has less fine-scale variability than case II (Table 10.1).

2. Concentration microscales are bounded upward by the half-wavelength scale: $\hat{\Delta}_i^c < \lambda_i$ and $\hat{\Delta}_2^c \leq \lambda_2/\pi$, the latter being expected, given the inequality (10.16) and given that the concentration field has to be periodic in $x_2$ here.
3. Concentration microscales are increasing functions of the local dispersivity. As an increase in the concentration microscales causes an increase in the VRT, local dispersion, to some extent, has a self-inhibiting role in dissipating concentration fluctuations.
4. For the same periodicity, increasing the fine-scale structure of the flow field will reduce the concentration microscales (i.e., speed up variance dissipation and dilution).

These properties of the concentration microscales are manifested in the calculated VRT, as shown in Figure 10.5. The VRT reflects the anisotropy of the concentration microscales, by definition (10.21), and the concentration microscales reflect the anisotropy of the hydraulic-conductivity microstructure (Figure 10.4). The VRT is also smaller for the case in which the hydraulic conductivity has more small-scale variations (Figure 10.5). Notwithstanding the increase in the concentration microscales with local dispersion (Figure 10.4), which reflects weakened mean squared concentration gradients due to local dispersion, the VRT is observed to be a decreasing function of the local dispersivity (Figure 10.5). Of course, a zero local dispersivity results in an infinite VRT.

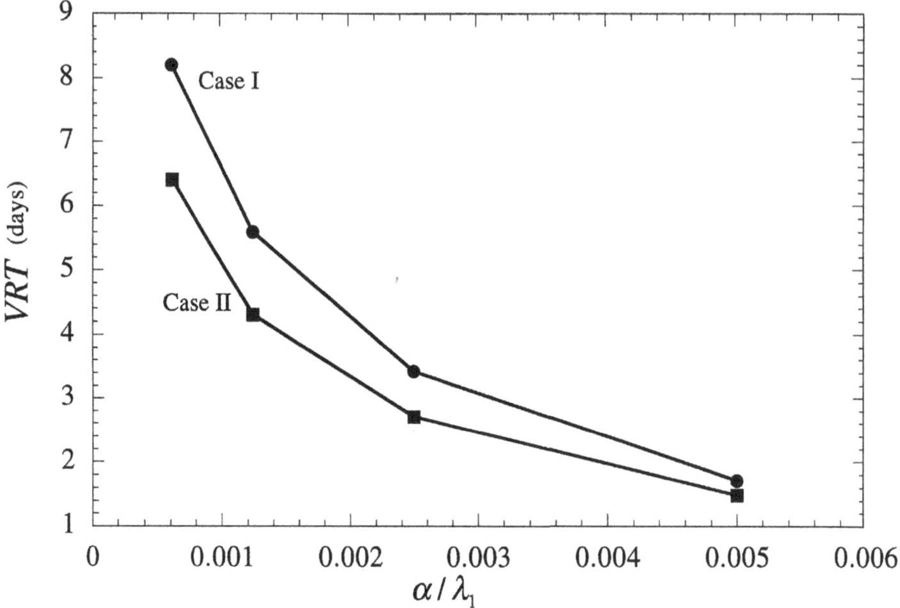


Figure 10.5. The time scale over which concentration fluctuations are destroyed by local dispersion, the VRT, is a decreasing function of the local dispersivity.

### *10.6.2 Dilution and Concentration Fluctuations*

From the simulations of Kapoor and Kitanidis (1996) and those reported in this chapter, the following picture emerges:

*At $t$ < VRT, Plumes Become Increasingly Irregular with Time.* The influence of local dispersion is not strongly manifest in the concentration field for $t <$ VRT, and the advective distortion of the solute body results in an increase in the raggedness of the concentration field with time. This is evidenced by an initially increasing CV (Figure 10.6) and initially falling reactor ratio (Figure 10.7). The peak concentration at any given time, $c_p(t)$, is also considerably underestimated by the peak of the mean concentration field, $\bar{c}_p(t)$, at early times (Figure 10.8). The results show that spatial second moments at early times will be poor indicators of dilution measures such as the reactor ratio, CV, and the peak concentration. In addition to the spatial second moments, an estimate of plume irregularity is needed for dilution assessment.

For the ZLD case, the VRT would be infinitely large, as concentration variance would not be destroyed at all. Therefore, the increase in the irregularity of the plume with time would continue indefinitely: The CV would grow unboundedly with time, and the reactor ratio would fall toward zero with time. However, local dispersion, no matter how small, qualitatively alters this picture, as discussed next.

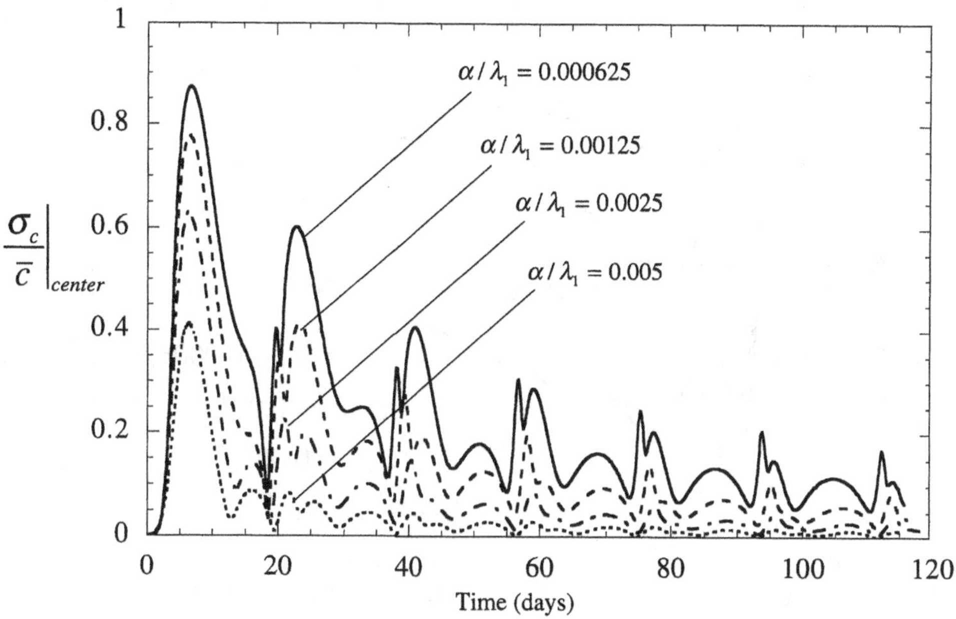


Figure 10.6a. Rise and fall of the concentration CV (case I). At large times ($t \gg$ VRT) the falling CV has an overall $t^{-1}$ behavior, predicted by Kapoor and Gelhar (1994a,b). Some other studies have predicted an unboundedly increasing CV, as they assumed zero local dispersion (ZLD).

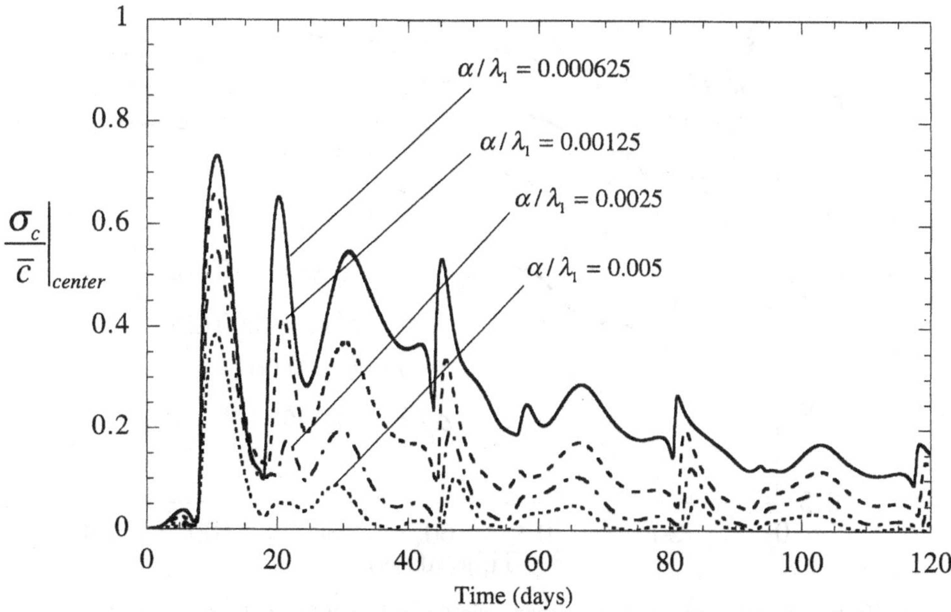


Figure 10.6b. Rise and fall of the concentration CV (case II). At large times ($t \gg$ VRT) the falling CV has an overall $t^{-1}$ behavior, predicted by Kapoor and Gelhar (1994a,b). Some other studies have predicted an unboundedly increasing CV, as they assumed ZLD.

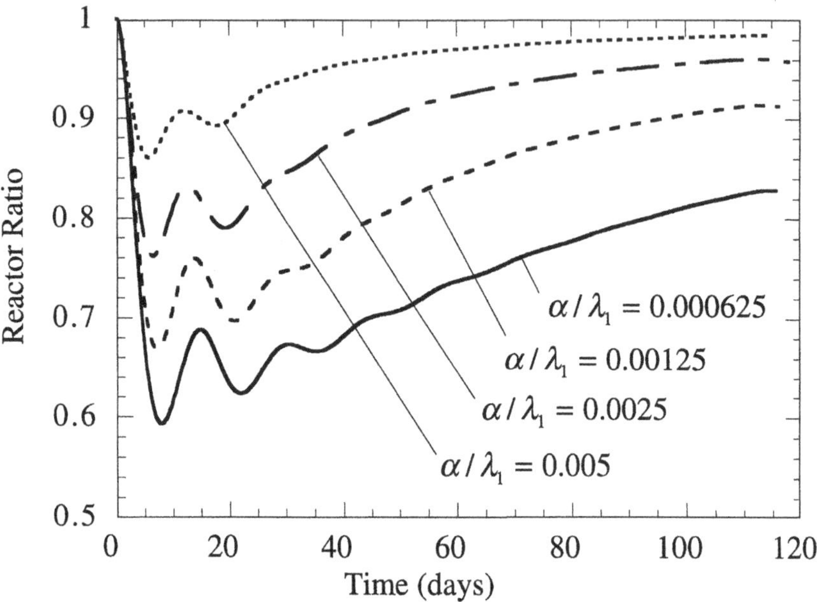


Figure 10.7a. Fall and recovery of the reactor ratio (case I). For the ZLD case, this ratio would asymptotically decrease toward zero.

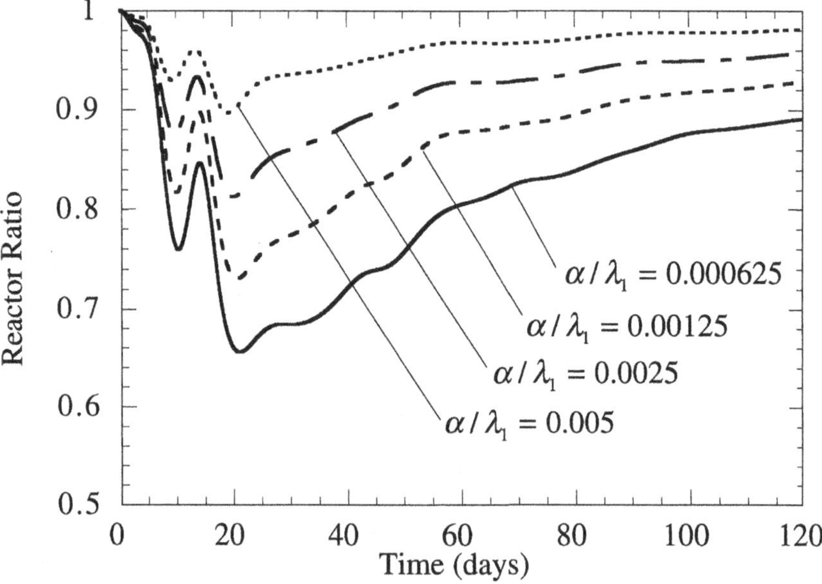


Figure 10.7b. Fall and recovery of the reactor ratio (case II). For the ZLD case, this ratio would asymptotically decrease toward zero.

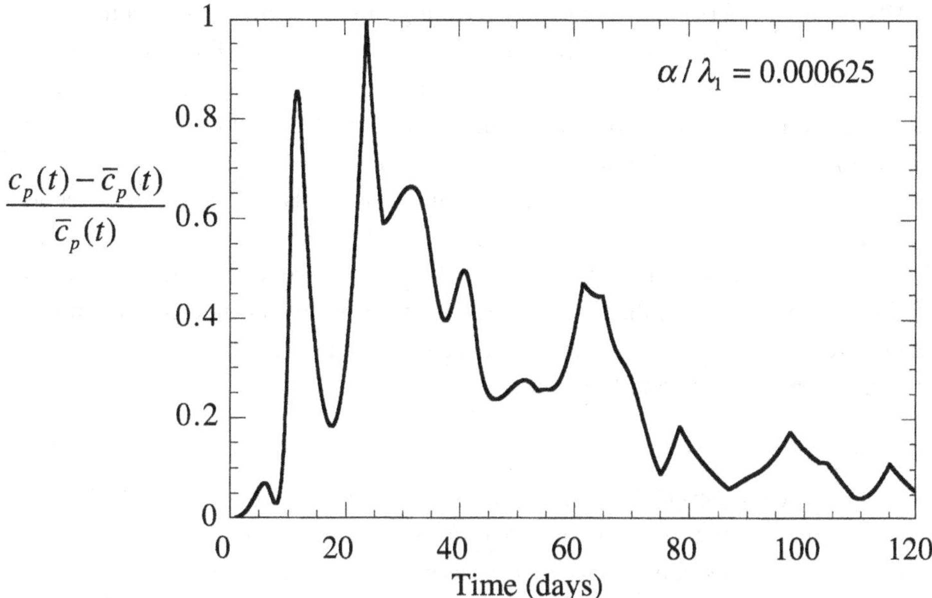


Figure 10.8. Underestimation of peak concentration by the mean (case II, $\alpha/\lambda_1 = 0.000625$). Initially the peak concentration ($c_p$) is increasingly underestimated by the peak of the cross-sectional average concentration ($\bar{c}_p$). However, at large times, $t >>$ VRT (VRT = 6.4 days for this case), they come close together. This is analogous to the CV decreasing toward zero (Figure 10.6) and the reactor ratio recovering toward unity (Figure 10.7) when $t >>$ VRT.

*At $t \gg$ VRT, Plumes Become Regular and More Uniform.* At large times, the CV heads toward its asymptotic value of zero, as shown in Figure 10.6, and the reactor ratio heads toward its maximum possible value of unity, as Figure 10.7 clearly demonstrates. The jumps in the CV at the center of mass (Figure 10.6) reflect that the center of mass is passing through zones of different velocity variability. Eventually, the peak concentration also comes close to the peak of the mean concentration (Figure 10.8). The larger the value of the VRT, the longer it takes for the irregularity of plumes to be dissipated. At the center of mass, the CV has a $t^{-1}$ fall at large times, according to the results shown in Figure 10.6. This type of decrease was theoretically predicted earlier by Kapoor and Gelhar (1994a,b) for random porous media with finite scales of heterogeneity. It has also been observed by the authors in direct numerical simulations for random porous media, and the approximate expressions for the recovery of the reactor ratio shown in Figure 10.7 have been developed (Kapoor and Kitanidis, in press). It appears that for a broad class of heterogeneous flows with finite scales of heterogeneity, at large times the CV at the center will decay with time. This is in qualitative contrast to the perpetually increasing CV predicted by some previous studies neglecting local dispersion.

Note that Figures 10.6 and 10.7 present results for four different values of the local dispersivity and therefore four different VRTs (Figure 10.5). In interpreting these figures it is necessary to remember that the smaller the local dispersivity, the larger the VRT, and vice versa. The physically meaningful time scale that one could use to render time dimensionless in Figures 10.6, 10.7, and 10.8 is the VRT, as that signifies the time scale of variance destruction. The mean advection time scale, $\lambda_1/v$, would be a less meaningful factor to render time dimensionless in these figures on concentration fluctuations, because it does not quantify the fluctuation destruction process. Plotting fluctuation measures versus time rendered dimension-less by the mean advection time scale ($\lambda_1/v$) can lead to the incorrect perception that a small local dispersion coefficient does not qualitatively alter the behavior of concentration fluctuations and dilution. In other words, if the simulations were prematurely terminated at a time that did not exceed the VRT, but possibly exceeded $\lambda_1/v$, the overall large-time decay of the fluctuation measures would not be observed.

*Heterogeneous Advection Enhances the Rate of Dilution by Local Dispersion.* As made plain by equation (10.6), local dispersion is essential for dilution, and advective transport is not. However, without variability in seepage velocity, the dilution rate would reduce to very small values over time as the plume spread out and concentration gradients diminished (Kitanidis, 1994). Although advection per se does not effect dilution, variability in velocity creates the conditions for local dispersion to become effective: Velocity variability distorts the isoconcentration lines and creates areas of steep concentration gradients (Figure 10.3b), thus enhancing dilution. This "stirring" or "blending" of the fluids controls the rate of dilution at large times ($t \gg$ VRT).

*VRT Can Be Quite Long.* The rates of change (or "kinetics") of dilution appear to be quantified well through the VRT. The numerical results presented here show that VRT varies approximately inversely with the local dispersivity (Figure 10.5) and can be quite long for realistically small local dispersivities. A long VRT means that the concentration variance is destroyed slowly, and therefore it can be significant at times relevant to remediation projects and, *a fortiori*, tracer tests.

Study of the data from field experiments at Borden and Cape Cod by Thierrin and Kitanidis (1994) indicates that the reactor ratio was far from converging to unity, even at the end of those experiments, and the concentration CV at the cape was large (Kapoor and Gelhar, 1994b), albeit not increasing with time, as would be the case under pure advection. Although difficulties in inferring concentration variances and dilution rates from the data available for those sites must be recognized, it appears that the VRT at those sites could be of the order of 100 days or more.

## 10.7 Conclusion

Studies of the issues of scale in the transport of nonreactive solutes in heterogeneous formations have focused on spatial second moments and their time rates of change (macrodispersion). Such studies have explained the scale dependence of dispersion coefficients and have shown that variability in flow velocity controls the rate of spreading of solute plumes in geologic formations. In applications, numerical models have employed large dispersivities that have predicted rapid reductions in peak concentrations. In this chapter we have argued that the rate of increase of spatial second moments for a solute body does not necessarily quantify the dilution of contaminant concentrations in a heterogeneous flow field, because spatial second moments are not informative about solute variability.

The dilution index, reactor ratio, and concentration CV quantify dilution and allow quantitative study of the mechanisms and parameters that control dilution. It has been shown how dilution, peak concentrations, and concentration fluctuations are controlled by the interactions of local dispersion and flow-field heterogeneities, which control the VRT. This connection is currently not being incorporated into practical assessments of concentrations of toxic substances in the subsurface, because of the widespread practice of employing large effective-dispersion coefficients to assess some sort of smoothly varying mean concentration. Our results show that the current approach of estimating concentrations based on effective-dispersion coefficients or spatial second moments can severely overestimate dilution. To remedy that, a solution to the concentration-variance equation (10.14) can be approximated, based on the concept of the VRT, which depends on the local dispersion coefficient and the heterogeneity microstructure, as shown in Figure 10.5. Through evaluation of the concentration variance, the mean concentration and the spatial second moments can be supplemented with information about the concentration CV and the reactor ratio. The characteristic VRT needs to be understood for a wide variety of porous media in order to develop practical guidelines for assessing dilution.

## Acknowledgments

Funding for this study was provided by a DOE contract for the study "Mixing Strategies for Enhanced In-Situ Bioremediation," and by the NSF under grant EAR-9204235, "Scale Dependence and Scale Invariance in Catchment Hydrologic Processes," at the Department of Civil Engineering, Stanford University.

## References

Ames, W. F. 1992. Numerical Methods for Partial Differential Equations. San Diego: Academic Press.

Bear, J. 1972. Dynamics of Fluids in Porous Media. New York: American Elsevier.

Dagan, G. 1982. Stochastic modeling of groundwater flow by unconditional and conditional probabilities. 2. The solute transport. *Water Resour. Res.* 18:835–48.

Dagan, G. 1984. Solute transport in heterogeneous porous formations. *J. Fluid Mech.* 145:151–77.

Dagan, G. 1987. Solute transport in groundwater. *Ann. Rev. Fluid Mech.* 19:183–215.
Dieulin, A., Matheron, A. G., de Marsily, G., and Beaudoin, B. 1981. Time dependence of an equivalent dispersion coefficient for transport in porous media. In: *Flow and Transport in Porous Media*, ed. A. Verruijt and F. B. J. Barends, pp. 199–202. Rotterdam: Balkema.
Eidsath, A., Carbonell, R. G., Whitaker, S., and Herrmann, L. R. 1983. Dispersion in pulsed systems. III. Comparison between theory and experiments in packed beds. *Chem. Eng. Sci.* 38:1803–16.
Gelhar, L. W. 1986. Stochastic subsurface hydrology from theory to applications. Water *Resour. Res.* (*suppl.*) 22:135S–45S.
Gelhar, L. W., and Axness, C. L. 1983. Three-dimensional stochastic analysis of macrodispersion in aquifers. *Water Resour. Res.* 19:161–80.
Gelhar, L. W., Gutjahr, A. L., and Naff, R. L. 1979. Stochastic analysis of macrodispersion in a stratified aquifer. *Water Resour. Res.* 15:1387–97.
Kapoor, V. 1997. Vorticity in three-dimensionally random porous media. *Trans. Porous Media* 26:103–19.
Kapoor, V., and Gelhar, L. W. 1994a. Transport in three-dimensionally heterogeneous aquifers: 1. Dynamics of concentration fluctuations. *Water Resour. Res.* 30:1775–8.
Kapoor, V., and Gelhar, L. W. 1994b. Transport in three-dimensionally heterogeneous aquifers: 2. Predictions and observations of concentration fluctuations. *Water Resour. Res.* 30:1789–801.
Kapoor, V., and Kitanidis, P. K. 1996. Concentration fluctuations and dilution in two-dimensionally periodic heterogeneous porous media. *Trans. Porous Media* 22:91–119.
Kapoor, V., and Kitanidis, P. K. In press. Concentration fluctuations and dilution in aquifers.
Kitanidis, P. K. 1988. Prediction by the method of moments of transport in heterogeneous formations. *J. Hydrol.* 102:453–73.
Kitanidis, P. K. 1992. Analysis of macrodispersion through volume-averaging: moment equations. *Stoch. Hydrol. Hydraul.* 6:5–25.
Kitanidis, P. K. 1994. The concept of the dilution index. *Water Resour. Res.* 30:2011–26.
Matheron, G., and de Marsily, G. 1980. Is transport in porous media always diffusive? A counterexample. *Water Resour. Res.* 16:901–17.
Mei, C. C. 1992. Method of homogenization applied to dispersion in porous media. *Trans. Porous Media* 9:261–74.
Neuman, S. P., Winter, C. L., and Newman, C. M. 1987. Stochastic theory of field-scale Fickian dispersion in anisotropic porous media. *Water Resour. Res.* 23:453–66.
Pickens, J. F., and Grisak, G. E. 1981. Scale dependent dispersion in a stratified granular aquifer. *Water Resour. Res.* 17:1191–211.
Schwartz, F. W. 1977. Macroscopic dispersion in porous media: the controlling factors. *Water Resour. Res.* 13:743–52.
Smith, L., and Schwartz, F. W. 1980. Mass transport. 1: Stochastic analysis of macrodispersion. *Water Resour. Res.* 16:303–13.
Sposito, G., and Barry, D. A. 1987. On the Dagan model of solute transport in groundwater: foundational aspects. *Water Resour. Res.* 23:1867–75.
Sudicky, E. A. 1986. A natural gradient experiment on solute transport in a sand aquifer: spatial variability of hydraulic conductivity and its role in the dispersion process. *Water Resour. Res.* 22:2069–82.
Thierrin, J., and Kitanidis, P. K. 1994. Solute dilution at the Borden and Cape-Cod groundwater tracer tests. *Water Resour. Res.* 30:2883–90.

## Appendix: Model Periodic Hydraulic Conductivity and Flow Field

To illustrate the basic features of dilution and concentration fluctuations, we consider a periodic saturated porous medium whose hydraulic conductivity $K$ varies as

$$\ln K = \ln K_G + \tilde{f},$$

$$\tilde{f}(x_1, x_2) = \sum_{k_1=1}^{k_1^m} \sum_{k_2=1}^{k_2^m} a(k_1, k_2) \sin\left(\frac{k_1 \pi x_1}{\lambda_1}\right) \cos\left(\frac{k_2 \pi x_2}{\lambda_2}\right)$$

The geometric mean of the hydraulic conductivity, $K_G$, is a constant. The number of modes of variability in the $i$th direction is denoted by $k_i^m$. The porous medium is of an infinite longitudinal ($x_1$) extent, and the vertical ($x_2$) thickness $H$ is an integral multiple of the vertical wavelength $2\lambda_2$ (Figure 10.3a). A reason for choosing this model to illustrate concepts of dilution and concentration fluctuations is the ease of solving the leading-order flow problem, including the no-flux boundary condition at $x_2 = 0$ and $H$. The variables with a superscript $A$ denote cell ($2\lambda_1 \times 2\lambda_2$) averages or, equivalently, depth–cell averages ($2\lambda_1 \times H$), and the variables with a tilde represent deviations around cell averages. The leading-order flow field $q_i(\mathbf{x}) = q_i^A + \tilde{q}_i(\mathbf{x})$ resulting from a mean hydraulic gradient $J$ in the $x_1$ direction, employing the "exponential generalization" for the effective hydraulic conductivity, is

$$\frac{\tilde{q}_1(x_1, x_2)}{K_G J} = \sum_{k_1=1}^{k_1^m} \sum_{k_2=1}^{k_2^m} \frac{a(k_1, k_2)\left(\frac{k_2 \pi}{\lambda_2}\right)^2}{\left(\frac{k_1 \pi}{\lambda_1}\right)^2 + \left(\frac{k_2 \pi}{\lambda_2}\right)^2} \sin\left(\frac{k_1 \pi x_1}{\lambda_1}\right) \cos\left(\frac{k_2 \pi x_2}{\lambda_2}\right)$$

$$\frac{\tilde{q}_2(x_1, x_2)}{K_G J} = -\sum_{k_1=1}^{k_1^m} \sum_{k_2=1}^{k_2^m} \frac{a(k_1, k_2)\left(\frac{k_1 \pi}{\lambda_1}\right)\left(\frac{k_2 \pi}{\lambda_2}\right)}{\left(\frac{k_1 \pi}{\lambda_1}\right)^2 + \left(\frac{k_2 \pi}{\lambda_2}\right)^2} \cos\left(\frac{k_1 \pi x_1}{\lambda_1}\right) \sin\left(\frac{k_2 \pi x_2}{\lambda_2}\right)$$

$$q_1^A = K_{\text{eff}} J, \qquad q_2^A = 0, \qquad K_{\text{eff}} = K_G e^{\beta},$$

$$\beta = \frac{1}{8} \sum_{k_1=1}^{k_1^m} \sum_{k_2=1}^{k_2^m} a^2(k_1, k_2) \frac{\left(\frac{k_2}{\lambda_2}\right)^2 - \left(\frac{k_1}{\lambda_1}\right)^2}{\left(\frac{k_1}{\lambda_1}\right)^2 + \left(\frac{k_2}{\lambda_2}\right)^2}$$

The details of derivation have been published (Kapoor and Kitanidis, 1996). In Table 10.1, where the parameters of the flow field are listed, $\sigma_{\tilde{f}}^2$ is the cell-averaged squared deviation of $\ln K$ around $\ln K_G$.

# 11

# Analysis of Scale Effects in Large-Scale Solute-Transport Models

ROGER BECKIE

## 11.1 Introduction

This chapter presents a framework in which to understand the scale dependence and interaction of scales in large-scale models of conservative solute transport in groundwater. We are interested in models of conservative solute transport at practical field scales, on the order of a few meters to many kilometers. In particular, we examine how the dynamics on different scales are represented and interact in models developed using large-eddy-simulation (LES) methods from geophysical fluid dynamics (Rogallo and Moin, 1984). LES methods provide an ideal framework for our analysis because they allow us to define the scale of our model precisely.

This chapter presents a synthesis of findings from both groundwater hydrology and computational fluid dynamics. Our principal contribution will be to bring these two fields together in applications to groundwater problems. We shall first review the conceptual and technical basis for application of LES methods to conservative solute transport. We shall then examine the effects of advection and dispersion on the variability in the concentration field. We shall next investigate the effects of smaller-scale advection on the larger-scale component of the concentration field. We shall conclude with a brief and speculative analysis of strategies to incorporate the effects of unresolved, smaller-scale processes in larger-scale models of solute transport in groundwater.

## 11.2 Overview of LES

The roots of LES are in geophysical and engineering computational fluid dynamics, where many problems of interest involve turbulence (Rogallo and Moin, 1984). Turbulent velocity fields, governed by the Navier-Stokes equations, vary over a wide and continuous range of length and time scales. It can be shown that the number of unknowns required for a numerical model to explicitly resolve a large-Reynolds-number turbulent velocity field on all significant scales exceeds the capacity of current computers, and likely any future computers (Frisch and Orszag, 1990). A typical numerical discretization is much too coarse to accurately resolve a turbulent velocity field on a scale that can be described by the so-called primitive (fine-scale) Navier-Stokes equations.

Modelers using LES recognize that whereas it may never be possible to resolve all significant scales of a turbulent flow with a numerical model, it may be possible to develop models of the larger-scale dynamics. Such large-scale models would not explicitly describe the smaller-scale details of the flow. The goal of LES methods is to develop larger-scale models from accepted smaller-scale primitive models.

An LES model of a physical process is developed in four steps. First, an appropriate primitive model of the physical process is specified. Second, the variables of the primitive model are decomposed into their larger- and smaller-scale components, which are called grid- and subgrid-scale variables. Grid-scale variables can be explicitly resolved on the particular numerical grid being used, but the subgrid-scale variables are too fine-scale to be explicitly resolved (Figure 11.1). The grid- and subgrid-scale variables are defined using a spatial filter that is compatible with the numerical discretization. Spatial filters are analogous to the time-domain filters used for signal processing. Third, a grid-scale model is defined by spatially filtering the primitive model. The resulting grid-scale model usually will contain new terms (not found in the primitive-model equations) that depend upon subgrid-scale variables. These terms are called subgrid-closure terms and can be interpreted as

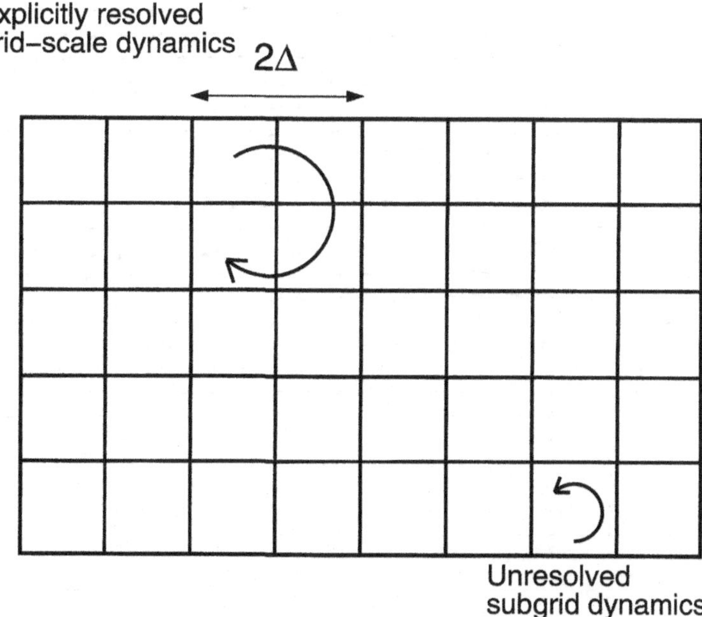


Figure 11.1. A numerical grid can explicitly resolve dynamics with scales greater than $2\Delta$, where $\Delta$ is the characteristic scale of the discretization. Subgrid-scale dynamics, by definition, cannot be resolved.

the effects of unresolved subgrid-scale dynamics upon explicitly resolved grid-scale dynamics. Fourth, the subgrid-closure terms are approximated using grid-scale variables. These approximations are called subgrid-closure models. The final result is a grid-scale model that can be solved on a coarser numerical grid.

These concepts can perhaps be best explained in physical terms with a simple transport example. Let us assume that at some fine scale, solute transport can be modeled exactly as a purely advective process. This is our primitive model. Given the "true" primitive velocity field (Figure 11.2), solute transport could be explicitly resolved in a numerical model at this scale by varying the concentration and velocity field in each grid block. Because the velocity field varies in space, solute released along one boundary in a fast part of the flow breaks through to the other boundary more quickly than solute released along the same boundary in slower parts of the flow. This leads to the purely advective primitive-model breakthrough curve shown in Figure 11.3.

Now consider modeling the same problem at a coarser grid scale. In a grid-scale model, only the larger-scale component of the primitive velocity field can be explicitly resolved (Figure 11.4). The grid-scale velocity field is much smoother and lacks the smaller-scale variability of the true velocity field. A purely advective model at this scale leads to the grid-scale breakthrough curve shown in Figure 11.3. The purely

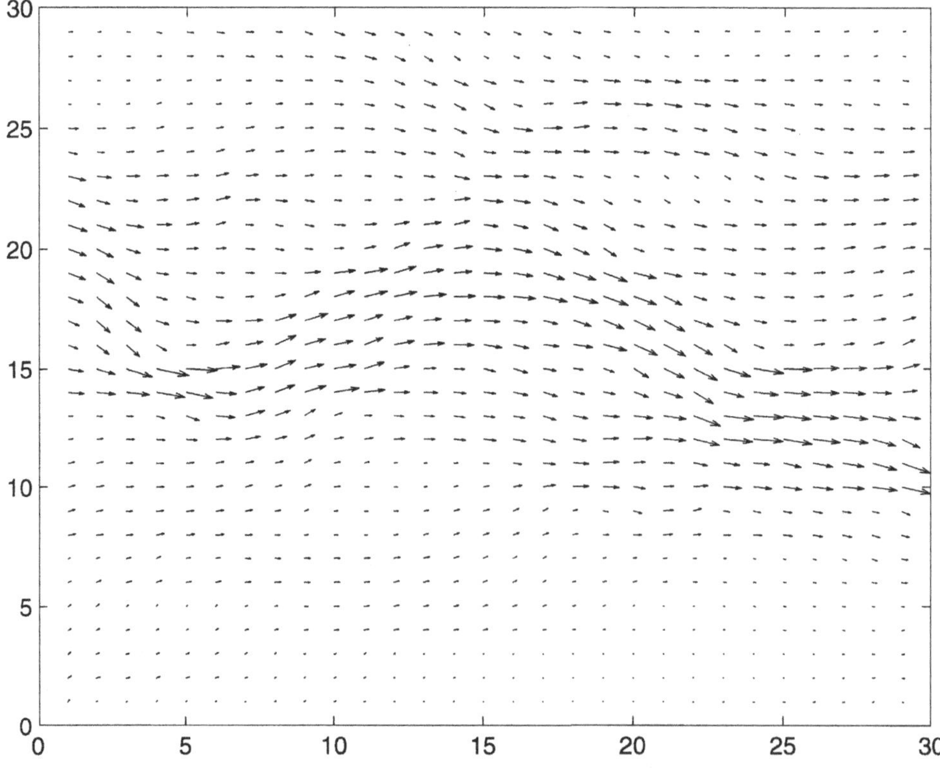


Figure 11.2. A hypothetical primitive velocity field.

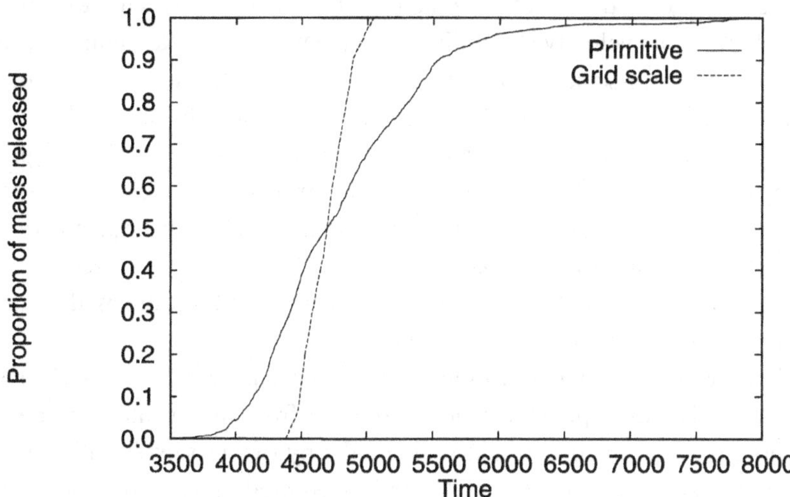


Figure 11.3. Breakthrough curves computed by advecting solute without dispersion or diffusion through a velocity field that is resolved at the primitive scale and through the same velocity field when it is resolved at a coarser grid scale.

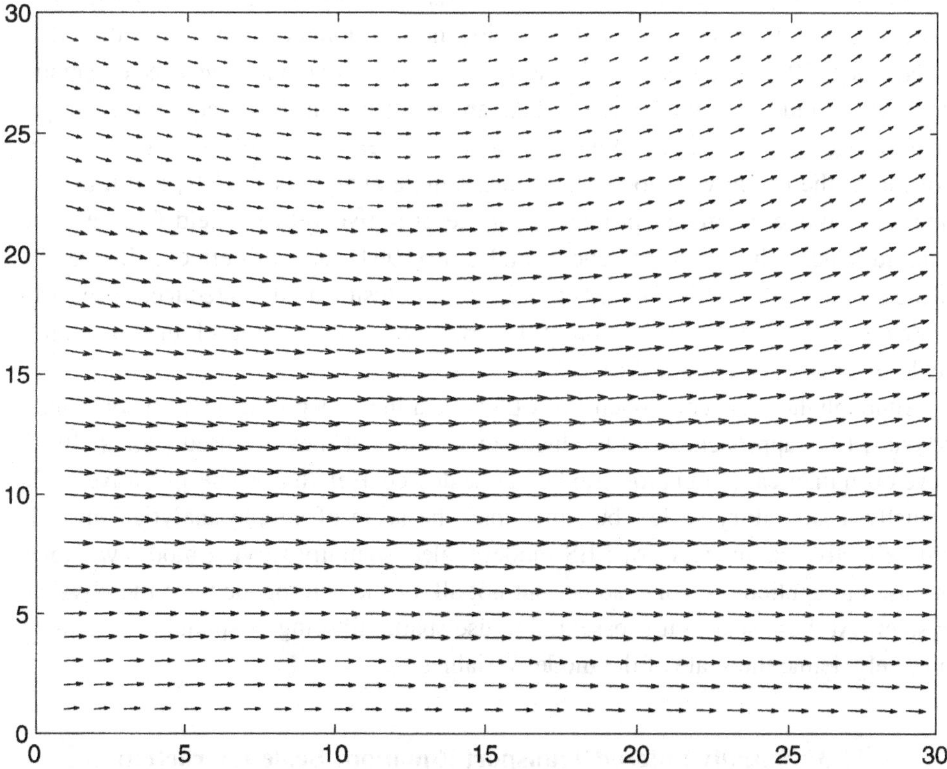


Figure 11.4. A coarser-scale representation of the hypothetical velocity field from Figure 11.1. The grid scale in this figure is approximately eight times greater than the scale of the primitive velocity field.

advective grid-scale model is not accurate because it does not capture the effects of smaller, subgrid-scale velocity variability on solute transport. Examining the curves, it appears that the grid-scale breakthrough curve is too sharp and could be made to more closely resemble the true, primitive-model breakthrough by dispersing the front. In other words, the effect of subgrid advection on the grid-scale breakthrough curve can be approximated by adding dispersion to the grid-scale model. The dispersion added to a grid-scale transport model in excess of that originally in the primitive model is called enhanced dispersion (Gelhar and Axness, 1983). The enhanced dispersion added to the grid-scale model is an approximation, a subgrid-closure model, for what is really just advection in the primitive model.

This simple example provides insight into the scale dependence of models of physical processes. The example shows how a process from the primitive model, such as advection, is represented in a grid-scale model by two processes: advection and dispersion. The amount of additional dispersion required in the grid-scale model increases as the proportion of the primitive velocity field that falls into the subgrid scales increases. As the grid scale is reduced, so that the velocity field is resolved on finer and finer scales, the amount of dispersion needed in the grid-scale model must diminish. The enhanced dispersion in a grid-scale model is thus scale-dependent. When the grid scale and the primitive-model scale coincide, no additional dispersion is required.

Neuman (1990) used essentially the same argument to explain why the dispersivities "measured" with numerical models are smaller than those required in simple analytical models. Indeed, numerical models can explicitly resolve a greater proportion of the primitive velocity field than simple analytical models, and thus in a numerical model a smaller proportion of the primitive velocity field falls into the subgrid scales. The results of Scheibe and Cole (1994) show this effect quite clearly. More generally, the burden placed upon a subgrid-closure model is reduced as the grid scale decreases and the proportion of the primitive-scale dynamics found at subgrid scales decreases (Beckie, 1996).

Although the basic components of LES appear in many problems from continuum physics, LES approaches can be distinguished in that they are used principally to develop numerical-simulation models at scales coarser than some primitive scale (usually a laboratory scale), but finer than the scale of simple analytical models with effective parameters. At LES-model scales, spatially varying model variables are used to explicitly resolve some, but not all, of the heterogeneity of the physical system. At these scales, it is essential to use spatial filtering or spatial averaging to precisely define the scale of the model variables.

## 11.3 Spatially Filtered Transport Equation: Scale Interactions

In this section we demonstrate spatial filtering of a primitive transport equation and show how a grid-scale-model equation with subgrid-closure terms results. We then investigate the scale interactions contained in the subgrid-closure terms and show how unresolved subgrid dynamics affect the resolved grid-scale dynamics through the closure terms. In the next section we shall quantify these effects with a mathematical analysis.

We begin with a primitive model to describe transport at an appropriate smaller scale. There is a hierarchy of models that describe solute transport on different scales (Bird, Steward, and Lightfoot, 1960; Chu and Sposito, 1980; Tompson and Gray, 1986; Bouchard and Georges, 1990). In our development, the primitive model can be formulated at any scale. Because our principal objective is to model transport at practical field scales, it is convenient to begin with a transport model defined on the scale of a core:

$$\frac{\partial C}{\partial t} + \frac{\partial}{\partial x_j}(V_j C) - \frac{\partial}{\partial x_j}\left(D_{jk}\frac{\partial C}{\partial x_k}\right) = 0 \qquad (11.1)$$

where $C(\mathbf{x}, t)$ is the core-scale solute-concentration field [M/L$^3$], $V_j(\mathbf{x}, t)$, with $j$ =1, 2, 3, is the core-scale average linear velocity of groundwater flow [L/T], $D_{jk}$ is the core-scale dispersion tensor [L$^2$/T], and where we have employed the summation convention for repeated indices. We take the relativist perspective of Baveye and Sposito (1984) and assume that the transport equation (11.1) describes the concentration field that would be measured with core-scale instrumentation. We assume that the primitive-model velocity field and dispersion tensor are those appropriate for this scale of description.

The spatial filters that define grid- and subgrid-scale variables in LES methods are selected to match the filtering properties of the numerical discretization. Indeed, a numerical grid can explicitly resolve model variables on spatial scales greater than $2\Delta$, where $\Delta$ is the characteristic discretization scale (Figure 11.1). Conceptually, resolving a primitive-model variable on a numerical grid is equivalent to spatially filtering the primitive-model variable.

Mathematically, if $V_1$ is the $x_1$ component of the primitive velocity field, then the grid-scale (filtered) velocity field $\bar{V}_1$ is expressed as a convolution integral (Leonard, 1974):

$$\bar{V}_1(\mathbf{x}) = \int \mathcal{G}(\mathbf{x} - \mathbf{x}'; \Delta) V_1(\mathbf{x}')\, d\mathbf{x}' \qquad (11.2)$$

where $\mathcal{G}(\mathbf{x}; \Delta)$ is a spatial-filter function with filter width $\Delta$, and $\mathbf{x} = (x_1, x_2, x_3)$ is the spatial-coordinate vector. The subgrid-scale velocity is then defined as

$$v_1(\mathbf{x}) = V_1(\mathbf{x}) - \bar{V}_1(\mathbf{x}) \qquad (11.3)$$

For notational convenience, we write primitive variables with uppercase letters, filtered variables with an overbar, and subgrid-scale variables with lowercase letters.

The three filter functions $\mathcal{G}$ that are customarily used for LES are selected to match the filtering effect of common numerical discretizations. Figure 11.5 illustrates the three filters in physical and Fourier space. The Fourier transformations of the spatial filters show the spatial frequencies $f = (f_1^2 + f_2^2 + f_3^2)^{1/2}$, or scales, that are passed by the filter (amplitude near unity) and those that are filtered out (amplitude near zero) (Bracewell, 1965, p. 46). Both the box and Gaussian filters (Figure 11.5a,b) mimic the filtering effect of a finite-difference discretization (Rogallo and Moin, 1984). The Fourier-space representation of the box filter shows that whereas it passes mostly low frequencies, it also passes and spuriously attenuates higher frequencies. Figure 11.5b shows that the Gaussian filter decays monotonically to zero at higher frequencies. The spectrally sharp filter (Figure 11.5c), appropriate for use in Fourier-spectral numerical methods (Canuto et al., 1987), complements the box filter in both physical space and Fourier space. The spectral filter passes only those spatial frequencies $f$ below a specified cutoff $f_c = 1/(2\Delta)$, where $\Delta$ is the cutoff length scale.

We spatially filter the primitive model, equation (11.1), to change from the primitive-model scale to the scale of the numerical grid, $\Delta$. We assume that the primitive-scale dispersion coefficient is smoothly varying, such that $\overline{D_{jk}(\partial C/\partial x_k)} \approx D_{jk}(\partial \bar{C}/\partial x_k)$. The grid-scale transport model is then

$$\frac{\partial \bar{C}}{\partial t} + \frac{\partial}{\partial x_j}\overline{(V_j C)} - \frac{\partial}{\partial x_j}\left(D_{jk}\frac{\partial \bar{C}}{\partial x_k}\right) = 0 \tag{11.4}$$

Except for the filtered advective term $\overline{(V_j C)} = \int \mathcal{G}(\mathbf{x}-\mathbf{x}'; \Delta) V_j(\mathbf{x}')C(\mathbf{x}')\, d\mathbf{x}'$, this grid-scale-model equation looks like the primitive transport equation written with grid-scale variables. The subgrid-closure terms are hidden in the filtered advective term, which is formulated with the primitive variables $V_j$ and $C$. The subgrid-scale components of $V_j$ and $C$ cannot be resolved by the numerical grid. The subgrid-closure terms can be exposed by decomposing the filtered advective term $\overline{(V_j C)}$ into grid-scale and subgrid-scale components using equations (11.2) and (11.3) (Beckie, Aldama, and Wood, 1994):

$$\begin{aligned}\overline{V_j C} &= \overline{\bar{V}_j \bar{C}} + \overline{\bar{V}_j c} + \overline{v_j \bar{C}} + \overline{v_j c} \\ &= \{1\} + \{2\} + \{3\} + \{4\}\end{aligned} \tag{11.5}$$

The essence of the transport-equation scale problem is contained in the decomposition of the filtered advective term in equation (11.5). The filtered advective term $\overline{(V_j C)}$ itself can be interpreted as the grid-scale effect of primitive-model advection. The right-hand side of equation (11.5) shows that this effect has four distinct components. Terms {2}, {3}, and {4} depend upon subgrid variables and are therefore subgrid-closure terms. The subgrid variables in these terms cannot be resolved on a grid-scale mesh and therefore must be approximated using grid-scale variables. We next

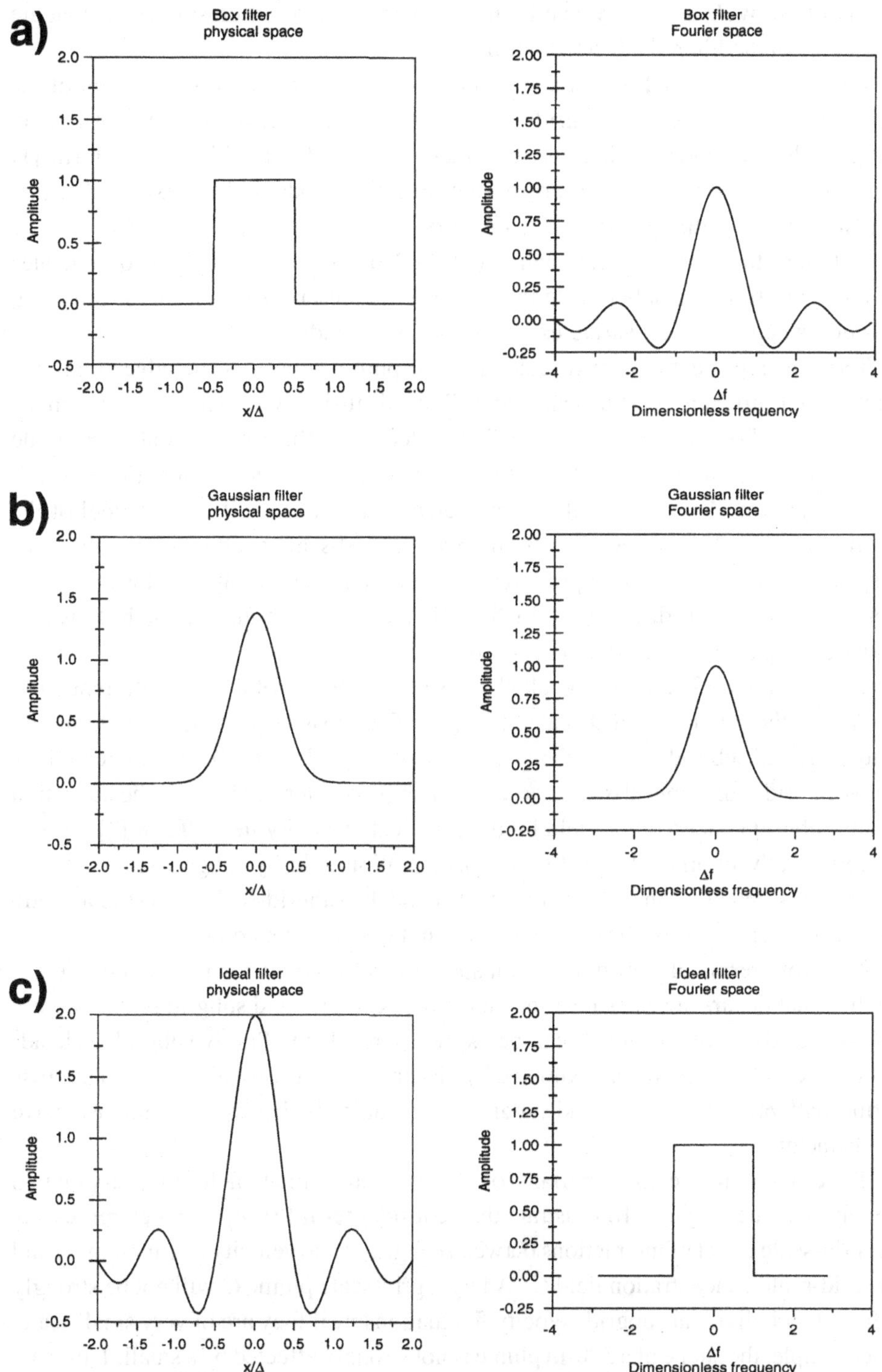


Figure 11.5. Physical- and Fourier-space plots of the one-dimensional forms of the three common spatial filters used for LES: (a) the box filter, (b) the Gaussian filter, and (c) the ideal filter.

attempt to provide some physical intuition about the subgrid-closure terms that are contained in the filtered advective term.

First, term {1}, which is not a subgrid-closure term, represents advection of the grid-scale solute field by the grid-scale velocity field. Term {1} does not require subgrid-closure modeling because it contains grid-scale variables only. Term {1} looks similar to the advective term in the primitive model (11.1), except that it is a nonlocal, or spatially integrated, quantity, as indicated by the overbar and the definition of the spatial filter, equation (11.2). Nonlocality is simply a consequence of changing from a smaller-scale to a larger-scale description. A nonlocal term at position **x** depends upon variables at positions surrounding **x**.

Term {4} represents the grid-scale effect of purely subgrid-scale advection. Advection at a sufficiently small scale "looks" like diffusion when viewed from a much larger scale. For example, what is called molecular diffusion at a continuum scale is in fact advection of molecules at the atomic scale. The purely subgrid-scale advection in term {4} is thus well approximated by a gradient-diffusion model of the form $\overline{v_j c} \approx D_{\text{eff}} \nabla \bar{C}$, where $D_{\text{eff}}$ is an effective-grid-scale coefficient or enhanced-dispersion coefficient. This approximation is an example of a subgrid-closure model. Term {4} is the classic dispersion term found in most Eulerian analyses of the transport equation (e.g., Bear, 1972, pp. 101, 604).

Terms {2} and {3} are the so-called cross terms and are of a much different character than the purely subgrid-scale term {4}. The cross terms, which contain both grid-scale and subgrid-scale variables, represent the grid-scale effects of interactions between grid- and subgrid-scale variables. The second term, $\overline{(\bar{V}_j c)}$, is the advection of the subgrid concentration field by the grid-scale velocity field. Term {3}, $\overline{(v_j \bar{C})}$, represents advection of the grid-scale concentration field by subgrid-scale velocity fluctuations. Whereas the advective term {1} and the subgrid-scale term {4} have tidy physical interpretations, the cross terms are not easily understood.

The cross terms of a grid-scale transport model account for primitive-scale advection that occurs at scales near the cutoff between grid and subgrid scales. Purely grid-scale advection, term {1}, can be explicitly resolved. Purely subgrid-scale advection, term {4}, is well approximated by dispersion in a grid-scale model. The cross terms, halfway between grid and subgrid scale, can be both advective and dispersive in character.

The cross terms are not important for systems with velocity fields that vary on two widely separated scales. To illustrate this, consider term {3}, $\overline{v_j \bar{C}}$, which represents the grid-scale effect of interactions between subgrid-scale velocity fluctuations $v_j$ and the grid-scale concentration field $\bar{C}$. A large, grid-scale plume $\bar{C}$ will not be strongly affected by individual subgrid velocity fluctuations $v_j$ if they are of very small scale. For example, the shape of a 500-m plume is not strongly affected by a small, 1-m scale heterogeneity. A similar argument can be made about term {2}, $\overline{\bar{V}_j c}$, the interaction between the grid-scale velocity field and the subgrid-scale concentration field.

The cross terms are most significant for systems with velocity fields that vary on a continuous range of scales. Intuitively, the larger a subgrid-scale velocity heterogeneity, the stronger its influence on the grid-scale concentration field. The largest subgrid-scale velocity heterogeneities are just smaller than the grid-scale $\Delta$, and the smallest grid-scale concentration fluctuations are just larger than the grid scale. Subgrid-scale velocity heterogeneities will thus most strongly affect the grid-scale concentration field at scales just larger than the grid scale. The cross terms are thus larger when the primitive velocity field varies on many scales than when the velocity field varies on widely separated grid and subgrid scales.

## 11.4 Concentration Variance and Entropy

In this section we quantify our intuitive investigation from the previous sections with a mathematical analysis of the grid-scale transport equation. We develop expressions that describe the time evolution of two different measures of the variability of the concentration field: the concentration variance, used in the fluid-mechanics literature (McComb, 1990, p. 473), and the concentration entropy, which is related to the dilution-index concept of Kitanidis (1994). We show that the processes of primitive-model dispersion and subgrid-scale advection tend to reduce the variability of the grid-scale concentration field with time, whereas grid-scale advection does not affect the concentration variability. Similarly, we investigate the effect of the subgrid-closure terms on the variability in the concentration field.

The variability of the primitive concentration field $C$ in a region $\Omega$ can be quantified with the global spatial variance $\sigma_C^2$,

$$\sigma_C^2(t) \equiv \frac{1}{\int_\Omega d\mathbf{x}} \int_\Omega [C(\mathbf{x}, t) - m_C]^2 \, d\mathbf{x} \tag{11.6}$$

$$\sigma_C^2(t) = \frac{1}{\int_\Omega d\mathbf{x}} \left[ \int_\Omega C^2(\mathbf{x}, t) \, d\mathbf{x} - 2m_C \int_\Omega C(\mathbf{x}, t) \, d\mathbf{x} + m_C^2 \int_\Omega d\mathbf{x} \right] \tag{11.7}$$

$$\sigma_C^2(t) = \frac{1}{\int_\Omega d\mathbf{x}} \int_\Omega C^2(\mathbf{x}, t) \, d\mathbf{x} - m_C^2 \tag{11.8}$$

where $m_C \equiv \int_\Omega C(\mathbf{x}, t) \, d\mathbf{x} / (\int_\Omega d\mathbf{x})$ is the spatial mean or average concentration in $\Omega$. A higher variance means greater variability of the concentration field in $\Omega$.

A local (nonintegral) measure of the concentration-field variability is the concentration variance $C^2(\mathbf{x}, t)$, which appears in equation (11.8). It is best understood through the definition of the global variance in equation (11.8). The global variance $\sigma_C^2$ equals the difference between the spatially averaged concentration variance and the square of the average concentration. When the concentration is everywhere the same, then the concentration variance is everywhere the same, and the global variance is zero. Both the global variance and the concentration variance are here defined as spatial statistics. They are not probabilistic quantities. We examine how advection on different scales affects the global variance by developing expressions for the concentration variance.

Similarly, the dilution index $E(t)$ of Kitanidis (1994), a global measure of plume dilution, is related to the local entropy of the concentration field $\mathcal{S}(\mathbf{x}, t)$ (defined later) in the same fashion as the global variance $\sigma_C^2$ is related to the concentration variance $C^2(\mathbf{x}, t)$. Dilution is the process by which solute is distributed through a larger volume of solvent, thereby reducing the maximum concentration (Kitanidis, 1994). The dilution index is a quantitative measure of mixing and dilution of a solute plume. Kitanidis (1994) used several instructive examples to build intuition about the concept. He showed that the dilution index grows as the solute plume diffuses and smooths. Any process that tends to distribute mass more evenly will tend to increase the dilution of the plume. The index is bounded by a maximum value, which in a finite domain corresponds to the case where the solute mass is distributed evenly through space, such that the solute concentration is everywhere the same. At the same time, the global variance $\sigma^2$ goes to zero when the solute mass is distributed evenly through space.

Kitanidis (1994) defined the dilution index using $P(\mathbf{x}, t)$, the concentration normalized by the total mass, $P(\mathbf{x}, t) = C(\mathbf{x}, t)/\int_\Omega C(\mathbf{x}, t)\, d\mathbf{x}$, as

$$E(t) = \exp\left[-\int_\Omega P(\mathbf{x}, t)\ln[P(\mathbf{x}, t)]\, d\mathbf{x}\right] \tag{11.9}$$

We define the local entropy of the concentration field as

$$\mathcal{S}(\mathbf{x}, t) = -P(\mathbf{x}, t)\ln[P(\mathbf{x}, t)] \tag{11.10}$$

This entropy is related to the dilution index through

$$E(t) = \exp\left[\int_\Omega \mathcal{S}(\mathbf{x}, t)\, d\mathbf{x}\right] \tag{11.11}$$

In keeping with the work of Kitanidis (1994), I believe that entropy as defined here can be best interpreted as a localized spatial measure of dilution, not as a thermodynamic or probabilistic quantity.

The importance of dilution or mixing of solvent and solute is that it controls the rate at which chemical reactions between solutes can progress. The rate at which a reaction between two solutes can progress depends upon the intrinsic reaction rate and the rate at which the solutes mix together. The rate at which solutes mix together can be quantified by either the rate at which dilution increases or the rate at which global variance decreases. A key result from Kitanidis (1994) is that the only mechanism that increases primitive-scale dilution is primitive-model dispersion and diffusion.

Similarly, we show next that primitive concentration variance can be reduced only by primitive-model dispersion and diffusion. In the remainder of this section, we examine the role of advection at different scales on the dilution of the grid-scale concentration field.

To develop an expression for the evolution of the primitive concentration variance, we first multiply the primitive transport equation (11.1) by $C(\mathbf{x}, t)$ and apply the chain rule of differentiation in reverse to yield

$$\frac{\partial C^2}{\partial t} + \frac{\partial}{\partial x_j}(V_j C^2) - \frac{\partial}{\partial x_j}\left(D_{jk}\frac{\partial C^2}{\partial x_k}\right) = -2\frac{\partial C}{\partial x_j} D_{jk} \frac{\partial C}{\partial x_k} \tag{11.12}$$

The result is the familiar advection–dispersion equation with a sink term. To relate this expression to the global variance, we integrate the equation over the domain $\Omega$ and apply the Gauss theorem to the divergence terms to yield

$$\frac{d\sigma_C^2}{dt} = -\frac{2}{\int_\Omega d\mathbf{x}} \int_\Omega \frac{\partial C}{\partial x_j} D_{jk} \frac{\partial C}{\partial x_k}\, d\mathbf{x} \tag{11.13}$$

expressing the rate at which the global concentration variance is reduced. Because the dispersion tensor is positive-definite, the right-hand side of equation (11.13) is always negative. These two results show that the sink term on the right-hand side of the concentration-variance equation (11.12) is responsible for the reduction in primitive global concentration variance $\sigma_C^2$. Kapoor and Gelhar (1994) provided a much more thorough analysis of concentration variance and coefficient of variation, but in a probabilistic framework. Because we use a spatial-averaging approach, our results are not directly comparable to theirs.

Very similar equations can be developed for the primitive entropy by multiplying the primitive transport equation (11.1) in terms of normalized concentrations $P(\mathbf{x}, t)$ by $-\ln[P(\mathbf{x}, t)]$:

$$\frac{\partial \mathcal{S}}{\partial t} + \frac{\partial}{\partial x_j}(V_j \mathcal{S}) - \frac{\partial}{\partial x_j}\left(D_{jk}\frac{\partial \mathcal{S}}{\partial x_k}\right) = \frac{1}{P}\frac{\partial P}{\partial x_j} D_{jk} \frac{\partial P}{\partial x_k} \tag{11.14}$$

where the right-hand side is always greater than or equal to zero. This is again the advection–dispersion equation, now with a source of entropy arising from primitive-model dispersion. An expression for the logarithm of the dilution index can be developed from this equation by integrating over the volume $\Omega$ and applying the Gauss theorem to the divergence terms to yield

$$\frac{d\ln(E)}{dt} = \int_\Omega \frac{1}{P}\frac{\partial P}{\partial x_j} D_{jk} \frac{\partial P}{\partial x_k} \tag{11.15}$$

Kitanidis (1994) developed the same equation and concluded that primitive-model dispersion is the only source of increased dilution.

If dispersion is the only means to destroy variance and increase entropy in a primitive model, what is the role of subgrid-scale advection in a larger, grid-scale model of transport? Intuitively, viewed from a large enough scale, smaller-scale subgrid advection "looks" like dispersion or diffusion. At some scale, subgrid advection tends to smooth out larger-scale concentration fluctuations, so we would expect subgrid advection to also drain variance from and supply entropy to the larger-scale concentration field.

Figures 11.6 and 11.7 provide an illustration of the effects of subgrid advection on the dilution and variance of a solute when viewed at two different scales. They show the same contaminant plume discretized on two different scales, after horizontal advection without dispersion or diffusion through a $256\Delta \times 256\Delta$ heterogeneous groundwater velocity field, where $\Delta$ is the scale of a grid block. The finer grid corresponds to the primitive model or "true" plume, and the coarser grid to some larger-scale representation of the same plume. The coarser-grid plume was obtained by averaging over the finer-grid plume using a spatial filter with a characteristic filter width of approximately 5 grid blocks.

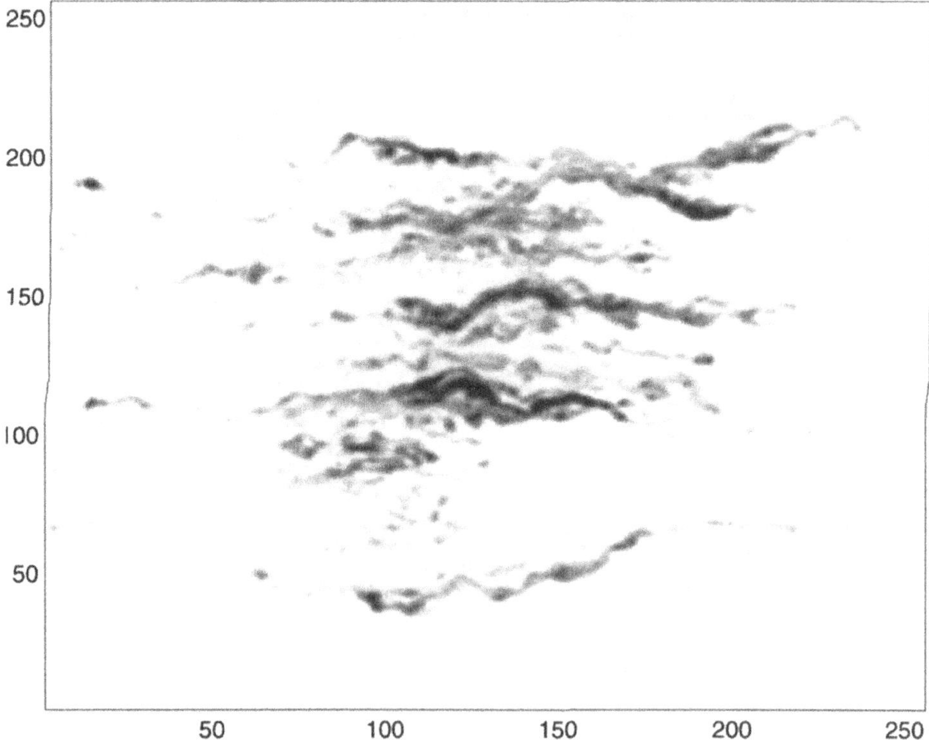


Figure 11.6. A solute plume represented at a fine scale.

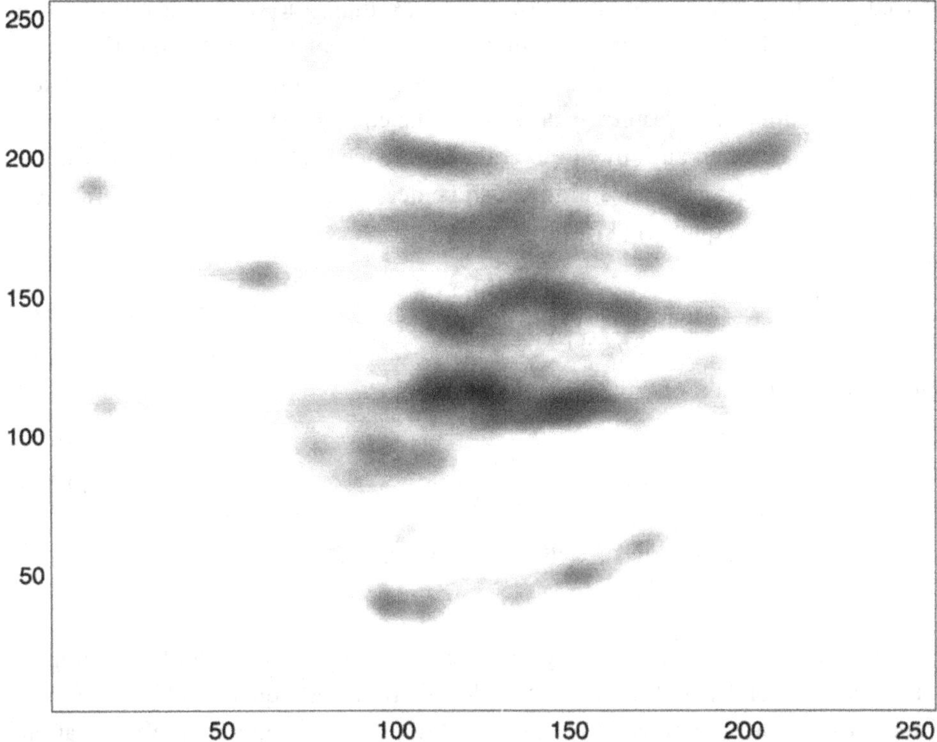


Figure 11.7. The same solute plume as in Figure 11.6 represented at a coarser scale. This plume was computed by spatially filtering the finer-scale plume using a spatial filter of width approximately $5\Delta$, where $\Delta$ is the width of a grid block.

Table 11.1 shows the variance and dilution of the coarse- and fine-grid plume computed using the discretized form of equation (11.8) and the discrete form of the dilution index from Kitanidis (1994). The variance and dilution are computed before and after horizontal advection. Initially the plume is distributed evenly in a box of approximately $48\Delta \times 225\Delta$. After advection, shown in Figures 11.6 and 11.7, the variance and dilution index of the fine-grid plume are unchanged, as expected from equations (11.13) and (11.15), while the variance of the coarse-grid plume has been

Table 11.1. *Variance and dilution for fine-grid and coarse-grid plumes*

| | Fine grid | | Coarse grid | |
|---|---|---|---|---|
| Parameter | Before | After | Before | After |
| Variance ($\Delta^4\sigma^2$) | $1.91 \times 10^3$ | $1.91 \times 10^3$ | $1.6 \times 10^3$ | $8.94 \times 10^2$ |
| Dilution ($E/\Delta^2$) | $1.08 \times 10^4$ | $1.08 \times 10^4$ | $1.15 \times 10^4$ | $1.85 \times 10^4$ |

reduced and the dilution index has grown. The example shows that primitive-scale advection can reduce the variance and increase the dilution of the coarser-grid representation, but not affect the variance or dilution of the primitive concentration field.

We can quantify these results by developing the equations that describe the evolution of the variance and entropy of the grid-scale concentration. An equation for the variance of the grid-scale concentration can be derived by multiplying a rearranged grid-scale transport equation (11.4) by $\bar{C}$:

$$\bar{C}\left[\frac{\partial \bar{C}}{\partial t} + \frac{\partial}{\partial x_j}(\bar{V}_j\bar{C}) - \frac{\partial}{\partial x_j}\left(D_{jk}\frac{\partial \bar{C}}{\partial x_k}\right) = \frac{\partial}{\partial x_j}(\overline{V_j C} - \bar{V}_j\bar{C})\right] \tag{11.16}$$

which, after application of the chain rule, becomes

$$\begin{aligned}\frac{\partial \bar{C}^2}{\partial t} &+ \frac{\partial}{\partial x_j}(\bar{V}_j\bar{C}^2) - \frac{\partial}{\partial x_j}\left(D_{jk}\frac{\partial \bar{C}^2}{\partial x_k}\right) \\ &= -2\frac{\partial \bar{C}}{\partial x_j}D_{jk}\frac{\partial \bar{C}}{\partial x_k} + 2\bar{C}\frac{\partial}{\partial x_j}(\overline{V_j C} - \bar{V}_j\,\bar{C})\end{aligned} \tag{11.17}$$

An equation for the entropy of the grid-scale concentration, $\mathcal{S}_{\bar{C}}$, can be developed in a similar fashion by multiplying the grid-scale transport equation (11.4) in terms of normalized concentrations $\bar{P}$ by $-\ln(\bar{P})$. After some manipulation, the equation becomes

$$\begin{aligned}\frac{\partial \mathcal{S}_{\bar{C}}}{\partial t} &+ \frac{\partial}{\partial x_j}(\bar{V}_j\mathcal{S}_{\bar{C}}) - \frac{\partial}{\partial x_j}\left(D_{jk}\frac{\partial \mathcal{S}_{\bar{C}}}{\partial x_k}\right) \\ &= \frac{1}{\bar{P}}\frac{\partial \bar{P}}{\partial x_j}D_{jk}\frac{\partial \bar{P}}{\partial x_k} + [1 - \ln(\bar{P})]\frac{\partial}{\partial x_j}(\overline{V_j P} - \bar{V}_j\bar{P})\end{aligned} \tag{11.18}$$

The grid-scale variance and entropy equations are identical with their primitive counterparts except for the appearance of new source terms $2\bar{C}(\partial/\partial x_j)(\overline{V_j C} - \bar{V}_j\bar{C})$ and $[1 - \ln(\bar{P})](\partial/\partial x_j)(\overline{V_j P} - \bar{V}_j\bar{P})$. As before for the primitive equations, the source terms can be interpreted as the effects of advection upon the variance and entropy of the grid-scale concentration field. These source terms will in turn appear in the equations for the global variance of the grid-scale concentration, $\sigma^2_{\bar{C}}$, and dilution index $E_{\bar{C}}(t)$ [cf. the primitive equations (11.13) and (11.15)].

The results of this section support our intuitive notions about subgrid advection. The additional source terms in the equations for the variance and entropy of the grid-scale concentration, equations (11.17) and (11.18), show that subgrid advection affects the variance and entropy of the grid-scale concentration field. For the most part, these source terms drain variance from and supply entropy to the grid-scale concentration field. In contrast, the primitive concentration field's variance and dilution are not affected by advection.

To fully explore the meaning of the grid-scale variance and entropy equations requires a more complicated mathematical analysis, which is beyond the scope of this chapter. The studies of the transport equation by Rose (1977) and of the Navier-Stokes equations in the context of turbulence by Leslie and Quarini (1979), Piomelli et al. (1991), and Zhou (1991) show that the grid-scale effect of subgrid advection (of either solute or momentum) is to transfer primitive variance and entropy between scales, with no effect on the global variance or dilution index. Variance can flow from larger to smaller scales (called forward scatter) (Piomelli et al., 1991) or from smaller to larger scales (called backscatter). Mathematically, the additional source terms in equations (11.17) and (11.18) can be either negative or positive and can either supply or drain variance from the grid-scale concentration field. This is in contrast to the source terms from the primitive concentration variance and entropy, equations (11.12) and (11.14), which always drain variance and supply entropy to the primitive concentration field. Even in the grid-scale model, the net flow of variance tends to be from the larger scales to the smaller scales, where variance is more aggressively consumed by the dispersive terms (Kraichnan, 1987). In a grid-scale model, the conservative transfer of variance and entropy between scales due to advection is interrupted at the grid scale. This transfer of variance from grid scale to subgrid scale (forward scatter), and the reverse (backscatter), must be reintroduced into the grid-scale model to accurately model the grid-scale dynamics. This is the role of subgrid modeling, the subject of the next section.

To conclude this section, it is worthwhile to briefly speculate about the implications of the additional grid-scale mixing provided by subgrid advection. The mixing of the grid-scale concentration field caused by subgrid advection leads to a more uniform grid-scale concentration field, but does not affect the mixing of the primitive concentration field. The smooth description of a grid-scale model may not be detailed enough for use in reactive-solute models, in particular when the reactions are relatively fast and the grid blocks are relatively large (e.g., Molz and Widdowson, 1988; Smith, Harvey, and Leblanc, 1991). The significant length scale for a reaction depends upon the reaction time scales compared with the time required for mixing. Fast reactions occur on small spatial scales before reactants have time to mix together at larger scales. Slower reactions allow the reactants time to mix over larger volumes, which enhances overall reaction progress. Thus, subgrid advection can be expected to enhance the larger-scale mixing that can affect slower reactions. Faster reactions likely are not promoted by larger-scale mixing caused by subgrid advection.

## 11.5 Subgrid-Closure Modeling Strategies for Groundwater Transport

We close this chapter with a speculative examination of methods that can be used to develop subgrid-closure models for the groundwater transport equation. We limit ourselves to a brief and incomplete review of this vast literature. Subgrid-closure modeling is closely related to constitutive theory from continuum mechanics (Eringen, 1980, p. 148). The principal goal of both is to provide a model for the larger-scale effects of unresolved, smaller-scale processes.

The problem of subgrid-closure modeling appears in the filtered transport equation as the problem of replacing the filtered advective term $\overline{V_j C}$ from equations (11.4) and (11.5) by some functional of grid-scale variables and parameters. Perhaps the simplest closure model is the Fickian closure, which is written with an enhanced or macrodispersion coefficient $D^*_{jk}$ as

$$\overline{V_j C} = \bar{V}_j \bar{C} - D^*_{jk} \frac{\partial \bar{C}}{\partial x_k} \tag{11.19}$$

or,

$$-D^*_{jk} \frac{\partial \bar{C}}{\partial x_k} = (\overline{\bar{V}_j \bar{C}} - \bar{V}_j \bar{C}) + \overline{\bar{V}_j c} + \overline{v_j \bar{C}} + \overline{v_j c} \tag{11.20}$$

The Fickian model has the advantage of simplicity. It presents the theoretical challenge of determining appropriate expressions for the macrodispersion coefficient.

Analysis of macrodispersion and the Fickian model has been a topic of considerable research in the groundwater literature, although most of it in a stochastic (versus spatial-filtering) framework; see, for example, Gelhar and Axness (1983), Dagan (1984, 1988), Sposito, Jury, and Gupta (1986), Neuman, Winter, and Newman (1987), and Sposito and Barry (1987). These stochastic results are not directly applicable to the problem of subgrid-closure modeling, for they do not account for the effect of the scale of the numerical grid on macrodispersion.

The more recent two-particle stochastic approaches of Dagan (1990, 1992), Rubin (1991), and Rajaram and Gelhar (1993) and the spatial-moments approach of Kitanidis (1988) have considered the effect of the plume scale on the magnitude of the dispersivity. These results show that the magnitude of the plume's dispersion coefficient grows as the plume grows in scale. Indeed, as the plume grows, it can experience larger-scale velocity variability. Conversely, velocity variability on scales larger than the plume act to sweep the entire plume, but do not spread it (Rajaram and Gelhar, 1993).

In these stochastic studies, the dispersion coefficient is a measure of the rate at which the second spatial moment of a plume grows. It is a global, Lagrangian dispersion coefficient that describes a single plume. To apply these models to a site with multiple plumes would require that each plume be described by its own dispersion coefficient. In contrast, the idea behind an LES spatial-filtering model is to develop a subgrid closure that is a property of each grid block of a numerical model.

Although they have received less attention, there are some alternatives to the Fickian model. Tompson and Gray (1986) and Tompson (1988) developed a second-order model for transport in which the dynamics of the dispersive flux [equivalent to the term $\overline{v_j c}$ from equation (11.5)] are governed by a separate equation instead of a simple Fickian constitutive model. Their method was inspired by the moment-equation approaches used in some turbulence models (Mellor and Yamada, 1974). Tompson and Gray (1986) developed their model by volume-averaging a pore-scale primitive model up to what is essentially a laboratory scale. Tompson (1988) showed that under simplifying assumptions, the second-order dispersive-flux model is equivalent to a time-dependent dispersion-coefficient model. It is not clear how to adapt these results for use as a subgrid closure in a field-scale transport model.

Graham and McLaughlin (1989) applied a strategy very similar to that of Tompson and Gray (1986) in a stochastic framework. They modeled the dispersive flux using higher-order statistical moments of the primitive-model variables. The statistical moments can then be used to make probability statements about the model variables. Their approach is based upon a stochastic framework and cannot be directly compared with spatial-filtering methods. Indeed, the goal of their work is not to rescale a primitive model, but to quantify the uncertainty in the primitive-model variables.

We close with two examples from the LES literature. Rose (1977) developed an LES model for transport in a time-invariant, spatially heterogeneous velocity field with a power-law spatial-correlation structure similar to fully developed turbulence. He applied renormalization-group (RNG) methods from statistical mechanics to derive a grid-scale transport model. The RNG method is an LES method that blends probabilistic averaging with spatial filtering. Spatial filters are used to define grid and subgrid scales. Subgrid-closure terms are then replaced by their statistical mean. The grid scales are treated deterministically. The resulting grid-scale model contains a grid-scale advection term, a Fickian macrodispersion term, a stochastic stirring term called "eddy advection," and a difficult-to-interpret term that Rose (1977) described as grid-scale advection of a nonlocal density. The eddy-advection term and the nonlocal term account for the strong coupling among the scales that are just larger and just smaller than the grid scale (i.e., the cross terms). This strong coupling means that the effects of dynamics on these scales cannot be replaced by their statistical-mean effect and an enhanced-dispersion coefficient. As the RNG model is derived using a spectrally sharp filter, it is most appropriately solved numerically with the spectral-numerical methods described by Canuto et al. (1987). The combination of spatial filtering and probabilistic averaging over the subgrid scales makes the RNG approach an attractive alternative.

Last, the expansion techniques of Leonard (1974), Aldama (1990, 1992), and Beckie et al. (1996) can be applied to create a subgrid closure for the filtered advective term. These expansion methods are based upon the Gaussian filter function. Leonard (1974) developed an expansion that approximates the nonlocal term $\overline{\bar{V}_j \bar{C}}$ by a local expression. Aldama (1990, 1992) extended that approach to the cross terms and the subgrid term $\overline{v_j c}$. The cross-term approximations are asymptotic as the ratio of $\Lambda$ (the dominant dynamic scale of the grid-scale variables) to $\Delta$ (the filter width) goes to zero: $\Lambda/\Delta \to 0$. This ratio can be thought of as a type of separation-of-scales requirement: When the energetic scales $\Lambda$ are much larger than the scales being filtered away, $\Delta$, the approximations will be accurate. The closure for the subgrid term $\overline{v_j c}$ is not accurate if the subgrid scales are too energetic (Beckie et al., 1996).

## 11.6 Conclusions

We present a spatial-filtering framework that can be used to understand and develop numerical models of solute transport in groundwater. We begin with a primitive-model description of solute transport on some convenient smaller scale. Then we use spatial filters to extract the larger-scale component of the primitive model. The spatial filters mimic the effect of a numerical discretization and allow us to define the grid-scale variables and dynamics (which are explicitly resolved by the numerical grid) and smaller, unresolved subgrid-scale variables and dynamics. Spatial filtering of a primitive transport model leads naturally to subgrid-closure terms that represent the effects of subgrid-scale dynamics on explicitly resolved grid-scale dynamics. In an LES model, the grid-scale effect of subgrid-scale dynamics is modeled using grid-scale variables and a subgrid-closure model. In grid-scale solute-transport models, the effect of subgrid-scale advection is often represented as a Fickian dispersion.

We investigate the effect of subgrid-scale advection upon the grid-scale concentration field by examining the concentration variance and entropy, which are local measures of the variability and dilution of the concentration field. We show that in a primitive-scale model, only primitive-model dispersion and diffusion can decrease the variance and increase the dilution of the concentration. However, the variance and entropy of the grid-scale concentration field are affected by subgrid advection. The role of subgrid advection is to drain variance and supply entropy to the grid scales in much the same fashion as primitive-model dispersion and diffusion drain variance and supply entropy to the primitive concentration field.

It should be recognized that our analysis is principally concerned with models and model scale and is not a study of the fundamental physics and scales of transport in natural porous media. Indeed, our point of departure is a primitive model that we accept as an accurate description at some scale. Our analysis leads to scale effects such as forward scatter and backscatter, which are not easily recognized in the primitive model. These scale effects should be considered to be properties of the model and model scale, for they do not appear in an appropriately fine-scale primitive model. These scale effects may not be important in models of many physical systems, particularly when the variability is confined to a range of small scales or when a large proportion of the primitive physics can be explicitly resolved.

## References


Aldama, A. A. 1990. *Filtering Techniques for Turbulent Flow Simulation*. Lecture Notes in Engineering, vol. 56. Berlin: Springer-Verlag.

Aldama, A. A. 1992. A subgrid scale theory for physical processes with quadratic nonlinearities: an a priori test for Burgers' flow. In: *Computational Methods in Water Resources IX*, vol. 2, ed. T. F. Russel, R. E. Ewing, C. A. Brebbia, W. G. Gray, and G. F. Pinder, pp. 91–8. Southampton: Computational Mechanics Publications.

Baveye, P., and Sposito, G. 1984. The operational significance of the continuum hypothesis in the theory of water movement through soils and aquifers. *Water Resour. Res*. 20:521–30.

Bear, J. 1972. *Dynamics of Fluids in Porous Media*. New York: Dover.

Beckie, R. 1996. Measurement scale, network sampling scale, and groundwater model parameters. *Water Resour. Res*. 32:65–76.

Beckie, R., Aldama, A. A., and Wood, E. F. 1994. The universal structure of the groundwater flow equations. *Water Resour. Res*. 30:1407–19.

Beckie, R., Aldama, A. A., and Wood, E. F. 1996. Modeling the large-scale dynamics of groundwater flow: 1. Theoretical development. *Water Resour. Res*. 32:1269–80.

Bird, R., Steward, W., and Lightfoot, B. 1960. *Transport Phenomena*. New York: Wiley.

Bouchard, J.-P., and Georges, A. 1990. Anomalous diffusion in disordered media: statistical mechanisms, models and physical applications. *Phys. Rep*. 195:127–293.

Bracewell, R. 1965. *The Fourier Transform and Its Applications*. New York: McGraw-Hill.

Canuto, C., Hussaini, M. Y., Quarteroni, A., and Zang, T. A. 1987. *Spectral Methods in Fluid Dynamics*. Berlin: Springer-Verlag.

Chu, S.-Y., and Sposito, G. 1980. A derivation of the macroscopic solute transport equation for homogeneous, saturated porous media. *Water Resour. Res*. 16:542–6.

Dagan, G. 1984. Solute transport in heterogeneous porous formations. *J. Fluid Mech*. 145:151–77.

Dagan, G. 1988. Time-dependent macrodispersion for solute transport in anisotropic heterogeneous aquifers. *Water Resour. Res*. 24:1491–500.

Dagan, G. 1990. Transport in heterogeneous porous formations: spatial moments, ergodicity, and effective dispersion. *Water Resour. Res*. 26:1281–90.

Dagan, G. 1992. Dispersion of a passive solute in non-ergodic transport by steady velocity fields in heterogeneous formations. *J. Fluid Mech*. 233:197–210.

Eringen, A. C. 1980. *Mechanics of Continua*. Huntington, NY: Robert E. Krieger.

Frisch, U., and Orszag, S. A. 1990. Turbulence: challenges for theory and experimentation. *Physics Today* 43:24–32.

Gelhar, L. W., and Axness, C. L. 1983. Three-dimensional stochastic analysis of macrodispersion in aquifers. *Water Resour. Res*. 19:161–80.

Graham, W., and McLaughlin, D. B. 1989. Stochastic analysis of nonstationary subsurface solute transport. 1. Unconditional moments. *Water Resour. Res*. 25:215–32.

Kapoor, V., and Gelhar, L. W. 1994. Transport in three-dimensionally heterogeneous aquifers. 1. Dynamics of concentration fluctuations. *Water Resour. Res*. 30:1775–88.

Kitanidis, P. K. 1988. Prediction by the method of moments of transport in a heterogeneous formation. *J. Hydrol*. 102:453–73.

Kitanidis, P. K. 1994. The concept of the dilution index. *Water Resour. Res*. 30:2011–26.

Kraichnan, R. H. 1987. Eddy viscosity and diffusivity: exact formulas and approximations. *Complex Systems* 1:805–20.

Leonard, A. 1974. Energy cascade in large-eddy simulations of turbulent fluid flows. *Adv. Geophys*. 18A:237–48.

Leslie, D. C., and Quarini, G. L. 1979. The application of turbulence theory to the formulation of subgrid modelling procedures. *J. Fluid Mech.* 91:65–91.
McComb, W. D. 1990. *The Physics of Fluid Turbulence*. Oxford University Press.
Mellor, G. L., and Yamada, T. 1974. A hierarchy of turbulence closure models for planetary boundary layers. *J. Atmos. Sci.* 31:1791–806.
Molz, F. J., and Widdowson, M. A. 1988. Internal inconsistencies in dispersion-dominated models that incorporate chemical and microbial kinetics. *Water Resour. Res.* 24:615–19.
Neuman, S. P. 1990. Universal scaling of hydraulic conductivities and dispersivities in geologic media. *Water Resour. Res.* 26:1749–58.
Neuman, S. P., Winter, C. L., and Newman, C. M. 1987. Stochastic theory of field-scale Fickian dispersion in anisotropic porous media. *Water Resour. Res.* 23:453–66.
Piomelli, U., Cabot, W. H., Moin, P., and Lee, S. 1991. Subgrid–scale backscatter in turbulent and transitional flows. *Phys. Fluids A* 7:1766–71.
Rajaram, H., and Gelhar, L. W. 1993. Plume scale-dependent dispersion in heterogeneous aquifers. 1. Lagrangian analysis in a stratified aquifer. *Water Resour. Res.* 29:3249–60.
Rogallo, R. S., and Moin, P. 1984. Numerical simulation of turbulent flows. *Ann. Rev. Fluid Mech.* 16:99–137.
Rose, H. A. 1977. Eddy diffusivity, eddy noise, and subgrid-scale modelling. *J. Fluid Mech.* 81:719–34.
Rubin, Y. 1991. Transport in heterogeneous porous media: prediction and uncertainty. *Water Resour. Res.* 27:1723–38.
Scheibe, T. D., and Cole, C. R. 1994. Non-Gaussian particle tracking: application to scaling of transport processes in heterogeneous media. *Water Resour. Res.* 30:2027–40.
Smith, R. L., Harvey, R. W., and Leblanc, D. R. 1991. The importance of closely spaced vertical sampling in delineating chemical and microbiological gradients in groundwater studies. *J. Contam. Hydr.* 7:285–300.
Sposito, G., and Barry, D. A. 1987. On the Dagan model of solute transport in groundwater: foundational aspects. *Water Resour. Res.* 23:1867–75.
Sposito, G., Jury, W. A., and Gupta, V. K. 1986. Fundamental problems in the stochastic convection–dispersion model of solute transport in aquifers and field soils. *Water Resour. Res.* 22:77–88.
Tompson, A. F. B. 1988. On a new functional form for the dispersive flux in porous media. *Water Resour. Res.* 24:1939–47.
Tompson, A. F. B., and Gray, W. G. 1986. A second-order approach for the modeling of dispersive transport in porous media. 1. Theoretical development. *Water Resour. Res.* 22:591–9.
Zhou, Y. 1991. Eddy damping, backscatter, and subgrid stresses in subgrid modeling of turbulence. *Phys. Rev. A* 43:7049–52.

# 12

# Scale Effects in Fluid Flow through Fractured Geologic Media

PAUL A. HSIEH

## 12.1 Introduction

Fractures result from mechanical breaks in intact geologic media such as rocks or compacted glacial tills. Although a fracture that is completely filled by minerals is still considered a fracture in the geologic sense, within the context of subsurface fluid flow we think of a fracture as a mechanical break that results in void space between the fracture walls. This void space is more or less planar – one of its dimensions (the aperture or distance between fracture walls) is much smaller than the other two (the extension of the fracture plane). When interconnected, fractures provide pathways for fluid flow through geologic media that would be significantly less permeable if the media were unfractured.

The geometry of subsurface fractures varies greatly, with fracture lengths ranging from less than a millimeter (e.g., a microcrack in a rock grain) to thousands of kilometers (e.g., a fault along a tectonic-plate boundary). Fracture apertures vary from minute "hairline" cracks, nearly imperceptible to the naked eye, to solution-enlarged channels wide enough for human exploration. Fractures can be highly interconnected in a densely fractured rock, or isolated and poorly connected in a sparsely fractured rock. Some fracture networks exhibit a nested pattern, with smaller fractures bounded by larger ones (Barton and Hsieh, 1989). Studies of the processes that create fractures over this broad range of scales constitute an active area of research in the earth sciences.

In a study of fragmentation, Turcotte (1986) found that the fragmentation process often results in a power-law or fractal distribution of fragment sizes. A fractal distribution is said to exhibit scale invariance because, except for a scaling factor, the frequency of larger fragments is the same as the frequency of smaller fragments. An implication of Turcotte's findings is that fracture patterns in nature might also exhibit scale invariance. In fact, fractal characteristics have been observed in studies of fractures patterns (e.g., Barton and Hsieh, 1989; Poulton, Mojtabai, and Farmer, 1990). Such observations have motivated the use of fractal-generation techniques such as iterative-function systems (Barnsley, 1988) to model fracture networks (e.g., Doughty et al., 1994; Acuna and Yortsos, 1995).

Whereas the network pattern is one factor that controls fluid flow in fractured media, an equally important factor is the ability of fractures to conduct fluid. Long et al. (1991) reported that in a study at the Fanay-Augeres mine in France, very few of the observed fractures were found to conduct fluid. Thus, even if a fracture network possesses scale invariance, that may not necessarily be true for hydraulic conductivity. Although major experiments have been carried out in the past decade to investigate flow and transport in fractured rocks (e.g., Cacas et al., 1990a,b; Abelin et al., 1991a,b; Dverstorp, Andersson, and Nordqvist, 1992; Tsang, Tsang, and Hale, 1996), those experiments were aimed primarily at site characterization and model validation. In experiments designed to study the relationship between scale and hydraulic conductivity (e.g., Guimera, Vives, and Carrera, 1995), the findings have been equivocal.

This chapter examines fluid flow through fractured geologic media at different length scales. We limit the discussion to single-phase, saturated, nearly isothermal conditions, where changes in hydraulic head are small compared with what is needed to alter the fracture system (e.g., we do not consider high-pressure injection and hydraulic fracturing). The emphasis is on field application. We begin with an overview of study approaches. A key issue that emerges from this overview is the characterization of heterogeneity or spatial variability. We discuss the concepts of "scale dependence" and "scale effect" from the perspectives of laboratory testing, borehole testing, and regional-scale modeling. Next, we review some field data to examine the relationship between hydraulic conductivity and measurement scale. The chapter concludes with a discussion of possible directions for future research.

## 12.2 Study Approaches

The study of fluid flow in fractured media is conventionally divided into two approaches: the continuum approach and the discrete-fracture approach. A recent book by the Committee on Fracture Characterization and Fluid Flow (1996, ch. 6) gives an extensive review of this topic. The continuum approach does not consider flow through every fracture. Instead, the fracture network is represented as if it were a granular porous medium. Fractures are analogous to pores; the intact rock blocks (assumed impermeable) are analogous to grains. In the continuum approach, the methods for analyzing flow in fractured media are identical with the methods for porous media. If the intact blocks themselves are also permeable, they can be represented by a "block continuum" that overlaps and interacts with the fracture continuum. This gives rise to the double-porosity model.

By contrast, the discrete-fracture approach considers flow through individual fractures. Each fracture in the network is specified by its location, shape, orientation,

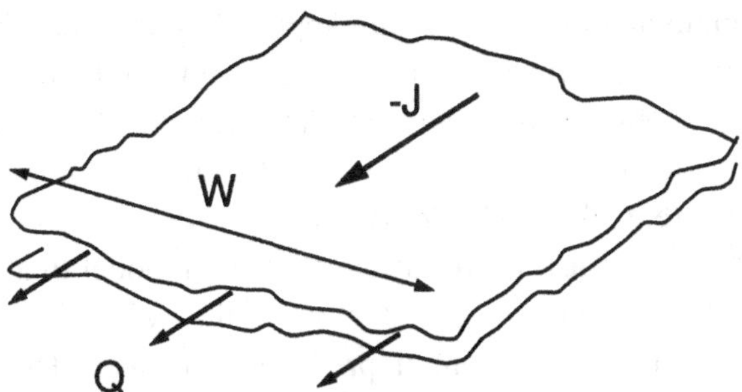


Figure 12.1. Discharge ($Q$) over a fracture width ($W$) under a hydraulic gradient ($J$). The fracture transmissivity is defined as $Q/(WJ)$.

spatial extent, and transmissivity (defined as the discharge per unit length of fracture under a unit hydraulic gradient) (Figure 12.1). In a two-dimensional discrete-fracture model, each fracture is represented by a line element. Fluid flow through the fracture network is analogous to electric-current flow in a resistor network, with a fracture analogous to a resistor. In three dimensions, each fracture is represented by a plane of finite extent, typically a circular or elliptical disk (Figure 12.2). Flow through the network is calculated by numerical methods that typically require substantial computational resources (e.g., Long, Gilmour, and Witherspoon, 1985). To reduce the computational requirement, attempts have been made to replace the three-dimensional network of disks by a three-dimensional network of pipes (e.g., Cacas et al., 1990a,b).

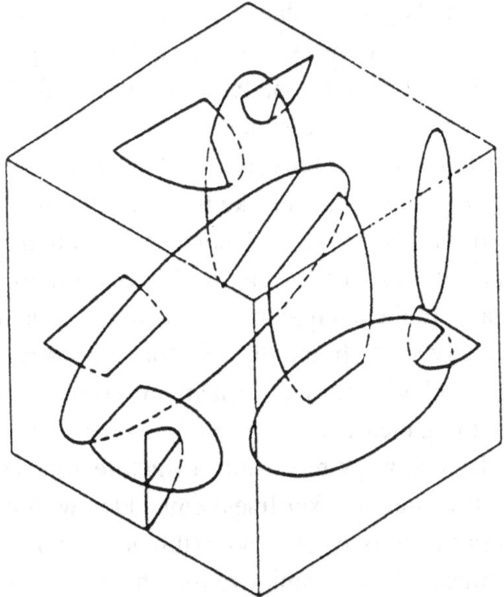

Figure 12.2. Example of a fracture network composed of disk-shaped fractures. (From Dershowitz and Einstein, 1988, with permission.)

Both the continuum and discrete-fracture approaches are amenable to deterministic or stochastic treatment. A deterministic-continuum model treats the hydraulic conductivity field as a deterministic function that is estimated by model calibration. To avoid nonuniqueness in calibration results, the function is commonly assumed to take on a simple form (e.g., the model domain might be divided into several homogeneous subregions). By contrast, a stochastic-continuum model treats the hydraulic-conductivity distribution as a random field in which the hydraulic conductivity fluctuates from point to point in a random, unpredictable manner. The random field is characterized by the mean and variance of the hydraulic conductivity and its correlation over varying distances. This information is then used to derive the mean, variance, and correlation for the hydraulic head and flow field.

For the discrete-fracture approach, generally there are insufficient data to specify the geometry and transmissivity of individual fractures. Practitioners of discrete-fracture modeling tend to avoid assuming a deterministic fracture pattern. Instead, a stochastic treatment is implemented by considering the fracture geometric parameters (spatial location, orientation, lateral extent, etc.) and transmissivity as random variables whose probability density functions are estimated from field data. By generating random geometric parameters and transmissivity from these probability density functions, a synthetic fracture network is created, and flow through the network can be solved by computer simulation. The variability in flow through different synthetic networks can be assessed by repeating the network generation and flow simulation many times.

Although the continuum and discrete-fracture approaches might appear diametrically opposed to each other, they in fact share the same underlying physical principles. Both approaches require that fluid mass be conserved and that flow be proportional to head gradient (Darcy's law). Additionally, for transient conditions, fluid storage occurs by elastic expansion or compression of fluid and matrix and by drainage of pore space as the water table rises or falls (Bear, 1972; Bear, Tsang, and de Marsily, 1993). For cases involving open-channel-like flow in solution-enlarged conduits in carbonate rocks, a nonlinear flow law (e.g., Scheidegger, 1960) might replace Darcy's law. The main difference between the two approaches is how they represent the structure of the model domain. In the continuum approach, the structure is characterized by a hydraulic-conductivity field. In the discrete-fracture approach, the structure is characterized by the network geometry and fracture transmissivity.

To illustrate the foregoing ideas, consider a simple, two-dimensional fracture network (Figure 12.3a). This network represents a fracture zone composed of denser and more transmissive fractures (thicker lines) embedded within a background network of sparser and less transmissive fractures (thinner lines). The fractures in the fracture zone are 100 times more transmissive than the fractures in the background network. The intact rock blocks are impermeable. To simulate flow through the network, we assign a hydraulic head of unity along the left boundary, a head of zero

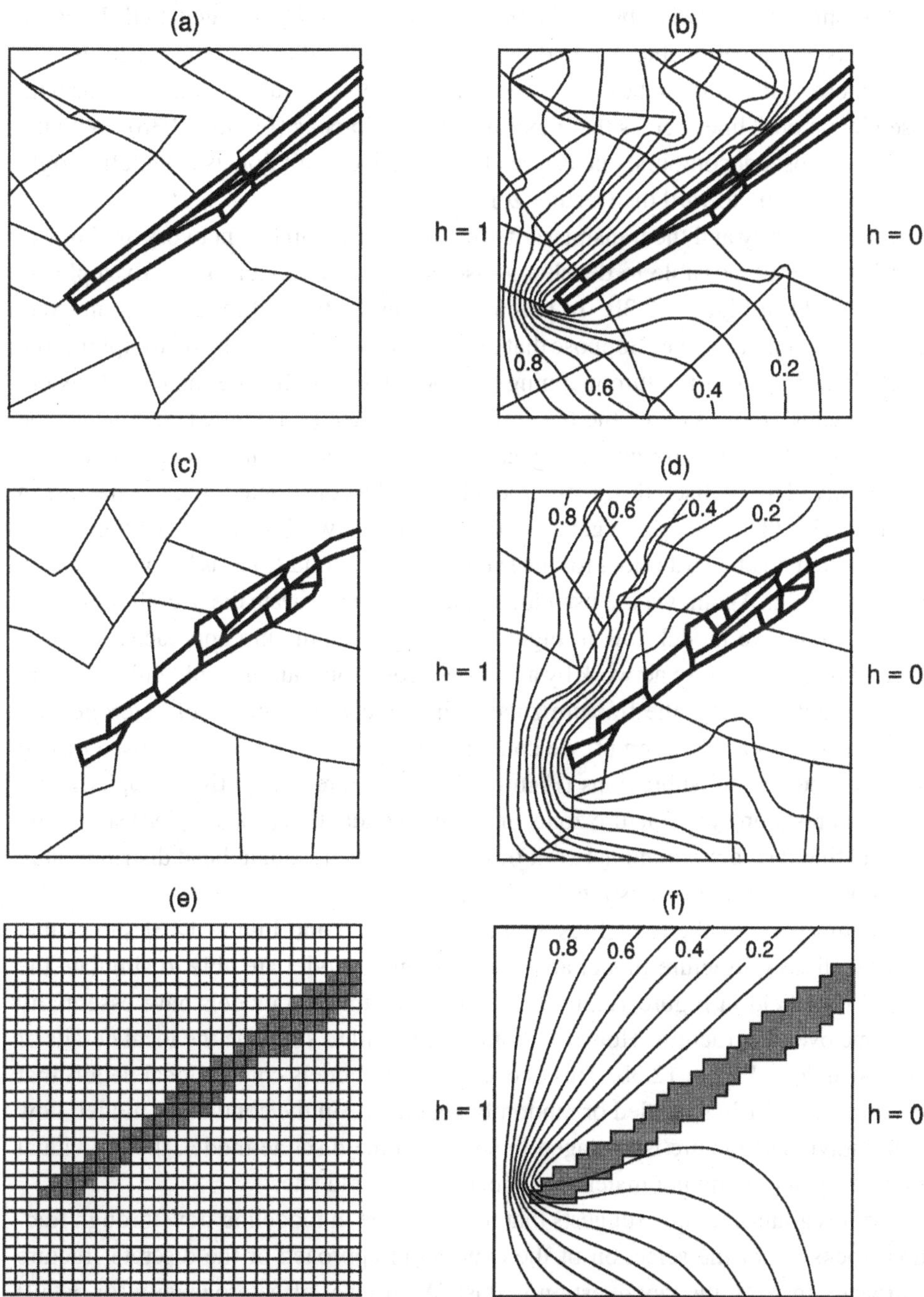


Figure 12.3. Simulation of flow in a fracture network: (a) the real fracture network; (b) distribution of hydraulic head in the real network; (c) discrete-fracture model; (d) distribution of hydraulic head in the discrete-fracture model; (e) continuum model; (f) distribution of hydraulic head in the continuum model.

along the right boundary, and a linearly varying head from unity to zero along the top and bottom. The head distribution in the network can readily be calculated. Because the blocks are impermeable, the hydraulic head is defined only in the fractures. For the purpose of illustration, however, we connect points of equal hydraulic heads with "pseudocontour" lines across the impermeable rock blocks (Figure 12.3b). With the aid of these pseudocontours, one can readily see that overall flow is from left to right and is converging toward the fracture zone.

If the geometry and the transmissivity of each fracture in this network are known, then it should be possible to develop a discrete-fracture model to exactly replicate the flow field in Figure 12.3b. In practice, fracture data are limited; thus an exact replication is unattainable. Suppose that by a combination of geophysical exploration, borehole testing, and fracture mapping it is possible to delineate the overall extent of the fracture zone, sample the transmissivity of selected fractures, and determine frequency distributions for geometric parameters such as fracture orientation, length, and density. Then a discrete-fracture model might be constructed, such as the one in Figure 12.3c. (The method to generate the fracture network is not of concern here.) The resultant head distribution, for the same boundary conditions as before, is shown in Figure 12.3d. Again, we use pseudocontour lines to illustrate the overall flow field.

How would we model flow in Figure 12.3b by a continuum approach? Clearly, representing the fracture network by a homogeneous continuum would fail to capture an important feature – the fracture zone. Thus, a certain amount of heterogeneity is necessary. As a first attempt, we construct a continuum model composed of two homogeneous model subregions (Figure 12.3e): a more conductive (stippled) subregion representing the fracture zone, and a less conductive (nonstippled) subregion representing the background network of fractures. The resultant head distribution in this continuum model is illustrated in Figure 12.3f.

A comparison of the model results in Figure 12.3d and Figure 12.3f shows that both the discrete-fracture model and the continuum model are able to simulate the overall flow field in Figure 12.3b. This is because the two models have essentially the same overall structure. However, both models fail to simulate the local details of the flow field in Figure 12.3b. This is because both models contain local details that are either arbitrarily specified or randomly generated. In the continuum model, flow in an individual fracture is averaged in space. In the discrete model, the individual fractures generally do not match those in an actual field site.

The foregoing example suggests that the key issue in modeling flow through fractured rocks is not the selection of the continuum approach or the discrete-fracture approach. Instead, the key question is this: What level of heterogeneity (or detail) should be incorporated into the model? The answer will depend on knowledge of the field site, the scale of interest, and the purpose of the investigation. For example, because propagation of hydraulic head is less sensitive to local heterogeneities than is movement of dissolved chemicals, a study to evaluate groundwater withdrawal might require less detailed knowledge of the subsurface than would a study to evaluate contaminant movement.

For model construction, it is useful to think of the heterogeneity in terms of resolved and unresolved features. In the foregoing example, the two resolved features are the fracture zone and the background fracture network. The unresolved features are the details of the fracture patterns, which are spatially variable and unknown. The continuum and discrete-fracture approaches can be viewed as different ways to deal with unresolved features. A deterministic-continuum model ignores the spatial variability in the unresolved features, and instead assigns hydraulic-conductivity values to model subregions, so that the average flow through a subregion is similar to the average flow through the unresolved features in the subregion. By contrast, a stochastic-continuum model mimics the variability in the unresolved features by using random hydraulic-conductivity fields. A deterministic-discrete-fracture model might represent the unresolved features by an arbitrary network pattern, although, as mentioned earlier, this practice is not favored. A stochastic-discrete-fracture model generates random, synthetic networks having geometric parameters and transmissivities that are inferred from statistical sampling of the unresolved features.

It should be noted that the continuum approach does not necessarily imply homogeneity. In the past, several studies have examined the conditions under which a fractured rock can be represented by an equivalent, homogeneous porous medium (e.g., Long et al., 1982). Perhaps as an unintended consequence, homogeneity has become associated with the continuum approach. For example, if pumping from a well causes greater drawdown in a distant observation well than in a nearby observation well, and all three wells lie on a straight line, so that the response cannot be explained by anisotropy, then the test result is sometimes characterized as "noncontinuum" or "anomalous" behavior. In fact, such a response indicates the presence of heterogeneity. A similar response can occur in a sedimentary aquifer that contains a highly permeable zone, such as a meandering string of gravel deposited by a stream in the geologic past (i.e., a paleostream channel). If the pumped well and the distant well are both drilled into a highly permeable zone, but the nearby observation well is drilled outside the highly permeable zone (Figure 12.4), then similar "anomalous" behavior is observed. This behavior can be readily simulated with a conventional groundwater model by assigning different hydraulic-conductivity values to different model subregions. A similar approach can be used to model flow in fractured media that contain highly transmissive fracture zones.

We conclude this section by commenting on the notion of representative elementary volume (REV), which traditionally has provided the theoretical foundation for continuum theory (Bear, 1972). In the classic theory, the size of the REV is defined over a scale range that is substantially larger than the scale of the microscopic heterogeneity (e.g., pore-level variation in void space), but substantially smaller than the scale of the macroscopic heterogeneity (e.g., porosity variation across a formation). By

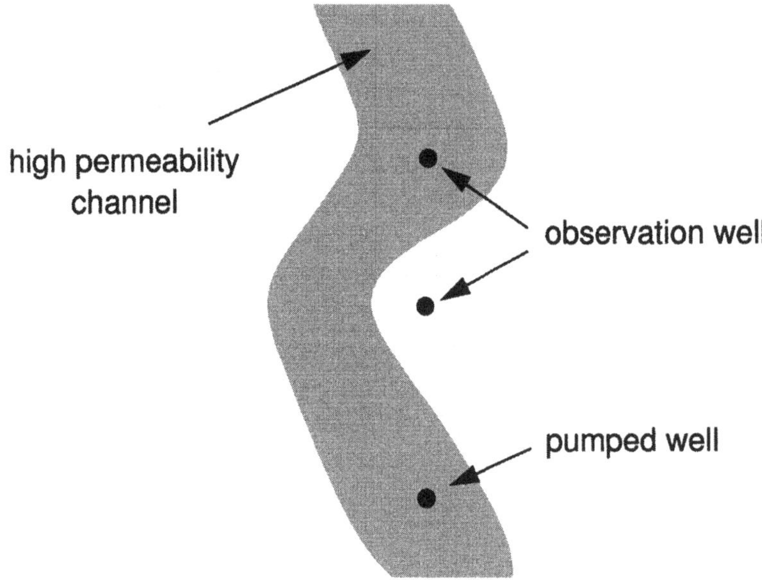


Figure 12.4. A pumped well and two observation wells in an aquifer containing a high-permeability channel.

averaging over the REV, microscopic heterogeneity is smoothed out, while macroscopic heterogeneity is left essentially unaffected. For granular porous media, an REV usually can be defined, because there exists a clear separation between microscopic and macroscopic scales. Within the "REV scale range," averaged quantities are independent of the size, shape, and orientation of the REV.

For fractured media, the heterogeneity typically occurs over a broad range of scales. Consequently, there may not exist a scale range over which an REV can be defined. Does the absence of an REV mean that the continuum approach is inapplicable, or that the REV concept is irrelevant? This question probably will remain a topic of debate. Alternatives to the REV concept have been proposed (e.g., Baveye and Sposito, 1984; Neuman, 1987). In the present discussion, we sidestep the issue of the REV and focus instead on heterogeneity. By incorporating an appropriate level of heterogeneity and detail into the flow model, the continuum approach has been successfully applied to simulate flow in fractured media (e.g., Carrera and Heredia, 1988).

## 12.3 Measurement and Scale Dependence

The presence of heterogeneity over a broad range of spatial scales leads directly to scale dependence. In a general sense, scale dependences means that what is observed on one scale might be different from what is observed on another scale. For example, at the scale of centimeters, a fractured medium might appear as an intact rock with no fracture, an intact rock containing one or more fractures, a mass of crushed fragments in a fault zone, or even a complete void within a solution-enlarged channel. As the scale of observation increases, closely spaced features that are correlated over larger distances tend to merge together to become a single large feature, and smaller, isolated features begin to lose their prominence. For example, a series of fractures arranged in a step-like (en echelon) fashion on an outcrop might appear as a single fault when viewed from the air. Conversely, a highly transmissive, through-going fracture at the scale of several meters might turn out to be a local "short circuit" at the kilometer scale.

From the perspective of flow in fractured media, different scales of observation correspond to different methods for determining hydraulic conductivity. These methods are often classified into three types: laboratory tests, borehole tests, and calibration of regional- or basin-scale modeling. Scale dependence in hydraulic conductivity means that the different methods of measurement might lead to different results, and thus different conceptualizations of the hydraulic-conductivity field.

Laboratory tests are typically performed on drill cores several centimeters in diameter and length. With sophisticated instruments, hydraulic conductivity as low as $10^{-17}\ \mathrm{m \cdot s^{-1}}$ has been measured (Trimmer et al., 1980). However, samples for laboratory tests are almost always taken from intact, unfractured portions of drill cores. Fractures in cores are often damaged during drilling, and they exist at a state of stress different from in situ conditions. Even if a fracture is undamaged, preparation of a fractured core for laboratory testing is difficult. A fracture with a wide opening (greater than several millimeters) is virtually impossible to reproduce in the laboratory. Thus, results of laboratory tests typically are representative of the intact rock.

Borehole tests are carried out by inducing a hydraulic stress (by pumping, injection, or pressurization) in a borehole and measuring the head or pressure response. If the response is measured in the same borehole, the test is a single-borehole test. If the response is measured in nearby observation boreholes, the test is a multiple-borehole test. Packers are often used to divide a borehole into sections for testing and for measurement. Analysis of the test response consists of two steps. The first step is to select a model to represent flow in the rock mass. The second step is to determine values for model parameters (such as hydraulic conductivity) such that the model-computed responses will match the test responses. Although fractured rocks are typically heterogeneous, most conventional models for borehole-test analysis assume that the rock is homogeneous. Consequently, the hydraulic conductivity derived with these models is an "equivalent" or "effective" property.

The volume sampled by a borehole test varies with the type of test. A single-borehole test roughly samples a cylindrical volume that extends vertically over the test interval (which can range from less than 1 m to more than 100 m) and radially outward to distances ranging from centimeters to tens of meters. By comparison, a multiple-borehole test samples a much larger volume, roughly that spanned by the network of boreholes. The dimensions of the sampled volume are typically tens of meters to more than 100 m.

Borehole testing methods are subject to practical constraints that limit their range of applicability. For example, the single-borehole slug test (Cooper, Bredehoeft, and Papadopulos, 1967), performed by rapidly raising or lowering the fluid level in a borehole and monitoring the subsequent recovery, is limited by test duration. When testing low-permeability rocks, the recovery period can be exceedingly long. When testing high-permeability rocks, the recovery can be too rapid for accurate measurement. In a typical program of field testing, the practical range of hydraulic-conductivity measurement for a slug test is roughly $10^{-5}$–$10^{-9}$ $\mathrm{m \cdot s^{-1}}$. By confining the test interval with packers and imposing a pressure pulse (Bredehoeft and Papadopulos, 1980), the lower limit can be extended to about $10^{-14}$ $\mathrm{m \cdot s^{-1}}$.

Single-hole injection and withdrawal (pumping) tests are limited by the flow capacity and by accuracy in flow-rate measurement. When testing low-permeability rocks, injection is preferred over withdrawal, because the flow rate may be too small to sustain steady pumping. For injection tests, the practical lower limit of hydraulic conductivity is about $10^{-10}$ $\mathrm{m \cdot s^{-1}}$ and is controlled by the ability to measure small flow rates. For both injection and pumping tests, the practical upper limit of hydraulic-conductivity measurement is nearly open-ended and depends on the availability of the water supply for injection or on the pump capacity.

For multiple-borehole tests, the limiting factor is the ability to detect head or pressure responses in observation wells. In a low-permeability environment, the hydraulic perturbation caused by injection or pumping attenuates rapidly with distance from the stressed borehole. Below a hydraulic conductivity of about $10^{-7}$ $\mathrm{m \cdot s^{-1}}$, the response in an observation borehole is nearly undetectable unless it is very close to the stressed borehole. Therefore, low-permeability regions are not likely candidates for multiple-borehole tests. This means that hydraulic-conductivity values determined by multiple-borehole tests tend to represent the higher-permeability regions.

At the regional or basin scale (from one to hundreds of kilometers), hydraulic conductivity can be inferred from calibration of numerical models. In principle, this approach is applicable over nearly the entire range of hydraulic-conductivity values exhibited by geologic material. For this approach to be successful, the calibration data must consist of hydraulic heads and groundwater flow rates. The latter can be estimated on the basis of groundwater recharge from precipitation, groundwater discharge to streams, or a basin-scale water-balance calculation, or possibly they can be inferred from the effects of groundwater flow in perturbing the geothermal gradient or affecting the transport of dissolved chemicals and minerals.

Several studies have compared published values for hydraulic conductivity at the laboratory, borehole, and regional scales (Brace, 1980, 1984; Clauser, 1992). For crystalline rocks, marked differences were observed between the laboratory- and borehole-derived values. Both Brace (1980) and Clauser (1992) recognized that the differences were due to the absence of fractures in laboratory samples. However, as the laboratory scale is smaller than the borehole scale, Brace (1984) also noted that "permeability depends on the size of the volume of rock being sampled." Clauser (1992) described this observation as a "scale effect," meaning that "the larger the experiment's scale, or characteristic volume, the greater the permeability." This notion appears to be broadly accepted in the hydrogeologic community (e.g., Garvin, 1986). However, what is called "scale effect" by Clauser might more reasonably be called "sampling bias." In other words, if fractured cores can be readily tested, then laboratory results should include both high and low hydraulic-conductivity values. A trend of increasing hydraulic conductivity from laboratory scale to borehole scale might no longer be evident.

If laboratory measurements are excluded, does the notion that "the larger the experiment's scale, ... the greater the permeability" still hold? A survey of case studies shows that some data support this notion, but other data contradict it. In a study of the Cretaceous Pierre Shale in South Dakota, Neuzil, Bredehoeft, and Wolff (1984) (Figure 12.5) obtained hydraulic conductivities in the range of $4 \times 10^{-12}$ to $3 \times 10^{-11}$ $\mathrm{m \cdot s^{-1}}$ from single-borehole slug tests using the pressure-pulse method. However, calibration of a regional flow model of the Pierre Shale and the underlying Dakota Aquifer yielded a shale hydraulic conductivity of about $10^{-9}$ $\mathrm{m \cdot s^{-1}}$. To explain the difference, Neuzil et al. (1984) postulated that the regional-scale hydraulic conductivity might be due to the presence of vertical, through-going fractures spaced on the order of 100 m to 1,000 m. The vertical orientation and large spacing would mean that those fractures would not likely be encountered by vertical boreholes.

Another case study provides an opposite example. In a study of heat and groundwater flow in the Uinta Basin of Utah, Willet and Chapman (1987) found that the geothermal gradient was depressed near the high-elevation northern flank of the basin (indicating recharge of colder water) and enhanced in the lower-elevation central part of the basin (indicating discharge of warmer water that had circulated to basin depths of several kilometers). The results from 159 single-borehole pumping tests conducted in the Tertiary rocks of the Duchesne River Formation yielded hydraulic-conductivity values in the range of $10^{-7}$–$10^{-2}$ $\mathrm{m \cdot s^{-1}}$, with a geometric-mean value of about $10^{-5}$ $\mathrm{m \cdot s^{-1}}$. By calibrating a coupled groundwater and heat-flow model, Willet and Chapman obtained a hydraulic conductivity of $5 \times 10^{-8}$ $\mathrm{m \cdot s^{-1}}$ for the same rocks. This regional-scale value is lower than the borehole-scale values. A possible explanation could be that the borehole tests were conducted in horizons that contained highly transmissive fractures, but those fractures were discontinuous on the regional scale.

The foregoing case studies suggest that a general trend relating hydraulic-conductivity to scale is not readily apparent. In the South Dakota (Pierre Shale) study, the borehole tests may have missed the higher-permeability features. In the Uinta Basin study, the boreholes may have undersampled the less permeable rocks. Therefore, under realistic, heterogeneous field conditions it is often difficult to infer larger-scale properties from smaller-scale observations, especially if features that control

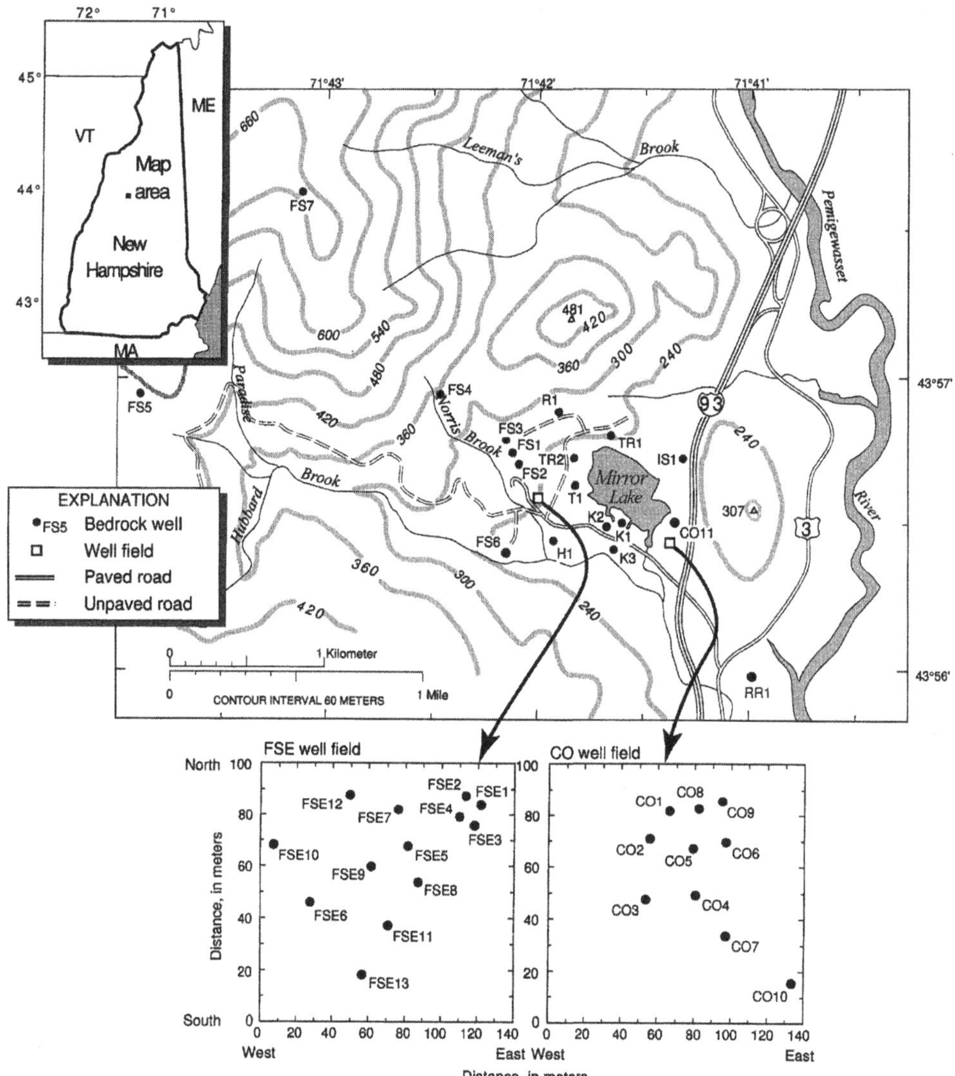


Figure 12.5. Map of the Mirror Lake fractured-rock research site.

large-scale behavior occur in very small portions of the study area (as in the Pierre Shale case). A multiscale investigation might be required to gain an understanding of flow in such settings.

## 12.4 Field Observations

If a general trend is not apparent, is there another type of relationship between hydraulic conductivity and scale? To examine this issue, we review findings from a recent study in the vicinity of Mirror Lake, New Hampshire (Shapiro and Hsieh,

1991). At the study site, glacial deposits (0–55 m) overlie a crystalline bedrock composed of schist that is extensively intruded by granite, pegmatite, and lesser amounts of lamprophyre (a fine-grained, volcanic-dike rock). For testing and sampling, 41 wells were drilled in an area of several square kilometers that included two well fields (Figure 12.5). One objective of the study was to compare the hydraulic conductivity determined at three scales: several meters, 100 m, and several kilometers.

At the scale of several meters, hydraulic conductivity was determined by single-borehole tests using packers to isolate 4–5-m-long test intervals. As an example, Figure 12.6 shows the distributions of hydraulic-conductivity values and fractures (as observed by borehole televiewer) in well FSE11, which was 85 m deep and was cased

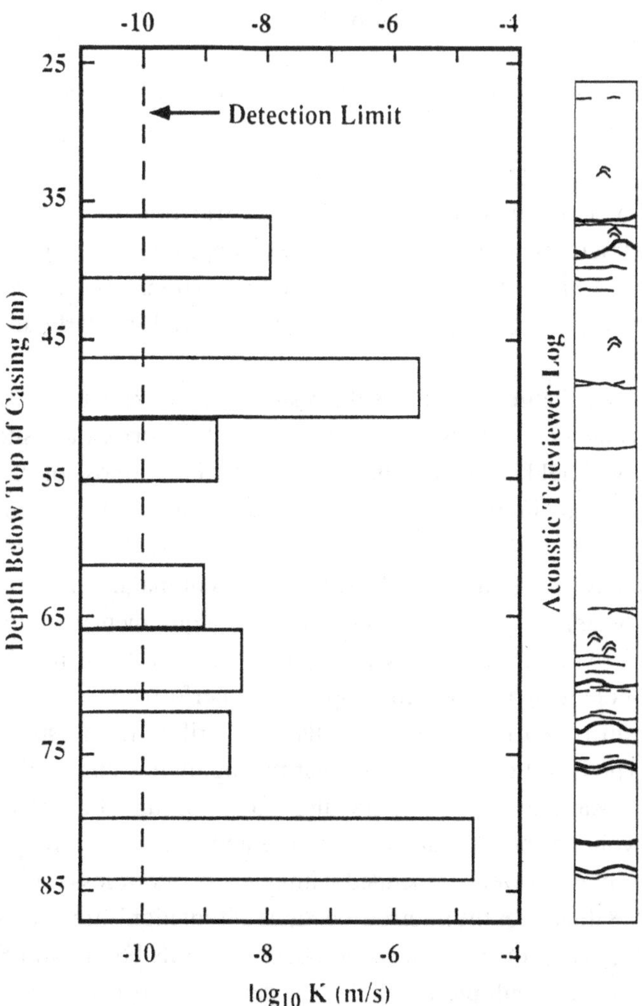


Figure 12.6. Left: Hydraulic conductivity determined by single-borehole packer tests. Right: Fractures observed by acoustic televiewer. Borehole FSE11, Mirror Lake fractured-rock research site.

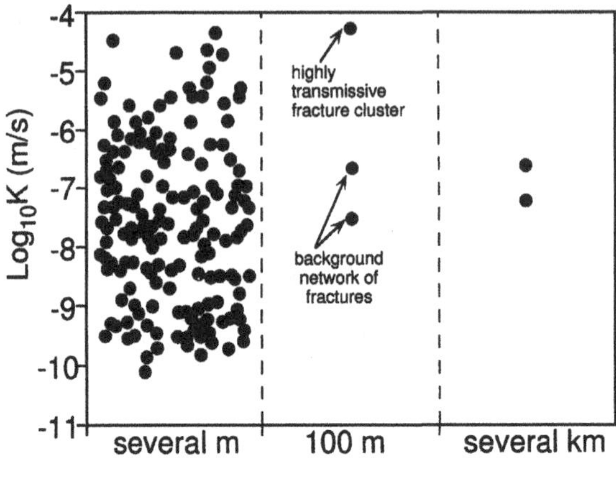


Figure 12.7. Distribution of hydraulic-conductivity values determined at scales of several meters (left panel), 100 m (center panel), and several kilometers (right panel). Mirror Lake fractured-rock research site.

over the upper 25 m. The hydraulic-conductivity distribution was highly nonuniform. Of the 25 or so fractures below the casing, several appeared to be significantly more transmissive than the rest. The presence of a few highly transmissive fractures among a background of less transmissive fractures was observed in nearly all the wells at the site.

The left panel of Figure 12.7 shows the results of approximately 200 single-hole hydraulic tests conducted in 11 wells distributed over the Mirror Lake site. (Only one well from each well field is included in this data set.) For purpose of illustration, the data points are spread out in the horizontal direction; thus the horizontal position of a point has no meaning. The test results show that at the scale of several meters, the hydraulic conductivity varied over at least five orders of magnitude, from below $1 \times 10^{-10}$ m · s$^{-1}$ (lower limit of measurement for the test equipment) to $5 \times 10^{-5}$ m · s$^{-1}$.

To determine hydraulic conductivity at the scale of 100 m, multiple-borehole tests and cross-hole geophysical tomography were performed at the FSE well field, an area of 120 m by 80 m, where 13 wells were drilled to investigate the upper 60 m of rock. The investigation revealed that the highly transmissive fractures connected to form local clusters. Each fracture cluster occupied a nearly horizontal, tabular-shaped volume approximately 1.5 m thick and extended laterally a distance of 20–50 m. These clusters were embedded within a network of less transmissive fractures. Figure 12.8 illustrates the inferred locations of four highly transmissive clusters, marked A through D, in the vertical section between wells FSE1 and FSE6.

Examination of the multiple-borehole test data showed that they could not be analyzed by borehole-test models that assumed homogeneity. Therefore, a three-dimensional finite-difference groundwater model was constructed to represent a block

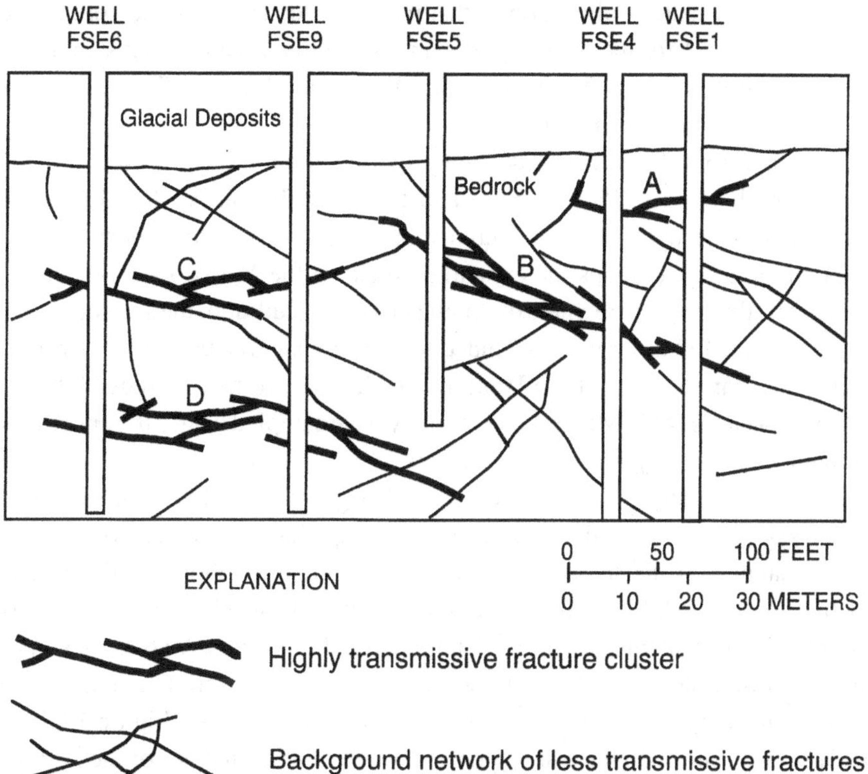


Figure 12.8. Vertical section in FSE well field illustrating inferred locations of highly transmissive fracture clusters and background network of less transmissive fractures. Mirror Lake fractured-rock research site.

of bedrock (120 m by 80 m by 60 m) beneath the well field. The model grid consisted of two different types of cells: those representing highly transmissive fracture clusters, and those representing the background network of less transmissive fractures. Calibrating the model simulation to match hydraulic-test data (drawdowns) yielded an isotropic hydraulic conductivity of $6 \times 10^{-5}$ m $\cdot$ s$^{-1}$ for the zone of more transmissive fracture clusters, and horizontal and vertical hydraulic conductivities of $3 \times 10^{-8}$ and $2 \times 10^{-7}$ m $\cdot$ s$^{-1}$ respectively, for the background network of less permeable fractures. These values are plotted on the middle panel of Figure 12.7.

On the scale of several kilometers, the hydraulic conductivity of the bedrock was determined by constructing a basin-scale model of an area 3 km by 3 km to a depth of 150 m below bedrock surface (Tiedeman, Goode, and Hsieh, 1997). Within that area, hydraulic heads were monitored at wells and piezometers, and groundwater discharge was estimated from base flow in streams. At the basin scale, the highly transmissive fracture clusters, such as those observed at the FSE well field, were not explicitly represented in the model grid. Instead, the model domain was divided into several large, homogeneous zones to represent hydraulic-conductivity variation at the drainage-basin scale. Calibration of that model yielded a hydraulic-conductivity of $3.2 \times 10^{-7}$ m · s$^{-1}$ for the bedrock beneath lower hillsides and valleys, and $6.3 \times 10^{-8}$ m · s$^{-1}$ for the bedrock beneath upper hillsides and hilltops. These values are plotted on the right panel of Figure 12.7.

The findings from the Mirror Lake site provide yet another contrast to the findings from South Dakota (Pierre Shale) and the Uinta Basin. At the Mirror Lake site, the basin-scale hydraulic conductivity falls within the range spanned by the single-borehole tests. Perhaps the crystalline rocks at the Mirror Lake site are less heterogeneous than the Pierre Shale of South Dakota. In addition, the single-borehole tests at the Mirror Lake site were extended to measure significantly lower hydraulic conductivities than the tests at the Uinta Basin. These two factors suggest that the single-borehole tests at the Mirror Lake site may have sampled nearly the entire range of hydraulic-conductivity variation in the rock.

When compared across measurement scales (Figure 12.7), data from the Mirror Lake site suggest a conductivity–scale relationship that can be described in the following manner. At the scale of several meters (single-borehole test), hydraulic-conductivity values exhibited large variations, because the test intervals ranged from those containing no fractures to those containing highly transmissive fractures. At the 100-m scale (multiple-borehole test), the intact rock blocks were too small to constitute distinguishable features. By contrast, the fracture clusters (Figure 12.8) still exerted a strong control on the test response. Consequently, analysis of the multiple-borehole tests identified two features – the fracture cluster and the background fracture network. At the scale of several kilometers, the hydraulic-conductivity values were similar to those for the background fracture network. This suggests that the fracture clusters there were too small to be distinguished at the basin scale. Thus the background fracture network controls basin-scale fluid flow.

## 12.5 Concluding Remarks

The Committee on Fracture Characterization and Fluid Flow (1996, p. 307) noted that "the hydraulic properties of rock masses are likely to be highly heterogeneous even within a single lithological unit if the rock is fractured. The main difficulty in modeling fluid flow in fractured rock is to describe this heterogeneity." Although there appears to be general agreement with this statement, the ongoing debates on the continuum approach versus the discrete-fracture approach, and on stochastic versus deterministic treatment, indicate that there is as yet no general agreement on how best to tackle the problem. The available evidence suggests that the heterogeneity in a fractured rock occurs over a broad range of scales. However, a general trend relating hydraulic conductivity to scale is not apparent.

If the findings at the Mirror Lake site can be generalized, they would suggest that at a certain scale range, the higher conductivity values might be correlated over larger distances than the lower conductivity values, but at another scale range the opposite might happen. If this conjecture holds, then stochastic models of fractured media will require a nested correlation structure that will involve different correlations for different spatial scales (e.g., Gelhar, 1986, fig. 8). In addition, different correlations might be assigned to different ranges of high hydraulic-conductivity values (Gomez-Hernandez and Srivastava, 1990). Whether or not such correlation structure implies scale invariance remains an open question for further research.

Whereas studies of fractured rocks can be advanced on many fronts, a key area of future research is to better understand the nature of heterogeneity. Improved methods in borehole testing and geophysical imaging could better define the hydraulic-conductivity distribution in a rock mass. To understand the nature of this heterogeneity, it is necessary to understand the diverse processes that control fracture permeability. Advances in the use of fracture-mechanics principles to simulate fracture-set formation (Renshaw and Pollard, 1994) could provide valuable insight into how stresses and tectonism control the development of fracture networks. In addition to physical processes, chemical processes such as weathering, dissolution, and infilling also influence fracture permeability. In different rock types and geologic settings, these mechanisms will lead to drastically different outcomes. Understanding the interactions of these processes will require the collaborative efforts of geologists, geophysicists, geochemists, and hydrologists.

## References


Abelin, H., Birgersson, L., Gidlund, J., and Neretnieks, I. 1991a. A large-scale flow and tracer experiment in granite. 1. Experimental design and flow distribution. *Water Resour. Res.* 27:3107–17.

Abelin, H., Birgersson, L., Moreno, L., Widen, H., Agren, T., and Neretnieks, I. 1991b. A large-scale flow and tracer experiment in granite. 2. Results and interpretation. *Water Resour. Res.* 27:3119–35.

Acuna, J. A., and Yortsos, Y. C. 1995. Application of fractal geometry to the study of networks of fractures and their pressure transient. *Water Resour. Res.* 31:527–40.

Barnsley, M. F. 1988. *Fractals Everywhere.* San Diego: Academic Press.

Barton, C. C., and Hsieh, P. A. 1989. *Physical and Hydrologic-Flow Properties of Fractures.* Field trip guidebook T385. Washington, DC: American Geophysical Union.

Baveye, P., and Sposito, G. 1984. The operational significance of the continuum hypothesis in the theory of water movement through soils and aquifers. *Water Resour. Res.* 20:521–30.

Bear, J. 1972. *Dynamics of Fluids in Porous Media.* New York: American Elsevier.

Bear, J., Tsang, C. F., and de Marsily, G. 1993. *Flow and Contaminate Transport in Fractured Rocks.* New York: Academic Press.

Brace, W. F. 1980. Permeability of crystalline and argillaceous rocks. *Int. J. Rock Mech. Mining Sci. and Geomech. Abstr.* 17:241–51.

Brace, W. F. 1984. Permeability of crystalline rocks: new in site measurements. *J. Geophys. Res.* 89:4327–30.

Bredehoeft, J. D., and Papadopulos, S. S. 1980. A method for determining the hydraulic properties of tight formations. *Water Resour. Res.* 16:233–8.

Cacas, M. C., Ledoux, E., de Marsily, G., Tillie, B., Barbreau, A., Durand, E., Feuga, B., and Peaudecerf, P. 1990a. Modeling fracture flow with a stochastic discrete fracture network: calibration and validation. 1. The flow model. *Water Resour. Res.* 26:479–89.

Cacas, M. C., Ledoux, E., de Marsily, G., Barbreau, A., Calmels, P., Gaillard, B., and Margritta, R. 1990b. Modeling fracture flow with a stochastic discrete fracture network: calibration and validation. 2. The transport model. *Water Resour. Res.* 26:491–500.

Carrera, J., and Heredia, J. 1988. *Inverse Modeling of Chalk River Block. NAGRA technical report 88-14*. Baden, Switzerland: National Cooperative for the Disposal of Radio Waste (NAGRA).

Clauser, C. 1992. Permeability of crystalline rocks. *EOS, Trans. AGU* 73:233–8.

Committee on Fracture Characterization and Fluid Flow. 1996. *Rock Fractures and Fluid Flow: Contemporary Understanding and Applications*. Washington, DC: National Academy Press.

Cooper, H. H., Bredehoeft, J. D., and Papadopulos, I. S. 1967. Response of a finite-diameter well to an instantaneous charge of water. *Water Resour. Res.* 3:263–9.

Dershowitz, W. S., and Einstein, H. H. 1988. Characterizing rock joint geometry with joint system models. *Rock Mech. and Rock Eng.* 21:21–51.

Doughty, C., Long, J. C. S., Hestin, K., and Benson, S. M. 1994. Hydrologic characterization of heterogeneous geologic media with an inverse method based on iterative function systems. *Water Resour. Res.* 30:1721–45.

Dverstorp, B., Andersson, J., and Nordqvist, W. 1992. Discrete network interpretation of field tracer migration in sparsely fractured rock. *Water Resour. Res.* 28:2327–43.

Garvin, G. 1986. The role of regional fluid flow in the genesis of the pine point deposit, Western Canada sedimentary basin – a reply. *Econ. Geol.* 81:1015–20.

Gelhar, L. W. 1986. Stochastic subsurface hydrology from theory to applications. *Water Resour. Res* 22:135S–45S.

Gomez-Hernandez, J. J., and Srivastava, R. M. 1990. ISIM3D: an ANSI-C three-dimensional multiple indicator conditional simulation program. *Comput. Geosci.* 16:395–440.

Guimera, J., Vives, L., and Carrera, J. 1995. A discussion of scale effects on hydraulic conductivity at a granitic site (El Berrocal, Spain). *Geophys. Res. Lett.* 22:1449–52.

Long, J. C. S., Gilmour, P., and Witherspoon, P. A., 1985. A model for steady fluid flow in random three-dimensional networks of disc-shaped fractures. *Water Resour. Res.* 21:1105–15.

Long, J. C. S., Karasaki, K., Davey, A., Peterson, J., Landsfeld, M., Kemeny, J., and Martel, S. 1991. An inverse approach to the construction of fracture hydrology models conditioned by geophysical data. *Int. J. Rock Mech. and Mining Sci. Geomech. Abstr.* 28:121–42.

Long, J. C. S., Remer, J. S., Wilson, C. R., and Witherspoon, P. A. 1982. Porous media equivalents for networks for discontinuous fractures. *Water Resour. Res.* 18:645–58.

Neuman, S. P. 1987. Stochastic continuum representation of fractured rock permeabilities as an alternative to the REV and fracture network concepts. In: *Rock Mechanics: Proceedings of the 28th U.S. Symposium*, ed. I. W. Farmer, J. J. K. Daemen, C. S. Desai, C. E. Glass, and S. P. Neuman, pp. 533–61. Rotterdam: Balkema.

Neuzil, C. E., Bredehoeft, D. D., and Wolff, R. G. 1984. Leakage and fracture permeability in the Cretaceous shales confining the Dakota Aquifer in South Dakota. In: *Proceedings of the First C. V. Theis Conference on Geohydrology*, ed. D. G. Jorgenson and D. C. Signor, pp. 113–20. Worthington, OH: National Water Well Association.

Poulton, M. M., Mojtabai, N., and Farmer, I. W. 1990. Scale invariant behavior of massive and fragmented rock. *Int. J. Rock Mech. Mining Sci. Geomech. Abstr.* 27:219–21.

Renshaw, C. E., and Pollard, D. D. 1994. Numerical simulation of fracture set formation: a fracture mechanics model consistent with experimental observations. *J. Geophys. Res.* 99:9353–72.

Scheidegger, A. E. 1960. *The Physics of Flow through Porous Media.* University of Toronto Press.

Shapiro, A. M., and Hsieh, P. A., 1991. Research in fractured-rock hydrogeology: characterizing fluid movement and chemical transport in fractured rock at the Mirror Lake drainage basin, New Hampshire. In: *U.S. Geological Survey Toxic Substances Hydrology Program – Proceedings of the Technical Meeting, Monterey, California, March 11–15, 1991*, ed. G. E. Mallard and D. A. Aronson, pp. 155–61. U.S. Geological Survey Water Resources Investigations report 91-4034. Washington, DC: USGS.

Tiedeman, C. R., Goode, D. J., and Hsieh, P. A. 1997. *Numerical Simulation of Ground-Water Flow through Glacial Deposits and Crystalline Bedrock in the Mirror Lake Area, Grafton County, New Hampshire.* U.S. Geological Survey professional paper 1572.

Trimmer, D., Bonner, B., Heard, H. C., and Duba, A. 1980. Effect of pressure and stress on water transport in intact and fractured gabbro and granite. *J. Geophys. Res.* 85:7059–71.

Tsang, Y. W., Tsang, C. F., and Hale, F. V. 1996. Tracer transport in a stochastic continuum model of fractured media. *Water Resour. Res.* 32:3077–92.

Turcotte, D. L. 1986. Fractals and fragmentation. *J. Geophys Res.* 91:1921–6.

Willet, S. D., and Chapman, D. S. 1987. Temperatures, fluid flow, and the thermal history of the Unita Basin. In: *Migration of Hydrocarbons in Sedimentary Basins*, ed. B. Doligez, pp. 533–51. Paris: Editions Technip.

# 13

# Correlation, Flow, and Transport in Multiscale Permeability Fields

SHLOMO P. NEUMAN and VITTORIO DI FEDERICO

## 13.1 Introduction

Flow and transport processes in natural soils and rocks traditionally have been described by means of partial differential equations (PDEs). These equations generally are taken to represent basic physical principles (conservation and constitutive laws) that operate on some macroscopic scale (theoretical support volume) at which the geologic medium can be viewed as a continuum. The precise nature of this theoretical macroscopic support scale remains generally unclear, though some derive comfort from associating it in the abstract with a "representative elementary volume" (REV). Unfortunately, the concept of an REV is equally difficult to define without ambiguity and to apply in practice. Flow and transport PDEs are local in the sense that all quantities (parameters; forcing functions, including initial, boundary, and source terms; dependent variables) that enter into them are defined at a single point $(\mathbf{x}, t)$ in space–time. Parameters such as permeability, porosity, and dispersivity are generally regarded as macroscopic medium properties that are well defined and thus can be determined (at least in principle) experimentally, and more or less uniquely, at any point $\mathbf{x}$ in the flow domain.

In reality, geologic media are heterogeneous and exhibit both discrete and continuous spatial variations on a multiplicity of scales, and therefore it can be anticipated that the flow and transport properties of these media will exhibit similar variations. Indeed, one manifestation of such variations is the observed and well-documented dependence of permeabilities and dispersivities on their scale of measurement (support volume). Figure 13.1 shows two superimposed profiles for hydraulic conductivity $K$ as determined by packer tests in a single borehole that penetrated fractured granitic rocks at the Finnsjön site in Sweden. One profile represents permeabilities obtained from measurements of steady-state pressure and flow rate during continuous injection of water into the borehole at intervals of 20 m, and the other at intervals of 2 m. Both sets of data give the appearance of random variation along the length of the borehole,

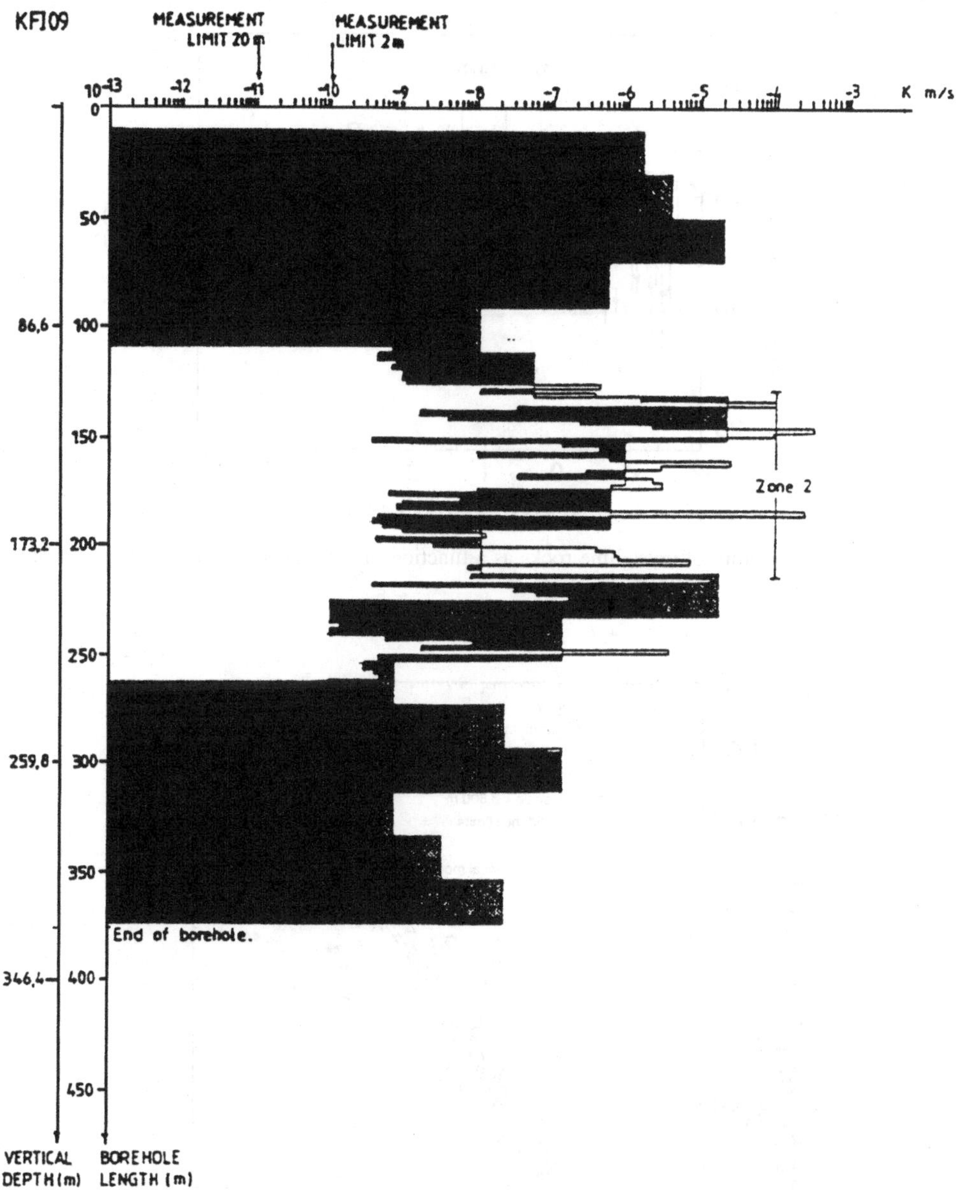


Figure 13.1. Hydraulic conductivity of fractured granitic rocks near Finnsjön, Sweden, as determined in 2-m and 20-m sections of a borehole. (From Andersson et al., 1988, with permission.)

but their values and statistics are different. Figure 13.2 summarizes laboratory and field permeability data from crystalline rocks at a variety of sites and suggests that these data vary with the scale of measurement. Figure 13.3 shows longitudinal dispersivities, derived by means of traditional advection–dispersion models from a variety of laboratory and field tracer studies worldwide, and suggests that these parameters

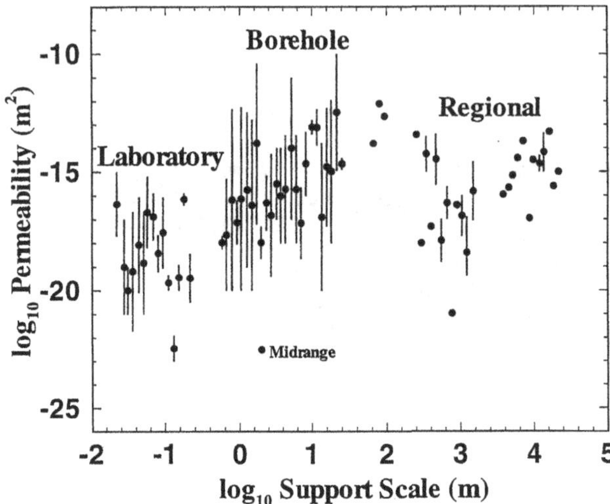


Figure 13.2. Permeability of crystalline rocks as a function of experimental scale. (Adapted from Clauser, 1992, and Neuman, 1994.)

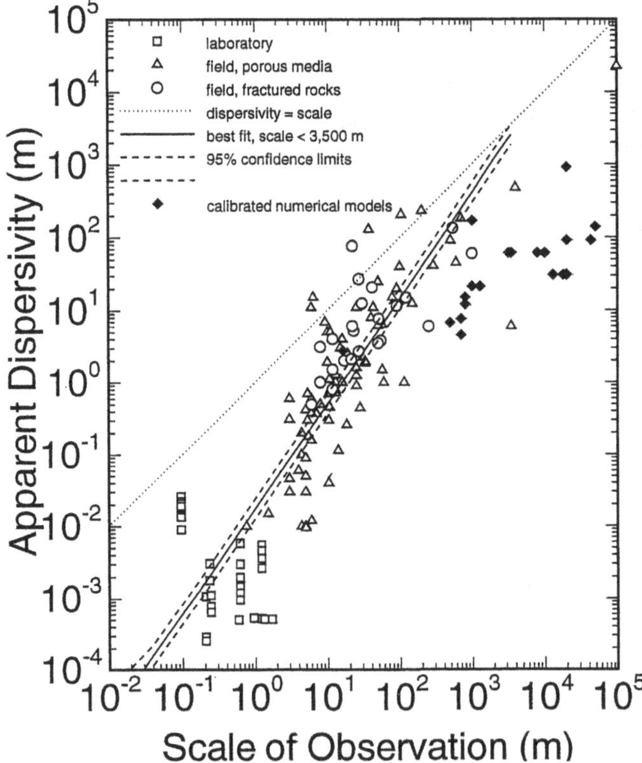


Figure 13.3. Apparent longitudinal dispersivities versus scale of observation. (From Neuman, 1995, with permission.)

also vary with the scale of observation. In other words, hydraulic conductivities and dispersivities seem not to be unique properties of the geologic medium, but to depend additionally on the support and observation scales.

Internal consistency of flow and transport PDEs requires that all quantities that enter into them (parameters, forcing terms, dependent variables) be defined on one unique support scale; consistency of these PDEs with data (without which the equations are not operational) requires that the latter be defined and measured on this same scale. To the extent that measurements and their statistics exhibit dependence on the support scale, it is necessary in theory, and advisable (though not always possible) in practice, to work with one consistent support scale, $\omega$. (In the example of Figure 13.1, the volume of rock affected by each 2-m or 20-m packer test varies from one interval to another and generally is unknown; one may have no choice but to consider, for working purposes, these two test intervals as the nominal length scales of two distinct $\omega$ scales.) The selection of a working $\omega$ depends not on the abstract question whether or not it constitutes an REV, but on the pragmatic question whether or not it allows in principle, and renders technically feasible as well as otherwise desirable, the definition and measurement of all relevant quantities on this scale throughout the domain of interest.

One key question of practical and theoretical interest is whether or not it is possible quantitatively to relate flow and transport parameters and associated statistics determined on one scale of measurement to those determined on another scale. Transforming such information from a small to a larger support scale is often referred to as upscaling, and transforming data and statistics from a large scale to a smaller scale is called downscaling. Only on rare occasions can upscaling or downscaling be properly accomplished by linear convolution (weighted averaging) or deconvolution of small- or large-scale data; one notable case where this may be possible is that of permeabilities in perfectly stratified porous media. For most other cases, there is a need to develop alternative ways of relating measurements and their statistics across disparate support scales. One approach that has led to useful insight into the scale dependence of subsurface flow and transport phenomena is their analysis within geostatistical and stochastic frameworks. This approach has been motivated by the recognition that subsurface data and especially permeability, which often exhibit seemingly erratic spatial fluctuations of the kind illustrated in Figure 13.1, nevertheless tend to be spatially autocorrelated as well as mutually cross-correlated. The key to understanding scale dependence in flow and transport seems to lie in an understanding of how permeabilities (and, to a lesser extent, other parameters) and their statistics (primarily spatial correlations) defined and measured on one scale affect related parameters and their statistics on larger scales. We shall pursue this question on the premise that all relevant flow and transport quantities are defined and measured on one consistent support scale, $\omega$.

Spatial fluctuations in permeability, such as those in Figure 13.1, are due in part to random errors of measurement and errors arising from indirect assessment of permeability based on measurements of pressure and flow rate. In large part, such fluctuations are caused by spatial variations in the makeup of $\omega$-scale soil and rock volumes that compose the geologic medium (i.e., by $\omega$-scale medium heterogeneity). Because permeabilities can be determined only at selected well locations and within discrete depth intervals, permeability values elsewhere in the subsurface remain unknown. As measured values tend to fluctuate erratically in space, they cannot be extended into untested portions of the subsurface with certainty. Hence the subsurface distribution of $\omega$-scale permeabilities is at once random and uncertain. The same is true of $\omega$-scale forcing terms, which can additionally fluctuate in an uncertain fashion with time. Upon introducing such random inputs into flow and transport PDEs, the latter become stochastic, and their $\omega$-scale solutions (in terms of heads, concentrations, fluxes, and velocities) are likewise random. Hence these solutions cannot be specified directly (other than as random samples, or realizations, of an infinite set or ensemble of equally likely and therefore nonunique solutions), but only in terms of $\omega$-scale ensemble statistics. Such theoretical, deterministic ensemble statistics represent the results of sampling and averaging of $\omega$-scale quantities, at a given point $(\mathbf{x}, t)$, across their infinite ensemble of equally likely values (realizations) in probability space (sampling and averaging across a random finite sample of these realizations yields sample statistics that are random variables). Constraining ensemble moments to be consistent with specific measurements renders them conditional on data; otherwise, if they are consistent with data statistics, but not with specific data values, they are unconditional. A conditional ensemble mean solution represents an optimum (minimum-variance) unbiased prediction of the unknown true $\omega$-scale solution at any point $(\mathbf{x}, t)$; the corresponding conditional ensemble variance is a measure of the associated prediction error.

In general, ensemble mean quantities satisfy nonlocal (integro-differential) rather than local (differential) flow equations (Neuman and Orr, 1993; Neuman et al., 1996; Tartakovsky and Neuman, 1998a, b) and transport equations (Neuman, 1993a; Zhang and Neuman, 1996). In these equations, ensemble mean (predicted) fluxes at any point $(\mathbf{x}, t)$ depend not only on mean gradients at this point but also on additional quantities at other points in space–time. In other words, the mean fluxes are nonlocal and therefore non-Darcian (in the case of flow) and non-Fickian (in the case of transport). It follows that the standard notions of permeability and dispersivity, as material properties (coefficients in Darcy's law and Fick's law) that are well defined at each point in space, generally lose meaning when one deals with ensemble-averaged quantities under uncertainty. This is especially true under conditioning, where the local (depending on one point) and nonlocal (depending on more than one point) parameters that control ensemble mean flow and transport are additionally functions of the quantity and quality (information content) of $\omega$-scale data. For standard notions of permeability and dispersivity to apply in the ensemble mean, one must restrict consideration to special situations without conditioning. The most common among these special situations is that of mean uniform steady-state flow in a statistically homogeneous permeability field. If the flow domain is additionally infinite (so large that, for practical purposes, its boundaries can be assumed to lie at infinite distances from the area of interest), one may achieve ergodicity, whereby the now-constant ensemble mean flux and hydraulic gradient also represent spatial averages of their random $\omega$-scale counterparts over a sufficiently large domain in one or more realizations. In this case, the effective hydraulic conductivity, which by definition relates the ensemble mean flux to the ensemble mean gradient, is equal to the equivalent hydraulic conductivity, which, by definition, relates the spatially averaged flux to the spatially averaged gradient; all of these quantities are deterministic. If the flow domain is finite, ergodic conditions may not be achieved, and spatial averages of flux and gradient, over one or a few realizations, may remain random; in this case, the deterministic ensemble mean flux and gradient represent not spatial averages of these quantities but only the ensemble means (i.e., best estimates) of such spatial averages. By the same token, the effective hydraulic conductivity that relates the ensemble mean flux to the gradient represents not any actual value, but only the ensemble mean (best estimate) of random equivalent hydraulic conductivities associated with random spatial averages of flux and gradient over a domain of restricted size. Similar ideas apply in principle to the more complex case of transport.

One important aspect of geostatistical analysis is the inference from such data of a (semi)variogram $\gamma(\mathbf{s})$, where $\mathbf{s}$ is a vector defining the spatial separation between any two points. The natural-log hydraulic conductivity ($Y = \ln K$) and transmissivity ($Y = \ln T$) data often appear to fit variogram models associated with a constant sill (variance) $\sigma^2$ and integral (spatial) autocorrelation scale $\lambda$. Figure 13.4 illustrates such behavior for 1-m-scale packer-test data from fractured granites at the Stripa mine in Sweden. This is commonly taken to imply that the data are representative of a statistically homogeneous random field $Y(\mathbf{x})$. Indeed, most stochastic analyses of subsurface flow and transport are based on the assumption that $Y(\mathbf{x})$ is statistically homogeneous. Such analyses show, among other things, that in an infinite domain subject to a uniform ensemble mean hydraulic gradient, the steady-state ensemble mean flux is related to this gradient via Darcy's law, with an effective hydraulic conductivity that depends primarily on the variance and integral scale of $Y(\mathbf{x})$. This effective hydraulic conductivity doubles as an equivalent conductivity provided that one averages the flux and gradient over a domain that is much larger than the integral scale of $Y(\mathbf{x})$; for an answer to the question of how much larger, refer to Paleologos, Neuman, and Tartakovsky (1996) and Tartakovsky and Neuman (1998c). By the same token, the mean spread of ensemble-averaged solute concentrations (sampled on the scale $\omega$) of an inert solute, under these same conditions, is controlled by an effective-dispersivity tensor whose longitudinal component first grows linearly with mean travel distance (or time) but later tends asymptotically to a constant (Fickian) value. The asymptotic Fickian regime is established when the mean travel distance

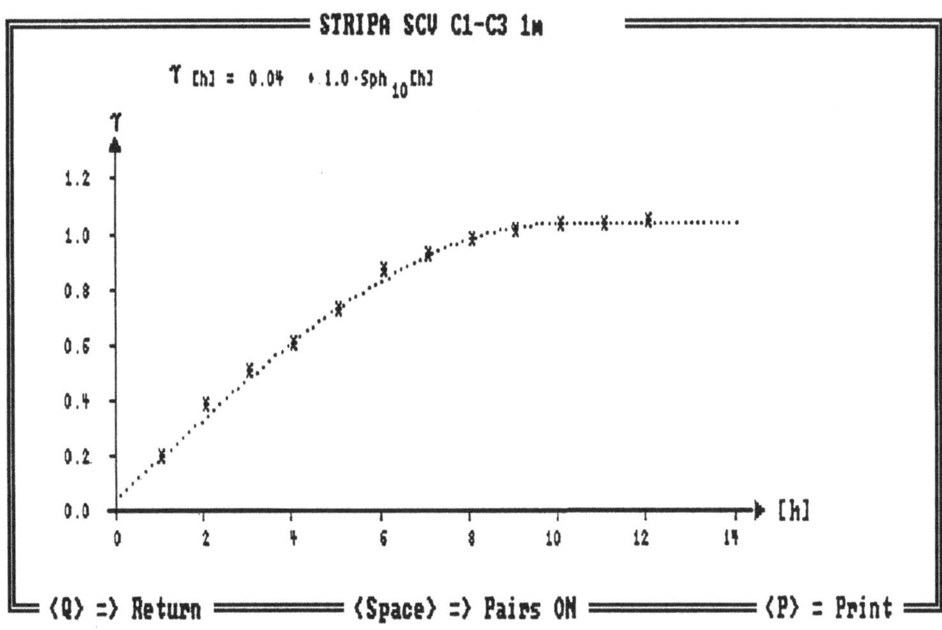


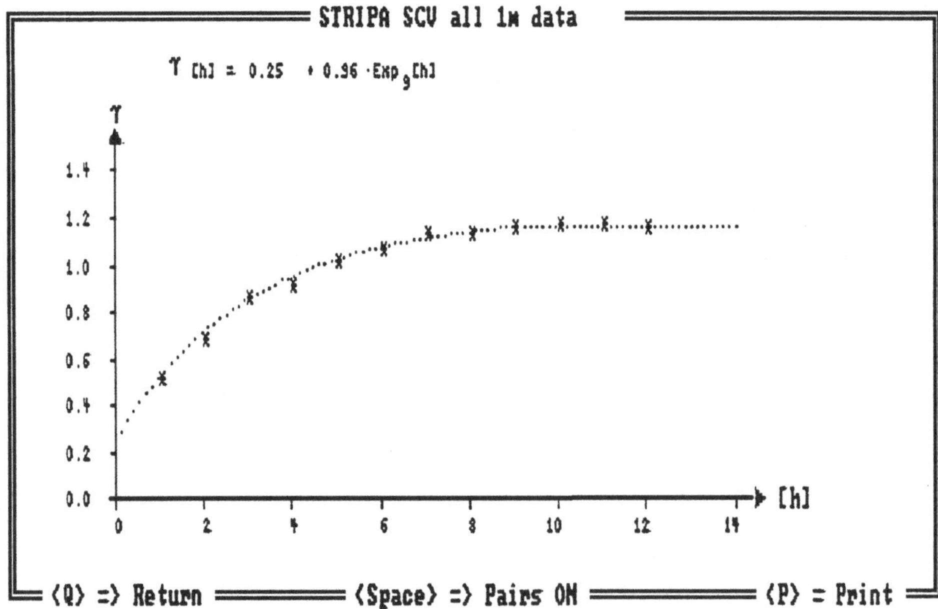


Figure 13.4. Variograms of natural-log hydraulic conductivities of fractured granites at Stripa, Sweden, as determined in 1-m sections of three (C, N, W) borehole sets. (From Winberg, 1991, with permission.)

becomes large compared with the integral scale of $Y(\mathbf{x})$; the corresponding effective longitudinal dispersivity depends on the variance and integral scale of $Y(\mathbf{x})$. Under ergodic conditions (for an answer to the question of when such conditions develop, refer to Dagan, 1990, 1991, and Zhang, Zhang, and Lin, 1996), the same dispersivity tensor describes the mean spread of an actual plume (in a single realization). In other words, a statistically homogeneous random $Y(\mathbf{x})$ field can, under the foregoing conditions, be associated with effective as well as equivalent hydraulic conductivity and longitudinal dispersivity that tend to constant values as the scale of observation increases.

We have seen earlier that, in reality, permeabilities (Figure 13.2) and longitudinal dispersivities (Figure 13.3) appear to vary continuously over a broad range of experimental scales. Likewise, the integral scale of $Y(\mathbf{x})$ (Figure 13.5) and, to a lesser extent, its variance (this will become clear later) appear to increase consistently with the size of the domain under investigation. These findings suggest that statistical homogeneity, as is often suggested by standard variogram analyses, may not be a true property of $Y(\mathbf{x})$, but rather an artifact of the scale of investigation and method of inference. It further suggests a possible link between this apparent scale dependence of $\lambda$ and $\sigma^2$ of $Y(\mathbf{x})$, on one hand, and that of larger-scale equivalent permeability and solute dispersivity on the other hand.

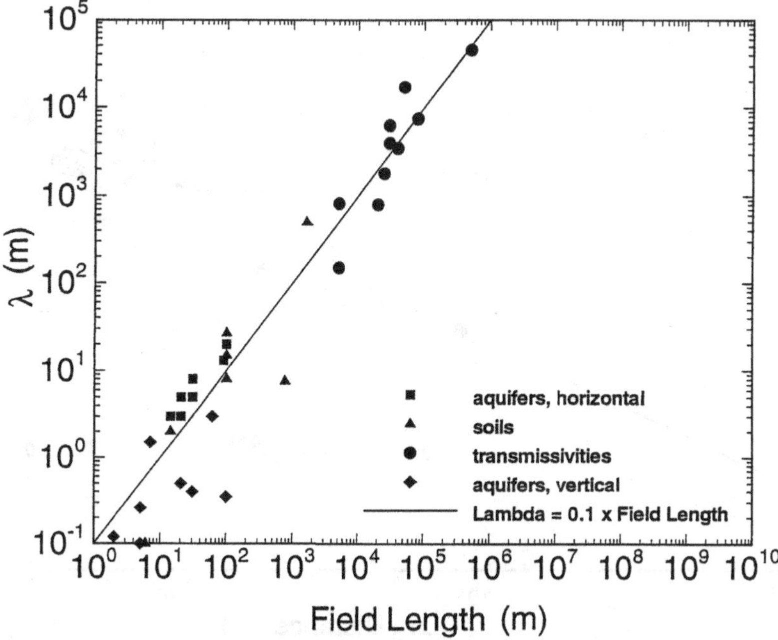


Figure 13.5. Correlation scales $\lambda$ of natural-log hydraulic conductivities and transmissivities at various sites versus field lengths. (From Neuman, 1994, with permission; data from Gelhar, 1993, table 6.1.)

Indeed, when sample variograms are plotted on logarithmic paper rather than on arithmetic paper (as had been the standard procedure until quite recently), the data often lie close to a straight line, which is not consistent with the assumption of homogeneity. Such behavior is being observed at an increasing number of sites (Neuman, 1995) on distance scales ranging from a few meters (Figure 13.6) to 100 km (Figure 13.7). The straight line is indicative of a nonstationary field with homogeneous spatial increments. If the field is statistically isotropic, a line with slope $2H$ represents a power variogram $\gamma(s) = as^{2H}$, where $s$ is separation distance (the magnitude of $\mathbf{s}$), $a$ is a constant, and the power $H$ is called the Hurst coefficient. Because the variogram scales as $\gamma(rs) = r^{2H}\gamma(s)$, the field is self-affine and, within the range $0 < H < 1$, constitutes a random field that exhibits fractal geometry (a random fractal), with dimension $D = d + 1 - H$, where $d$ is the Euclidean (topologic) dimension (Voss, 1985). If the field is additionally Gaussian, it constitutes fractional Brownian motion (fBM) (Mandelbrot and Van Ness, 1968). When

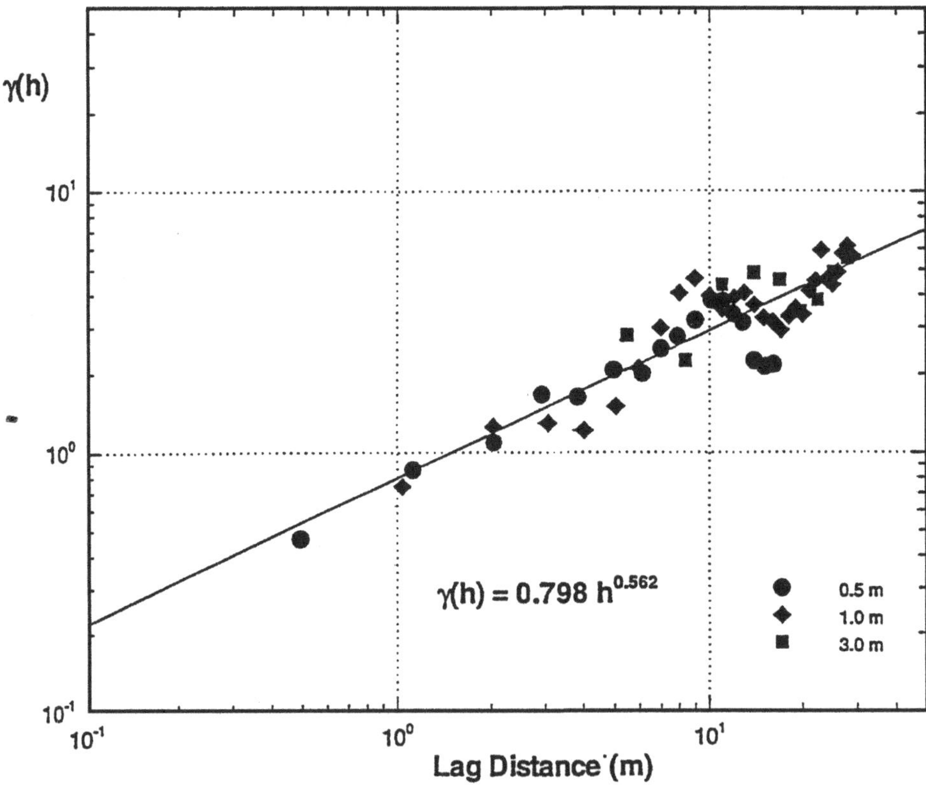


Figure 13.6. Log-log variogram of natural-log air permeabilities of unsaturated fracture tuffs near Superior, Arizona, as determined in various size sections of several boreholes. (From Guzman and Neuman, 1996, with permission.)

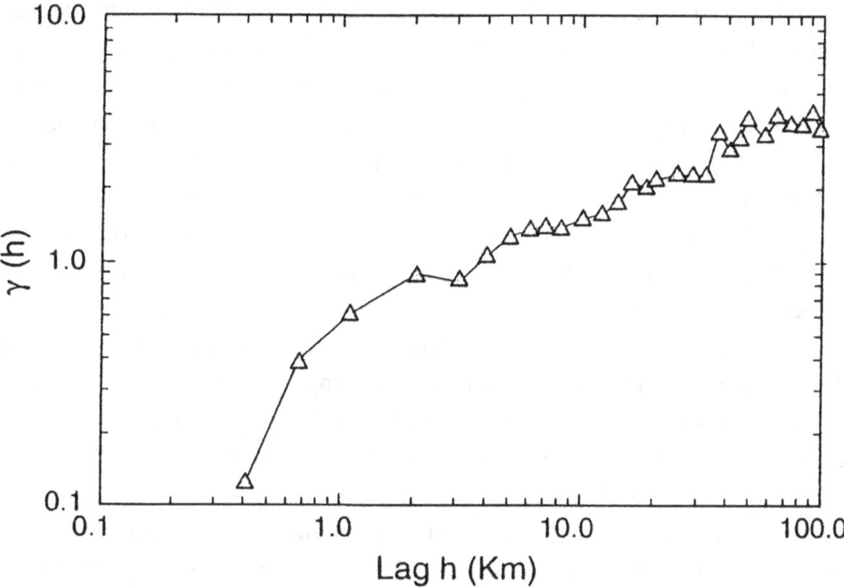


Figure 13.7. Log-log variogram of natural-log transmissivities in a sandstone aquifer. (From Desbarats and Bachu, 1994, with permission.)

$0.5 < H < 1$, random spatial increments in field values are positively correlated, so that positive and negative deviations from the mean tend to persist over distance, a phenomenon known as persistence. When $0 < H < 0.5$, the increments are negatively correlated, so that positive and negative deviations from the mean tend to alternate rapidly, a phenomenon called antipersistence. When $H = 0.5$, the increments are uncorrelated, and the field represents Brownian motion (BM).

The Hurst coefficient of $Y(\mathbf{x})$ is not the same at each site, though it has been found to lie near the midrange of $0 < H < 0.5$ in several recent studies quoted by Neuman (1995) and in the study by Molz and Boman (1995). Within this range, the increments are negatively correlated and relatively noisy, exhibiting antipersistent behavior. Nevertheless, when one juxtaposes apparent values of $\sigma^2$ and $\lambda$ from many different sites, inferred from $Y$ data by assuming that the underlying field is homogeneous, one finds that they fit a generalized power model with $H \approx 0.25$ (Neuman, 1994), as discussed in more detail later. Such generalized behavior was deduced earlier by Neuman (1990) from the observed scale dependence (Figure 13.3) of juxtaposed apparent dispersivities reported for a large number of tracer studies worldwide. This, too, will be discussed in more detail later.

Neuman (1990) has shown that any random field with homogeneous increments can be viewed as an infinite hierarchy of mutually uncorrelated homogeneous fields (modes) characterized by exponential autocovariance functions and variances that increase as some power of scale. He has noted that accounting deterministically for large-scale spatial variability is equivalent to filtering out low-frequency modes from this hierarchy. Here we extend these ideas by demonstrating that both the power variogram and associated spectra of a random field with homogeneous isotropic increments can be constructed as weighted integrals from zero to infinity (an infinite hierarchy) of either exponential or Gaussian variograms and spectra of uncorrelated homogeneous isotropic modes. We then analyze the effect of filtering out (truncating) high- and low-frequency modes from this infinite hierarchy in the real and spectral domains. We show that a low-frequency cutoff renders the truncated hierarchy statistically homogeneous, with a spatial autocovariance function that varies monotonically with separation distance in a manner not too dissimilar from that of its constituent modes. The integral scales of the lowest- and highest-frequency modes (cutoffs) are related, respectively, to domain and sample (support) sizes. Taking this relationship to be one of proportionality renders our expressions for integral scale and variance dependent on domain size in a manner consistent with observations.

The traditional approach has been one of truncating power spectral densities, rather than the hierarchy of spectra associated with exponential or Gaussian autocovariance functions as we propose. This approach has been used previously: by Ababou and Gelhar (1990) to analyze the scale dependence of effective hydraulic conductivity and longitudinal dispersivity in one and three dimensions, by Kemblowski and Wen (1993) to study dispersion in perfectly stratified soils, by Dagan (1994) to explore the scale dependence of effective dispersion during nonergodic transport, and by Kemblowski and Chang (1993) and Chang and Kemblowski (1994) in the context of unsaturated flow. Rajaram and Gelhar (1995) adopted a power-law spectral density in their analysis of a "relative dispersivity," which is defined a priori so as to include a low-wave-number filter whose aim is to exclude the influence of velocity variations at scales larger than the plume. We show that truncated power spectra yield autocovariance functions that oscillate about zero with finite (in two and three dimensions) or vanishing (in one dimension) integral scales.

We continue by describing how our hierarchical theory allows bridging across scales. We show that this can be done at a given locale by calibrating a truncated variogram model to data observed on a given support in one domain and predicting the autocovariance structure of the corresponding field in domains that are either smaller or larger. We further suggest that one can also venture to bridge across both domain scales and locales by adopting the generalized Hurst coefficient $H = 0.25$ and other generalized variogram parameters.

Next we discuss flow and transport in isotropic, multiscale, random log hydraulic-conductivity fields formed by the weighted superposition of mutually uncorrelated exponential or Gaussian modes between a lower cutoff and an upper cutoff. In particular, we present expressions for the effective hydraulic conductivity in one, two, and three dimensions, followed by first-order expressions for pre-asymptotic and asymptotic effective dispersivities in two dimensions. We conclude by examining the effect of conditioning on asymptotic and pre-asymptotic transport. Much of the material that follows has been reported elsewhere by Neuman (1994, 1995) and Di Federico and Neuman (1995, 1997, 1998a, b)

## 13.2 Theoretical Background

Consider a power variogram of the form

$$\gamma(s) = \frac{1}{2}\langle[Y(x+s) - Y(x)]^2\rangle = C_0 s^{2H} \qquad (0 < H < 1) \tag{13.1}$$

where $s$ is distance (lag), $x$ is position in one-dimensional space, $\langle\ \rangle$ indicates the ensemble mean (expectation), $C_0$ is a constant, and $H$ is the Hurst coefficient. The associated one-dimensional random field $Y(x)$ is then self-affine (Yaglom, 1987) and devoid of a characteristic correlation scale. The random field $Y(x)$ is statistically nonhomogeneous, but it possesses homogeneous spatial increments. When these increments are multivariate Gaussian, the field constitutes fBM (Mandelbrot and Van Ness, 1968) that can be constructed as a moving average of white noise from $-\infty$ to $x$ (Bras and Rodríguez-Iturbe, 1985). When $0 < H < 1/2$, the increments are negatively autocorrelated and thus antipersistent; when $1/2 < H < 1$, the increments are positively autocorrelated and persistent; when $H = 1/2$, the increments are uncorrelated, and $Y(x)$ degenerates into ordinary BM.

The variogram (13.1) is associated with a power-law spectral density $S(k)$ of the form [Yaglom, 1987, eq. (4.263), p. 407; his $f(k)$ is our $S(k)$]

$$S(k) = \frac{C_1}{k^{1+2H}} \qquad (0 < H < 1); \qquad C_1 = C_0\frac{\Gamma(1+2H)\sin(\pi H)}{\pi} \tag{13.2}$$

where $k$ is wavenumber, and $\Gamma(x)$ is the gamma function (Abramowitz and Stegun, 1972, p. 255). The spectral density (13.2) is associated with a nondifferentiable random field of stationary increments (Yaglom, 1987, p. 401) and infinite variance (Yaglom, 1987, p. 105) and a variogram that satisfies [Yaglom, 1987, eq. (4.250a), p. 400]

$$2\gamma(s) = 4\int_0^\infty [1 - \cos(ks)]S(k)\,dk \tag{13.3}$$

Substitution of the spectral density (13.2) into equation (13.3) yields equation (13.1).

In a $d$-dimensional Euclidean domain $R^d$, the location $\mathbf{x}$ and displacement $\mathbf{s}$ are vectors. Then $Y(\mathbf{x})$ is an isotropic self-affine random field provided that

$$\gamma(s) = \frac{1}{2}\langle[Y(\mathbf{x}+\mathbf{s}) - Y(\mathbf{x})]^2\rangle = C_0 s^{2H} \qquad (0 < H < 1;\ s = |\mathbf{s}|) \tag{13.4}$$

The corresponding spectral densities are given by [Yaglom, 1987, eq. (4.119), p. 358; eq. (4.362), p. 441]

$$S^*(k) = \frac{\varphi^*(k)}{[2\pi^{d/2}/\Gamma(d/2)]k^{d-1}} \tag{13.5}$$

$$\varphi^*(k) = \frac{A}{k^{1+2H}} \qquad (0 < H < 1); \qquad A = 2C_0 \frac{2^{2H+1} H\Gamma[(d+2H)/2]}{\Gamma(d/2)\Gamma(1-H)} \tag{13.6}$$

where $k = |\mathbf{k}|$, $\mathbf{k}$ being the wavenumber vector. The relationship between the spectral density so defined and the variogram is given by [Yaglom, 1987, eq. (4.358a), p. 440; eq. (4.109), p. 355]

$$2\gamma(s) = \int_0^\infty [1 - \Lambda_d(ks)]\varphi^*(k)\, dk; \tag{13.7}$$

$$\Lambda_2(x) = J_0(x), \qquad \Lambda_3(x) = \frac{\sin(x)}{x}$$

where $J_0$ is a Bessel function of the first kind and zero order (Abramowitz and Stegun, 1972, p. 358). In effect, equations (13.5) and (13.6) written for $d = 1$ yield, for $S^*(k)$, twice the value given by equation (13.2); furthermore, relations between one-dimensional and two/three-dimensional spectral densities (Yaglom, 1987, pp. 359–60) yield $S^*(k)$ values that are half those given by equations (13.5) and (13.6). To be fully consistent, we define

$$\varphi(k) = \frac{\varphi^*(k)}{2}; \qquad S(k) = \frac{S^*(k)}{2} \tag{13.8}$$

For $d = 3$, these coincide respectively with the spectral densities $E(k)$ and $F(k)$ defined by Monin and Yaglom (1975, pp. 99–100).

## 13.3 Isotropic Multiscale Random Field as a Weighted Superposition of Mutually Uncorrelated Modes

### *13.3.1 Superposition of Modes in a Real Domain*

*Exponential Modes.* Consider an infinite hierarchy of mutually uncorrelated, statistically homogeneous and isotropic random fields (modes), each of which is associated with an exponential variogram

$$\gamma(s, \lambda) = \sigma^2(\lambda)[1 - \exp(-s/\lambda)] \tag{13.9}$$

and dimensionless variance

$$\sigma^2(n) = \frac{C}{n^{2H}} \tag{13.10}$$

where $\lambda$ is integral scale, $n = 1/\lambda$ is mode number, and $C$ is a constant of dimensions $[\mathrm{L}^{-2H}]$, so that the variance decreases as a power of the mode number. Integrating a continuous hierarchy of such modes over all possible scales, and weighting each contribution by a factor $1/n$, gives

$$\gamma(s) = \int_0^\infty \gamma(s, n)\frac{dn}{n} \tag{13.11}$$

Substituting equations (13.9) and (13.10) into equation (13.11) and evaluating yields [Gradshteyn and Ryzhik, 1994, eq. (3.551.1), p. 403]

$$\gamma(s) = C_0 s^{2H} \tag{13.12}$$

where

$$C_0 = C\frac{\Gamma(1-2H)}{2H} \qquad \left(0 < H < \frac{1}{2}\right) \tag{13.13}$$

is a constant proportional to $C$ and having the same dimensions. The weighted superposition of exponential modes is thus a power variogram of the form (13.1). Our derivation and result are almost identical with those of Neuman (1990), except that he included the weight $1/n$ in his definition of $\sigma^2(n)$, so that $C$ and $C_0$ acquired disparate dimensions. A somewhat related superposition of temporal autocovariance functions (as opposed to spatial variograms), which leads to power behavior at large times (equivalent to large spatial separations), was described earlier by Philip (1986).

Consider now the case

$$\gamma(s, n_l) = \int_{n_l}^\infty \gamma(s, n)\frac{dn}{n} \tag{13.14}$$

where integration is performed with a lower cutoff $n_l = 1/\lambda_l$, so that all modes of integral scale larger than $\lambda_l$ are filtered out (excluded). Then [Gradshteyn and Ryzhik, 1994, eq. (3.381.3), p. 364] (see also Appendix B in this chapter)

$$\gamma(s, n_l) = \frac{C_0}{\Gamma(1-2H)n_l^{2H}}[1 - \exp(-n_l s) + (n_l s)^{2H}\Gamma(1-2H, n_l s)] \tag{13.15}$$

where $\Gamma(a, x)$ is the incomplete gamma function [Abramowitz and Stegun, 1972, eq. (6.5.3), p. 260]; in the limit as $n_l \rightarrow 0$, equation (13.15) reduces to equation (13.12). The variogram (13.15) defines a homogeneous field associated with a constant variance

$$\sigma^2(n_l) = \frac{C_0}{\Gamma(1-2H)n_l^{2H}} \tag{13.16}$$

an autocovariance $C(s, n_l) = \sigma^2(s, n_l) - \gamma(s, n_l)$ given by

$$C(s, n_l) = \frac{C_0}{\Gamma(1-2H)n_l^{2H}} \left[\exp(-n_l s) - (n_l s)^{2H}\Gamma(1-2H, n_l s)\right] \tag{13.17}$$

and a corresponding finite integral scale [Gradshteyn and Ryzhik, 1994, eq. (6.455.1), p. 690]

$$I(n_l) = \frac{1}{\sigma^2(n_l)} \int_0^\infty C(s, n_l)\, ds = \frac{2H}{1+2H} \frac{1}{n_l} = \frac{2H}{1+2H} \lambda_l \tag{13.18}$$

Because these results hold for $0 < H < 1/2$, the integral scale $I(n_l)$ of the truncated multiscale random field is proportional to (and always smaller than) the integral scale $\lambda_l$ of the lowest mode retained, the constant of proportionality increasing with $H$. Note also that for $n_l s \ll 1$, or equivalently $s \ll \lambda_l$, $\gamma(n_l, s) \simeq C_0 s^{2H}$ still holds, despite the truncation. This implies that the cutoff has no effect on lags much smaller than the integral scale of the smallest active mode.

If the hierarchy of modes is truncated both below a lower cutoff $n_l$ and above an upper cutoff $n_u$ according to

$$\gamma(s, n_l, n_u) = \int_{n_l}^{n_u} \gamma(s, n) \frac{dn}{n} \tag{13.19}$$

then one obtains, in analogy to equation (13.15),

$$\gamma(s, n_l, n_u) = \gamma(s, n_l) - (s, n_u) \tag{13.20}$$

When $n_l \rightarrow 0$ and $n_u \rightarrow \infty$, equation (13.20) reduces to equation (13.12). The variance, autocovariance, and integral scale associated with equation (13.20) are, respectively,

$$\sigma^2(n_l, n_u) = \sigma^2(n_l) - \sigma^2(n_u) \tag{13.21}$$

$$C(s, n_l, n_u) = C(s, n_l) - C(s, n_u) \tag{13.22}$$

$$I(n_l, n_u) = \frac{2H}{1+2H} \frac{n_u^{1+2H} - n_l^{1+2H}}{n_l n_u (n_u^{2H} - n_l^{2H})} \tag{13.23}$$

where $n_l = 1/\lambda_l$ and $n_u = 1/\lambda_u$, $\lambda_l$ and $\lambda_u$ being, respectively, the integral scales of the lowest and highest modes retained in the superposition. Note that equations (13.20)–(13.23) differ little from equations (13.15)–(13.18) when $n_u \gg n_l$; in particular, for given $H$, $C_0$, and $n_l$, the introduction of a higher cutoff $n_u$ brings about a reduction in variance and an increase in integral scale.

*Gaussian Modes.* Consider now modes having Gaussian variograms

$$\gamma(s,\lambda) = \sigma^2(\lambda)[1 - \exp(-\pi s^2/4\lambda^2)] \tag{13.24}$$

where $\lambda$ is again integral scale. Substituting equations (13.10) and (13.24) into equation (13.11) yields, after introduction of a new variable $x = n^2$ [Gradshteyn and Ryzhik, 1994, eq. (3.551.1), p. 403; Abramowitz and Stegun, 1972, eq. (6.1.17), p. 256],

$$\gamma(s) = C_0' s^{2H} \tag{13.25}$$

where

$$C_0' = C\frac{\Gamma(1-H)}{2H}\left(\frac{\pi}{4}\right)^H \qquad (0 < H < 1) \tag{13.26}$$

is a constant proportional to $C$ and having the same dimensions. The weighted superposition of Gaussian modes is thus a power variogram of the same form as equation (13.12), but is valid over a wider range of $H$ values. The ratio $C_0/C_0'$ between the coefficients in equations (13.13) and (13.26) is larger than unity (1.12 for $H = 0.10$, 3.40 for $H = 0.40$), implying that for given $C$ and $H$, the sum of exponential modes increases more rapidly with separation distance than does the sum of Gaussian modes.

Superimposing Gaussian modes with a lower cutoff $n_l = 1/\lambda_l$ according to equation (13.14) gives, following a change of variable $x = n^2$ [Gradshteyn and Ryzhik, 1994, eq. (3.381.3), p. 364; eq. (8.356.2), p. 951] (see also Appendix B of this chapter),

$$\gamma(s, n_l) = \frac{C_0'}{\Gamma(1-H)(\pi/4)^H n_l^{2H}} \times\left[1 - \exp\left(-\frac{\pi}{4}n_l^2 s^2\right) + \left(\frac{\pi}{4}n_l^2 s^2\right)^H \Gamma\left(1-H, \frac{\pi}{4}n_l^2 s^2\right)\right] \tag{13.27}$$

Because $\Gamma(1-H, 0) = \Gamma(1-H)$ [Gradshteyn and Ryzhik, 1994, eq. (8.354.2), p. 950], equation (13.27) reduces to equation (13.25) in the limit as $n_l \to 0$. The variance, autocovariance, and integral scale [Gradshteyn and Ryzhik, 1994, eq. (6.455.1), p. 690] associated with equation (13.27) are, respectively,

$$\sigma^2(n_l) = \frac{C_0'}{\Gamma(1-H)(\pi/4)^H n_l^{2H}} \tag{13.28}$$

$$C(s, n_l) = \frac{C_0'}{\Gamma(1-H)(\pi/4)^H n_l^{2H}} \times \left[\exp\left(-\frac{\pi}{4}n_l^2 s^2\right) - \left(\frac{\pi}{4}n_l^2 s^2\right)^H \Gamma\left(1-H, \frac{\pi}{4}n_l^2 s^2\right)\right] \tag{13.29}$$

$$I(n_l) = \frac{1}{\sigma^2(n_l)}\int_0^\infty C(s, n_l)\, ds = \frac{2H}{1+2H}\frac{1}{n_l} = \frac{2H}{1+2H}\lambda_l \tag{13.30}$$

The latter is identical with equation (13.18) corresponding to exponential modes, even though equations (13.27)–(13.29) differ from equations (13.21)–(13.23). From equation (13.28) it follows that equation (13.27) remains approximately equal to equation (13.24) when $n_l s \ll 1$.

If the hierarchy of modes is truncated both below a lower cutoff $n_l$ and above an upper cutoff $n_u$ according to equation (13.14), then the variogram, variance, autocovariance, and integral scale are given respectively by equations (13.20)–(13.23).

The autocorrelation functions $\rho(s, n_l) = C(s, n_l)/\sigma^2(s, n_l)$ corresponding to exponential and Gaussian modes with a lower cutoff are plotted in Figures 13.8 and 13.9, respectively, as functions of dimensionless distance $s/I$ at various values of $H$. Because the variance [equations (13.16) and (13.28)] and integral scale

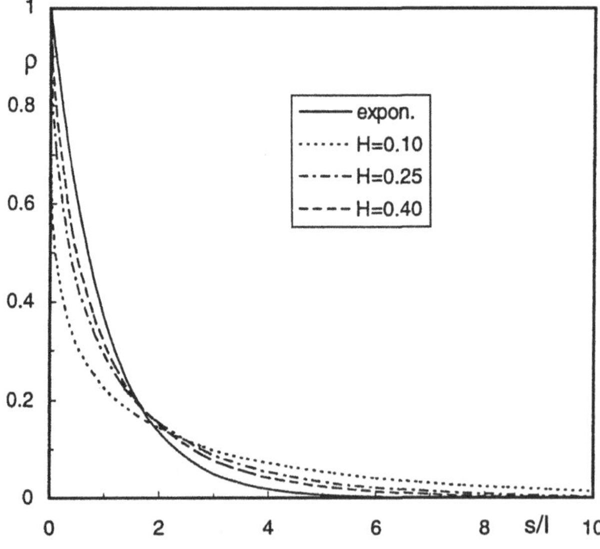


Figure 13.8. Autocorrelation function corresponding to exponential modes with lower cutoff versus dimensionless distance $s/I$ for various values of $H$.

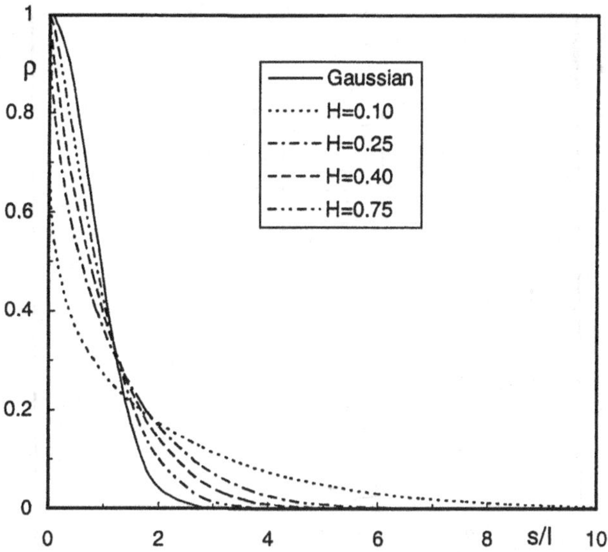


Figure 13.9. Autocorrelation function corresponding to Gaussian modes with lower cutoff versus dimensionless distance $s/I$ for various values of $H$.

[equations (13.18) and (13.30)] depend on $H$, we vary $C_0$ or $C_0'$ and $n_l$ with $H$ so as to maintain the values of $\sigma^2$ and $I$ constant. Exponential and Gaussian models with the same variance and integral scale are included for comparison. As $H$ increases from 0 to 0.5, the multiscale autocorrelation function in Figure 13.8 approaches the exponential model; as $H$ increases from 0 to 1, the multiscale autocorrelation function in Figure 13.9 approaches the Gaussian model.

Figures 13.10 and 13.11 show the multiscale autocorrelation functions $\rho(s, n_l, n_u) = C(s, n_l, n_u)/\sigma^2(s, n_l, n_u)$ with both lower and upper cutoffs for exponential and Gaussian modes, respectively, versus $s/I$ at $H = 0.25$ for various $\beta = n_l/n_u$, where $0 \leq \beta \leq 1$; the case without upper cutoff corresponds to $\beta = 0$. As $\beta$ increases, a wider range of high modes is excluded, and the resulting truncated autocorrelation functions approach the corresponding exponential or Gaussian models.

### *13.3.2 Superposition of Modes in a Spectral Domain*

All results obtained in the real domain have direct counterparts in the spectral domain. Our starting point is the spectral density corresponding to a given autocovariance function; the spectral density and autocovariance are related through the Fourier-transform pair

$$S(\mathbf{k}) = \frac{1}{(2\pi)^d} \int_{-\infty}^{\infty} C(\mathbf{s}) e^{-i\mathbf{k}\cdot\mathbf{s}} d\mathbf{s} \tag{13.31}$$

$$C(\mathbf{s}) = \int_{-\infty}^{+\infty} S(\mathbf{k}) e^{i\mathbf{k}\cdot\mathbf{s}} d\mathbf{k} \tag{13.32}$$

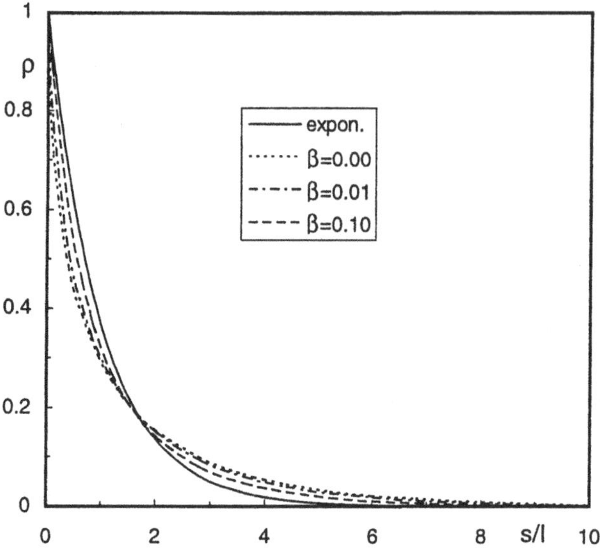


Figure 13.10. Autocorrelation function corresponding to exponential modes with lower and upper cutoffs versus dimensionless distance $s/I$ for $H = 0.25$ and various $\beta = n_l/n_u$, $0 \leq \beta \leq 1$.

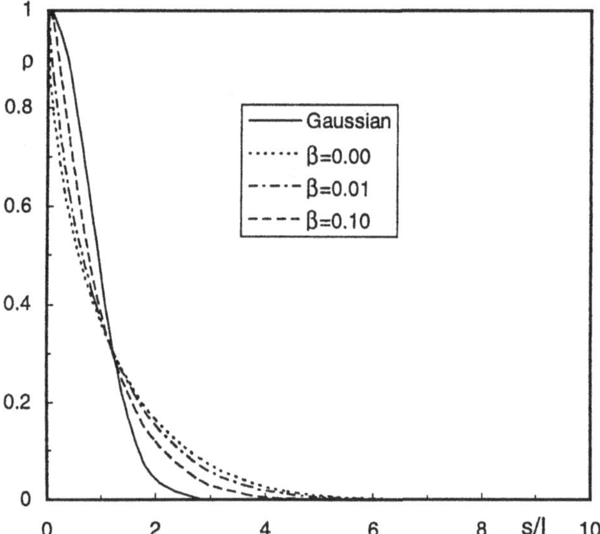


Figure 13.11. Autocorrelation function corresponding to Gaussian modes with lower and upper cutoffs versus dimensionless distance $s/I$ for $H = 0.25$ and various $\beta = n_l/n_u$, $0 \leq \beta \leq 1$.

*Exponential Modes.* For isotropic exponential autocovariance, the $d$-dimensional spectral density is [Yaglom, 1987, eq. (4.126), p. 362]

$$S(k) = \frac{\sigma^2 \lambda^d \Gamma[(d+1)/2]}{\pi^{(d+1)/2}(1 + k^2\lambda^2)^{(d+1)/2}} \tag{13.33}$$

Upon setting $\lambda = 1/n$, the superposition of a continuous hierarchy of individual spectra over all possible scales, weighted by $1/n$, is accomplished via

$$S(k) = \int_0^\infty S(k, n)\frac{dn}{n} \tag{13.34}$$

Defining the variance in equation (13.33) according to equation (13.10), and integrating equation (13.33) according to equation (13.34), following a change of variable $x = k/n$, yields the following spectral density [Gradshteyn and Ryzhik, 1994, eq. (3.241.2), p. 340; eq. (3.241.4), p. 341], valid for $0 < H < 1/2$:

$$S(k) = C_0 \frac{2^{2H} H\Gamma[(d+2H)/2]}{\pi^{d/2}\Gamma(1-H)k^{d+2H}} \tag{13.35}$$

where identities involving the gamma function [Gradshteyn and Ryzhik, 1994, eq. (8.331) and (8.335.1), p. 946] were employed. These power spectral densities coincide with those predicted by equations (13.2) and (13.5), (13.6) and (13.8), demonstrating that superposition leads to the same results in real and in spectral domains.

Introducing a lower cutoff by setting the lower limit of integration in equation (13.34) equal to $n_l$, and changing variable $x = k^2/n^2$, results in the following spectral density [Gradshteyn and Ryzhik, 1994, eq. (3.194.1), p. 333]:

$$S(k, n_l) = \frac{C_0}{\Gamma(1-2H)} \frac{2H\Gamma[(d+1)2]}{\pi^{(d+1)/2}(d+2H)n_l^{d+2H}} F_{2,1}\left(\frac{d+1}{2}, \frac{d+2H}{2}, \frac{d+2+2H}{2}, -\frac{k^2}{n_l^2}\right) \tag{13.36}$$

where $F_{2,1}$ is the Gauss hypergeometric function (Abramowitz and Stegun, 1972, p. 556). Adding an upper cutoff $n_u$ leads to the following expression valid in $d$ dimensions:

$$S(k, n_l, n_u) = S(k, n_l) - S(k, n_u) \tag{13.37}$$

*Gaussian Modes.* The $d$-dimensional spectral density associated with an isotropic Gaussian autocovariance (13.24) is [Yaglom, 1987, eq. (4.134), p. 364]

$$S(k) = \frac{\sigma^2\lambda^d}{\pi^d} \exp\left(-\frac{k^2\lambda^2}{\pi}\right) \tag{13.38}$$

The weighted superposition of a continuous hierarchy of these individual spectra over all possible scales is again accomplished via equation (13.34). Introducing the change of variable $x = 1/n^2$ yields a spectral density [Gradshteyn and Ryzhik, 1994, eq. (3.381.4), p. 364], valid for $0 < H < 1$, given by equation (13.35) upon replacement of $C_0$ by $C_0'$.

Introducing a lower cutoff in the superposition yields the following $d$-dimensional spectral density [Gradshteyn and Ryzhik, 1994, eq. (3.381.1), p. 364]:

$$S(k, n_l) = \frac{C_0'}{\Gamma(1-H)} \frac{2^{2H} H}{\pi^{d/2} k^{d+2H}} \gamma'\left(\frac{d+2H}{2}, -\frac{n_l^2 k^2}{\pi}\right) \tag{13.39}$$

where $\gamma'(a, x) = \Gamma(a) - \Gamma(a, x)$ is the incomplete gamma function [Abramowitz and Stegun, 1972, eq. (6.5.3), p. 260]. Adding an upper cutoff $n_u$ leads to an expression formally identical with equation (13.37).

We confirmed all the foregoing spectral-density expressions, for both exponential and Gaussian modes, by obtaining them for various $H$ values through evaluation of the Fourier transform of corresponding autocovariance functions in one, two, and three dimensions by means of Mathematica.

## 13.4 Truncated Power Spectral Densities

In this section we consider random fields having power spectral densities of the functional forms given by equations (13.2), (13.5), and (13.6), with a sharp lower cutoff such that

$$\begin{aligned} S(k) &= \frac{A_d}{k^{d+2H}} \qquad (k \geq k_l) \\ S(k) &= 0 \qquad (k < k_l) \end{aligned} \tag{13.40}$$

where $A_d$ $(d = 1, 2, 3)$ is the spectral density for unit wavenumber. The spectral density $S(\mathbf{k})$ and autocovariance $C(\mathbf{s})$ are related through the Fourier-transform pair, equations (13.31) and (13.32). Truncated random fields with the foregoing cutoff have finite variance and autocovariance. In particular, the $d$-dimensional variance is given by (Appendix A)

$$\sigma^2(k_l) = \frac{\pi^{d/2} A_d}{\Gamma(d/2) H k_l^{2H}} \tag{13.41}$$

and the corresponding autocovariance function by (Appendix A)

$$\begin{aligned} C(s, k_l) = \frac{\pi^{d/2} A_d}{\Gamma(d/2) H k_l^{2H}} \Bigg[ & F_{1,2}\left(-H; \frac{d}{2}, 1-H; -\frac{k_l^2 s^2}{4}\right) \\ & - \frac{\Gamma(d/2)\Gamma(1-H)}{2^{2H}\Gamma[(d+2H)/2]} (k_l s)^{2H} \Bigg] \end{aligned} \tag{13.42}$$

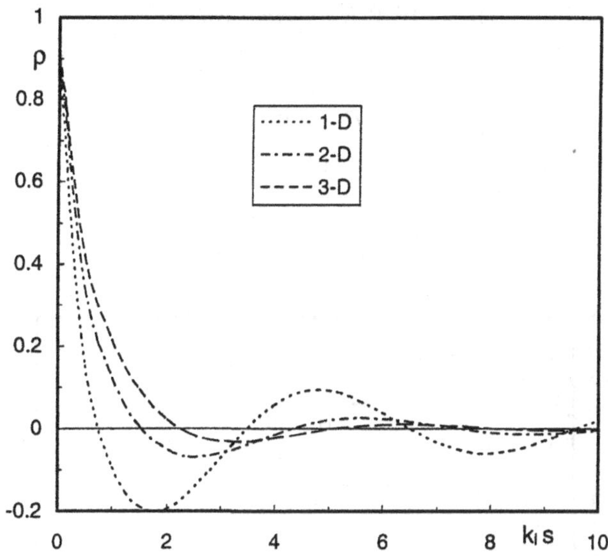


Figure 13.12. Autocorrelation function corresponding to power spectral density with lower cutoff versus dimensionless distance $k_l s$ in one, two, and three dimensions for $H = 0.25$.

where $F_{p,q}$ is the generalized hypergeometric function (Erdélyi et al., 1953)

$$F_{p,q}(a_1, \ldots, a_p; b_1, \ldots, b_q; Z) = \sum_{k=0}^{\infty} \frac{(a_1)_k \cdots (a_p)_k Z^k}{(b_1)_k \cdots (b_q)_k k!} \tag{13.43}$$

$$(a_0) = 0, \qquad (a)_k = a(a+1)\cdots(a+k-1) = \frac{\Gamma(a+k)}{\Gamma(a)} \tag{13.44}$$

The autocorrelation functions $\rho(s, k_l) = C(s, k_l)/C(0, k_l)$ corresponding to equation (13.42) for $H = 0.25$ are plotted in Figure 13.12 versus the dimensionless distance $k_l s$. They are seen to oscillate asymptotically about zero, with amplitudes that diminish as the dimensionality $d$ of the domain increases from 1 to 3. Figure 13.13 shows that in one dimension the amplitude is larger for larger values of $H$. The corresponding variograms are simply $\gamma(s, k_l) = \sigma^2(k_l) - C(s, k_l)$. As such, they approach their sill in an oscillatory fashion at an asymptotic rate that increases with the cutoff frequency $k_l$. The corresponding integral scale is

$$I(k_l) = \int_0^\infty \rho(s, k_l)\, ds = \frac{1}{k_l} \int_0^\infty \rho(r)\, dr = \frac{1}{k_l} f_1(H) \tag{13.45}$$

where $r = k_l s$. It is difficult to evaluate $I(k_l)$ analytically for autocovariance functions (13.42) because of their binomial form. Instead, we evaluated $f_1(H)$ numerically for $H = 0.10$ and $0.25$ by means of the Newton-Cotes integration routine built

Table 13.1. *Numerical values for* $f_1(H)$

| Dimensions | $H = 0.10$ | $H = 0.25$ |
|---|---|---|
| 1 | 0.00 | 0.00 |
| 2 | 0.17 | 0.33 |
| 3 | 0.26 | 0.52 |

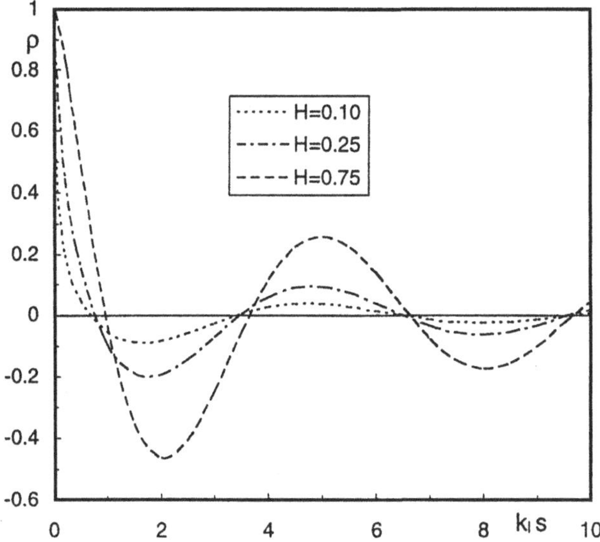


Figure 13.13. Autocorrelation function corresponding to power spectral density with lower cutoff versus dimensionless distance $k_l s$ in one dimension for various values of $H$.

into Maple V. The results are listed in Table 13.1. They suggest that the autocovariance functions expressed by equation (13.42) are associated with either zero or finite integral scales; an indirect argument that supports this in two and three dimensions has been developed elsewhere (Di Federico and Neuman, 1995).

If one considers both a lower cutoff and an upper cutoff in the spectral domain, the spectral density takes the form

$$S(k) = \frac{A_d}{k^{d+2H}} \qquad (k_u \geq k \geq k_l) \tag{13.46}$$

$$S(k) = 0 \qquad (k < k_l \quad \text{or} \quad k > k_u)$$

Accordingly, the variance is given by

$$\sigma^2(k_l, k_u) = \sigma^2(k_l) - \sigma^2(k_u) \tag{13.47}$$

The corresponding autocovariance function, obtained by means of Mathematica, is

$$C(s, k_l, k_u) = \frac{\pi^{d/2} A_d}{\Gamma(d/2) H k_l^{2H}} \left[ \frac{1}{k_l^{2H}} F_{1,2}\left( -H; \frac{d}{2}, 1 - H; -\frac{k_l^2 s^2}{4} \right) - \frac{1}{k_u^{2H}} F_{1,2}\left( -H; \frac{d}{2}, 1 - H; -\frac{k_u^2 s^2}{4} \right) \right] \tag{13.48}$$

The associated variograms $\gamma(s, k_l, k_u) = \sigma^2(k_l, k_u) - C(s, k_l, k_u)$ oscillate asymptotically about a constant sill; when $k_u \gg k_l$, they differ little from those involving only a lower cutoff. The one-dimensional version of our variogram is almost identical with that recently developed by Yordanov and Nikolaev (1994), except for a minor difference stemming from different definitions of the Fourier transform.

We see that the traditional approach of truncating power spectral densities yields autocovariance functions that oscillate about zero with finite (in two and three dimensions) or vanishing (in one dimension) integral scales.

Recently Dagan (1994) considered two-dimensional steady-state flow in saturated multiscale porous media. He derived linearized pre-asymptotic and asymptotic approximations for the nonergodic longitudinal-dispersion coefficient in log-conductivity fields that were either stationary with a power covariance $C(s) = as^\beta$, $-1 \leq \beta \leq 0$, or nonstationary with a power variogram $\gamma(s) = as^\beta$, $0 \leq \beta \leq 2$. This dispersion coefficient defines the mean spread about the actual centroid of a plume, as opposed to the ergodic dispersion coefficient, which defines the mean spread about the mean centroid. To derive such a nonergodic dispersion coefficient for a log-conductivity field associated with a power variogram, Dagan (1994, sec. 5) proposed to introduce a low-frequency (infrared) cutoff in the associated power spectrum so as to render the log-conductivity field homogeneous, with a constant variance. His theory did not require specifying the variance explicitly. As to the variogram, Dagan assumed that regardless of the low-frequency cutoff, it retained its original power form $\gamma(s) = as^\beta$. Based on that, Dagan concluded that the nonergodic longitudinal-dispersion coefficient in two dimensions tends asymptotically to a Fickian constant as mean travel distance increases when $0 \leq \beta < 1$, but continues to grow without bound when $1 \leq \beta \leq 2$, in which case its growth rate cannot be faster than linear.

The actual effects of truncating a two-dimensional power spectrum below some frequency $k_l$ on the associated variance and variogram are given, respectively, by our equations (13.41) and (13.42). In Dagan's notation ($\beta = 2H$), the corresponding variogram is written as

$$\gamma(s) = \frac{2\pi A_2}{\beta k_l^\beta} \left[ \frac{\Gamma(1 - \beta/2)}{2^\beta \Gamma(1 + \beta/2)} (k_l s)^\beta + 1 - F_{1,2}\left( -\beta/2; 1, 1 - \beta/2; -\frac{k_l^2 s^2}{4} \right) \right] \tag{13.49}$$

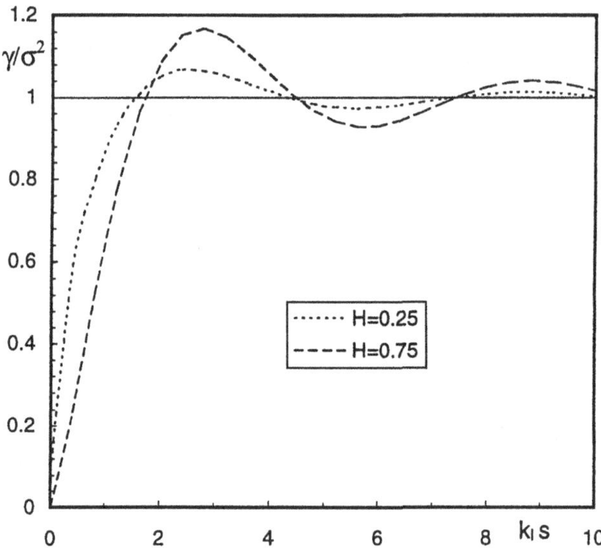


Figure 13.14. Dimensionless variogram corresponding to power spectral density with lower cutoff versus dimensionless distance $k_l s$ in two dimensions for $H = 0.25$ and $0.75$.

Figure 13.14 shows a plot of this variogram, normalized by the variance, as a function of $k_l s$ for $H = \beta/2 = 0.25$ and $0.75$. It is evident that the variogram does not follow a power, but oscillates asymptotically around its sill. We mentioned earlier that a similar variogram was obtained by Yordanov and Nikolaev (1994) in one dimension, including both upper and lower cutoffs. They proposed that one could approximate it by a power law within a restricted interval of lag $s$ values, which, however, depends critically on the infrared cutoff frequency $k_l$, the exponent of their approximate power being other than $\beta$. According to Yordanov and Nikolaev (1994), in one dimension the exponent exceeds $\beta$ when $0 \leq \beta < 1$ $(0 \leq H < 0.5)$ and is less than $\beta$ when $1 < \beta < 2$ $(0.5 < H < 1)$.

More importantly, because the introduction of a low-frequency cutoff renders the integral scale of the covariance either finite or zero for all admissible $\beta$, transport becomes asymptotically Fickian and ergodic (Dagan, 1990) for all $0 < \beta < 2$, including transport within the range $1 \leq \beta < 2$, which Dagan (1994) thought would remain permanently pre-asymptotic. Hence there is no need to distinguish between ergodic and nonergodic asymptotic dispersions; such a distinction is of interest at best during the pre-asymptotic stages of plume dispersion.

## 13.5 Verification of Scaling Relations

To verify our multiscale model directly, we consider values of spatial autocorrelation (integral) scale and variance, inferred by various researchers from natural-log hydraulic conductivity and transmissivity data at diverse sites on the assumption that the corresponding fields are statistically homogeneous and isotropic, as recently summarized by Gelhar (1993, table 6.1). Gelhar's table also lists a characteristic length for each domain within which conductivity or transmissivity was sampled. Figure 13.5 is a plot of reported integral scales versus characteristic lengths of the various field sites (domains). It suggests that integral scales tend to increase linearly with domain size, at a generalized rate of about 1/10, over domains that range in length scale from a few meters to several hundred kilometers. This is consistent with our truncated hierarchical model provided only that one postulates a fixed ratio $\mu$ between the length scale of any sampling window superimposed on the infinite hierarchy and the integral scale of the lowest-frequency mode that is filtered out by this window. Thus, if $\omega$ is the length scale of the data support (sample size) and $\Omega$ is the length scale of the domain being sampled, then the sampling window has a length scale bounded from below by $\omega$ and from above by $\Omega$, so as to introduce an upper frequency cutoff $n_u = 1/\lambda_u = 1/\mu\omega$, with associated integral scale $\lambda_u = \mu\omega$, and a lower frequency cutoff $n_l = 1/\lambda_l = 1/\mu\Omega$, with associated integral scale $\lambda_l = \mu\Omega$. In the common case where $\Omega \gg \omega$ and thus $n_u \gg n_l$, it is appropriate to work with the limiting case $n_u \to \infty$. Then, according to equation (13.18) or (13.30), the integral scale of the filtered field is given by

$$I(\lambda_l) = \alpha\lambda_l, \qquad \alpha = \frac{2H}{1+2H} \tag{13.50}$$

where $0 < \alpha < 1/2$ for exponential modes and $0 < \alpha < \frac{2}{3}$ for Gaussian modes. Hence

$$I(\lambda_l) = \alpha\mu\Omega \tag{13.51}$$

which, as we saw, is supported by the data in Figure 13.5.

We mentioned earlier that Neuman (1990, 1995) had derived, on the basis of juxtaposed apparent-dispersivity data from tracer studies at many diverse sites, a generalized power variogram for natural-log hydraulic conductivities having the simple form $\gamma(s) \approx cs^{1/2}$, where $c \approx 0.027$. His interpretation seems to hold (Neuman, 1993b, 1995) despite the fact that some of the field-derived dispersivities may not be entirely reliable (Gelhar, Welty, and Rehfeldt, 1992). The corresponding variogram will be identical with equation (13.12) or (13.25) provided that one adopts the generalized Hurst coefficient $H \approx 0.25$. If we take the variogram to consist of exponential modes, then, according to equation (13.13), $C_0 \approx 0.027$, and $C \approx 0.027/(2\pi^{1/2}) = 0.0076$. If we take it to consist of Gaussian modes, then, according to equation (13.26), $C_0 \approx 0.027$, and $C \approx 0.027(4/\pi)^{0.25}/[2\Gamma(\frac{3}{4})] = 0.0117$. It follows that $C$ has a generalized value close to 0.01. Figure 13.5 suggests that for juxtaposed hydraulic conductivity and transmissivity data from many diverse sites, $\alpha\mu \approx 0.1$. According to equation (13.50), $H \approx 0.25$ corresponds to a generalized value of $\alpha \approx \frac{1}{3}$, and so we obtain a generalized value of $\mu \approx \frac{1}{3}$.

Equations (13.16) and (13.28), together with our postulate $\lambda_l = \mu\Omega$, imply that when $\Omega \gg \omega$ the variance of a truncated hierarchy increases as a power of domain size according to $\sigma^2 \propto \Omega^{2H}$, and so $\sigma^2\lambda \propto \Omega^{1+2H}$. For juxtaposed natural-log hydraulic conductivities and transmissivities from many sites, we expect $H \approx 0.25$, and hence $\sigma^2\lambda \propto \Omega^{1.5}$. Indeed, Figure 13.15 shows that when the hydraulic-conductivity data (squares and triangles) from Gelhar (1993, table 6.1) are juxtaposed on a log-log plot of $\sigma^2\lambda$ versus $\lambda$ (or, equivalently, $\Omega$), they are scattered closely about a (dashed) line having the predicted slope of 1.5; the corresponding transmissivity data (circles) are scattered closely about another (solid) line having the same slope and an intercept corresponding to the value predicted by Neuman (1990) (on the basis of tracer-study data) of $c \approx 0.027$ (see the preceding paragraph). The offset between the two lines was attributed by Neuman (1994) to a reduction in variance obtained when hydraulic conductivities were averaged over the vertical to yield transmissivities. This explanation is in turn qualitatively consistent with equation (13.21), which, together with our postulate $\lambda_u = \mu\omega$, predicts a reduction in variance with support scale according to

$$\sigma^2 \propto (\Omega^{2H} - \omega^{2H}) \tag{13.52}$$

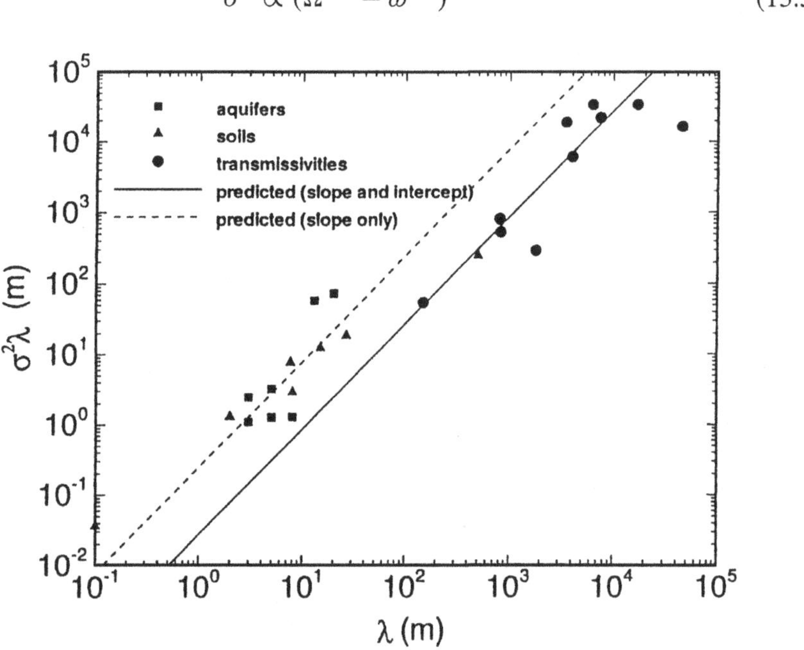


Figure 13.15. Product of variance $\sigma^2$ and horizontal correlation scale $\lambda$ of natural-log hydraulic conductivities and transmissivities at various sites versus $\lambda$. Predictions correspond to $\gamma(s) = c\sqrt{s}$. (From Neuman, 1994, with permission; data from Gelhar, 1993, table 6.1.)

## 13.6 Bridging Across Scales

Our hierarchical theory allows bridging across scales by predicting the effect of viewing a multiscale random field, defined and measured on a given support scale, through a larger window defined by the domain under investigation. More precisely, our theory predicts the effects of domain size on the integral scale, variance, and covariance of such a multiscale random field. At a given locale, one can, in principle, calibrate our truncated variogram model (estimate its parameters $C$, $H$, $\mu$) to sample data observed on a given support length scale $\omega$ (which, with $\mu$, defines the upper frequency cutoff $n_u$) in one domain of length scale $\Omega$ (which, with $\mu$, defines the lower cutoff $n_l$), and then predict the autocovariance structure of the corresponding field in domains that are either smaller ($<\Omega$) or larger ($>\Omega$). One can also venture (we suspect with less predictive power) to transfer such predictions from one locale to another by introducing the generalized parameters $C \approx 0.0076$ for exponential modes, $C \approx 0.0117$ for Gaussian modes, $H \approx 0.25$, and $\mu \approx \frac{1}{3}$ into our truncated variogram models. Selection between alternative models (exponential modes, Gaussian modes with or without lower cutoffs, etc.) is also possible in principle by means of formal model discrimination criteria, coupled with methods such as maximum-likelihood cross-validation, as proposed by Samper and Neuman (1989).

## 13.7 Effective Hydraulic Conductivity

It is common in the stochastic groundwater literature to treat log hydraulic conductivity as a statistically homogeneous Gaussian field. For an unbounded flow domain subject to a uniform mean hydraulic gradient, the effective conductivity corresponding to a statistically isotropic field has been conjectured by Matheron (1967) to be

$$K_{\text{eff}} = K_g \exp\left[\sigma^2\left(\frac{1}{2} - \frac{1}{d}\right)\right] \tag{13.53}$$

where $K_g$ is the geometric mean conductivity. This result is rigorously valid for the one-dimensional case ($d = 1$), where $K_{\text{eff}}$ is the harmonic mean $K_h = K_g \exp(-\sigma^2/2)$, and for the two-dimensional case ($d = 2$), where $K_{\text{eff}} = K_g$. Its validity in the three-dimensional case ($d = 3$) has been demonstrated numerically for $\sigma^2$ as large as 7 by Neuman and Orr (1993), and for somewhat smaller values of $\sigma^2$ by Ababou (1988), Desbarats (1992), and Dykaar and Kitanidis (1992).

The effective conductivity for a multiscale isotropic field consisting of exponential modes is obtained by substituting the variance (13.21), coupled with equation (13.16), into equation (13.53). The results for one, two, and three dimensions are, respectively,

$$K_{\text{eff}} = K_g \exp\left(-\frac{\sigma^2}{2}\right) = K_g \exp\left[-\frac{C_0}{2\Gamma(1-2H)}\left(\frac{1}{n_l^{2H}} - \frac{1}{n_u^{2H}}\right)\right] \quad (13.54)$$

$$K_{\text{eff}} = K_g \quad (13.55)$$

$$K_{\text{eff}} = K_g \exp\left(\frac{\sigma^2}{6}\right) = K_g \exp\left[-\frac{C_0}{6\Gamma(1-2H)}\left(\frac{1}{n_l^{2H}} - \frac{1}{n_u^{2H}}\right)\right] \quad (13.56)$$

These expressions are closely related to the expression derived by Neuman (1994) for the effective conductivity of a block of length scale $\Omega$ without considering truncation. The two results become identical if we consider $C_0$ to be the same in both, set $\beta = 1$ in Neuman's expression to eliminate boundary influence, consider $n_u \to \infty$ and $n_l = 1/\lambda_l$ (where $\lambda_l$ is the integral scale of the lowest mode retained in the superposition), and write

$$\Omega^{2H} = \frac{\lambda_l^{2H}}{\Gamma(1-2H)} \quad (13.57)$$

The same can be done with Gaussian modes by using equation (13.28) instead of equation (13.16).

In a somewhat related study, Ababou and Gelhar (1990) considered a statistically isotropic log permeability field with variance $\sigma^2$ proportional to $\ln \Omega$. They concluded that in such a field the effective permeability (not its logarithm) increases systematically as a power of $\Omega$ in three dimensions, decreases in one dimension, and shows no systematic variation with $\Omega$ in two dimensions.

Figure 13.2 shows only ranges of measured permeability, which are insufficient to validate quantitatively the theoretical scaling relations (13.54)–(13.57). However, the figure does provide qualitative support for these relations. The laboratory and borehole permeability data appear to exhibit a systematic increase with the reported scale of measurement. No such trend is indicated by the regional data. This could be due to a combination of factors: The available sample may be too small to indicate a trend; some of the data correspond to flow regimes that are nearly two-dimensional; some of the data are derived from calibrated numerical models that account explicitly for medium heterogeneity and thus filter out large-scale, low-frequency variations from the hierarchy.

### 13.8 First-Order Theory of Flow and Advective Transport

Di Federico and Neuman (1995, 1998a, b) developed a first-order theory of flow and advective transport in isotropic, multiscale, random log conductivity fields formed by the weighted superposition of mutually uncorrelated exponential and Gaussian modes with lower and upper cutoffs in two and three dimensions. We present some of their results for exponential modes in two dimensions.

Consider steady-state flow satisfying the continuity equation and Darcy's law:

$$\nabla \cdot \mathbf{q}(\mathbf{x}) = 0, \qquad \mathbf{q}(\mathbf{x}) = -K(\mathbf{x})\nabla h(\mathbf{x}) \tag{13.58}$$

where $\mathbf{q}$ is Darcy flux, and $h$ is hydraulic head. The seepage velocity is given by $\mathbf{u}(\mathbf{x}) = \mathbf{q}(\mathbf{x})/\phi$, where $\phi$ is porosity (assumed to be constant). We treat the natural logarithm $Y(\mathbf{x}) = \ln K(\mathbf{x})$ of hydraulic conductivity as a statistically homogeneous, multivariate Gaussian random field, uniquely defined by its constant ensemble mean (expected value) $\langle Y \rangle$, variance $\sigma^2$, and spatial covariance function $C(\mathbf{s})$, where $\mathbf{s} = \mathbf{x} - \chi$ is a separation vector. We further assume that $Y(\mathbf{x})$ results from a weighted superposition of exponential modes with a lower cutoff $n_l$, so that $\sigma^2$, $C(\mathbf{s})$, and the associated finite integral scale are given, respectively, by equations (13.16)–(13.18). This allows us to derive explicit expressions for the cross-covariance $C_{Yh}(\mathbf{x}, \chi)$ between $Y$ and $h$, the head variogram $\gamma_h(\mathbf{x}, \chi)$, the cross-covariance $C_{uY}(\mathbf{x}, \chi)$ between any component of $\mathbf{u}$ and $Y$, the cross-covariance $C_{uh}(\mathbf{x}, \chi)$ between any component of $\mathbf{u}$ and $h$, and the cross-covariance and autocovariance $C_{uu}(\mathbf{x}, \chi)$ between any two components of $\mathbf{u}$ [Di Federico and Neuman, 1995, e.g. (3.42)–(3.50) and figs. 6–14].

Consider an indivisible particle of inert solute that has finite mass but zero volume. At time $t = 0$ the particle is known to be located at the origin of the coordinates. It is then swept by the random groundwater velocity $\mathbf{u}(\mathbf{x})$ to describe a time-dependent, random trajectory $\mathbf{X}(t)$. Under steady-state uniform mean flow, the mean displacement is $\langle \mathbf{X}(t) \rangle = \langle \mathbf{u} \rangle t$, where

$$\langle \mathbf{u} \rangle = \frac{K_G}{\phi}\mathbf{J} = \mathbf{U} \tag{13.59}$$

and $\mathbf{J} = -\nabla \langle h(\mathbf{x}) \rangle$ is the constant mean hydraulic gradient. We write $\mathbf{X}(t) = \langle \mathbf{X}(t) \rangle + \mathbf{X}'(t)$ and note that $X_{ij}(t) = \langle X_i'(t) X_j'(t) \rangle$ is the covariance of particle displacement. Explicit first-order expressions for $X_{11}$ and $X_{22}$ are given for the case of zero local dispersion by Di Federico and Neuman [1995, e.g. (3.55) and (3.56) and figs. 15 and 16]. From these, it is easy to derive corresponding expressions for the macrodispersion coefficients:

$$D_{ij}(t) = \frac{1}{2}\frac{dX_{ij}}{dt} \tag{13.60}$$

The longitudinal-macrodispersion coefficient is

$$D_{11}(r) = U\frac{\sigma^2}{4(1+2H)(2+H)(1+H)n_l} \tag{13.61}$$
$$\times\left[8H(2+H)(1+H) + 12H(1+2H)(1+H)\frac{1}{r^3}\right.$$
$$- 12H(1+2H)(1+H)\frac{e^{-r}}{r^3} - 12H(1+2H)(1+H)\frac{e^{-r}}{r^2}$$
$$- 6H(1+2H)(2+H)\frac{1}{r} + 6H(1+2H)\frac{e^{-r}}{r}$$
$$\left.- 6He^{-r} + 3re^{-r} - 3r^{1+2H}\Gamma(1-2H, r)\right]$$

and the transverse coefficient is

$$D_{22}(r) = U\frac{\sigma^2}{4(2+H)(1+H)n_l} \tag{13.62}$$
$$\times\left[-12H(1+H)\frac{1}{r^3} + 12H(1+H)\frac{e^{-r}}{r^3} + 12H(1+H)\frac{e^{-r}}{r^2}\right.$$
$$+ 2H(2+H)\frac{1}{r} + 2H(1+2H)\frac{e^{-r}}{r}$$
$$\left.- 2He^{-r} + re^{-r} - r^{1+2H}\Gamma(1-2H, r)\right]$$

where $U$ is the magnitude of $\mathbf{U}$, and $r = n_l s$, $s$ being the magnitude of the displacement vector $\mathbf{s}$. Asymptotically, as $r \to \infty$, all terms, except the first, on the right-hand side of equation (13.61) vanish, and so

$$D_{11}(\infty) = U\sigma^2\frac{2H}{(1+2H)n_l} = U\frac{2HC_0}{(1+2H)\Gamma(1-2H)n_l^{1+2H}} \tag{13.63}$$

Note that this Fickian asymptote can be recast as $D_{11}(\infty) = U\sigma^2 I$, which yields the asymptotic (Fickian) longitudinal dispersivity

$$\alpha_{11}(\infty) = \frac{D_{11}(\infty)}{U} = \sigma^2 I \tag{13.64}$$

We saw earlier that $\sigma^2 I$ grows as a power $1 + 2H$ of distance; hence equation (13.64) implies that the same is true of $\alpha_{11}(\infty)$. Considering that the apparent longitudinal dispersivities in Figure 13.3 were obtained by fitting tracer experimental data to Fickian models, we now have a theoretical explanation (and prediction) of their observed supralinear increase with mean travel distance (or time). This is true for the open symbols in Figure 13.3; the solid symbols were obtained from numerical models that accounted explicitly (deterministically) for grid-scale medium heterogeneity and thus fall outside the scope of the theory discussed here. Note that our theoretical prediction as to how $\alpha_{11}(\infty)$ scales with distance corresponds exactly to, and thus confirms, the scaling relation derived semiempirically by Neuman (1990) from Figure 13.3.

The first-order transverse-dispersion coefficient in equation (13.62) tends to zero at large mean travel distance (or time).

Figures 13.16 and 13.17 show how dimensionless $D_{11}$ and $D_{22}$ (normalized with respect to $U\sigma^2 I$ so that they also represent dimensionless dispersivities), respectively, vary with the dimensionless distance $s/I$ as $H$ takes on different values. Because the integral scale $I = (2H)/[(1 + 2H)n_l]$ and the variance $\sigma^2 = C_0/[\Gamma(1 - 2H)n_l^{2H}]$ depends on $H$, we vary $C_0$ and $n_l$ with $H$ so as to maintain the values of $I$ and $\sigma^2$ constant. The longitudinal dispersivity for the multiscale model is seen to be always inferior to that for an exponential model with a similar integral scale. The difference increases as $H$ decreases, and the rate at which $D_{11}$ approaches its asymptote goes down. At a travel distance of 80 integral scales, dimensionless $D_{11}$ for $H = 0.10$ is still 0.93, whereas the exponential model gives 0.98. The greatest relative difference between the two models occurs in the intermediate range of a few tens of integral scales. The transverse dispersivity shows lower rising limbs and peaks as $H$ decreases; the peaks become lower and the descending limbs are offset to the right.

When an upper truncation is present, its effect is insignificant if $n_u \gg n_l$, and the foregoing still holds. Figures 13.18 and 13.19 show what happens to dimensionless

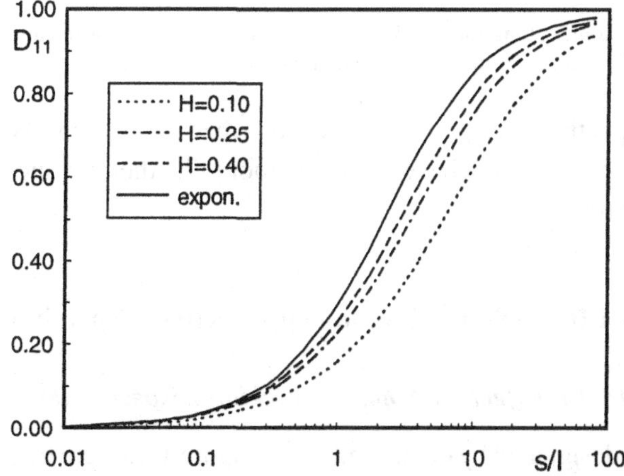


Figure 13.16. Dimensionless longitudinal dispersion versus dimensionless distance as a function of $H$ in the presence of a lower cutoff, compared with exponential model.

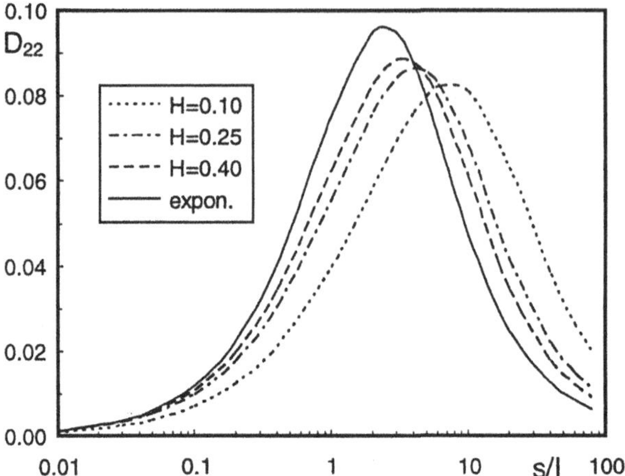


Figure 13.17. Dimensionless transverse dispersion versus dimensionless distance as a function of $H$ in the presence of a lower cutoff, compared with exponential model.

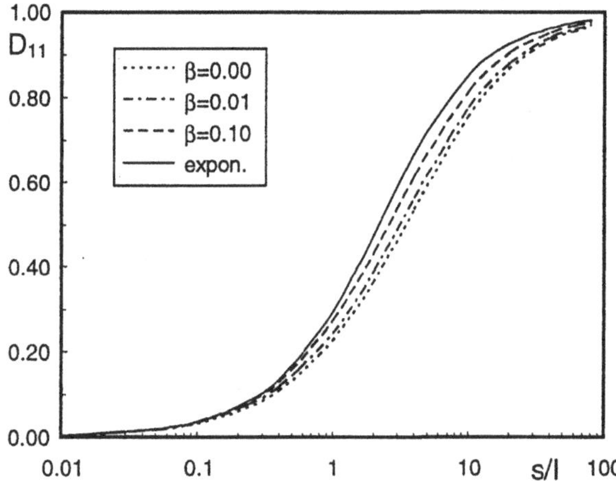


Figure 13.18. Dimensionless longitudinal dispersion versus dimensionless distance as a function of cutoff ratio $\beta$ when $H = 0.25$, compared with exponential model.

$D_{11}$ and $D_{22}$, respectively, as $\beta = n_l/n_u$ varies when $H = 0.25$. We see that as $\beta$ increases, both dispersivities behave more and more like their counterparts based on the exponential models.

## 13.9 Effect of Conditioning on Effective Dispersivity

### *13.9.1 Effect on Apparent Fickian Dispersivities*

The dispersivities designated by open symbols in Figure 13.3 were obtained by means of Fickian (mostly analytical) models that consider the permeability (or transmissivity) to be uniform in each tracer study. Dispersivities designated by solid symbols

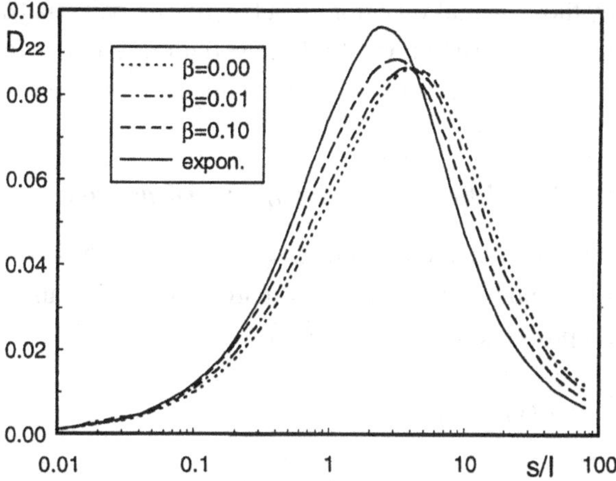


Figure 13.19. Dimensionless transverse dispersion versus dimensionless distance as a function of cutoff ratio $\beta$ when $H = 0.25$, compared with exponential model.

were obtained from the calibration of numerical models against hydraulic and concentration data corresponding to large-scale plumes. Such models account explicitly for intermediate-scale (larger than the support scale $\omega$, smaller than the flow domain $\Omega$) medium heterogeneity by allowing the permeability (or transmissivity) to vary spatially in a deterministic manner. In the language of stochastic theory, such models are conditioned (though not in a rigorous manner) on site-specific measurements. What effect does this have on the computed apparent Fickian dispersivities?

The apparent dispersivities from calibrated, explicitly heterogeneous models are seen to increase much more slowly with scale than is predicted by equation (13.64). Applying regression to these dispersivities yields a straight line with a slope of 0.54 (not shown in the figure). The null hypothesis that the slope is unity or larger must be rejected at the 99% confidence level. Hence the recorded dispersivities from calibrated models are inconsistent with the generalized power variogram and the associated fBM model, which we found to be consistent with the rest of the data. The two data sets constitute samples from different populations and must be analyzed separately. This explains why we have not included data from calibrated (conditional) models in our earlier discussion of unconditional fields.

The calibrated dispersivities in Figure 13.3 grow approximately in proportion to $\sqrt{s}$, and the discrepancy between them and values predicted by equation (13.64) grows in proportion to $s$. Neuman (1990) attributed this reduction in scale partly to a nonlinear effect and partly to conditioning on (accounting explicitly and deterministically for) site-specific data. Such conditioning renders the model better able to resolve spatial variations in the advective velocity field. It introduces an upper bound (cutoff) on the correlation scales that affect dispersivities, causing the latter to diminish as information about the spatial variability of the log permeabilities improves in quantity and quality. The diminution in apparent longitudinal dispersivity occurs because only those spatial variations in advective velocity that are not resolved explicitly have an effect on dispersion. As these variations are now limited to higher frequencies (and shorter wavelengths), so are the dispersivities reduced.

### *13.9.2 Effect on Preasymptotic Dispersion*

Following Neuman (1995), let us consider steady-state advection in a random velocity vector field $\mathbf{v}(\mathbf{x})$. The field has homogeneous spatial increments and is characterized by a tensor (dyadic) power-law semivariogram:

$$\gamma(s) = b_0 \rho s^{2H} \qquad (0 < H < 1) \tag{13.65}$$

where $b_0$ is a constant and $\rho$ is a constant dyadic. As such, $\mathbf{v}(\mathbf{x})$ is a self-affine random field with fractal dimension $D = d + 1 - H$. Its ensemble mean and variance are generally undefined. We define them conditionally by assuming, without loss of generality, that the velocity is known (from hydraulic and/or tracer data) to be $\mathbf{v}_0$ at some point $\mathbf{x}_0$. Hence $\mathbf{v}_0$ is the conditional ensemble mean of $\mathbf{v}$.

At time $t_0$, a mass $M_0$ of tracer is introduced into the stream at another point $\mathbf{y}_0$. We associate this mass with an indivisible particle that advects with the random velocity $\mathbf{v}(\mathbf{x})$ without diffusion or local dispersion. Let $\tau = t - t_0$ be the particle residence time at any $t > t_0$, so that its conditional mean displacement from the origin $\mathbf{y}_0$ is $\mathbf{h} = \mathbf{v}_0\tau$. Let $\beta$ be the angle between $\mathbf{h}$ and the separation vector $\mathbf{r} = \mathbf{y}_0 - \mathbf{x}_0$. Then the conditional mean distance $s$ of the particle from $\mathbf{x}_0$ is given by $s^2 = r^2 - 2rh\cos\beta + h^2$, where $r$ and $h$ are the respective magnitudes of $\mathbf{r}$ and $\mathbf{h}$. The corresponding displacement vector is $\mathbf{s} = \mathbf{r}+\mathbf{h}$. The velocity of the particle after a displacement $\mathbf{s}$ is $\mathbf{v}(\mathbf{x}_0 + \mathbf{s}) = \mathbf{v}_0 + \Delta\mathbf{v}(\mathbf{s})$, where $\Delta\mathbf{v}(\mathbf{s})$ is a velocity increment. As velocity increments are homogeneous, the conditional velocity cross-covariance (not autocovariance!) is simply $\mathbf{V}(\mathbf{v}) = \mathbf{V}(\Delta\mathbf{v}) = \gamma(s)$. If $r = 0$ (the particle originates at the conditioning point), $\mathbf{s} = \mathbf{h}$, and the velocities experienced by the traveling particle have a cross-covariance dyadic that grows indefinitely as a simple power of the residence time $\tau$,

$$\mathbf{V}(\tau) = b_1 \rho \tau^{2H} \tag{13.66}$$

where $b_1 = b_0 v_0^{2H}$. The same happens for $r > 0$ following a displacement $\mathbf{h}$ large enough to satisfy $h \gg r$. We now restrict ourselves to $r$ values that are small enough, or (equivalently) $\tau$ values that are large enough, to satisfy equation (13.66).

It is most important to note that $\mathbf{v}(\mathbf{x})$ is autocorrelated spatially over all distance scales; it has an infinite integral scale $\lambda_\infty$. Hence the particle velocities are autocorrelated temporally at all times. In other words, the particle velocities are associated with an infinite (Lagrangian, conditional) correlation time $T = \lambda_\infty / v_0$.

Let $p(\mathbf{v})$ be the conditional univariate probability density function (PDF) of $\mathbf{v}$. It is well established (Batchelor, 1952a, pp. 349–50; Neuman, 1993a, app. A) that, in a velocity field with finite Lagrangian correlation scale $T < \infty$, the conditional ensemble mean concentration due to the foregoing instantaneous point source is given at early time $\tau \ll T$ by

$$\langle c(\mathbf{y}, \tau)\rangle = \frac{M_0}{\phi\tau^d} p\left(\frac{\mathbf{y}}{\tau}\right) \tag{13.67}$$

where $\mathbf{y} = \mathbf{x} - \mathbf{y}_0$ is a radius vector measured from the particle origin. Whereas in general $\langle c(\mathbf{y}, \tau)\rangle$ satisfies a nonlocal (space–time integro-differential) conditional mean transport equation (Neuman, 1993a), the solution for early times, equation (13.67), satisfies a local (partial differential) equation (Neuman, 1993a, app. A). The recognition that $\langle c(\mathbf{y}, \tau)\rangle$ is local when $\tau \ll T$ plays an important role in the analytical-numerical computational method recently proposed for conditional advective transport by Zhang and Neuman (1995), who cited Monte Carlo studies showing that the early time period during which equation (13.67) is valid usually exceeds $0.1T$ by a sizable margin and may last as long as $0.5T$.

In a fractal velocity field, $T$ is infinite, so it follows that $\langle c(\mathbf{y}, \tau)\rangle$ is local and given by equation (13.67) at all times $\tau < \infty$. This simple but far-reaching fact appears to have gone unnoticed in the literature. Our simple demonstration that it must be valid in a fractal velocity field implies that mean advective transport in such a field is inherently local, rather than nonlocal as proposed by some authors. This is true provided that the actual (random) concentration $c$ is taken to be governed by a local advective-transport equation, as considered by Neuman (1993a), rather than by a nonlocal equation.

The local transport equation satisfied by $\langle c(\mathbf{y}, \tau)\rangle$ in equation (13.67) is not the standard advection–dispersion equation (e.g., Neuman, 1993a, app. A). As such, it is neither Fickian nor quasi-Fickian and allows $\langle c(\mathbf{y}, \tau)\rangle$ to have a non-Gaussian spatial profile [note, however, that $\langle c(\mathbf{y}, \tau)\rangle$ is spatially Gaussian if $\mathbf{v}$ is univariate Gaussian, a point to which we shall return later]. It is nevertheless well known that one can describe its mean spread by a dispersion tensor (dyadic):

$$\mathbf{D}(\tau) = \frac{1}{2}\frac{d\Omega(\tau)}{d\tau} \tag{13.68}$$

where $\Omega(\tau)$ is the second spatial moment of $\langle c(\mathbf{y}, \tau)\rangle$ about its conditional center of mass,

$$\Omega(\tau) = \frac{\phi}{M_0}\int (\mathbf{y} - \mathbf{v}_0\tau)(\mathbf{y} - \mathbf{v}_0\tau)^T \langle c(\mathbf{y}, \tau)\rangle d\mathbf{y} \tag{13.69}$$

Because $\phi\langle c(\mathbf{y}, \tau)\rangle / M_0$ is the probability density function of particle displacements

(Batchelor, 1949, 1952b; Dagan, 1987), $\Omega(\tau)$ also represents the covariance of particle displacements $\mathbf{y}$ about its conditional mean displacement $\mathbf{h} = \mathbf{v}_0\tau$. Substitution of equation (13.67) into equation (13.69), followed by substitution of the result into equation (13.68), and the change of variables $\mathbf{z} = \mathbf{y}/\tau$, yields

$$\mathbf{D}(\tau) = \frac{1}{2}\frac{d}{d\tau}[\tau^2\mathbf{V}] \tag{13.70}$$

which is known in the literature to hold for $\tau \ll T$. Because in our case $T$ is infinite, both equations (13.67) and (13.70) hold for all $\tau < \infty$. This also appears to have remained unnoticed in the literature.

In the special case where $\mathbf{v}(\mathbf{x})$ is a homogeneous field with mean $\mathbf{v}_0$, variance $\sigma^2$, and a finite Lagrangian correlation time $T < \infty$, $\mathbf{V}$ is a constant, and equation (13.70) reduces to the classic result for early time (Taylor, 1921)

$$\mathbf{D}(\tau) = \mathbf{V}\tau \qquad (\tau \ll T) \tag{13.71}$$

The corresponding dispersivity tensor, defined as

$$\alpha = \frac{\mathbf{D}}{v_0} \tag{13.72}$$

can be expressed in the form

$$\alpha(h) = C_v^2 \rho h \tag{13.73}$$

where $h$ is the mean displacement, $C_v = \sigma/v_0$ is a velocity coefficient of variation, and $\rho$ is the cross-correlation dyadic of velocity components. Note that $\alpha(h)$ in such a homogeneous velocity field grows in proportion to $h$ during the early pre-asymptotic regime (Figures 13.15–13.19).

Only in the case of a finite $T$ is an asymptotic regime ever reached. In our fractal case where $T$ is infinite, the transport remains always pre-asymptotic, as given by equations (13.67) and (13.70). Recalling that we consider only cases where $r$ is small enough, and/or (equivalently) $\tau$ is large enough, so that $\mathbf{h} \approx \mathbf{s}$ and equation (13.66) holds, then substituting the latter into equation (13.70) and evaluating the derivative yields a power-law growth for the dispersivity dyadic:

$$\mathbf{D}(\tau) = (1 + H) b_1 \rho \tau^{1+2H} \tag{13.74}$$

By virtue of equation (13.72), the dispersivity then grows as a power of the mean travel distance according to

$$\alpha(h) = (1 + H) \tilde{C}_v^2 \rho h^{1+2H} \tag{13.75}$$

where $\tilde{C}_v = \sqrt{b_0}/v_0$ is a velocity coefficient of variation.

Because $H$ is theoretically confined to the range $0 \leq \omega \leq 1$, $\alpha$ in equation (13.75) grows with $h$ at a rate faster than linear. We saw earlier that this predicted growth rate is strongly supported by the apparent longitudinal dispersivities (open symbols) in Figure 13.3; however, it differs fundamentally from growth rates commonly predicted in the literature on dispersion in fractal media (e.g., Adler and Thovert, 1993; Muralidhar and Ramkrishna, 1993; Sahimi, 1993; Glimm et al., 1993; Dagan, 1994). For example, Sahimi (1993, p. 32) concluded that longitudinal dispersivity varies as $\tau^{2H-1}$, where, contrary to his assertion, $H$ is the same as in this chapter. The latter implies a rate of growth slower than linear with either $\tau$ or $h$. We suspect that at least part of this difference stems from a focus by some authors on an asymptotic transport regime, which we believe never develops unless low-frequency components of the fractal velocity field are filtered out, as we have done earlier. Such artificial filtering of "infrared" components is one of the factors that led Dagan (1994, p. 333) to conclude that "the ensemble mean of the effective dispersion coefficient cannot grow faster than linearly with the travel distance; the result is in variance with that of Neuman (1990)." Both our unconditional asymptotic and conditional pre-asymptotic theories in this chapter demonstrate that Dagan's conclusion is invalid.

Under what circumstances does the velocity field form a random fractal with a power-law cross-covariance, such as in equation (13.66)? Consider a statistically homogeneous and isotropic Gaussian (natural) log permeability field with variance $\sigma_y^2$ and an exponential autocovariance function. Then for mean uniform flow parallel to the $x_1$ coordinate, the velocity cross-covariance is, to first order in $\sigma_y^2$, diagonal, with nonzero components (Dagan, 1989, p. 257)

$$V_{11} \simeq \frac{3}{8}\sigma_y^2 K_g^2 \frac{J^2}{\phi^2} \simeq \frac{3}{8}\sigma_y^2 v_0^2, \qquad V_{22} \simeq \frac{1}{8}\sigma_y^2 v_0^2 \tag{13.76a}$$

in two dimensions and

$$V_{11} \simeq \frac{8}{15}\sigma_y^2 v_0^2, \qquad V_{22} = V_{33} \simeq \frac{1}{15}\sigma_y^2 v_0^2 \tag{13.76b}$$

in three dimensions. Here $v_0$ is the magnitude of the mean velocity, $V_{11}$ is the variance of the longitudinal velocity component $v_1$, and $V_{22}$ and $V_{33}$ are the variances of the transverse components $v_2$ and $v_3$, respectively.

Consider now a log permeability field with homogeneous increments characterized by a power-law variogram $\gamma(h) = ch^{2H}$, with $0 < H \leq \frac{1}{2}$. We recall that any such field can be viewed as the superposition of an infinite hierarchy of mutually uncorrelated, statistically homogeneous fields characterized by exponential autocorrelation functions. Hence, for $\gamma(h) < 1$, the variances of velocities encountered by a traveling particle are given to first order by

$$V_{11}(h) \simeq \frac{3}{8}cv_0^2 h^{2H}, \qquad V_{22}(h) \simeq \frac{1}{8}cv_0^2 h^{2H} \tag{13.77a}$$

in two dimensions and

$$V_{11}(h) \simeq \frac{8}{15} c v_0^2 h^{2H}, \qquad V_{22}(h) = V_{33}(h) \simeq \frac{1}{15} c v_0^2 h^{2H} \tag{13.77b}$$

in three dimensions. The velocity thus behaves approximately as a fractal for mean travel distances $h$ that satisfy $\gamma(h) < 1$.

It follows from equations (13.66), (13.70), (13.72), and (13.77) that for $\gamma(h) < 1$, conditional pre-asymptotic dispersivity is a diagonal tensor with scale-dependent components given to first order by

$$\alpha_{11}(h) \simeq \frac{3}{8} c(1 + H) h^{1+2H}, \qquad \alpha_{22}(h) \simeq \frac{1}{8} c(1 + H) h^{1+2H} \tag{13.78a}$$

in two dimensions and

$$\alpha_{11}(h) \simeq \frac{8}{15} c(1 + H) h^{1+2H}, \qquad \alpha_{22}(h) = \alpha_{33}(h) \simeq \frac{1}{15} c(1 + H) h^{1+2H} \tag{13.78b}$$

in three dimensions. Here $\alpha_{11}$ is the longitudinal dispersivity parallel to $v_1$, and $\alpha_{22}$ and $\alpha_{33}$ are transverse dispersivities parallel to $v_2$ and $v_3$, respectively.

## 13.10 Summary

We have shown that both the power (semi)variogram and associated spectra of a random field with homogeneous isotropic increments can be constructed as weighted integrals from zero to infinity (an infinite hierarchy) of either exponential or Gaussian variograms and spectra of uncorrelated homogeneous isotropic random fields (modes). We have investigated mathematically the effects of filtering out (truncating) high- and low-frequency modes from this infinite hierarchy in one-, two-, and three-dimensional real and spectral domains. Our analytical results show that a low-frequency cutoff renders the truncated hierarchy statistically homogeneous, with a positive spatial autocovariance function that decays monotonically with separation distance in a manner not too dissimilar from that of its constituent (exponential or Gaussian) modes.

The integral scale of the lowest-frequency mode (cutoff) is related to the length scale of a sampling window defined by the domain under investigation. The integral scale of the highest-frequency mode (cutoff) is related to the length scale of the data support (volume of measurement). Taking each relationship to be one of proportionality renders our expressions for the integral scale and variance of a truncated field dependent on window and support scales in a manner consistent with observations. In particular, we have predicted and confirmed on the basis of field hydraulic-conductivity and transmissivity data that the integral scale of a truncated multiscale hierarchy increases linearly with the length scale of a superimposed window (domain of investigation) provided that the latter is much larger than the length scale of data support. We likewise have predicted and confirmed that the variance of a hierarchy so truncated increases as a power $2H$ of window (domain) scale.

The traditional approach has been one of truncating power spectral densities, rather than the hierarchy of spectra associated with exponential or Gaussian autocovariance functions as we propose. We have shown that truncated power spectra yield autocovariance functions having zero or finite integral scales that oscillate about zero.

Our hierarchical theory allows bridging across scales by predicting the effect of viewing a multiscale random field, defined and measured on a given support scale, through a larger window defined by the domain under investigation. At a given locale, one can, in principle, calibrate a truncated variogram model to sample data observed on a given support scale in one domain, and then predict the autocovariance structure of the corresponding field in domains that are either smaller or larger. One can also venture (we suspect with less predictive power) to transfer such predictions from one locale to another by introducing the generalized variogram parameters [derived by Neuman (1990, 1994, 1995) and ourselves on the basis of juxtaposed hydraulic and tracer data from many sites] $C \approx 0.0076$ for exponential modes, $C \approx 0.0117$ for Gaussian modes, $H \approx 0.25$, and $\mu \approx \frac{1}{3}$ into our truncated variogram models. Selection between alternative models is possible in principle by means of formal model discrimination criteria, coupled with methods such as maximum-likelihood cross-validation, as proposed by Samper and Neuman (1989).

Our hierarchical theory allows solution of problems of flow and transport in multiscale random log permeability fields. We have presented expressions for effective hydraulic conductivities in such fields under uniform mean flow and expressions for first-order pre-asymptotic as well as asymptotic longitudinal and transverse macrodispersivities under advective transport. Our theory predicts that apparent longitudinal dispersivities, derived from tracer data by means of standard Fickian models, should grow as a power $1 + 2H$ of mean travel distance or time. This confirms theoretically an earlier semiempirical interpretation of the observed dispersivity scale effect by Neuman (1990).

Juxtaposed hydraulic and tracer data from many sites and geologic environments worldwide seem to exhibit generalized scaling behaviors that are captured well, and consistently, by our theory. There is evidence in the tracer data that accounting explicitly for site-specific medium heterogeneity by conditioning is tantamount to filtering out large-scale, low-frequency modes from the multiscale hierarchy of log permeabilities. The effect of such filtering is to render the apparent dispersivity dependent on information about the advective velocity field. As the quantity and quality of such data increase, the rate at which the apparent dispersivity increases with scale slows down.

A theoretical analysis has been presented for pre-asymptotic advective transport in a steady-state random velocity field with homogeneous increments. As the mean and variance of such a field are undefined, the theory is conditioned on knowledge of the velocity at some point $\mathbf{x}_0$. If a tracer is introduced at another point $\mathbf{y}_0$, then its conditional mean dispersion is local at all times. Its conditional mean concentration and variance are given explicitly by well-established expressions, which, however, have not previously been recognized as being valid in fractal fields. Once the conditional mean travel distance $s$ of the tracer becomes large compared with the distance between $\mathbf{y}_0$ and $\mathbf{x}_0$, the corresponding dispersion and dispersivity tensors grow in proportion to $s^{1+2H}$, where $0 < H < 1$. This supralinear rate of growth is similar to the rate we have established by means of an unconditional theory of transport in a truncated fractal hydraulic-conductivity field; both are consistent with juxtaposed apparent longitudinal dispersivities obtained by standard methods of interpretation from tracer behavior observed in a variety of geologic media under varied flow and transport regimes.

A self-affine natural-log permeability field gives rise to a self-affine velocity field while $s$ is sufficiently small to ensure that the variance of the log permeabilities (which grows as a power of $s$) will remain nominally less than unity.

## Acknowledgments

This work was supported in part by USIA-CIES, the Fulbright Scholarship Program, and the U.S. Nuclear Regulatory Commission under contract NRC-04-95-038.

## References

Ababou, R. 1988. Three-dimensional flow in random porous media. Ph.D. dissertation, Ralph Parsons Laboratory, Massachusetts Institute of Technology.

Ababou, R., and Gelhar, L. W. 1990. Self-similar randomness and spectral conditioning: analysis of scale effect in subsurface hydrology. In: *Dynamics of Fluids in Hierarchical Porous Media*, ed. J. H. Cushman, pp. 393–428. San Diego: Academic Press.

Abramowitz, M., and Stegun, I. A. 1972. *Handbook of Mathematical Functions.* New York: Dover.

Adler, P. M., and Thovert, J.-F. 1993. Fractal porous media. *Trans. Porous Media* 13:41–78.

Andersson, J.-E., Ekman, L., Gustafsson, E., Nordqvist, R., and Tiren, S. 1988. *Hydraulic Interference Tests and Tracer Tests Within the Brändan, Area, Finnsjön Study Site, Fracture Zone Project – Phase 3.* Technical report 89-12. Stockholm: Swedish Nuclear Fuel and Waste Management Co. SKB.

Batchelor, G. K. 1949. Diffusion in a field of homogeneous turbulence. I. Eulerian analysis. *Aust. J. Sci. Res. A* 2:437–50.

Batchelor, G. K. 1952a. Diffusion in a field of homogeneous turbulence. II. The relative motion of particles. *Proc. Cambridge Philos. Soc.* 48:345–63.

Batchelor, G. K. 1952b. The effect of homogeneous turbulence on material lines and surfaces, *Proc. R. Soc. London A* 213:349–66.

Bras, R. L., and Rodríguez-Iturbe, I. 1985. *Random Functions and Hydrology*. Reading, MA: Addison-Wesley.

Chang, C. M., and Kemblowski, M. W. 1994. Unsaturated flow in soils with self-similar hydraulic conductivity distribution. *Stoch. Hydrol. Hydraul.* 8:281–300.

Clauser, C. 1992. Permeability of crystalline rocks. *EOS, Trans. AGU* 73:233, 237–8.

Dagan, G. 1987. Theory of solute transport by groundwater. *Ann. Rev. Fluid Mech.* 19:183–215.

Dagan, G. 1989. *Flow and Transport in Porous Formations*. Berlin: Springer-Verlag.

Dagan, G. 1990. Transport in heterogeneous porous formations: spatial moments, ergodicity, and effective dispersion. *Water Resour. Res.* 26:1281–90.

Dagan, G. 1991. Dispersion of a passive solute in non-ergodic transport by steady velocity fields in heterogeneous formations. *J. Fluid Mech.* 233:197–210.

Dagan, G. 1994. The significance of heterogeneity of evolving scales to transport in porous formations. *Water Resour. Res.*, 30:3327–36.

Desbarats, A. J. 1992. Spatial averaging of transmissivity in heterogeneous fields with flow toward a well. *Water Resour. Res.* 28:757–67.

Desbarats, A. J., and Bachu, S. 1994. Geostatistical analysis of aquifer heterogeneity from the core to the basin scale: a case study. *Water Resour. Res.* 30:673–84.

Di Federico, V., and Neuman, S. P. 1995. *Effect of Filtering on Autocorrelation, Flow and Transport in Random Fractal Fields*. Technical report HWR-95-060, Department of Hydrology and Water Resources, University of Arizona.

Di Federico, V., and Neuman, S. P. 1997. Scaling of random fields by means of truncated power variograms and associated spectra. *Water Resour. Res.* 33:1075–85.

Di Federico, V., and Neuman, S. P. 1998a. Flow in multiscale log conductivity fields with truncated power variograms, *Water Resour. Res.* (in press).

Di Federico, V., and Neuman, S. P. 1998b. Transport in multiscale log conductivity fields with truncated power variograms, *Water Resour. Res.* (in press).

Dykaar, B. B., and Kitanidis, P. K. 1992. Determination of the effective hydraulic conductivity for heterogeneous porous media using a numerical spectral approach. 2. Results. *Water Resour. Res.* 28:1167–78.

Erdélyi, A., Magnus, W., Oberhettinger, F., and Tricomi, F. G. 1953. *Higher Transcendental Functions*, vol. 1. New York: McGraw-Hill.

Gelhar, L. W. 1993. *Stochastic Subsurface Hydrology*. Englewood Cliffs, NJ: Prentice-Hall.

Gelhar, L. W., Welty, C., and Rehfeldt, K. R. 1992. A critical review of field-scale dispersion in aquifers. *Water Resour. Res.* 28:1955–74.

Glimm, J., Lindquist, W. B., Pereira, F., and Zhang, Q. 1993. A theory of macrodispersion for the scale-up problem. *Trans. Porous Media* 13:97–122.

Gradshteyn, I. S., and Ryzhik, I. M. 1994. *Tables of Integrals, Series, and Products*, ed. A. Jeffrey. San Diego: Academic Press.

Guzman, A., and Neuman, S. P. 1996. Field air injection experiments. In: *Apache Leap Tuff INTRAVAL Experiments, Results and Lessons Learned*, ed. T. C. Rasmussen, S. C. Rhodes, A. Guzman, and S. P. Neuman, pp. 52–94. NUREG/CR-6096. Washington, DC: U.S. Nuclear Regulatory Commission.

Kemblowski, M. W., and Chang, C. M. 1993. Infiltration in soils with fractal permeability distribution. *Ground Water* 31:187–92.

Kemblowski, M. W., and Wen, J.-C. 1993. Contaminant spreading in stratified soils with fractal permeability distribution. *Water Resour. Res.* 29:419–25.

Mandelbrot, B. B., and Van Ness, J. W. 1968. Fractional Brownian motions, fractional noises and applications. *SIAM Rev.* 10:422–37.

Matheron, G. 1967. *Elements Pour une Theorie des Millieux Poreux*. Paris: Masson et Cie.

Molz, F. J., and Boman, G. K. 1995. Further evidence of fractal structure in hydraulic conductivity distribution. *Geophys. Res. Lett.* 22:2545–8.

Monin, A. S., and Yaglom, A. M. 1975. *Statistical Fluid Mechanics*, vol. 2, ed. J. L. Lumley. Cambridge, MA: MIT Press.

Muralidhar, R., and Ramkrishna, D. 1993. Diffusion in pore fractals: a review of linear response models. *Trans. Porous Media* 13:79–95.

Neuman, S. P. 1990. Universal scaling of hydraulic conductivities and dispersivities in geologic media. *Water Resour. Res.* 26:1749–58.

Neuman, S. P. 1993a. Eulerian-Lagrangian theory of transport in space-time nonstationary velocity fields: exact nonlocal formalism by conditional moments and weak approximations. *Water Resour. Res.* 29:633–45.

Neuman, S. P. 1993b. Comment on "A critical review of data on field-scale dispersion in aquifers" by L. W. Gelhar, C. Welty, and K. R. Rehfeldt. *Water Resour. Res.* 29:1863–5.

Neuman, S. P. 1994. Generalized scaling of permeabilities: validation and effect of support scale. *Geophys. Res. Lett.* 21:349–52.

Neuman, S. P. 1995. On advective transport in fractal velocity and permeability fields. *Water Resour. Res.* 31:1455–60.

Neuman, S. P., and Orr, S. 1993. Prediction of steady state flow in nonuniform geologic media by conditional moments: exact nonlocal formalism, effective conductivities, and weak approximation. *Water Resour. Res.* 29:341–64.

Neuman, S. P., Tartakovsky, D., Wallstrom, T. C., and Winter, C. L. 1996. Correction to "Prediction of steady state flow in nonuniform geologic media by conditional moments: exact nonlocal formalism, effective hydraulic conductivities, and weak approximation" by S. P. Neuman and S. Orr. *Water Resour. Res.* 32: 1479–80.

Paleologos, E. K., Neuman, S. P., and Tartakovsky, D. 1996. Effective hydraulic conductivity of bounded, strongly heterogeneous porous media. *Water Resour. Res.* 32:1333–41.

Philip, J. R. 1986. Issues of flow and transport in heterogeneous porous media. *Trans. Porous Media* 1:319–38.

Rajaram, H., and Gelhar, L. W. 1995. Plume-scale dependent dispersion in aquifers with a wide range of scales of heterogeneity. *Water Resour. Res.* 31:2469–82.

Sahimi, M. 1993. Fractal and superdiffusive transport and hydrodynamic dispersion in heterogeneous porous media. *Trans. Porous Media* 13:3–40.

Samper, F. J., and Neuman, S. P. 1989. Estimation of spatial covariance structures by adjoint state maximum likelihood cross validation. 1. Theory. *Water Resour. Res.* 25:351–62.

Tartakovsky, D. M., and Neuman, S. P. 1998a. Transient flow in bounded randomly heterogeneous domains, 1. Exact conditional moment equations and recursive approximations. *Water Resour. Res.* 34(1):1–12.

Tartakovsky, D. M., and Neuman, S. P. 1998b. Localization of conditional mean equations and temporal nonlocality effects. *Water Resour. Res.* 34(1):13–20.

Tartakovsky, D. M., and Neuman, S. P. 1998c. Transient effective hydraulic conductivities under slowly and rapidly varying mean gradients in bounded three-dimensional random media. *Water Resour. Res.* 34(1):21–32.

Taylor, G. I. 1921. Diffusion by continuous movements. *Proc. London Math. Soc.* 2:196–214.

Voss, R. F. 1985. Random fractals: characterization and measurement. In: *Scaling Phenomena in Disordered Systems*, ed. R. Pynn and A. Skjeltorp. NATO ASI series 133.

Winberg, A. 1991. *Analysis of Spatial Correlation of Hydraulic Conductivity Data from the Stripa Mine*. Stripa project technical report 91-28. Stockholm: Swedish Nuclear Fuel and Waste Management Co. SKB.

Yaglom, A. M. 1987. *Correlation Theory of Stationary and Related Random Functions. I: Basic Results*. Berlin: Springer-Verlag.
Yordanov, O. I., and Nikolaev, N. I. 1994. Self-affinity of time series with finite domain power spectrum. *Phys. Rev. E* 49:2517–20.
Zhang, D., and Neuman, S. P. 1995. Eulerian-Lagrangian analysis of transport conditioned on hydraulic data: 1. Analytical-numerical approach. *Water Resour. Res.* 31:39–51.
Zhang, D., and Neuman, S. P. 1996. Effect of local dispersion on solute transport in randomly heterogeneous media. *Water Resour. Res.* 32:2715–23.
Zhang, Y.-K., Zhang, D., and Lin, J. 1996. Non-ergodic solute transport in three-dimensional heterogeneous isotropic aquifers. *Water Resour. Res.* 32:2955–63.

## Appendix A

To evaluate variances and autocovariances via equation (13.32) for truncated power-law densities, we used the following relations for isotropic, real-valued correlation functions, valid for one, two, and three dimensions, respectively [Yaglom, 1987, eq. (2.71), p. 104; eq. (4.118)–(4.118a), p. 358]:

$$C(s) = 2\int_0^\infty \cos(ks)S(k)\,dk \tag{13.A1}$$

$$C(s) = 2\pi\int_0^\infty J_0(ks)kS(k)\,dk \tag{13.A2}$$

$$C(s) = 4\pi\int_0^\infty \frac{\sin(ks)}{ks}k^2S(k)\,dk \tag{13.A3}$$

where $J_0$ is a Bessel function of first kind and zero order (Abramowitz and Stegun, 1972, p. 358). With the spectral density (13.40), equations (13.A1)–(13.A3) yield, for zero lag in one, two, and three dimensions, the variances given by equation (13.41). The autocovariances given by equation (13.42) were obtained via equations (13.A1)–(13.A3) with the help of the symbolic software Mathematica. The variances (13.47) and autocovariances (13.48), for lower and upper cutoffs, were obtained as above upon integrating from $k_l$ to $k_u$.

## Appendix B

Integrals (13.14) and (13.27) are of the form reported by Gradshteyn and Ryzhik [1994, eq. (3.381.3), p. 364], where a typographical error is present. The correct form of the integral is

$$\int_u^\infty x^{v-1}\exp(-\mu x)\,dx = \mu^{-v}\Gamma(v, \mu u) \qquad (u > 0,\ \mu > 0) \tag{13.B1}$$

as indicated in the original reference (Erdélyi et al., 1953).

# 14

# Conditional Simulation of Geologic Media with Evolving Scales of Heterogeneity

YORAM RUBIN and ALBERTO BELLIN

## 14.1 Introduction

Stochastic models are frequently used for analysis of physical phenomena. Many physical and chemical processes in the surface and subsurface environments are controlled by spatially distributed parameters of quite complex and intricate structure. The complexity of the spatial structures, on the one hand, and the limited amount of data usually available for capturing and reconstructing them, on the other, inevitably have led to the use of stochastic techniques, with which the problems of extensive heterogeneity and data scarcity can be treated quantitatively. Stochastic generation of random fields, subject to some a-priori-described spatial laws, is the main vehicle for stochastic modeling. Random-field generation is hardly a new area in stochastic modeling in general, and particularly in geostatistics. Still, in reviewing the state of the art of existing techniques and the conceptual problems that arise as a result (Sposito, Jury, and Gupta, 1986; Dagan, 1994), it is realized that major challenges still remain. First, it seems that even the recent rapid increases in computing power have not kept up with the increasing demand for computing power posed by current-generation techniques. The spatial random-function models in demand for simulation are becoming more and more complex. They are defined by more intricate structures and by many scales of heterogeneity; they are, in general, statistically anisotropic and nonstationary. There is also an increasing demand for three-dimensional simulators. Such specifications can generally be accommodated by very large computing resources, and consequently the popular use of random-field generators has been limited to simple, low-dimension, single-scale, stationary isotropic models. A second challenge is the level of sophistication involved in exhausting the information available through measurements. In the simplest case, the heterogeneity is defined by a single scale of heterogeneity, and the measurements are defined over support volumes that are much smaller than this scale and hence can be viewed as point measurements. Most real-life cases are much more complex: The heterogeneity is defined by several scales of heterogeneity, and different types of measurements are available that are defined by different supports. Techniques for simulations that are conditional to measurements are needed, and in particular there is a need for techniques that can handle measurements with variable supports.

Of particular interest in this chapter is the heterogeneity characterized by evolving scales of heterogeneity. The existence of this type of heterogeneity in groundwater aquifers and petroleum reservoirs was discussed by Gelhar (1986). Direct evidence for evolving-scale heterogeneity and the apparently unbounded growth in the variability that is associated with such heterogeneity was supplied by Hewett (1986), and more recently by Neuman (1990), Painter and Paterson (1994), Molz and Boman (1995), and Liu and Molz (1996). The motivation to focus on this type of variability is the growing recognition that evolving-scale variability is more common than previously assumed and that further understanding of this type of variability will require low-cost random-field generators.

Discussions of the significance of this type of heterogeneity for flow and transport processes in groundwater aquifers have been published by Sposito et al. (1986); Cushman and Ginn (1993); Glimm et al. (1993); Dagan (1994), and Bellin et al. (1996). In those works, the effects of evolving-scale heterogeneity on flow and transport phenomena were investigated theoretically. There are, however, only very limited field data linking evolving-scale heterogeneity and transport processes in groundwater aquifers, and such data as there are show the clear imprimatur of the former on the later. In fact, it has only recently been suggested by Engesgaard et al. (1996), following an analysis of large-scale transport experiments in Denmark, that dispersivities do not show scale dependence beyond a certain travel distance, and they have suggested that dispersivities be represented by error bars, rather than circles, as has been the practice in many diagrams depicting scale dependence. By that method, some of the trends that had been suggested have become less obvious. In conclusion, to take advantage of the growing interest in evolving-scale hetrogeneity, more field data in support of such models are needed, particularly data that can show the imprint of evolving-scale heterogeneity on transport.

Figure 14.1 shows a semivariogram model of the log-conductivity, as hypothesized by Gelhar (1986), and it shows an increase without bound as the scale of observation increases. It is reasonable to expect that as the scale of observation increases, the variability of the hydrogeologic properties will follow suit, because additional types of variability are introduced by the larger variety of geologic processes that affect the hydrogeologic properties. Similar arguments were introduced by Wheatcraft and Tyler (1988), and they emphasized that evolving-scale heterogeneity does not necessarily imply self-similarity.

Figure 14.1 shows disparities between the various scales of heterogeneity. These disparities are manifested in the form of sills that characterize the transitions between the different observation scales. Such a disparity between scales is favorable for geostatistical simulations because it allows simulation of the variability at one scale

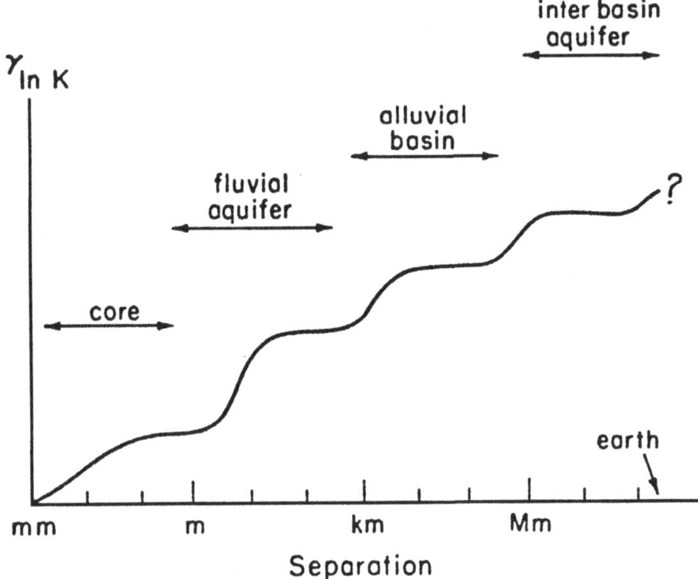


Figure 14.1. Hypothetical ln $K$ semivariogram illustrating the notion of scale-dependent correlation scales. (From Gelhar, 1986, with permission.)

while ignoring the larger-scale variability, which can be represented locally as a deterministic trend. An outstanding study of evolving-scale heterogeneity has been published by Desbarats and Bachu (1994), who examined the spatial variability of the conductivity in a sandstone aquifer. Semivariogram analysis of the log-conductivity (Figure 14.2) shows a steady increase in the spatial variability over length scales ranging from 400 m to 100 km. The semivariogram in Figure 14.2 does not appear to reach any sill. Extrapolating the semivariogram to a larger scale suggests that a sill cannot be reached, and that apparently supports the concept of unbounded variability. Like any extrapolation, this suggestion will remain speculative, because it can never be proved. The semivariogram in Figure 14.2 does not show a similar pattern of increasing variability at all scales, and although it represents evolving scales of heterogeneity, it suggests multifractal variability (Feder, 1988). This conclusion can be drawn from the observation that the experimental points showed in Figure 14.2 can be interpolated using two straight lines of different slopes. Regardless of whether or not the variability grows unboundedly, the field studies mentioned earlier support the assumption of scale-dependent variability, but they do not support the model of scale disparities. In the absence of disparity between heterogeneity scales, they must all be recognized in the geostatistical simulation process.

In surface hydrology (Rodríguez-Iturbe and Rinaldo, 1997), geomorphology (Chase, 1992; Rodríguez-Iturbe et al., Marani, Rigon and Rinaldo, 1994), and meteorology (Lovejoy and Schertzer, 1990; Gupta and Waymire, 1990) it has been found that many spatial phenomena are characterized by increasing variability as the scale

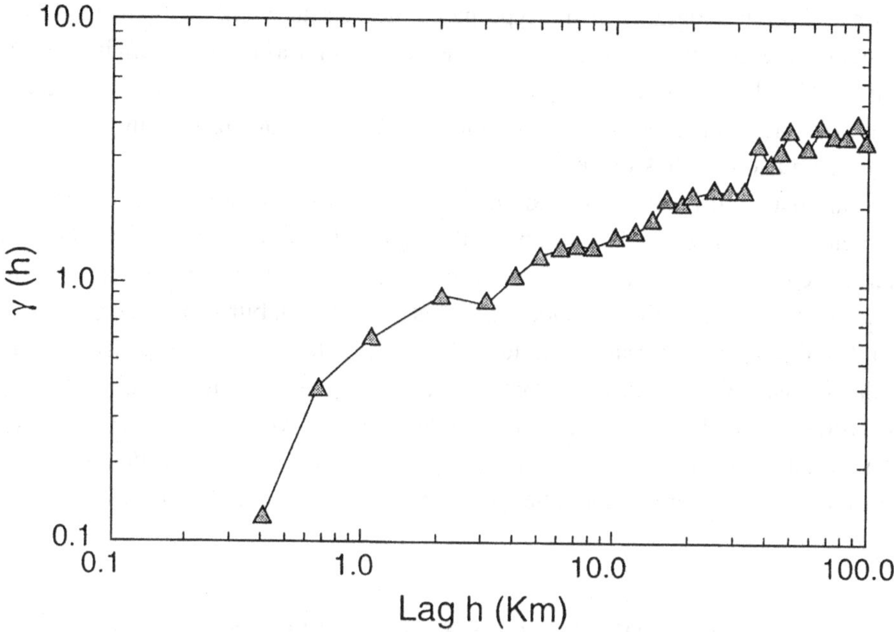


Figure 14.2. Global omnidirectional semivariogram of log-transmissivity. (From Desbarats and Bachu, 1994, with permission.)

of the sampled domain increases. A particular case of evolving-scale heterogeneity is the case of self-similarity. The biggest asset of self-similarity is its simplicity: The type of scale-dependent variability detected over domains of limited extent that are parts of a larger domain is speculated to persist over the entire domain, and consequently one can speculate on the nature of the transport of a contaminant plume, for example, once it travels beyond charted territory.

Self-similarity implies a power-law semivariogram, and hence a linear semivariogram on a log-log scale is the test for self-similarity. Self-similarity stems from the observation that a semivariogram $\gamma(r) \propto r^{\beta}$ satisfies $\gamma(ar) \propto a^{\beta}\gamma(r)$, and hence the increments are characterized by the similarity transformation $\Delta f(ar) \propto a^{\beta/2}\Delta f(r)$. The implication is that complex correlation structures can be summarized quite succinctly, and inference is made simple. From our inspection, however, self-similarity does not always exist at all scales, as evidenced by Figure 14.2, and practically it cannot be proved to exist at all scales. Hence the simulators that are needed should not be limited to the same power law at all scales.

To get a better sense of the applicability of the concept of self-similarity and the associated universal power-law behavior in groundwater, we can rely on the studies by Neuman (1994, 1995), where he showed that a random field that is composed of an infinite hierarchy of mutually uncorrelated, statistically homogeneous and isotropic random fields with similar correlation structures (e.g., all the fields are characterized by exponential semivariograms) and scale-dependent variance is characterized by power-law semivariograms of the type mentioned earlier. The mathematical constraints on the geologic processes that lead to spatial variability as envisioned by Neuman (1994, 1995) may be quite restrictive, but perhaps can be substantiated by numerical simulations of the geologic processes, following the suggestions of Koltermann and Gorelick (1996).

In summary, work is still needed on developing low-cost, sophisticated random-field generators, and special attention should be given to simulations of evolving-scale heterogeneity in general, and self-similar heterogeneity in particular, because of the growing recognition of the significance of this type of variability. This chapter deals with different aspects of random-field generation, with particular emphasis on self-similar fields. We present a random-field generator and use it as a basis for our discussion. Some difficulties in simulating fields with evolving-scale heterogeneity are highlighted, as are the benefits from conditional simulations as a function of the exponents of the power law and other points that need further attention and additional research.

## 14.2 The HYDRO_GEN Random-Field Generator

### *14.2.1 General Approach*

The spatial structure of a distributed attribute $z(\mathbf{x})$, where $\mathbf{x}$ denotes the spatial coordinate, is modeled as a random space function (RSF) $Z(\mathbf{x})$. Here and subsequently, boldface letters denote vectors, capital letters denote RSFs, and lowercase letters denote realizations of $Z(\mathbf{x})$. $Z(\mathbf{x})$ is characterized by its probability density function (PDF) or, as an alternative, by its statistical moments, such as the expected value

$$\langle Z(\mathbf{x})\rangle = m_z(\mathbf{x}) \tag{14.1}$$

(with $\langle\ \rangle$ being the expectation operator) and the spatial covariance (defined for every $\mathbf{r}$)

$$\begin{aligned} C_z(\mathbf{x}, \mathbf{x}') &= \langle [Z(\mathbf{x}) - m_z(\mathbf{x})][Z(\mathbf{x}') - m_z(\mathbf{x}')]\rangle \\ &= C_z(\mathbf{r} = \mathbf{x} - \mathbf{x}') \end{aligned} \tag{14.2}$$

with the last equality applicable in case of stationary domains. In case $Z(\mathbf{x})$ is either Gaussian or log-Gaussian, its entire distribution is defined by equations (14.1) and (14.2). In this study we limit ourselves to $Z$ that is Gaussian or can be transformed to Gaussian by a simple transformation (Journel and Alabert, 1989).

In the case of self-similar fields, statistical information is given only for the increments. Rather than working with $Z$, we rely on the spatial structure of the increments $Z'(\mathbf{x}) = Z(\mathbf{x}) - m_z(\mathbf{x})$. In this case, $Z$ can be defined through the spatial structure of the fluctuations $Z'(\mathbf{x}) = Z(\mathbf{x}) - m_z(\mathbf{x})$. The fluctuations are defined by a stationary mean

$$\langle Z(\mathbf{x}) - m_z(\mathbf{x}) \rangle = 0 \tag{14.3}$$

and the semivariogram

$$\gamma_z(\mathbf{x}, \mathbf{x}') = \frac{1}{2}\langle [Z'(\mathbf{x}) - Z'(\mathbf{x}')]^2 \rangle \tag{14.4}$$

This model is analogous to the one defined by equations (14.1) and (14.2) only when the semivariogram reaches a sill at a finite separation distance; otherwise the two foregoing models define fields of completely different natures. Equations (14.1) and (14.2) define fields whose heterogeneity is characterized by a finite integral scale, but fields defined by equations (14.3) and (14.4) do not possess a finite length scale if the semivariogram grows without bound. Semivariograms like equation (14.4) that scale as $\gamma_z(r) = ar^{2H}$, $r = |\mathbf{r}|$, and $0 < H < 1$, represent self-similar or fractal fields (Neuman, 1990).

The algorithm HYDRO_GEN developed by Bellin and Rubin (1996) is intended to generate replicates of $Z$ fields whose spatial statistics are defined by equations (14.1) and (14.2) or, alternatively, by equations (14.3) and (14.4). The focus of this chapter is on fast and efficient generation of many conditional realizations of nonstationary RSFs defined by power-law semivariograms. Whereas $Z$ is generally a continuous function, the proposed algorithm consists in generating the $Z$ field discretely over a predetermined arbitrary grid. The grid can be of variable density and of arbitrary geometry. In the following section, the generating technique for producing a single replicate of the $Z$ field is described.

### *14.2.2 Field-generating Technique*

The generation technique resembles the sequential-indicator simulation (SIS) method, proposed by Gòmez-Hernañdez and Srivastava (1990) and applied to self-similar random fields by Omre, Solna, and Tjelmeland (1992). HYDRO_GEN and SIS have in common some general concepts, but HYDRO_GEN introduces significant modifications that make it faster and more accurate.

Our starting point is the generation of a realization $z$ of the random field at the node $\mathbf{x}_0$, where local data are not available. This is accomplished using a standard random generator, with the unconditional mean and the unconditional variance $\sigma_z^2$ used as target statistics. For self-similar fields (say $Z$) the variance is estimated using the value of the semivariogram (14.4) with lag equal to the field dimension. Once $z(\mathbf{x}_0)$ is generated, it is considered as a datum, and it will be used to condition the $Z$ values that will be generated subsequently at neighboring nodes.

At the next step, generation of a realization at a nearby point $\mathbf{x}_1$ is considered. This time $Z(\mathbf{x}_1)$ is conditioned on the previously generated $z(\mathbf{x}_0)$ using the Gaussian conditioning procedure (Mood and Graybill, 1963). The procedure continues with the generation of a realization at the third and subsequent grid nodes while conditioning at each step on a selected number of previously generated $z$ values (Bellin and Rubin, 1996). In the general case we generate a realization for a generic $Z(\mathbf{x}_N)$ conditional to the previously generated $(N-1)$ data. The conditional expected value of $Z(\mathbf{x}_N)$ is given by

$$\langle Z^c(\mathbf{x}_N)\rangle = \sum_{j=1}^{N-1} \lambda_j(\mathbf{x}_N) z(\mathbf{x}_j) \tag{14.5}$$

and the conditional variance is given by

$$\left(\sigma_z^c\right)^2(\mathbf{x}_N) = \mu(\mathbf{x}_N) + \sum_{j=1}^{N-1} \lambda_j(\mathbf{x}_N)\gamma_z(\mathbf{x}_N, \mathbf{x}_j) \tag{14.6}$$

Once these last two statistics are computed, $z(\mathbf{x}_N)$ can be generated from any standard generator that uses these two statistics as target statistics. The steps applied to the generic $\mathbf{x}_N$ are repeated for each of the nodes over the grid. The only difference between the successive nodes is in the number of data used for conditioning and hence in the number of interpolation coefficients $\lambda_j$, which keeps increasing. Note that a conditional variance can be defined for fractal fields, despite the fact that an unconditional variance does not exist.

The coefficients $\lambda_j$ [equations (14.5) and (14.6)] provide the solution of the following linear system (Dagan, 1989):

$$\begin{cases} \displaystyle\sum_{j=1}^{N-1} \lambda_j(\mathbf{x}_N)\gamma_z(\mathbf{x}_j, \mathbf{x}_q) + \mu(\mathbf{x}_N) = \gamma_z(\mathbf{x}_N, \mathbf{x}_q) \\ \qquad\qquad\qquad\qquad (q = 1, \ldots, N-1) \\ \displaystyle\sum_{j=1}^{N-1} \lambda_j(\mathbf{x}_N) = 1 \end{cases} \tag{14.7}$$

The repetitive solution of equation (14.7) (i.e., for each new node) is computationally the most demanding step in the algorithm. The order of the linear system (14.7) grows linearly with $N$, and its repetitive solution for each new $\mathbf{x}_N$ can slow down the computations considerably. To reduce the computational burden, the pattern of moving over the grid and generating $z(\mathbf{x}_N)$ is kept fixed and is repeated in each realization. That allows the use of a search neighborhood that is fixed both in the number of conditional points and in their geometry. Savings stem from the fact that the interpolation coefficients $\lambda_j$ [equation (14.5)] do not depend on the actual $z$ values, but rather on the distance $\mathbf{x}_k - \mathbf{x}_j$, $k = 1, \ldots, N-1$ (Journel and Huijbregts, 1978) and on the data configuration in the search neighborhood; hence the coefficients can be computed once and stored for further use, because the data configuration in our method is kept fixed. The savings in computational effort become particularly significant when many realizations of the random field are required. When applied to stationary RSFs in a finite-integral-scale environment, the computational burden can be further alleviated by conditioning only on a limited-size search neighborhood, with dimensions depending on the integral scale (Bellin and Rubin, 1996). Nodes at distances larger than a few integral scales from $\mathbf{x}_N$ can be excluded from the set of conditional nodes without an appreciable effect on the reproduction of the target statistics. Unfortunately, this is not true for fields with evolving scales of heterogeneity, which require conditioning on the full set of previously generated $z$ values. Nevertheless, even here the sets of coefficients are computed only once and stored for use in subsequent realizations. The HYDRO_GEN concept can now be presented as an algorithm, with the following steps (referring to a generic point $\mathbf{x}_N$, and given the data set of all previously generated $z_j$, $j < N$):

1. Compute $\langle Z^c(\mathbf{x}_N) \rangle$ and $(\sigma_z^c)^2(\mathbf{x}_N)$, conditional to the set of previously generated data, using equations (14.5) and (14.6).
2. Generate a realization $z(\mathbf{x}_N)$ using the conditional mean and variance computed at step 1 as target statistics.
3. Add $z(\mathbf{x}_N)$ to the data base to be used in the subsequent generations.
4. Move to the next node according to an a-priori-determined scheme, increase $N$ by 1, and go to step 1. The pattern we use for moving along the grid is along rows, moving row by row.

HYDRO_GEN is not limited in principle to power-law semivariograms and can accommodate models where the heterogeneity evolves with different patterns over different ranges of scales.

The data used for conditioning can be of two types: (1) realizations of $Z(\mathbf{x}_i)$, $i < N$, that were generated at the previous steps, and (2) actual field data. When data of the first and second types are used, the generated fields are said to be conditional fields, meaning that spatial statistics are reconstructed and that the generated fields are constrained by measurements. The inclusion of actual field data modifies the interpolation coefficients as well as the conditional variance. The procedure used for generating realizations of the random field conditional to measurements is discussed in Appendix A.

### 14.2.3 Multistage Grid Refinement

An important issue in random-field generation is that of grid density. High-density grids are often adopted in order to obtain good reproduction of rapidly varying RSFs, such as those representing the hydraulic properties of geologic formations. High-density grids increase the computational burden, both at the field-generation stage and at the stage of the numerical simulation of the physical/chemical processes that occur in the generated domain. Trying to alleviate the computational burden, advantage can be taken of the fact that usually high resolution is required over only a limited portion of the field. To allow for high, spatially variable resolution, our method introduces multistage grid refinement. The grid can be refined to any a-priori-determined level of discretization, and the refinement can be executed over any subsection of the grid, as dictated by the problem.

Grid refinement is performed as follows. First, realizations of $Z$ at the nodes of a coarse initial grid are obtained following the method outlined in Section 4.2.2. Then subsections of the grid are refined by generating additional $Z$ values at the additional nodes. Because large-scale spatial correlation is already taken care of at the stage of coarse-grid generation, the additional $Z$ values are computed using equations (14.5) and (14.6), but conditioned only on the nearest neighbors, utilizing the screening effect often employed and observed in geostatistical applications. Figure 14.3 shows an example of the refinement for a generic coarse-grid cell. The refinement consists in two steps, indicated as stage 1 and stage 2. First, a realization of $Z$ at the cell's center is generated conditional to the $Z$ values at the four coarse-grid neighborhood points marked by filled circles in Figure 14.3a. This step is followed by the generation of $Z$ values at the four nodes on the cell boundaries marked by open squares in Figure 14.3b. This is accomplished by conditioning on the four nearest nodes symmetrically located

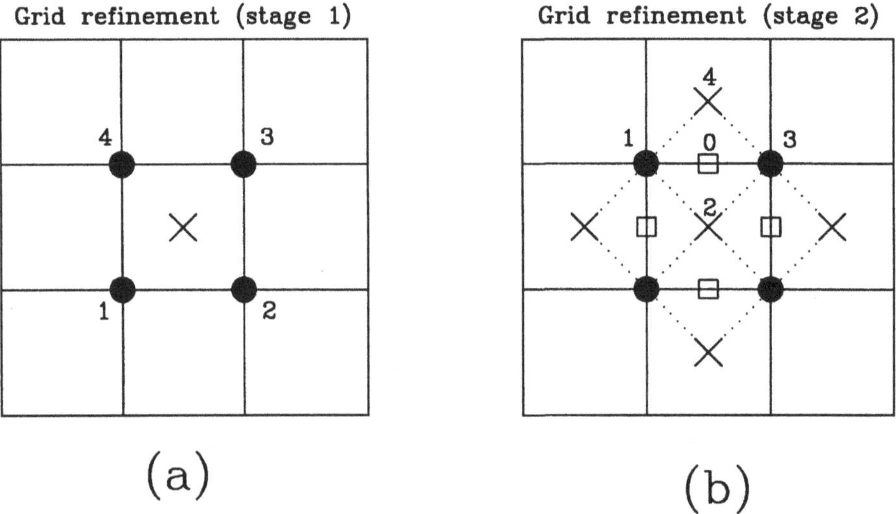


Figure 14.3. (a) Example of first stage of the refinement; (b) example of second stage of the refinement.

around the node under consideration. Figure 14.3b shows an example of the second-stage refinement. According to the refinement procedure, the generation at node 0 is conditional to the nodes indicated by the numbers 1, 2, 3, and 4. The desired level of refinement can be obtained by applying repeatedly the foregoing two stage refinements, with an ever-decreasing grid spacing.

The computational burden involved in generating fine-grid, large-scale fields is alleviated by using the multistage approach, which obviates the need for inversion of huge matrices. Furthermore, by refining the grid in a systematic manner, and working with fixed spatial-data configurations for the conditioning set, the additional effort becomes practically nil, because all it takes is a one-time inversion of a small matrix for each level of refinement. Application of the HYDRO_GEN method to irregular and nonrectangular grids, such as those encountered in finite-element discretization of irregular domains, has been discussed by Bellin and Rubin (1996).

The HYDRO_GEN package is available free of charge at the following www site: http://www.ing.unitn.it/~bellin/Hydro_ge.htn.

### *14.2.4 Applications of HYDRO_GEN to Self-similar Fields*

In this section we present applications HYDRO_GEN to self-similar fields. A more general modification of the code for general models of evolving-scale heterogeneity has not been pursued, although it does not pose conceptual difficulties. Neuman (1990) considered the power-law model

$$\gamma_z(r) = C_0 r^{2H} \tag{14.8}$$

to represent the semivariogram for self-similar log-conductivity fields, and we adopt this model for applications. Note that the $Z$ variable described by equation (14.8) does not have a finite variance, and its integral scale is infinite. HYDRO_GEN (Bellin and Rubin, 1996) can accommodate such a semivariogram and has been used to generate fields characterized by this type of semivariogram.

The properties of fields defined by equation (14.8) are demonstrated in Figure 14.4, which shows images of zero-mean Gaussian fields, generated using equation (14.8) as a model of spatial variability, with various values of $H$. These images show dramatically different behaviors, depending on whether $H$ is smaller or larger than 0.5.

Fields with semivariograms of the type given by equation (14.8) show long-run correlations. This property is an outcome of the fact that the correlation between increments at nodes $\mathbf{x}-\mathbf{r}$ and $\mathbf{x}+\mathbf{r}$ is given by the following scale-invariant relationship [Feder, 1988]:

$$\rho = \frac{\langle [Z(\mathbf{x}) - Z(\mathbf{x}-\mathbf{r})][Z(\mathbf{x}+\mathbf{r}) - Z(\mathbf{x})] \rangle}{\langle [Z(\mathbf{x}+\mathbf{r}) - Z(\mathbf{x})]^2 \rangle} = 2^{2H-1} - 1 \tag{14.9}$$

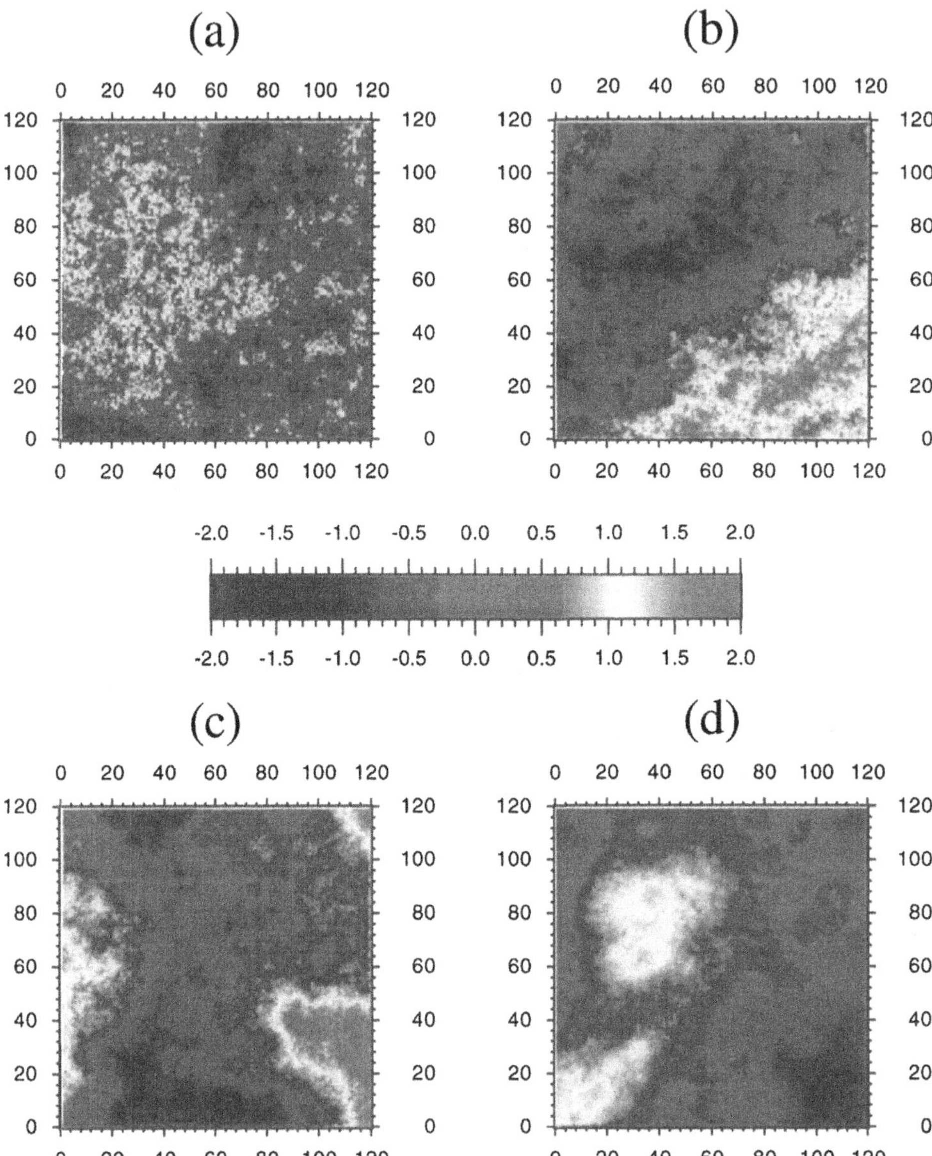


Figure 14.4. Examples of self-similar fields: (a) $H = 0.2$; (b) $H = 0.4$; (c) $H = 0.6$; (d) $H = 0.8$. To view this figure in color, as well as Figures 14.7 through 14.11, see the color section following page 418.

The derivation of equation (14.9) is given in Appendix B. We note from equation (14.9) that the correlation between increments is either positive or negative depending on whether $H > 0.5$ or $H < 0.5$, regardless of the separation distance $r$. The special case $H = 0.5$ corresponds to uncorrelated Gaussian noise. When $H$ is larger than 0.5, the increments of $Z$ are positively correlated. The probability of observing increments of the same sign over large distances increases as $H$ increases and the field becomes smoother in appearance. This type of behavior is typical of many geophysical time series, landscape topography, and river networks (Voss, 1985; Yaglom, 1987; Rodríguez-Iturbe and Rinaldo, 1997). For $H < 0.5$ the increments are negatively correlated: The amplitude of the small-scale variability and the probability of observing increments of opposite sign over short distances increase as $H$ decreases. The field then assumes its grainy, roughed appearance. This variability is typical of soil parameters such as bulk density, layer thickness, and percentage of fines or coarse material (Burrough, 1983; Molz and Boman, 1995; Liu and Molz, 1996). The effects of $H$ on the simulated fields can be seen in Figure 14.5, which shows the ensemble-averaged semivariogram and its histogram at different lags. The results suggest that no bias is present in the $H = 0.8$ case, but a slight one is found in the $H = 0.2$ case, in the form of reduced variability. This bias is an aliasing effect that is particular to the $H < 0.5$ case and can be eliminated by working with a smaller grid, which will capture the small-scale variability. In both cases, the spread in the semivariogram is large. This spread is caused by the correlation structure of equation (14.8), as discussed earlier: The fields we generate may be large in dimension, but still they are small if we recall that we are dealing with infinite integral scales.

Estimation and modeling in self-similar fields pose special challenges. Estimation of the semivariogram in the earth sciences assumes stationarity and ergodicity. However, the assumption of ergodicity is of limited value, especially when dealing with the large lags of the semivariogram, given the infinite integral scale. Hence, power-law behavior can be demonstrated over limited distances, but because of the lack of ergodicity it can be conjectured only over larger distances. Such an illusive

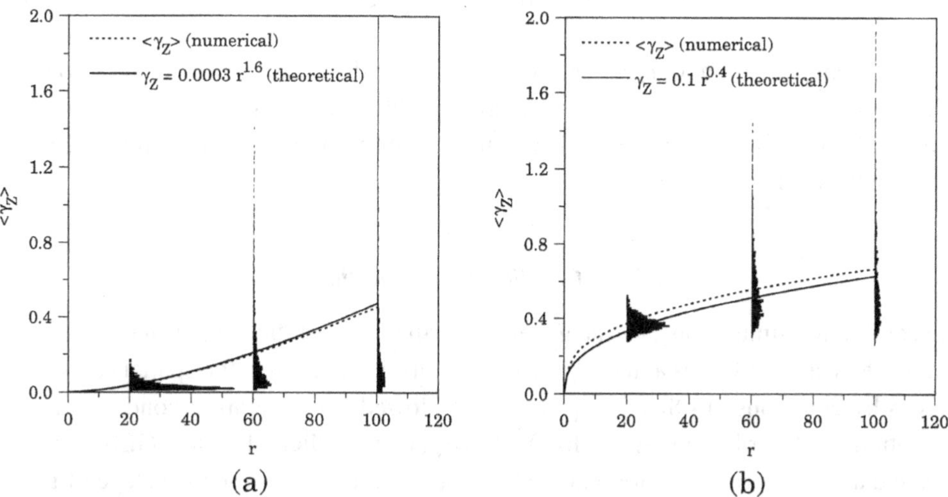


Figure 14.5. Expected value of the reconstructed semivariogram: (a) $H = 0.8$ and $C_0 = 0.0003$; (b) $H = 0.2$ and $C_0 = 0.1$. In all cases the semivariogram PDF is shown for $r = 20, 60$, and 100.

behavior opens the door for alternative interpretations of the semivariogram structure: For example, an integral scale that is larger than the domain of observation can also lead to power-law behavior at the smaller lags (Rubin, 1995).

The question regarding the large-lag behavior of the semivariogram becomes less acute if we recall that the near-orgin behavior of the semivariogram is dominant in point-value estimation. However, the fundamental issue in semivariogram model estimation lies in the dichotomy between the ensemble statistics of equation (14.8), on the one hand, and the lack of ergodicity in the field, on the other. In other words, to sample the ensemble statistics predicted by equation (14.8), a spatial process must occur over an infinite area, which is impractical. A complementary issue is how to model processes of limited spatial extent in a fractal field, because the variability of equation (14.8) will never be met in the finite-size domain and will not be sampled by processes defined by a limited spatial extent. This issue was recently discussed by Ababou and Gelhar (1990), Dagan (1994), and Bellin et al. (1996) in the context of transport of solutes in porous media. Those studies have shown that the finite scale of the flow-and-transport domain imposes an upper cutoff on the scales of variability that are sampled and creates a profound difference in the nature of the process from the description obtained without accounting for the finite scale of the process. Another issue of concern is that of point-value estimation. If a power-law semivariogram is adopted, unconditional point-value estimation is not a viable option because of the infinite variance of $Z$. This difficulty can be avoided through conditional estimation. As an example, consider the case of estimating $Z$ at $\mathbf{x}$, conditional on a single measurement located at $\mathbf{x}_0$. If a simple kriging estimator is chosen, the estimation variance at $\mathbf{x}$, $\sigma^2(\mathbf{x})$, is

$$\sigma^2(\mathbf{x}) = \gamma_z(\mathbf{x}, \mathbf{x}_0) = C_0(|\mathbf{x} - \mathbf{x}_0|)^{2H} \tag{14.10}$$

which becomes infinite only when $\mathbf{x}$ is at an infinite distance from $\mathbf{x}_0$. Hence conditioning is a means for making the estimation problem tractable in self-similar fields and for limiting the infinite variability of the ensemble statistics to the finite one that represents local conditions.

### *14.2.5 Conditional Simulations*

In conditional simulations, all the generated realizations of the $Z$ fields have in common the measured values at their known locations, in addition to the spatial statistics, given by equations (14.8). The procedure followed for generating conditional realizations is described in Appendix A. As explained earlier, HYDRO_GEN is built around a coarse-grid generation at the first stage, followed by a second stage of grid refinement. Because of the infinite correlation length in self-similar fields, we applied the conditioning procedure at $\mathbf{x}_N$ on all the previously generated nodes, in addition to the measurements within a limited search neighborhood defined by a circle of radius $r_c$. At the grid-refinement stage, however, only the four nearest coarse-grid nodes and the measurements at shorter distances from the generation node were used, in order to simplify and reduce the computational effort. By doing that, we did not compromise accuracy, because we have found out that the screening effect, whereby the nearby measurements carry the maximum effect, overrides the infinite correlation length.

Figure 14.6 shows the standard deviation of the estimation error, which is defined as the difference between estimated and actual values of $Z$, $\delta_z = \langle Z^c(\mathbf{x})\rangle - z(\mathbf{x})$, normalized by the standard deviation of the actual values of $Z$, as a function of the radius of the measurement search neighborhood $r_c$ used at the coarse-grid generation stage. The actual, benchmark values of $Z$ are chosen from an unconditional field generation with the semivariogram given by equation (14.8). Note that the case $r_c = 0$ indicates unconditional generation. We see that with a limited search radius of about $r_c = 20$, which is roughly one-sixth of the domain size, results are obtained that are similar to those obtained with larger radii. By that, we have realized an enormous benefit in terms of the computational effort.

For demonstration, consider Figure 14.7a, which is an unconditional field obtained using equation (14.8), with $C_0 = 0.0003$ and $H = 0.8$, and will be used as a base case. Figure 14.7b shows the conditional mean field $\langle Z(\mathbf{x})^c\rangle$ obtained through conditioning on six uniformly distributed measurements taken from Figure 14.7a. The measurements are used only for conditioning, and the semivariogram is assumed

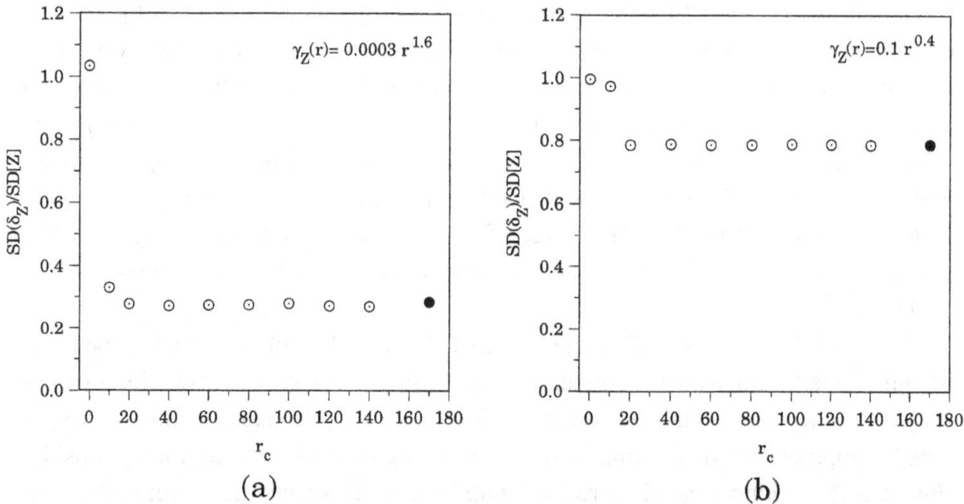


Figure 14.6. Standard deviation of the estimation error, SD[$\delta_Z$], normalized by the standard deviation of the reference field, SD[$Z$]: (a) $H = 0.8$ and $C_0 = 0.0003$; (b) $H = 0.2$ and $C_0 = 0.1$. The filled circle indicates conditioning on all the measurements at both the coarse stage and the refinement stage, and the dotted circles represent the results of conditioning on measurements inside the circle of radius $r_c$ at the coarse-grid stage and only on the four nearest grid nodes at the refinement stage.

to be known deterministically and given by equation (14.8). The conditional mean field captures some of the more significant aspects of the base-case field, but leaves a lot of room for improvement. To what extent a single conditional realization, or the conditional mean, reproduces the base-case field is a question that will be further addressed in the following. Figure 14.7c shows the amount of information at each node in the conditional realization in the form of information entropy $E$, which is defined as follows (Shannon, 1948):

$$E = -\int f_z^c \ln f_z^c \, dZ \tag{14.11}$$

where $f_z^c$ is the PDF of $Z$ conditional to the measurements. Because we are conditioning only on $Z$ measurements, $f_z^c$ assumes the following expression:

$$\begin{aligned} f_z^c(\mathbf{x}) &= f_z(\mathbf{x}|Z(\mathbf{x}_1), \ldots, Z(\mathbf{x}_{N'})) \\ &= \frac{1}{\sqrt{2\pi}\sigma_z^c(\mathbf{x})} \exp\left[ -\frac{[Z(\mathbf{x}) - \langle Z^c(\mathbf{x})\rangle]^2}{2\sigma_z^{c^2}} \right] \end{aligned} \tag{14.12}$$

where $N'$ is the number of measurements. Substituting equation (14.12) into equation (14.11) and integrating yields

$$E = 0.5 + \ln\left[\sqrt{2\pi}\sigma_z^c(\mathbf{x})\right] \tag{14.13}$$

Equations (14.11) and (14.13) reaffirm that low entropy means low uncertainty. The entropy assumes the value $E = -\infty$ at the measurement nodes, where the $Z$ values are known deterministically, whereas at nodes in unconditional simulations the entropy assumes its maximum value, which is given by equation (14.13), where $\sigma_z^c$ is replaced by the unconditional standard deviation. Figure 14.7c shows that the entropy is smaller when approaching the measurement nodes and increases as we move from them. Note that in order to reduce the range of $E$ values to be represented by the color scale, the value $E = -\infty$ at the measurement nodes is not shown in Figures 14.7–14.11.

Figure 14.7d shows the absolute value of the actual error $\delta_z$, which appears to be quite small, and with no noticeable trend. In any case, the estimation error is finite, and the conditional mean appears to be a very reasonable estimate. To get a better appreciation of the impact of conditioning on each measurement, consider Figure 14.8, which repeats Figure 14.7 with 81 equally spaced measurements. This time, the conditional mean (Figure 14.8b) better resembles the base case, and the entropy is lower compared with Figure 14.7. Figure 14.8c shows that appreciable improvement in accuracy has been obtained throughout the field and that the 81 measurements used represent a significant increase in the amount of information, as compared with the original 6 measurements. The striking finding is that while the power-law semivariogram leads to infinite variance in unbounded domains, the other aspect of that behavior is the infinite integral scale, which allows reduction of point-estimation uncertainty using a relatively small number of measurements.

Figure 14.9 uses $C_0 = 0.1$, $H = 0.2$, and six measurements. The smooth variations of $Z$ in the $H > 0.5$ case (Figure 14.7a) stand in contrast to the grainy, discontinuous nature of $Z$ in the $H < 0.5$ case (Figure 14.9a). Inspection of Figures 14.7c and 14.9c reveals that in the case $H > 0.5$, more information is transferred from the measurements to the conditional simulations. Furthermore, for $H < 0.5$ the entropy increases faster with distance from the measurement nodes than in the case of $H > 0.5$. Inspection of Figures 14.7d and 14.9d reveals that the estimation errors in the $H < 0.5$ case are spread over a much larger range, with a significant part being above the maximum value of 0.6 observed in Figure 14.7d. Figure 14.10 repeats Figure 14.9, but with 20 measurements, and Figure 14.11 repeats Figure 14.7, also with 20 measurements. We see from these figures that 20 measurements in the $H > 0.5$ case are very effective, that the conditional mean is very close to the base case, and that the entropy is small. In the $H < 0.5$ case, the increase in the number of measurements is much less rewarding.

In order to assess the global amount of information transferred from the measurements to the conditional simulations, we computed the spatial mean $\bar{E}$ and standard deviation SD$[E]$ of the entropy. The value of the entropy at the measurement nodes is not included in the computation. The results of the computations are shown in Table 14.1 for different Hurst coefficients and measurements. Inspection of Table 14.1 reaffirms that in the case $H > 0.5$ more information is extracted from the measurements. Furthermore, increasing the number of measurements from 6 to 20 results in appreciable reduction of the mean entropy only in the case $H > 0.5$.

This point is further demonstrated in Figure 14.12, which shows the PDFs for the differences between estimated and actual $Z$ for $H = 0.8$ and $H = 0.2$, for different numbers of conditioning points. The estimation errors are normalized by the standard deviation of $Z$ in order to remove the effects of the different variabilities in the

Table 14.1.

| $H$ | $N'$ | $\bar{E}$ | SD$[E]$ |
|---|---|---|---|
| 0.8 | 6 | −0.231 | 0.488 |
| 0.8 | 20 | −0.839 | 0.496 |
| 0.8 | 81 | −1.468 | 0.443 |
| 0.2 | 6 | 1.043 | 0.126 |
| 0.2 | 20 | 0.889 | 0.137 |

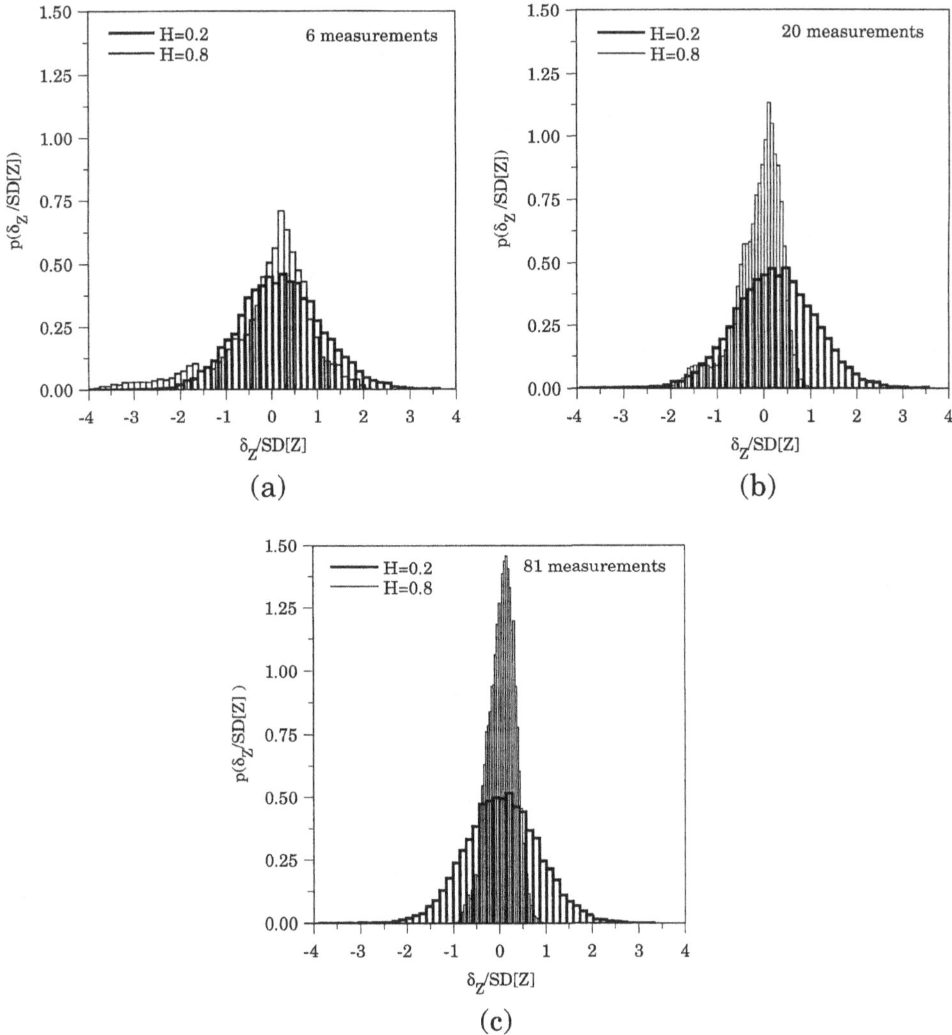


Figure 14.12. PDF for the normalized error $\delta_z/\mathrm{SD}[Z]$ for conditioning on (a) 6 measurements, (b) 20 measurements, and (c) 81 measurements.

two fields and to concentrate only on the effects of the increase in the number of conditioning points. We find that in the $H = 0.8$ case the PDF reduces in width and becomes less diffuse at a much faster rate than in the $H = 0.2$ case. In the $H = 0.2$ case, the increase in the number of measurement point from 6 to 81 has little effect on the estimation-error PDF. These effects can all be attributed to the exponent $H$. For $H < 0.5$, the fast ascent of the semivariogram near the origin implies a faster decay of the correlation, and hence a limited, localized effect of each of the measurements. It is expected that in the presence of a nugget, an exponent $H > 0.5$ may not be sufficient to guarantee a high degree of effectiveness of the samples.

A basic question that needs to be addressed prior to any Monte Carlo study is its ability to capture and model the unsampled heterogeneity. In fact, before embarking on a demanding Monte Carlo study, the sufficiency of estimation, rather than simulation, should be assessed. One way to eliminate the risk of failure of the Monte Carlo study is, of course, to increase the investment in data. As the number of measurements increases, the semivariogram (14.8) inferred from the data will represent the field variability more accurately. In fortunate situations of an ergodic data base, the inferred semivariogram will also capture the underlying spatial-variability process. Paradoxically, as the number of measurements increases, accurate modeling of the semivariogram becomes a minor issue because of the large degree of constraining applied by the measurements. Some degree of freedom will still remain with the small-scale variability, and the significance of that issue as a motivation for a Monte Carlo study needs to be addressed on a case-by-case basis. The real difficulty arises when only a small number of measurements is available. What then are the chances for conducting a meaningful Monte Carlo study in the case of self-similar variability? To asses this issue, but only in a cursory way, consider Figure 14.13. The striking result here is that whereas for both the $H = 0.8$ and $H = 0.2$ cases a limited number of measurements suffices to capture the variability of the single base-case realization, in the $H = 0.2$ case this number suffices to capture and reproduce the underlying stochastic process, but in the $H = 0.8$ case there is a wide gap between the statistics of the simulations and the underlying process represented by the semivariogram (14.8). The surprise here is that whereas in the $H = 0.2$ case we have found considerable

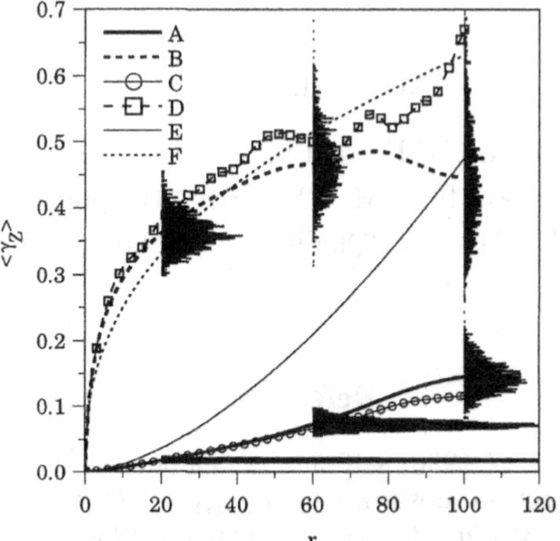


Figure 14.13. Expected values of the spatial semivariogram for both $H = 0.8$ and $H = 0.2$, with the semivariogram PDF at the separation distances $r = 20$, 60, and 100 and conditioning on 81 measurements. Legend: A, expected value of the spatial semivariogram for $H = 0.8$; B, expected value of the spatial semivariogram for $H = 0.2$; C, semivariogram of the reference field for $H = 0.8$; D, semivariogram of the reference field for $H = 0.2$; E, theoretical semivariogram for $H = 0.8$ ($\gamma_z = 0.0003r^{1.6}$); F, theoretical semivariogram for $H = 0.2$ ($\gamma_z = 0.1r^{0.4}$).

difficulties in point estimation, because of the "antipersistence" it is quite safe in term of capturing the stochastic process. The antipersistence implies that with a limited number of measurements there are far better chances to sample the many scales of variability and heterogeneity in the $H = 0.2$ case than in the $H = 0.8$ case. Pending further verification, we tend to conclude that $H > 0.5$ implies estimation, whereas $H < 0.5$ implies simulation, as aids in the decision-making process.

## 14.3 Summary

This chapter discussed some aspects of self-similar RSFs and techniques for generation of fields with evolving scales of heterogeneity. The shape of the semivariogram function characterizing the self-similarity and the number of measurements are reviewed in terms of their influence on estimation and simulation. We have found that the shape of the semivariogram, particularly its near-origin behavior, has profound effects on both estimation and simulation.

Variability in self-similar fields scales proportionally to $r^{2H}$. For $H < 0.5$, point estimation is very poor and requires a large number of measurements for a significant reduction in the estimation error. On the other hand, $H < 0.5$ ensures that a small number of measurements is sufficient for accurate inference of the semivariogram structure. In the case $H > 0.5$, measurements carry an influence over very large distances, making point estimation very rewarding. At the same time, inference of the semivariogram is very much error-prone. In all cases, however, conditioning is efficient in terms of making the estimation variance finite in fields with apparent infinite variability.

## Acknowledgments

This study was supported by NSF grant 9304481 to the first author. The second author acknowledges support from MURST 40% "Trasporto di inquinanti nei corpi idrici naturali" and MURST 40% "Qualità delle acque e controllo in tempo reale nelle reti fognarie."

## References

Ababou, R., and Gelhar, L. W. 1990. Self-similar randomness and spectral conditioning: analysis of scale effects in subsurface hydrology. In: *Dynamics of Fluids in Hierarchical Porous Media*, ed. J. H. Cushman, pp. 393–428. San Diego: Academic Press.

Barnes, R. J., and Watson, A. G. 1992. Efficient updating of kriging estimates and variances. *Math. Geol.* 24:129–33.

Bellin, A., Pannone, M., Fiori, A., and Rinaldo, A. 1996. On transport in porous formations characterized by heterogeneity of evolving scales. *Water Resour. Res.* 32:3485–96.

Bellin, A., and Rubin, Y. 1996. HYDRO_GEN: a spatially distributed random field generator for correlated properties. *Stoch. Hydrol. Hydraul.* 10:253–78.
Burrough, P. A. 1983. Multiscale sources of spatial variation in soil. 1. The application of fractal concepts to nested levels of soil variation. *J. Soil Sci.* 34:577–97.
Chase, C. G. 1992. Fluvial landsculpting and the fractal dimension of topography. *Geomorphology* 5:39–57.
Cushman, J. H., and Ginn, T. R. 1993. On dispersion in fractal porous media. *Water Resour. Res.* 29:3513–15.
Dagan, G. 1989. *Flow and Transport in Porous Formations*. Berlin: Springer-Verlag.
Dagan, G. 1994. The significance of heterogeneity of evolving scales to transport in porous formations. *Water Resour. Res.* 30:3327–36.
Desbarats, A. J., and Bachu, S. 1994. Geostatistical analysis of aquifer heterogeneity from the core scale to the basin scale: a case study. *Water Resour. Res.* 30:673–84.
Engesgaard, P., Jensen, K. H., Molson, J., Frind, E. O., and Olsen, H. 1996. Large-scale dispersion in a sandy aquifer: simulation of subsurface transport of environmental tritium. *Water Resour. Res.* 32:3253–66.
Feder, J. 1988. *Fractals*. New York: Plenum Press.
Gelhar, L. W. 1986. Stochastic subsurface hydrology from theory to applications. *Water Resour. Res.* 22:135S–45S.
Glimm, J., Lindquist, W. B., Pereira, F., and Zhang, Q. 1993. A theory of macrodispersion for the scale-up problem. *Trans. Porous Media* 13:97–122.
Gómez-Hernández, J. J., and Srivastava, R. M. 1990. Isim3d: an ANSI-C three-dimensional multiple indicator conditional simulation program. *Comput. Geosci.* 16:395–440.
Gupta, V. K., and Waymire, E. 1990. Multiscaling properties of spatial rainfall and river flow distributions. *J. Geophys. Res.* 95:1999–2009.
Hewett, T. A. 1986. Fractal distributions of reservoir heterogeneity and their influence on fluid transport. In: *Proceedings of the 61st Annual Technical Conference of the Society of Petroleum Engineers*. Paper SPE 15386.
Journel, A. G., and Alabert, F. A. 1989. Non-Gaussian data expansion in the earth sciences. *Terra Nova* 1:123–34.
Journel, A. G., and Huijbregts, C. J. 1978. *Mining Geostatistics*. New York: Academic Press.
Koltermann, C. E., and Gorelick, S. M. 1996. Heterogeneity in sedimentary deposits: a review of structure-imitating, process-imitating, and descriptive approaches. *Water Resour. Res.* 32:2617–58.
Liu, H. H., and Molz, F. J. 1996. Discrimination of fractional Brownian movement and fractional Gaussian noise structures in permeability and related property distributions with range analyses. *Water Resour. Res.* 32:2601–5.
Lovejoy, S., and Schertzer, D. 1990. Multifractals, universality classes and satellite and radar measurements of cloud and rain fields. *J. Geophys. Res.* 95:2021–34.
Molz, F. J., and Boman, G. K 1995. Further evidence of fractal structure in hydraulic conductivity distributions. *Geophys. Res. Lett.* 22:2545–8.
Mood, A. M. F., and Graybill, F. A. 1963. *Introduction to the Theory of Statistics*. New York: McGraw-Hill.
Neuman, S. P. 1990. Universal scaling of hydraulic conductivities and dispersivities in geologic media. *Water Resour. Res.* 26:1749–58.
Neuman, S. P. 1994. Generalized scaling of permeabilities: validation and effect of support scale. *Geophys. Res. Lett.* 21:349–52.
Neuman, S. P. 1995. On advective transport in fractal permeability and velocity fields. *Water Resour. Res.* 31:1455–60.

Omre, H., Solna, K., and Tjelmeland, H. 1992. Simulation of random functions on large lattices. In: *Geostatistics Tròia '92*. vol. 1, ed. A. Soares, pp. 179–200. Dordrecht: Kluwer.
Painter, S., and Paterson, L. 1994. Fractional Lèvy motion as a model for spatial variability in sedimentary rock. *Geophys. Res. Lett.* 21:2857–60.
Rodríguez-Iturbe, I., Marani, M., Rigon, R., and Rinaldo, A. 1994. Self-organized river basin landscapes: fractal and multifractal characteristics. *Water Resour. Res.* 30:3531–9.
Rodríguez-Iturbe, I., and Rinaldo, A. 1997. *Fractal River Basins: Chance and Self-organization*. Cambridge University Press.
Rubin, Y. 1995. Flow and transport in bimodal heterogeneous formations. *Water Resour. Res.* 31:2461–8.
Shannon, C. E. 1948. A mathematical theory of communication. *Bell System Tech. J.* 27:379–423, 623–56.
Sposito, G., Jury, W. A., and Gupta, V. K. 1986. Fundamental problems in the stochastic convection–dispersion model of solute transport in aquifers and field soils. *Water Resour. Res.* 22:77–88.
Voss, R. F. 1985. Random fractals: characterization and measurement. In: *Scaling Phenomena in Disordered Systems*, ed. R. Pynn and A. Skjeltorp, pp. 1–11. New York: Plenum Press.
Wheatcraft, S. W., and Tyler, S. W. 1988. An explanation of scale-dependent dispersivity in heterogeneous aquifers using concepts of fractal geometry. *Water Resour. Res.* 24:566–78.
Yaglom, A. M. 1987. *Correlation Theory of Stationary and Related Random Functions, Basic Results*, vol. 1. Berlin: Springer-Verlag.

## Appendix A: Generation Conditional to Measurements

Assume that $N'$ field measurements are available at the positions $\tilde{\mathbf{x}}_i, i = 1, \ldots, N'$. The straightforward procedure consisting in adding the $N'$ measurement to the set of $N-1$ previously generated $z$ values at nodes $\mathbf{x}_i$, $i = 1, \ldots, N-1$, and solving the system (14.7) for the resulting $N + N' - 1$ conditional nodes is not pursued here because it would eliminate all the savings realized from using fixed sets of conditioning coefficients. Instead, we follow the method of Barnes and Watson (1992), which is superior computationally. Taking advantage of the properties of partitioned matrices, the conditional variance can be expressed as follows:

$$\langle Z^c(\mathbf{x}_N|N-1+N') = \langle Z^c(\mathbf{x}_N \mid N-1)\rangle\rangle + [\mathbf{r}_{N'} - \hat{\mathbf{r}}_{N'}]^T \mathbf{D}^{-1}[\mathbf{z}_{N'} - \hat{\mathbf{z}}_{N'}] \tag{14.A1}$$

where $\langle Z^c(\mathbf{x}_N \mid N-1)\rangle$ is the conditional mean obtained using only the $N-1$ previously generated $z$ values (prior to the availability of the $N'$ measurements), $\hat{\mathbf{z}}_{N'}$ is the $(N' \times 1)$ vector of the conditional means computed at the measurement nodes $\tilde{\mathbf{x}}_i$, $i = 1, \ldots, N'$, using only the $N-1$ samples. The element $(i, j)$ of the matrix

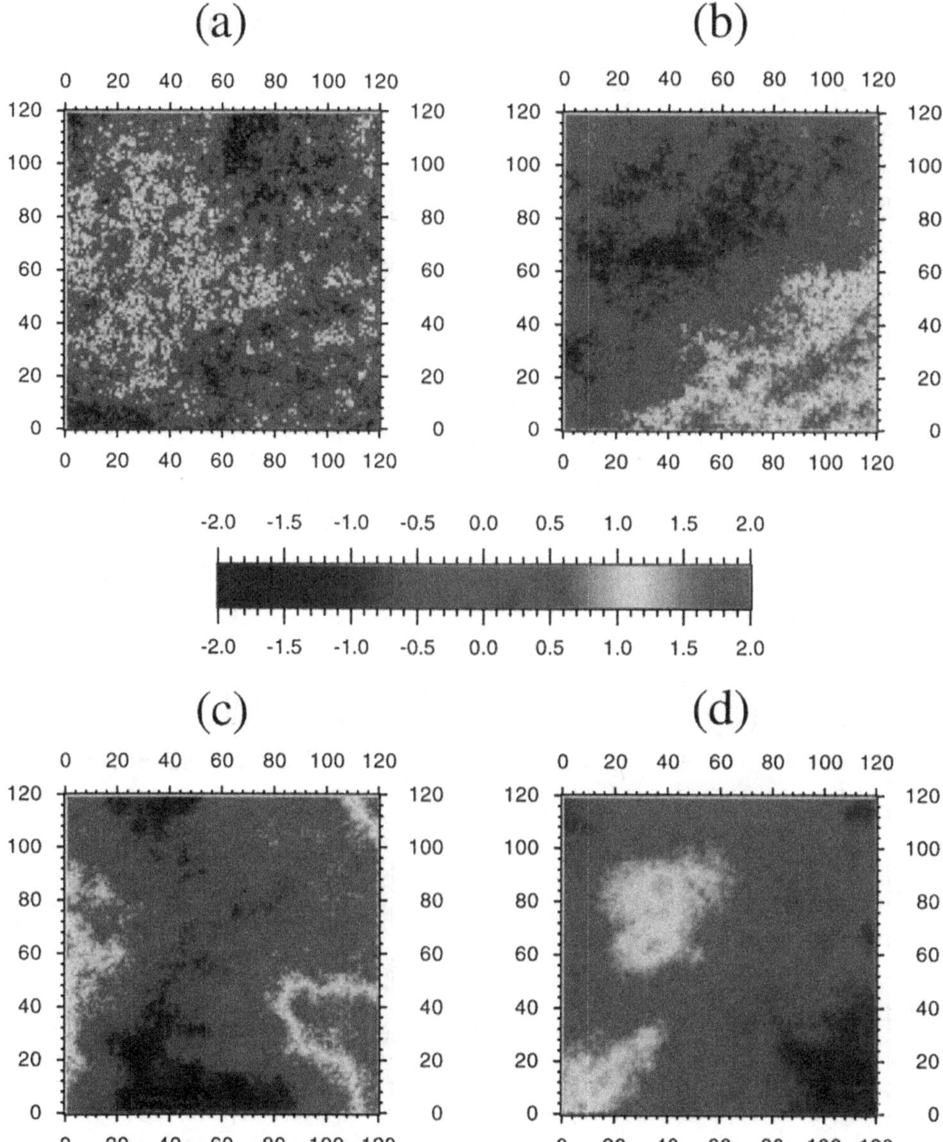


Figure 14.4. Examples of self-similar fields: (a) $H = 0.2$; (b) $H = 0.4$; (c) $H = 0.6$; (d) $H = 0.8$.

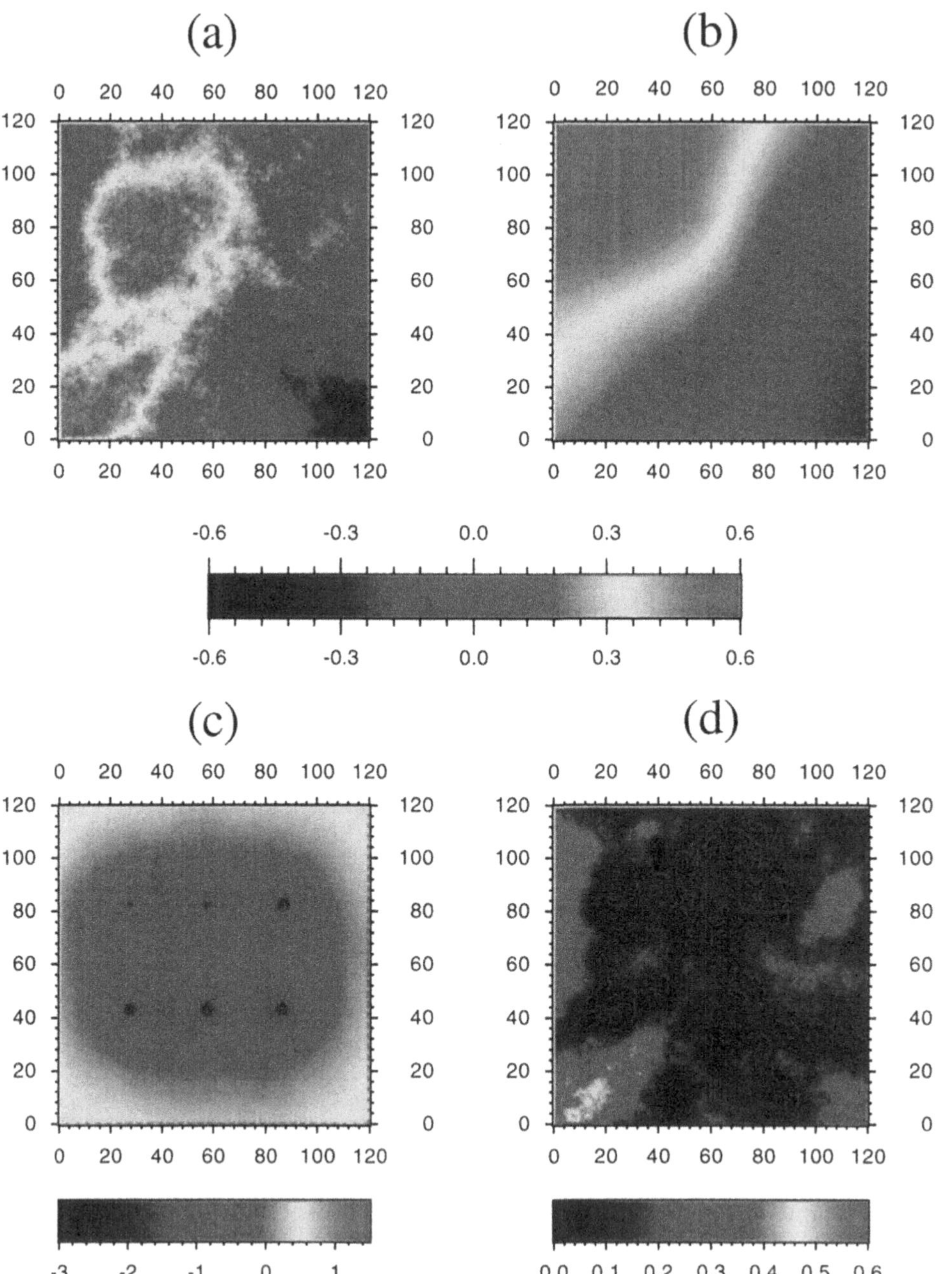


Figure 14.7. Color-scale representation of the following quantities: (a) reference field; (b) conditional mean; (c) information entropy; (d) absolute value of the actual error $\delta_z$ for the case $H = 0.8$ and $C_0 = 0.0003$ conditional on six measurements uniformly distributed over the domain.

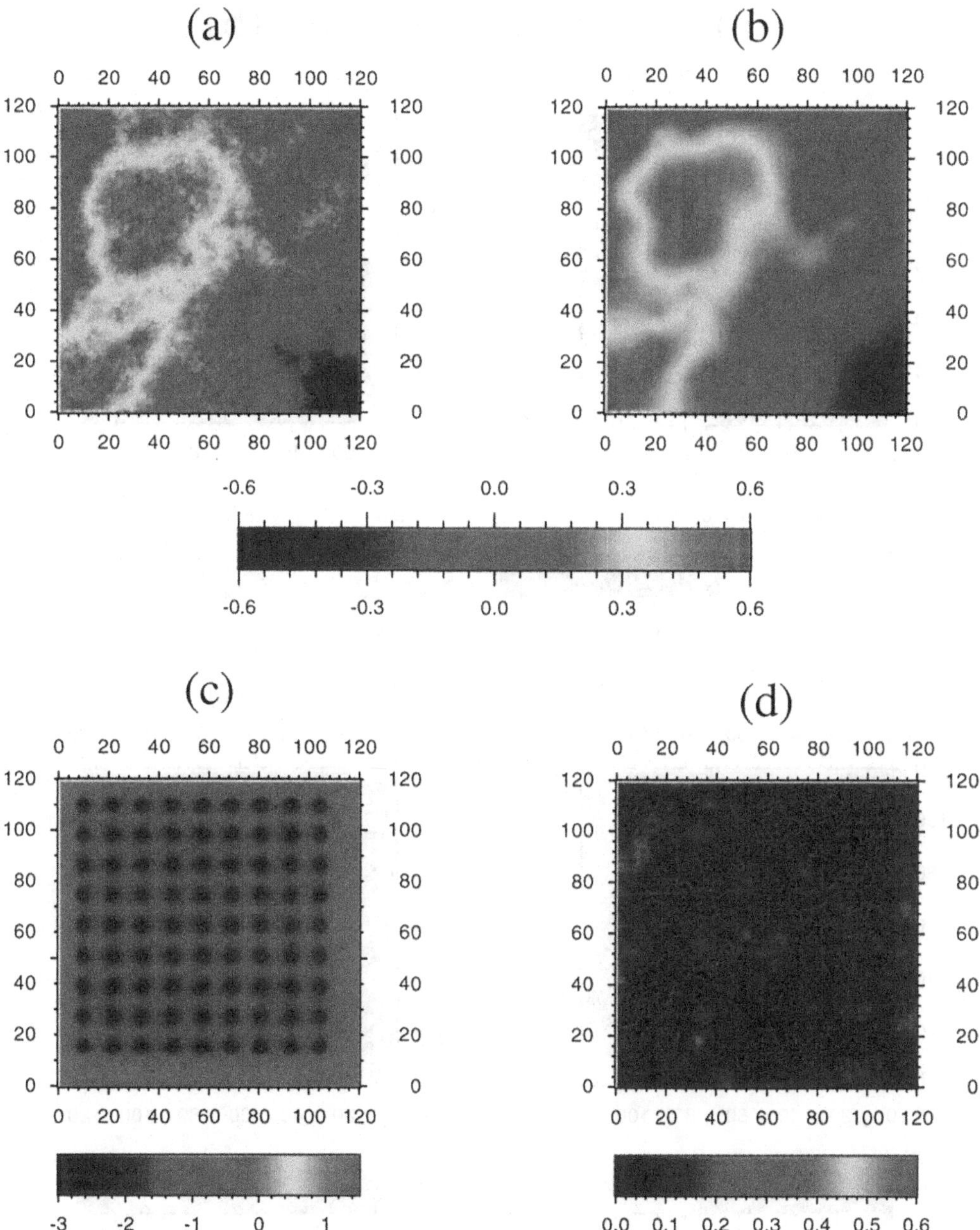


Figure 14.8. Color-scale representation of the following quantities: (a) reference field; (b) conditional mean; (c) information entropy; (d) absolute value of the actual error $\delta_z$ for the case $H$ = 0.8 and $C_0$ = 0.0003 conditional on 81 measurements uniformly distributed over the domain.

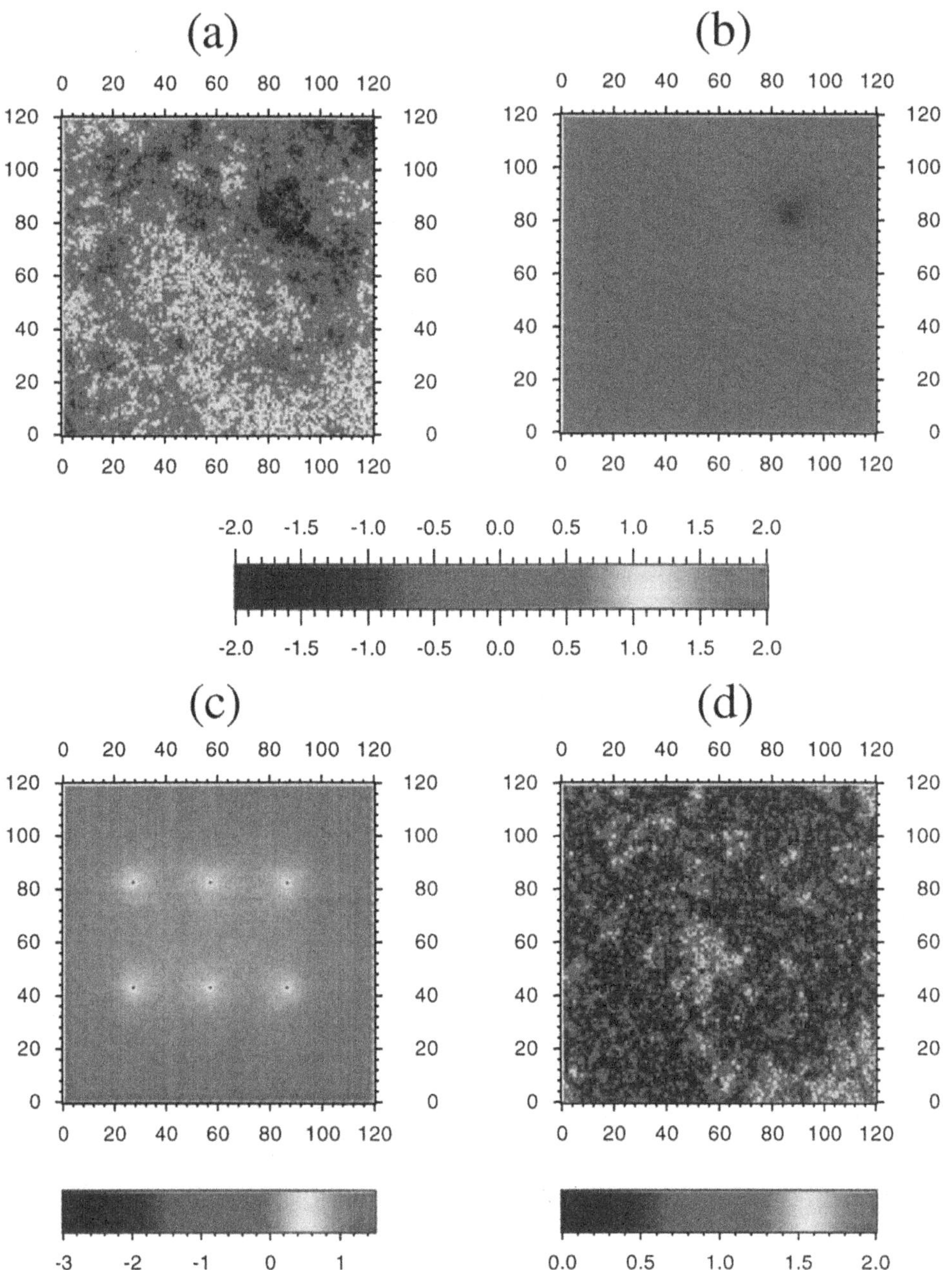


Figure 14.9. Color-scale representation of the following quantities: (a) reference field; (b) conditional mean; (c) information entropy; (d) absolute value of the actual error $\delta_z$ for the case $H$ = 0.2 and $C_0$ = 0.1 conditional on six measurements uniformly distributed over the domain.

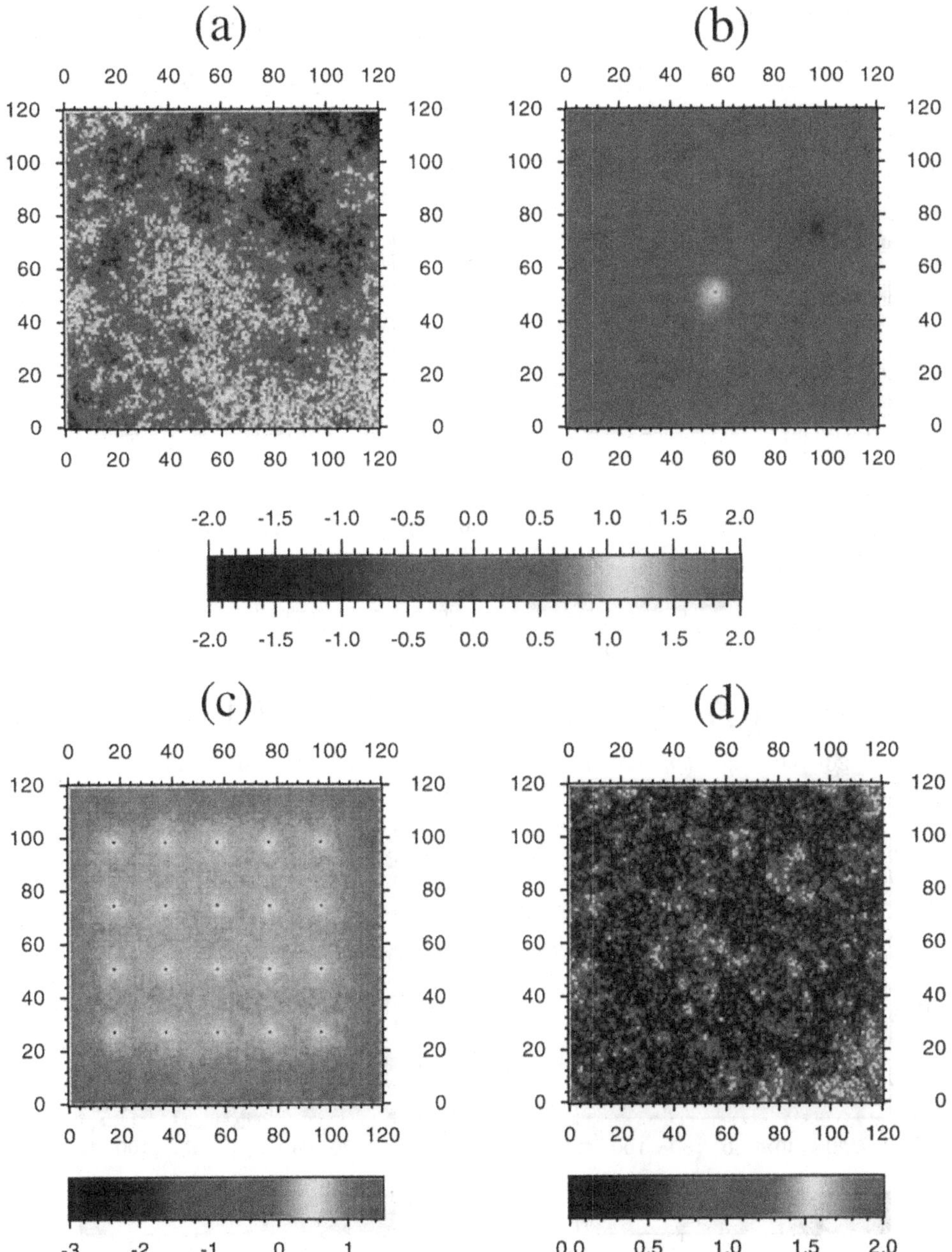


Figure 14.10. Color-scale representation of the following quantities: (a) reference field; (b) conditional mean; (c) information entropy; (d) absolute value of the actual error $\delta_z$ for the case $H = 0.2$ and $C_0 = 0.1$ conditional on 20 measurements uniformly distributed over the domain.

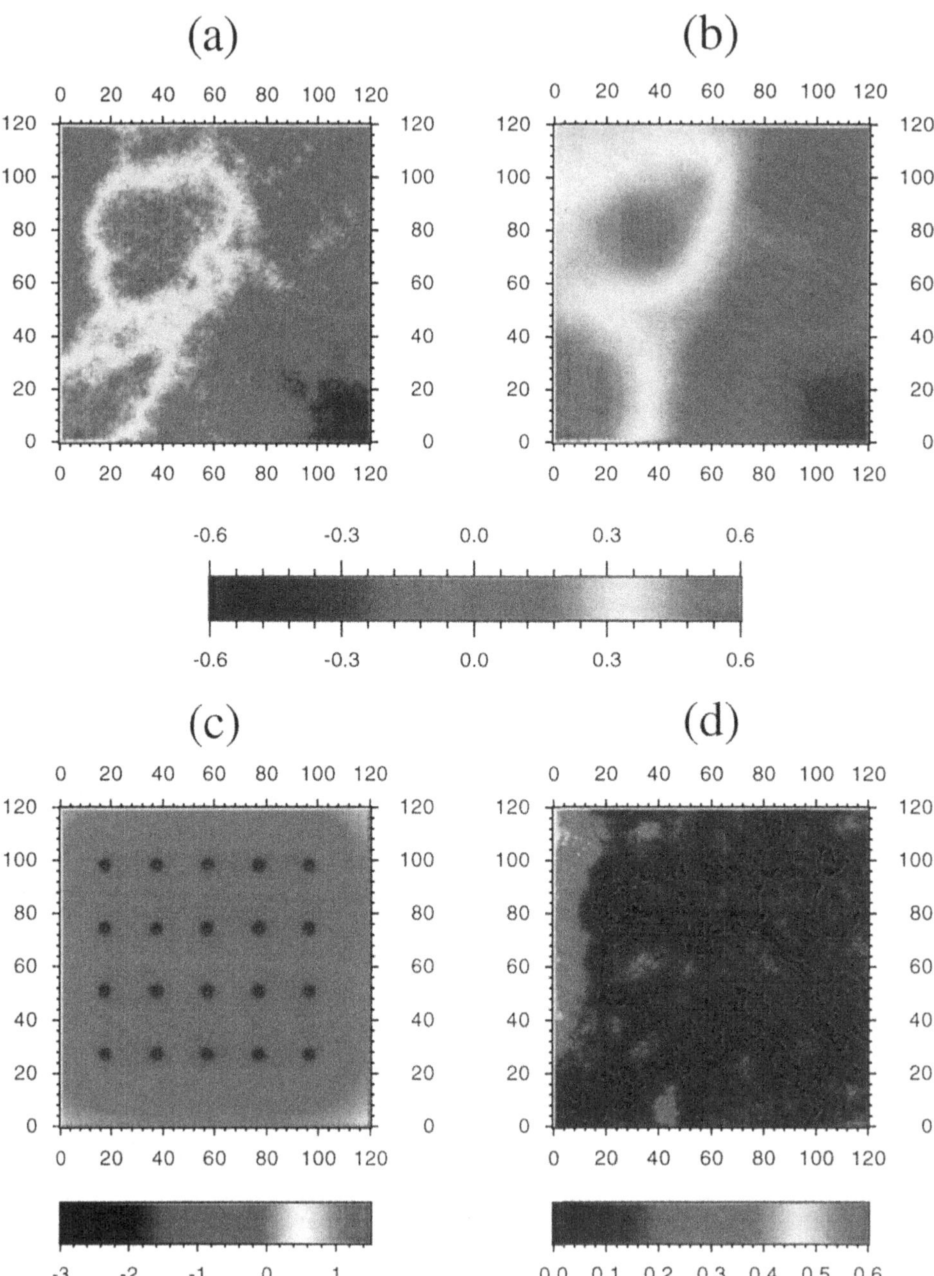


Figure 14.11. Color-scale representation of the following quantities: (a) reference field; (b) conditional mean; (c) information entropy; (d) absolute value of the actual error $\delta_z$ for the case $H = 0.8$ and $C_0 = 0.0003$ conditional on 20 measurements uniformly distributed over the domain.

**D** is given by

$$D_{ij} = \gamma_z(\tilde{\mathbf{x}}_i, \tilde{\mathbf{x}}_j) - \sum_{k=1}^{N-1} \lambda_k(\tilde{\mathbf{x}}_i)\gamma_z(\mathbf{x}_k, \tilde{\mathbf{x}}_j) - \mu(\tilde{\mathbf{x}}_i) \qquad (i, j = 1, \ldots, N') \tag{14.A2}$$

and the $i$th elements of the vectors multiplying the matrix **D** in equation (14.A1) are given by

$$[\mathbf{r}_{N'} - \hat{\mathbf{r}}_{N'}]_i = \gamma_z(\mathbf{x}_N, \tilde{\mathbf{x}}_i) - \sum_{j=1}^{N-1} \lambda_j(\tilde{\mathbf{x}}_i)\gamma_z(\mathbf{x}_N, \mathbf{x}_j) - \mu(\tilde{\mathbf{x}}_i) \qquad (i = 1, \ldots, N') \tag{14.A3}$$

$$[\mathbf{z}_{N'} - \hat{\mathbf{z}}_{N'}]_i = z(\tilde{\mathbf{x}}_i) - \sum_{j=1}^{N-1} \lambda_j(\tilde{\mathbf{x}}_i)z(\mathbf{x}_j) \qquad (i = 1, \ldots, N') \tag{14.A4}$$

where $\lambda_j(\tilde{\mathbf{x}}_i),\ j = 1, \ldots, N-1$, are the interpolation coefficients computed for the node $\tilde{\mathbf{x}}_i$ using the $N-1$ original samples, and $z(\tilde{\mathbf{x}}_i),\ i = 1, \ldots, N'$, are the field measurements. To obtain the conditional mean (14.A1), the updating procedure requires inversion of the matrix **D** of dimensions equal to the number of measurements, computation of the two vectors (14.A3) and (14.A4), and computation of a vector matrix and a vector-vector product. The updated conditional variance is then given by

$$\left(\sigma_z^c\right)^2(\mathbf{x}_N \mid N-1+N') = \left(\sigma_z^c\right)^2(\mathbf{x}_N \mid N-1) + [\mathbf{r}_{N'} - \hat{\mathbf{r}}_{N'}]^T \mathbf{D}^{-1} [\mathbf{r}_{N'} - \hat{\mathbf{r}}_{N'}] \tag{14.A5}$$

where $(\sigma_z^c)^2(\mathbf{x}_N \mid N-1+N')$ is the variance at node $\mathbf{x}_N$ conditioned on the $N-1$ original samples $z$ and on the $N'$ measurements, and $(\sigma_z^c)^2(\mathbf{x}_N \mid N-1)$ is the variance conditional only on the $N-1$ original sample. Once the updated mean and variance are computed, they become the updated target statistics for node $\mathbf{x}_N$, and a new value of $Z(\mathbf{x}_N)$ is generated, replacing the previous one, which was generated at the unconditional stage, based on equations (14.5) and (14.6).

## Appendix B: Derivation of the Correlation $\rho$

Consider the variance of the increment of $Z$ between nodes $\mathbf{x} - \mathbf{r}$ and $\mathbf{x} + \mathbf{r}$:

$$\langle [Z(\mathbf{x}+\mathbf{r}) - Z(\mathbf{x}-\mathbf{r})]^2 \rangle = 2\gamma_z(2r) \tag{14.B1}$$

which can be rewritten as

$$2\gamma_z(2r) = \langle\{[Z(\mathbf{x}+\mathbf{r}) - Z(\mathbf{x})] + [Z(\mathbf{x}) - Z(\mathbf{x}-\mathbf{r})]\}^2\rangle$$
$$= 4\gamma_z(r) + 4\rho\gamma_z(r) \tag{14.B2}$$

Comparing the left- and right-hand-side terms, we obtain

$$\rho = \frac{\gamma_z(2r)}{2\gamma_z(r)} - 1 \tag{14.B3}$$

which for the semivariogram given by equation (14.8) assumes the following final expression:

$$\rho = 2^{2H-1} - 1 \tag{14.B4}$$

# Index